Advances in ORGANOMETALLIC CHEMISTRY

VOLUME 42

Advances in Organometallic Chemistry

EDITED BY

F. GORDON A. STONE

DEPARTMENT OF CHEMISTRY
BAYLOR UNIVERSITY
WACO, TEXAS

ROBERT WEST

DEPARTMENT OF CHEMISTRY
UNIVERSITY OF WISCONSIN
MADISON, WISCONSIN

VOLUME 42

ACADEMIC PRESS
San Diego London Boston New York
Sydney Tokyo Toronto

This book is printed on acid-free paper. 

Academic Press
a division of Harcourt Brace & Company
525 B Street, Suite 1900, San Diego, California 92101-4495, USA
http://www.apnet.com

Academic Press Limited
24-28 Oval Road, London NW1 7DX, UK
http://www.hbuk.co.uk/ap/

International Standard Book Number: 0-12-031142-9

PRINTED IN THE UNITED STATES OF AMERICA
98 99 00 01 02 03 QW 9 8 7 6 5 4 3 2 1

Contents

Organometallic Complexes in Nonlinear Optics I: Second-Order Nonlinearities

IAN R. WHITTALL, ANDREW M. McDONAGH, MARK G. HUMPHREY, and MAREK SAMOC

Catalytic Dehydrocoupling: A General Strategy for the Formation of Element–Element Bonds

FRANÇOIS GAUVIN, JOHN F. HARROD, and HEE GWEON WOO

Contributors

Numbers in parentheses indicate the pages on which the authors' contributions begin.

FRANÇOIS GAUVIN (363), Chemistry Department, Bishops University, Lennoxville, Quebec, Canada J1M 1Z7

JOHN F. HARROD (363), Chemistry Department, McGill University, Montréal, Quebec, Canada H3A 2K6

MARK G. HUMPHREY (291), Department of Chemistry, Australian National University, Canberra, ACT 0200, Australia

WILLIAM M. JONES (147), Department of Chemistry, University of Florida, Gainesville, Florida 32611

JERZY KLOSIN (147), Catalysis Laboratory, The Dow Chemical Company, Midland, Michigan 48674

ANDREW M. McDONAGH (291), Department of Chemistry, Australian National University, Canberra, ACT 0200, Australia

HIROSHI OGINO (223), Department of Chemistry, Graduate School of Science, Tohoku University, Sendai 980-77, Japan

MAREK SAMOC (291), Australian Photonics Cooperative Research Centre, Laser Physics Centre, Research School of Physical Sciences and Engineering, Australian National University, Canberra, ACT 0200, Australia

HIROMI TOBITA (223), Department of Chemistry, Graduate School of Science, Tohoku University, Sendai 980-77, Japan

KENTON H. WHITMIRE (1), Department of Chemistry, Rice University, Houston, Texas 77005

IAN R. WHITTALL (291), Department of Chemistry, Australian National University, Canberra, ACT 0200, Australia

HEE GWEON WOO (363), Chonnam National University, Kwangju 500-757, Korea

ADVANCES IN ORGANOMETALLIC CHEMISTRY, VOL. 42

Main Group–Transition Metal Cluster Compounds of the Group 15 Elements

KENTON H. WHITMIRE

Department of Chemistry
Rice University
Houston, Texas 77005

0065-3055/98 $25.00

I

INTRODUCTION

Main group elements (E) and transition metal (M) fragments interact with one another in interesting and often unexpected fashions. The bonding patterns that are observed may give rise to structures with no direct counterpart among homometallic cluster compounds of either the main group elements or the transition metals. These unexpected situations arise as a result of the differing orbital requirements of the *p*-block vs *d*-block elements, as well as mismatches in size, electronegativities, substituent properties, and so on. For the heavier elements, relativistic effects also appear to be important.

Our interest in this area stemmed largely from a desire to understand how the incorporation of main group elements into transition metal clusters would affect their stability and ensuing chemistry. At the time we began working on these systems in 1982, there was considerable interest in developing clusters as models for catalysts and for use as catalysts themselves. This relationship was well recognized, as metallic clusters were showing a striking resemblance in their structures and stoichiometric transformations to those of metal particles and metal surfaces. An inherent problem was that many homometallic transition metal clusters degraded to mononuclear species under pressures of CO and the benefit of using a cluster was lost. There was, however, a developing body of information suggesting that incorporation of a main group element fragment would considerably stabilize a cluster framework. This has proven to be true in many cases, but the conditions and limitations of this stability are still not fully explored. Another aspect of the chemistry relating to catalysis is that in many commercial catalyst preparations, main group elements are often added as promoters to give higher yields or better selectivities, but how the main group elements interact with the transition metals was, and still is, largely unknown so that their function in the catalyst is speculative. It is clear from the chemistry that has been developed on E–M clusters that there are many factors involved. The E fragments may simply serve to block certain reaction sites, or they may modify the structural and electronic properties of attached transition metals. They may also serve as sites of reactivity themselves. More recently, there has been a developing interest in creating large clusters of metals and alloy materials. When the particles become large enough their physical properties (electronic, optical, magnetic, etc.) begin to resemble those of the related solid-state phases. The onset of this phenomenon is of considerable interest. Furthermore, particles on a nanometer scale—the so-called quantum dots—are predicted to have unusual and interesting properties as the size of the delocalized wave functions exceeds that of the particle.

This article focuses on transition metal compounds containing fragments based on the group 15 elements (N, P, As, Sb, and Bi), but it should not be forgotten that this chemistry must be set in the context of cluster compounds of the other main group elements. There are many similarities, but also important differences, between the behavior of these elements and those in groups 13, 14, and 16. A complete picture can only be obtained by a careful consideration of the entire set of related compounds. For example, compounds will often be found that have very similar structural arrangements for a whole range of main group elements, but the synthetic routes that are necessary to make them may be completely different, or the ensuing chemical reactivity (stability, redox behavior, etc.) may follow divergent patterns.

It was difficult to set the limits for the types of compounds to be coverred in this article. The boundaries between "cluster" chemistry and other types of organometallic and inorganic chemistry are diffuse, and opinions vary considerably about what is and what is not a cluster compound. To establish some reasonable boundaries, we focus on complexes that contain E with on more than one nonmetal substituent. Thus, E may have a lone pair of electrons or an R group such as hydride, halide, alkyl, or aryl. The lone pair of electrons may serve as a donor to an external metal fragment, E: → M. The rationale behind this selection is that, in cluster compounds, E atoms will generally contribute all valence electrons to cluster bonding except for one external pair, which may be either a bonding or a nonbonding pair. Such fragments will show the most "vertex-like" behavior when incorporated into the cluster, as opposed to ER_2 fragments and (most definitely) ER_3, which are more ligand-like. ER_3 compounds are, of course, the classical ligands having a lone pair of electrons that can be donated to a metal center; these compounds fall under the classification of traditional coordination chemistry. From a scientific point of view, this distinction is not completely satisfactory, as there are a number of similarities as well as chemical relationships between the compounds chosen for discussion and those for omission (e.g., many reactions leading to E–M clusters often begin with metal complexes having ER_2 and/or ER_3 groups). A more practical consideration was the necessity to limit the number of complexes—the number meeting the prevoius criteria is already very large, and including those with ER_2 fragments (especially with E = N or P) would have made the study overwhelming.

Undoubtedly, other researchers working in this field would choose a different set of boundary conditions. Many have actually done so, as the reader will note in a number of review papers that have dealt with various aspects of the interactions between main group elements and transition metals.[1–25] Although some of these may be considered old by current standards, they offer differing, and perhaps insightful, perspectives as well as providing a historical development of the field. Several of these reviews focus on transition-metal cluster chemistry, but contain sections on mixed E–M compounds, and so are included in the list to allow the reader to put E–M cluster compounds into the context of transition metal clusters in general. To direct the reader who is new to this area, the titles of these reviews are included in these reference citations.

The chemistry of E–M compounds lies at the interface of two distinct areas of cluster chemistry. The chemistry of the low-valent, organometallic clusters of the transition metals developed largely from the pioneering work of W. Hieber, who reported many of the early studies on E–M clusters as well as those on homometallic transition-metal carbonyl species.[25] At the other end of the spectrum, it has been known since the late 1800s, from the work of Johannis, Zintl, and other researchers, that reduction of the main group elements in liquid ammonia or amine solvents leads to the formation of polyatomic cluster anions now commonly known as Zintl ions.[26–38] These species often revert to alloy phases rather than forming discrete compounds upon solvent removal. Zintl ion chemistry has been reviewed and the bonding in the cluster anions and the related solid state phases discussed.[39,40]

Much of the early work on E–M clusters was conducted from the viewpoint of the transition metal, and so the E fragment was often considered simply as a ligand. One observation of this review is that the E function is often so integrally involved in the cluster framework and bonding that in many cases it is most appropriate to consider it as a cluster vertex. As more attention is paid to this area, more examples of this type of influence on structure and reactivity are being reported. E can be a site of reaction just as M is. Given the development of transition metal cluster chemistry and also of that of the Zintl ions, it is not surprising that there have been a number of studies using the main group cluster anions as starting points for syntheses of E–M cluster compounds. It is useful for comparison purposes to consider the bonding in the solid-state phases and discrete cluster compounds that are made up only of E.

In this article, R denotes an organic radical such as Me, Et, Pr, or Ph. X is used to designate halides and pseudohalides, whereas Y is employed to represent a substituent that may be either X or R. For simplicity in some of the structural drawings, M is used to denote a transition metal with its associated ligands. The open arrow ($\Rightarrow$) is used when it is not desirable or possible to completely balance a chemical equation. In addition to those commonly employed, the following abbreviations are used:

Cp^{1M}	C_5H_4Me	Cp^{M4E}	C_5Me_4Et
Cp^{4P}	$C_5H^iPr_4$	Cp^{1B}	$C_5H_4^tBu$
Cp^{2B}	1,3-$C_5H_3^tBu_2$	Fp*	Cp*Fe$(CO)_2$Fp*
Fp^{1M}	$(C_5H_4Me)Fe(CO)_2$	Fp^{1B}	$(C_5H_4^tBu)Fe(CO)_2$
np_3	$N(CH_2CH_2PPh_2)_3$		

II

STRUCTURE AND BONDING PATTERNS

A. *General Considerations*

The discussion of structure and bonding is organized according to increasing metal connectivity at E for simple compounds with one E function. This includes molecules where more than one such E function is present. Ultimately this leads to high coordination geometries at E, which are only found when E is an interstitial atom. After the presentation of molecules with isolated E functions, compounds that have $E_x (x > 1)$ showing direct E–E bonding are presented in order of increasing x. A table format has been chosen to summarize the molecules belonging to each particular class. In some cases, molecules may belong to more than one classification and attempts have been made to place these in as many locations as are applicable. Some complicated molecules that do not fit easily into simple classification schemes have been placed in tables according to the local geometry about the E atom in the compound.

The Group 15 elements have five valence electrons that may be involved in bonding to other elements. Lewis dot structures for the most common arrangements are shown as **1–3**. As can be readily seen, these fragments may provide three or four electrons for bonding when serving as a cluster vertex. It is also possible for E to adopt an interstitial geometry, in which case it is considered to donate all five electrons to cluster bonding.

R M

E E E

1 **2** **3**

B. *Open Compounds of the Type $ER_x(ML_n)_y$*

The simplest E–M compounds are electron precise (obey the octet rule at E and the 18-electron rule at M) and can be viewed as derivatives of EY_3 (**4**) in which the Y groups are successively replaced by 17-electron metal fragments (**5**–**7**). Compounds of types **6** and **7** are summarized in Tables I and II. The E–M interactions can be considered as conventional covalent bonds for these molecules. All of the examples given in the two tables can be taken to have E^{3+} bonded to two or three 18-electron $[ML_n]^-$ units.

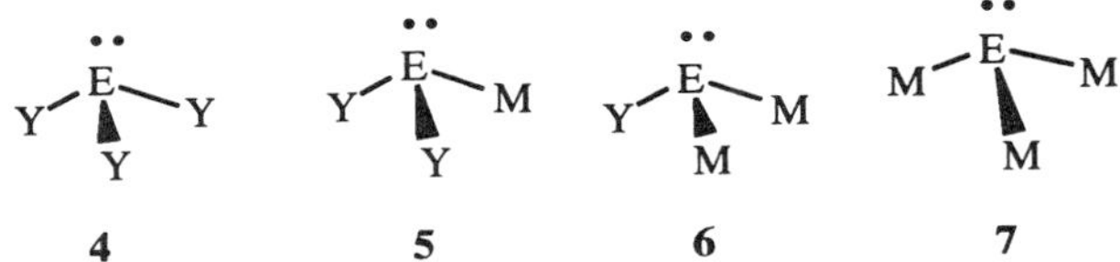

Two variations may be noted on this type of structure. The first is that two ER fragments can bridge two nonbonded metal centers, as exemplified by $Fe_2(CO)_8(BiR)_2$ (**8**, no Fe–Fe bond). The second is found only for the heaviest member of the Group 15 family and involves the formation of Lewis acid interactions between molecules to generate hypervalent bonding at Bi. Thus, the molecules $[Fp^x]_2BiCl$ (Fp^x = Cp or substituted Cp) exist in the solid state as cyclic trimers (**9**),[50–52] whereas $[Cp(CO)_3Mo]_2BiI$ and $[Cp(CO)_2Mn]_2BiCl$ exist as dimers.[46,61] For $[Cp(CO)_3Mo]_2BiBr$ the inter-

TABLE I

ELECTRON PRECISE PYRAMIDAL $(L_nM)_2EY$ COMPOUNDS, **6**

Compound	Ref.
$\{Cp(CO)_3M\}_2PCl$ (R = Mo, W}	41
$\{Cp(CO)_3M\}_2SbCl$ (M = Mo, W)	42
$\{Cp(CO)_2Fe\}_2SbX$ (X = Cl, Br, I)	42, 43
$\{(CO)_3(PPh_3)Co\}_2BiCl$	44
$\{Cp(CO)_3M\}_2BiX$ (M = Cr, Mo, W; X = Me, halide)	45–48
$\{Cp(CN^tBu)(CO)_2Mo\}_2BiCl$	48
$\{Fp^x\}_2Bi(NO_3)$ (Fp^x = Fp, Fp^{1M})	49
$\{Cp(CO)_2Ru\}_2Bi(NO_3)$	49
$\{Cp(CO)_2Ru\}_2BiCl$	49
$\{Cp^x(CO)_3M\}_2Bi(NO_3)$ (M = Cr, Mo, W; Cp^x = Cp, Cp^{1M})	49
$[\{Fp^x\}_2BiCl]_3$ (Fp^x = Fp, Fp^{1M})	50–52
$Fe_2(CO)_8(BiR)_2$ (R = Me, Et, CH_2Ph, Ph, ^{i}Pr, ^{n}Bu, ^{t}Bu, ^{i}Bu, $(CH_2)_3Cl$, $(CH_2)_4Br$)	53–56

TABLE II

$M_3(\mu_3\text{-E})$ COMPLEXES WITH NO M–M BONDS, **7**

Compound	Ref.
$P\{M(CO)_3Cp\}_3$ (M = Mo, W)	41
$As\{Mn(CO)_3\}_3$	41
$Sb\{M(CO)_3Cp\}_3$ (M = Mo, W)	42
$Sb\{Co(CO)_3PPh_3\}_3$	57
$Sb\{Mn(CO)_5\}_3$	41
$Sb\{Fp\}_3$	42
$Bi\{Cr(CO)_3Cp\}_3$	47
$Bi\{Mo(CO)_3Cp\}_3$	48
$Bi\{Mn(CO)_5\}_3$	58
$Bi\{Fp^x\}_3$ (Fp^x = Fp, Fp^{1M})	50
$Bi\{Fe(CO)_2(NO)PPh_3\}_3$	59
$Bi\{Fe(CO)_3(NO)\}_3$	59
$Bi\{Ru(CO)_2Cp\}_3$	49
$Bi\{Co(CO)_4\}_3$	60
$Bi\{Co(CO)_3SbPh_3\}_2\{Co(CO)_4\}$	60
$Bi\{Co(CO)_3L\}_3$ (L = PPh_3, $AsPh_3$)	57, 59, 60

molecular interactions form a polymeric chain network in the solid state.[46] Nitrate ion can also serve as a bridge between Bi centers.[49] Lewis acid interactions of this sort are common for Bi, but as yet unknown for the lighter Group 15 elements in this class of compounds.

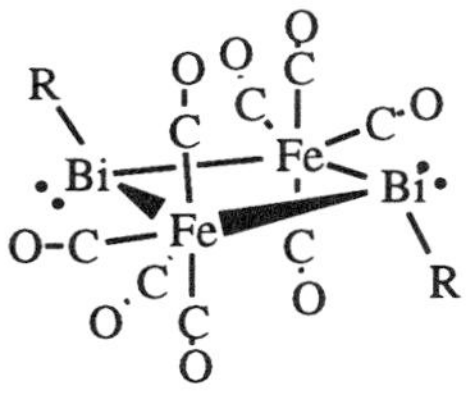

8

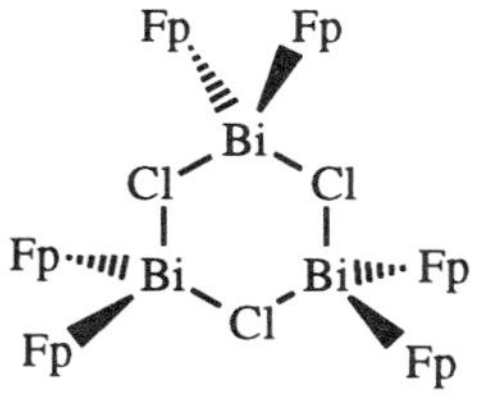

9

As is obvious from the drawings, the E atoms in these simple molecules retain a lone pair of electrons may be used for a dative bond to another ML_n fragment (**10**–**13**). Data on complexes of types **12** and **13** are summarized in Tables III and IV, respectively. The dilemma that arises is that from the electron counting procedures one would expect two types of bonds—covalent and dative (see Ref. 62 for a discussion of the differences between dative and covalent bonding)—for complexes **11**–**13**, but the ML_n fragments

TABLE III

$M_3(\mu_3\text{-EY})$ COMPLEXES WITH NO M–M BONDS, **12**

Compound	Ref.
$[\{(CO)_4Fe\}_3AsH]^{2-}$	66–68
$[\{Cp(CO)_2Fe\}_3AsOH]^{+}$	69
$[\{(CO)_4Fe\}_3SbH]^{2-}$	66
$[\{(CO)_4Fe\}_3SbCl]^{2-}$	70, 71
$[\{Cp(CO)_2Fe\}_3SbX]^{+}$ (X = Cl, Br)	59, 72
$[\{(CO)_4Fe\}_3BiCl]^{2-}$	70
$[\{(CO)_4Fe\}_3BiR]^{2-}$ (R = ^{i}Pr, ^{i}Bu, ^{n}Bu, ^{t}Bu, $(CH_2)_3Cl$, $(CH_2)_4Br$)	54

are indistinguishable. This discussion is further complicated by variation in the nature of ML_n. As long as ML_n is a 17-electron fragment that readily forms electron-pair bonds with one of the valence electrons on E, the situation is straightforward. It is possible to consider the metal fragments as being charged, and this makes the assignment more obscure. For example, a Lewis dot diagram for $[Bi\{Fe(CO)_4\}_4]^{3-}$ can easily be drawn that allows the bismuth atom to have an octet of electrons and the iron groups to obey the 18-electron rule (**14**). But is bismuth present as Bi^{3-} [with neutral $Fe(CO)_4$ fragments], Bi^{5+} [with $Fe(CO)_4^{2-}$ fragments], or some intermediate value? The two extreme views at least have the advantage that all the metal carbonyl fragments are equivalent, which agrees with the bond parameters for this regular tetrahedral molecule. It would appear, based on empirical observations of chemical reactivity, that the bonding is highly covalent.

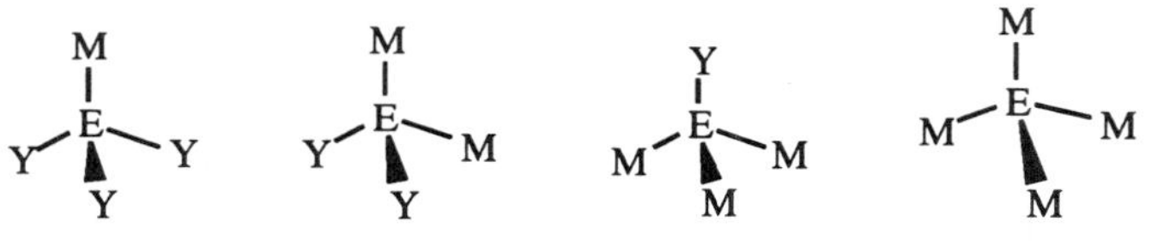

10 **11** **12** **13**

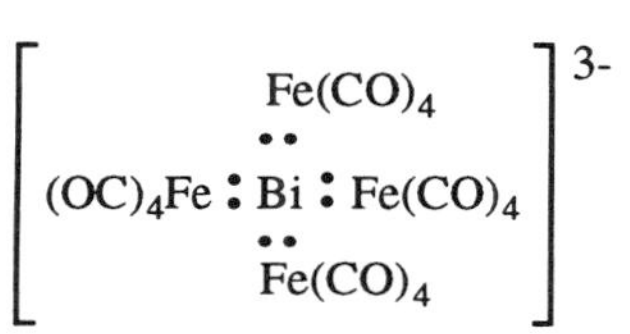

$$\left[(OC)_4Fe : Bi\{Fe(CO)_4\}_3 \right]^{3-}$$

14

TABLE IV

M_4E COMPLEXES WITH NO M–M BONDS, **13**

Compound	Ref.
$[Sb\{Fe(CO)_4\}_4]^{3-}$	70
$[Bi\{Fe(CO)_4\}_4]^{3-}$	73, 74
$[Bi\{Co(CO)_4\}_4]^-$	64, 75
$[\{Cp(CO)_2Fe\}_4As]X$ (x = Cl, BPh_4)	69
$[Sb\{Fe(CO)_3NO\}_4]^+$	59
$[Sb\{Co(CO)_3PPh_3\}_4]^+$	59

This is also true for the molecules $[Fp_3AsOH]^+$, $[Fp_3SbX]^+$, $[Sb\{Fe(CO)_3(NO)\}_4]^+$, and $[Sb\{Co(CO)_3PPh_3\}_4]^+$, where the choice of the E in the +5 oxidation state is appealing given the overall charge on the complex. As with $[Bi\{Fe(CO)_4\}_4]^{3-}$, the placement of reducing metal carbonyl anions next to oxidizing main group element cations would not be expected to be a stable situation. The Sb Mössbauer spectrum for $[Sb\{Fe(CO)_4\}_4]^{3-}$ suggests that the Sb atom is in the +1 oxidation state, agreeing with the conclusion that the bonding is highly covalent.[63] An unusual molecule in this class is $[Bi\{Co(CO)_4\}_4]^-$, in which the bismuth is formally a 10-electron center.[64,65] This hypervalency is reflected in the very long Bi–Co bonds (ca. 0.2 Å greater than for other "normal" Bi–Co bond distances).

Complexes may also be constructed of a fusion of open fragments. For **7** this can lead to the well-known cubane class of complexes **15** (Table V). The cubane core is a structure found naturally in biological systems having E = S and metal fragments based upon Fe and/or Mo. Organometallic

TABLE V

E_4M_4 CUBANE AND CUBANE-LIKE STRUCTURES, **15–17**

Compound	Ref.
$\{Cp^*Ti\}_4N_4$	76
$[CpCo]_4P_4$	77
$Co_4(PPh_3)_4(PPh)_4$	78
$\{Cp^xNi\}_4P_4$ ($Cp^x = Cp^{1M}, Cp^{1B}$)	79
$\{Cp^xNi\}_3P_5$ ($Cp^x = Cp^{1M}, Cp^{1B}$)	79
$Fe_4(CO)_{12}(AsMe)_4$	80, 81
$[CpMoAs]_4$	82
$Sb_4Co_4(CO)_{12}$	83
$Bi_4Co_4(CO)_{12}$	84

clusters most often have 24 electrons associated with the core structure, giving rise to 12 edge-localized bonds. This electron count and structural pattern holds for $\{Cp^*Ti\}_4N_4$, $Fe_4(CO)_{12}(AsMe)_4$, $Sb_4Co_4(CO)_{12}$, and $Bi_4Co_4(CO)_{12}$. $[CpCo]_4P_4$ and $Co_4(PPh_3)_4(PPh)_4$ differ in their electron counts, with the former having 20 core electrons and the latter having only 12. For each pair of electrons missing, an M–M bond is formed, causing a distortion of the basic cubane structure. This principle of M–M bond formation can be generally applied for the whole range of E–M clusters. Thus, $[CpCo]_4P_4$ is found to have two Co–Co bonds and $(PPh_3)_4Co_4(PPh)_4$ has six, consistent with the structures shown as **16** and **17**. Perhaps a better view of **17** is to consider it a metal tetrahedron capped on all four faces with PPh groups.

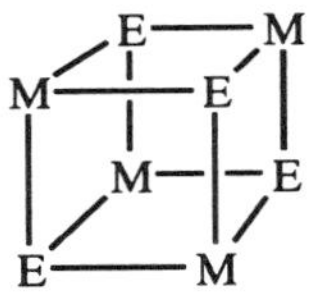

15 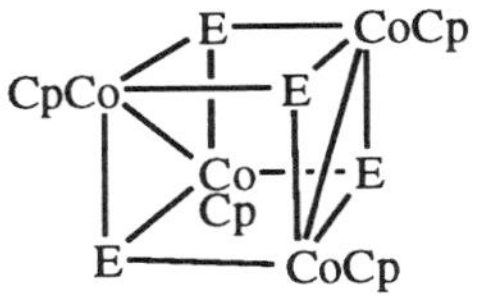 **16** 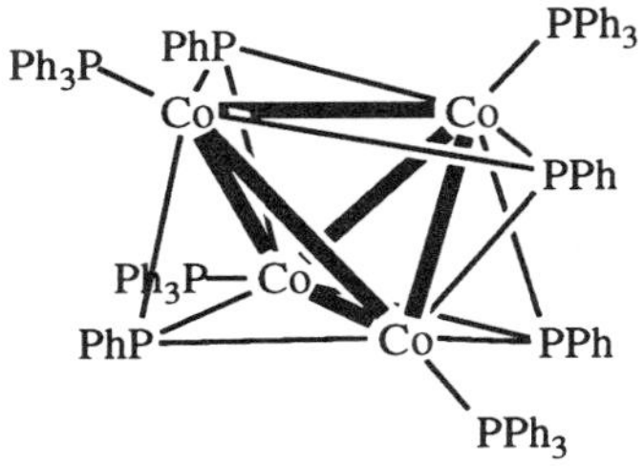 **17**

C. *Electron Precise $ER_x(ML_n)_y$ Compounds with M–M Bonding*

As illustrated by the cubane structure, the next step in molecular complexity is to introduce M–M bonding into the simpler structural types already noted. In each of the cases with two or more ML_n fragments, the removal of two electrons via either oxidation or loss of a two-electron donor ligand results in the formation of M–M bonds (**18**–**29**, Tables VI–XIII), just as the reverse addition may result in M–M bond cleavage. Multiple loss of two electrons or two-electron-donor ligands results in creation of more M–M bonds, leading to *closo* structures, of which the symmetrical, tetrahedral types **27**, **28**, and **29** are by far the most common. They possess six skeletal bonding pairs of electrons, coinciding with the number of bonds in the tetrahedron for a localized bonding approach. The tetrahedron can also, however, be derived from vertex removal from a trigonal bipyramid ($N = 5$), where six skeletal electron pairs ($N + 1$) are predicted by cluster-counting formalisms. Thus, a tetrahedron may also be considered as a *nido*-trigonal bipyramid. The difference between this view and the former is

TABLE VI

COMPOUNDS CONTAINING AN X–EM$_2$ FRAGMENT WITH AN M–M BOND, **18**

Compound	Ref.
$Fe_2(CO)_8Sb\{CH(SiMe_3)_2\}$	87, 88
$\{W_2(CO)_{10}\}Sb\{CH(SiMe_3)_2\}$	89
$[CpRu(CO)]_2(\mu\text{-}CO)(\mu\text{-}AsR)$ (R = Me, Ph)	90

that here the bonding electrons are *delocalized* over the metal framework. This formalism is predicated upon the metal vertices using three orbitals for cluster bonding, but King[85] writes that the four-connected vertices can also be allowed to use four orbitals, making the compounds electron precise. He has discussed this idea for PR vertices in the context of tetrahedral and trigonal bipyramidal molecules. A spectroscopic and redox investigation of $Co_3(CO)_9PPh$ and $Co_2Fe(CO)_9PPh$ supported the contention that the HOMO of the all-Co compound (49-electron cluster) and the LUMO of the Co_2Fe compound (48 electrons) were essentially all metal-*d* orbital in character.[86]

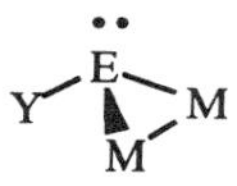

18 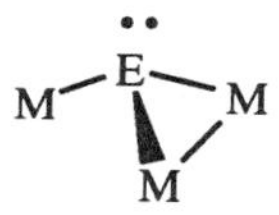 **19** 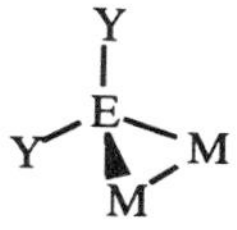 **20** 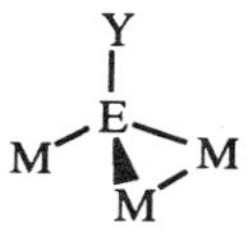 **21** 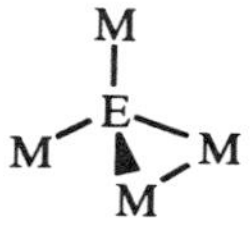 **22**

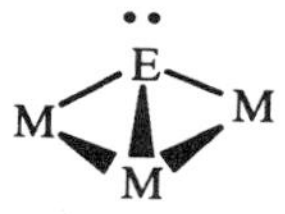

23 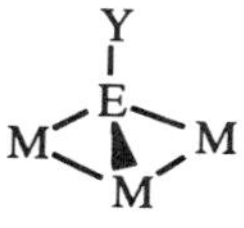 **24** 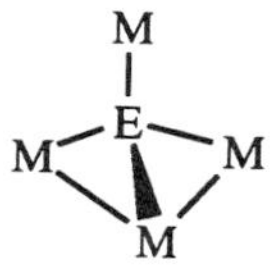 **25** 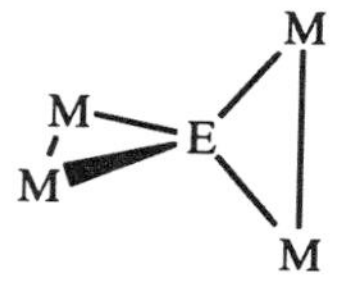 **26**

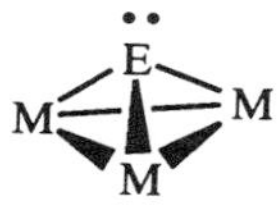

27 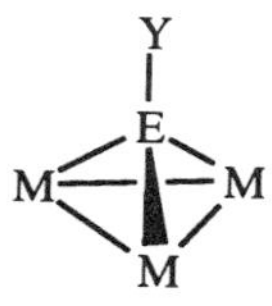 **28** 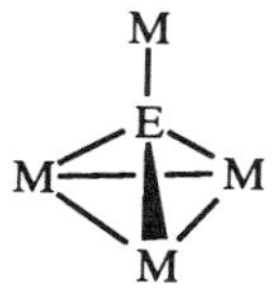 **29**

TABLE VII

M_3EY COMPOUNDS WITH ONE M–M BOND, **21**

Compound	Ref.
$\{W_2(CO)_{10}\}\{W(CO)_5\}SbR$ (R = ^{t}Bu, Cl)	44
$\{Bi_2W_2(CO)_8\}\{W(CO)_5\}\{Bi(Me)\}$	89
$\{Fe_2(CO)_5L_3\}\{Fe(CO)_3L\}PR$ (R = Ph, C_6H_4OMe-p, ^{t}Bu; L = CO, $P(OMe)_3$, CN^tBu)	91
$[\{HFe_2(CO)_7\}\{Fe(CO)_4\}PMe]^-$	92
$Ru_4(CO)_{13}(PPh)_2$	93
$[HOs_3(CO)_{10}(NCMe)][HOs_3(CO)_{10}]PH$	94, 95
$[HOs_3(CO)_{11}][HOs_3(CO)_{10}]PH$	94
$[\{Os_3(CO)_{11}\}\{HOs_3(CO)_{10}\}PH]^-$	94
$[\{HOs_3(CO)_9(\mu, \eta^2\text{-OCOMe})\}\{HOs_3(CO)_{10}\}PH]^-$	94
$[H_2Os_3(CO)_9(\mu, \eta^2\text{-OCOMe})][HOs_3(CO)_{10}]PH$	94
$[Co_2(CO)_5]_2(P^iPr)_3$	96
$[(C_8H_{12})_2(Cl)Ir_2]_2(PhPPPh)(PPh)_2$	96
$\{FeCo(CO)_6(PMe_2Ph)\}\{Co(CO)_3(PMe_2Ph)\}AsMe$	97

The "loss" of electrons/ligands with concomitant M–M bond formation is often purely conceptual, as the actual loss/addition process may never be observed directly, but it does serve as a very useful guiding principle in predicting chemical reactivity. Chemical oxidation/reduction reactions or ligand addition/removal may not lead simply to the anticipated derivative. Either these reactions may not occur at all, or the outcome may be a more complicated fragmentation or rearrangement. That is not to say that these electron loss/addition relationships are completely fanciful. There are some well-documented cases where the electron or ligand addition/loss processes proceed cleanly with M–M bond cleavage/formation as expected.

D. *Electronically Unsaturated Molecules*

The compounds described to this point have generally been electron precise, obeying the octet rule at E and the 18-electron rule at M. Exceptions

TABLE VIII

EM_4 COMPLEXES WITH ONE M–M BOND, **22**

Compound	Ref.
$[Et_4N][Sb\{Fe_2(CO)_8\}\{Fe(CO)_4\}_2]$	98
$[Et_4N][Sb\{Fe_2(CO)_8\}\{Fe(CO)_4\}\{Cr(CO)_5\}]$	53
$[Et_4N][Bi\{Fe_2(CO)_8\}\{Fe(CO)_4\}\{Cr(CO)_5\}]$	53

TABLE IX

EM_3, REM_3, and EM_4 COMPOUNDS WITH TWO M–M BONDS: **23**, **24**, and **25**

Compound	Ref.
$Cp_3(CO)_4(O)Mo_3N$	99
$Mo_4(O^iPr)_{12}(N)_2$	100
$Os_3(CO)_9(\mu$-X)NR (X = Cl, I; R = Ph, Tol)	101
$HOs_4(CO)_{12}(\mu_3$-$MPPh_3)\{NC(O)Me\}$ (M = Cu, Au)	102
$CrCo_2(CO)_{11}PH$	103
$[Fe_3(CO)_{10}(XCN)(PR)]^-$ (X = S, Se, Te; R = ^{i}Pr, ^{t}Bu)	104
$Fe_3(CO)_8(\mu$-$TeC_6H_2Me_3$-2,4,6$)_2(P^iPr)$	104
$Fe_3(CO)_8L_3PR$ (R = Ph, C_6H_4OMe-*p*, ^{t}Bu; L = CO, $P(OMe)_3$, CN^tBu)	91
$Fe_3(CO)_9(\mu_3$-S)(PR) R = Me, 4-$MeOC_6H_4$)	105
$HFe_3(CO)_9\{\mu$-P(Me)R\}(PR) (R = ^{i}Pr, CH_2Ph, Me, ^{t}Bu)	106
$Fe_3(CO)_8(\mu$-$P(H)Ph)_2(PPh)$	107
$Fe_2Co(CO)_9(\mu$-PHR)(PR) (R = Me, ^{t}Bu, Ph, Tol)	108
$CpMnFe_2(CO)_8L(PPh)$ (L = CO, PPh_3)	109
$Ru_4(CO)_{13}(PPh)$	102, 110, 111
$Ru_4(CO)_{13}(PPh)_2$	93
$H_2Ru_4(CO)_{12}(PPh)$	78, 111
$Ru_3(CO)_7(\mu$-$CHPPh_2)$(dppm)(PPh)	112
$Ru_3(CO)_9\{C_6H_4Cr(CO)_3\}\{PC_6H_5Cr(CO)_3\}$	113
$HOs_3(CO)_7[Fe(C_5H_4P^iPr_2)(C_5H_4)]P^iPr$	114
$Os_3(CO)_9[\mu_3$-$(C_5H_3)FeCp][P(C_5H_4)FeCp]$	115
$Os_3(CO)_9[C_6H_4][P(C_5H_4)FeCp]$	115
$[HOs_4(CO)_{12}\{\mu_3$-NC(O)(Me)\}$]^-$	102
$HOs_4(CO)_{12}(\mu_3$-$MPPh_3)\{\mu_3$-NC(O)(Me\} (M = Cu, Au)	102
$Ni_5(CO)_5(P_3{}^tBu_3)(\mu_4P^tBu)(\mu_3$-$P^tBu)_2$	116
$Cp(CO)_4(PPh_3)_2MoPt_2As$	117

to this in cluster chemistry can arise when metals that are stable with lower than 18-electron counts, such as Pd or Pt, are incorporated into the cluster, the cluster electron count often reflecting the lower electron demand by the metal. Another exception is the compound $HRu_3(CO)_9\{NS(O)MePh\}$, which has two fewer electrons than expected.[170] It appears to be analogous

TABLE X

SPIROCYCLIC EM_4 AND RELATED COMPOUNDS WITH TWO M–M BONDS, **26**

Compound	Ref.
$(CO)_8Fe_2AsFe_2(CO)_6Cl$	118
$\{Cp(CO)_2MoC(H)(PPh_3)As\}Mo_2(CO)_4Cp_2$	119
$[Fe_2(CO)_8(\mu_4$-$As)][Fe_2(CO)_6]$	120
$[\{Cp_2(CO)_2Co_2\}_2As][BF_4]$	121
$Fe_2(CO)_6Sb_2[Fe_2(CO)_8]_2$	122, 123
$Ru_2(CO)_8Bi\{HRu_3(CO)_{10}\}$	124

TABLE XI

TETRAHEDRAL M_3E COMPLEXES, **27**

Compound	Ref.
$Cp_3(CO)_6M_3P$ (M = Mo, W)	125
$Co_3(CO)_9P$	126
$Cp_3(CO)_6M_3As$ (M = Mo, W)	119, 127, 128
$(\mu_3\text{-HC})Fe_3(CO)_9As$	129
$[\{Me_3PAu\}_3NR]^+$ (R = Me, ^{t}Bu, CH_2Ph, $SiMe_3$)	130
$[\{Ph_3PAu\}_3PTol]^+$	131
$(\mu_3\text{-HC})Fe_3(CO)_9Sb$	129
$[PPN][Fe_3(CO)_{10}Sb]$	132
$[Et_4N]_2[(OC)_3Mo(OMe)_3Mo_3(CO)_9Bi]$	133
$[Fe_3(CO)_{10}Bi]^-$	73, 132
$(\mu_3\text{-MeOC})Fe_3(CO)_9Bi$	134
$H_3Fe_3(CO)_9Bi$	134
$(\mu_3\text{-HC})Fe_3(CO)_9Bi$	129
$(Cp^x)_3Fe_3(CO)_3Bi$ (Cp^x = Cp, Cp^{1M})	50, 135
$H_3Ru_3(CO)_9Bi$	136, 137
$H_3Ru_3(CO)_{9-x}(L)_xBi$ (L = PPh_3, $As(Tol)_3$, $SbPh_3$, $P(OMe)_3$; x = 1–3; L = PEt_3, x = 1, 2)	138
$H_3Os_3(CO)_9Bi$	136
$Co_3(CO)_9Bi$	65, 139
$Ir_3(CO)_9Bi$	140

to $H_2Os_3(CO)_{10}$, where M–M multiple bonding stabilizes the lower electron count.

Lower electron counts may in general be stabilized by multiple bonding, and a number of "inidene" complexes showing localized or delocalized E–M multiple bonding have been reported. The structural types (**30–35**) are summarized in Table XIV. These complexes arise when a lone pair of electrons on E is donated to an adjacent electron-deficient metal center [e.g., ML_n = $CpMn(CO)_2$, Cp_2^*Zr] or when a saturated metal center contributes electron density to an electron-deficient E fragment [e.g., ML_n = $Fe(CO)_4$, $M(CO)_5$ (M = Cr, Mo, W)]. Multiple bond character may be delocalized as in **34**, **35**, and **36**, or localized as in **37** or **38**. The most common structural type encountered to date has been **34**. In the case of X = Cl, addition of $CF_3SO_3SiMe_3$ results in elimination of Me_2SiCl to give $[\{L_nM\}_2E][CF_3SO_3]$, which has been shown for $[\{Cp^*(CO)_2Mn\}_2As][BF_4]$ to have a linear Mn–As–Mn linkage (**33**). This is presumably true for the bismuth analogue as well, but the X-ray structure of that complex has not yet been reported. Another apparent anomaly is complex **39**, which appears to retain partial Mn–Bi multiple bonding as evidenced by the Mn–Bi bond

TABLE XII

TETRAHEDRAL YEM_3 COMPLEXES, **28**

Compound	Ref.
$Cp^*_3(CO)_8Mo_3Co_2(\mu_4\text{-}N)(\mu_3\text{-}NH)$	141
$(Cp^{1M})MoFe_2(CO)_7(NO)(NH)$	142
$[(Cp^{1M})_3Mn_3(NO)_3(NOH)]^+$	143
$[(Cp^{1M})_3Mn_3(NO)_3(NH)]^+$	143
$[Cp\{Cp^{1M}\}_2MnFe_2(CO)_2(NO)(NR)]^-$ (R = OMe, OH, H)	144
$[\{Cp^{1M}\}Cp_2MnFe_2(CO)_2(NO)(NR)]^+$ (R = OMe, OH, H)	144
$[\{Cp^{1M}\}_3MnFe_2(CO)_2(NO)(NH)]^+$	144
$Fe_3(CO)_{10}NH$	145
$[Fe_3(CO)_8(NO)(NH)]^-$	146
$Fe_3(CO)_{10}NSiMe_3$	147–149
$Fe_3(CO)_{10}NCH_2Ph$	150
$Fe_3(CO)_{10}NEt$	151
$[HFe_3(CO)_9NPh]^-$	152, 153
$[Fe_3(CO)_9NPh]^{2-}$	153
$H_2Fe_3(CO)_9NR$ (R = Et, CH_2Ph)	154
$FeRu_2(CO)_{10}NH$	155
$Fe_2Ru(CO)_{10}NPh$	156
$FeRu_2(CO)_{10}NPh$	156
$FeRu_2(CO)_9\{P(OMe)_3\}NH$	155
$Ru_3(CO)_{10}NH$	155
$Ru_3(CO)_{10}NPh$	157–161
$Ru_3(CO)_{10}(NOMe)$	162, 163
$Ru_3(CO)_{10}(NOH)$	162
$[Ru_3(CO)_9(X)NPh]^-$ (X = H, Cl, Br, I, CN)	156, 164, 165
$[Ru_3(CO)_9\{C(O)R\}NPh]^-$ (R = Me, Ph)	156
$HRu_3(CO)_9\{C(O)R\}NPh$ (R = Me, Ph)	156
$H_2Ru_3(CO)_9NH$	166, 167
$H_2Ru_3(CO)_9NPh$	158, 168
$HRu_3(CO)_8(MeOC_6H_3NNC_6H_4OMe)(NC_6H_5)$	161
$H_2Ru_3(CO)_9NCH_2Ph$	169
$HRu_3(CO)_9\{NS(O)MePh\}$	170
$Ru_3(CO)_7(C_6H_6)(NPh)$	171
$FeCo_2(CO)_9(NH)$	167
$FeCo_2(CO)_9(NPh)$	171
$[CoRu_2(CO)_9NPh]^-$	156
$HCoRu_2(CO)_9NPh$	156
$FeRu_3(CO)_{10}\{P(OMe)_3\}_2NH$	155
$H_2Os_3(CO)_9NR$ (R = Ph, ^{n}Bu, CH_2Ph, cy)	172, 173
$Os_3(CO)_{10}NR$ (R = Ph, Tol)	101, 156
$[Os_3(CO)_9(I)NPh]^-$ (R = Ph, Tol)	101
$Cp_3Ni_3N^tBu$	174
$(P_2{}^tBu_2)Cr_3(CO)_{10}P^tBu$	175
$WFe_2(CO)_7(\mu\text{-}CTol)(NOH)$	176
$WFe_2(CO)_7(\mu\text{-}CTol)(NTol)$	176
$H_2Fe_3(CO)_9PR$ (R = Ph, p-C_6H_4OMe, cy, ^{t}Bu, CH_2CH_2CN, ^{i}Pr, Me, Tol)	92, 177–180

(*continues*)

TABLE XII (*continued*)

Compound	Ref.
$H_2Fe_3(CO)_9PSi(^iPr)_3$	181
$[HFe_3(CO)_9PSi(^iPr)_3]^-$	181
$HFe_3(CO)_9(\mu\text{-}P(H)SiMe_3)(PSiMe_3)$	182
$[HFe_3(CO)_9PR]^-$ (R = Me, ^{t}Bu, Ph, Tol)	181
$[H_2Fe_3(CO)_9P]C_6H_4[PFe_3(CO)_9H_2]$	183
$[H_2Fe_3(CO)_9P]C_6H_4\text{-}C_6H_4[PFe_3(CO)_9]H_2$	183
$[(Ph_3PAu)_2Fe_3(CO)_9P]C_6H_4[PFe_3(CO)_9(AuPPh_3)_2]$	183
$[Fe_3(CO)_9PR]^{2-}$ (R = Ph, cy, Me, Tol, ^{t}Bu)	92, 184
$Fe_3(CO)_{10}PR$	
(R = Me, ^{i}Pr, CH_2Ph, $CH(SiMe_3)_2$, ^{t}Bu, NEt_2, Ph, $C_6H_2Me_3$-2,4,6, C_6H_4OMe-4)	104, 178, 179 185–194
$Fe_3(CO)_{10-x}L_xPR$ (R = alkyl, aryl; x = 0–2; L = phosphite, isocyanide)	91
$Fe_3(CO)_{10}PNEt_2$	195
$Fe_3(CO)_9(PR_2Me)PR$ (R = Pr, CH_2Ph)	185
$[HFe_3(CO)_9(PH)]^-$	181
$H_2Fe_3(CO)_9(PC_6H_4OMe\text{-}4)$	196
$H_2Fe_3(CO)_8(PR_2Me)(PR)$ (R = Me, ^{i}Pr, CH_2Ph)	185
$Fe_3(CO)_9\{C(R)OR'\}(P^tBu)$ (R = Ph, Me; R′ = OEt, OMe)	197
$Fe_3(CO)_9\{=C(R)OR'\}(P^tBu)$ (R = Ph, Me; R′ = OEt, OMe)	197
$Fe_3(CO)_9(\mu\text{-}SnMe_2)PPh$	92
$HFe_3(CO)_9(\mu\text{-}AgL_2)PR$ (R = Ph, Tol; L = PMe_3, PPh_3)	92
$Fe_3(CO)_9(\mu\text{-}AgL_2)_2PR$ (R = Ph, Tol; L = PMe_3, PPh_3)	92
$Fe_3(CO)_9(\mu\text{-}AuL)_2PR$ (R = Ph, Tol; L = PMe_3, PPh_3)	92
$Fe_3(CO)_9(\mu\text{-}CuL)_2PR$ (R = Ph, Tol; L = PMe_3, PPh_3)	92
$HFe_3(CO)_9(\mu\text{-}R(R)Me)(PR)$ (R = ^{i}Pr, CH_2Ph, Me, ^{t}Bu)	106
$CpMFeCo(CO)_8(PR)$ (M = Mo; R = Ph, ^{t}Bu; M = W, R = ^{t}Bu)	198
$CoMCo_2(CO)_6\{(CO)_4FeAsMe_2\}$ (M = Mo, W)	198
$FeCo_2(CO)_8\{AsMe_2Mo(CO)_3Cp\}(PR)$ (R = Me, Ph)	198
$CpMoFeCo_2(CO)_8\{\mu\text{-}AsMe_2\}(PR)$ (R = Me, Ph, ^{t}Bu)	198
$CpWFeCo_2(CO)_8\{\mu\text{-}AsMe_2\}(P^tBu)$ (R = Me, Ph, ^{t}Bu)	198
$CpMnFe_2(CO)_8(PPh)$	199
$Cp(CO)_2MnFe_2(CO)_6PPh$	177, 199
$FeCo_2(CO)_9PH$	183
$FeCo_2(CO)_9PR$ (R = Me, Ph, ^{t}Bu)	200
$HFe_2Co(CO)_9PR$ (R = Me, Ph, ^{t}Bu)	200
$HFeRuCo(CO)_9PMe$	200
$HRu_2Co(CO)_9PMe$	200
$[FeCo_2(CO)_9P]C_6H_4[PFeCo_2(CO)_9]$	183
$[FeCo_2(CO)_9P]C_6H_4\text{-}C_6H_4[PFeCo_2(CO)_9]$	183
$[HFe_2Co(CO)_9P]C_6H_4[PFe_2Co(CO)_9H]$	183
$[HFe_2Co(CO)_9P]C_6H_4\text{-}C_6H_4[PFe_2Co(CO)_9H]$	183
$H_2Ru_3(CO)_9PPh$	111, 201, 202
$[HRu_3(CO)_9PR]^-$ (R = Ph, ^{t}Bu)	203
$H_2Ru_3(CO)_8(PPh_3)PPh$	204
$H_2Ru_3(CO)_9P(C_6H_4X\text{-p})$ (X = Ome, Br)	205
$H_2M_3(CO)_8\{PPhH_2\}PPh$ (M = Ru, Os)	201

TABLE XII (*continued*)

Compound	Ref.
$HRu_5(CO)_{10}(\mu\text{-}PPh_2)(\mu_3\text{-}PPh)(\mu_4\text{-}PPh)$	206
$Ru_6(CO)_{12}(\mu_3\text{-}PPh)_2(\mu_4\text{-}PPh)_2$	207
$Ru_6(CO)_{12}(\mu_3\text{-}PPh)_2(\mu_4\text{-}PPh)_3$	207
$Cr(C_6H_4)Ru_3(CO)_{11}P^tBu$	208
$FeCo_2(CO)_{9-x}L_x(PR)$ (L = PMe_2Ph; x = 1, 2; R = Me, Ph, ^tBu)	200
$M_4(CO)_{13}PPh$ (M = Ru, Os)	209, 210
$HRu_4(CO)_9\{\mu\text{-}P(N^iPr_2)_2\}(PN^iPr_2)$	211
$H_2Ru_6(CO)_{12}(\mu_3\text{-}P^tBu)_3$	212
$Ru_6(CO)_{11}(\mu_3\text{-}P^tBu)_3(\mu_4\text{-}P^tBu)$	213
$Ru_7(CO)_{14}(\mu_3\text{-}P^tBu)_2(\mu_4\text{-}P^tBu)_2$	213
$H_3Ru_9(CO)_{20}(\mu_3\text{-}P^tBu)_3(\mu_7\text{-}P)$	213
$H_2RhRu_3(CO)_{10}(PEt_3)PPh$	214
$[Co_3(CO)_9P]C_6H_4[PCo_3(CO)_9]$	183
$[Co_3(CO)_9P]C_6H_4\text{-}C_6H_4[PCo_3(CO)_9]$	183
$Co_3(CO)_9\{PCH(SiMe_3)_2\}$	215
$Ir_4(CO)_6(PPh_3)_4(\mu_3\text{-}Ph)$	216
$FeCo_2(CO)_9AsMe$	81, 97
$CpMoFeCo(CO)_8AsMe$	97
$H_2Ru_3(CO)_9AsPh$	205
$H_2Ru_3(CO)_8(AsPh_3)(AsPh)$	217
$H_2Ru_3(CO)_7(AsPh_3)_2(AsPh)$	217
$(\mu\text{-}H)_2Ru_3(CO)_8(\mu_2\text{-}NSNAs^tBu_2)(\mu_3\text{-}As^tBu)$	218
$HOs_3(CO)_8(NSNAs^tBu_2)(As^tBu)$	219, 220
$Os_3(CO)_9(\mu_3\text{-}C_6H_3Me)(AsTol)$	221, 222
$Os_3(CO)_8(\mu_3\text{-}C_6H_3Me)(AsPh_3)(AsTol)$	222
$(PPh_3)_4Co_4(\mu_3, \eta^3\text{-}As_3)(\mu_3\text{-}As)_3$	223
$[PPN][HFe_3(CO)_9Sb^tBu]$	132

distances (2.47 Å), even though the formation of the halide bridges alleviates the electron deficiency at Bi. This is based on the supposition that Bi is in the +1 oxidation state. A standard Lewis dot diagram can be drawn for this molecule that places 18 electrons at Mn and 8 electrons at Bi, so the driving force for the formation of the multiple bond interaction is not clear. In this scenario, the Bi atom acts as a two-electron donor to each Mn atom, leading to a charge structure described as $[Cp(CO)_2Mn]^{2-}$ and Bi^{5+}. Bismuth(V) would have a naturally smaller radius, and this might be the cause of the short Mn–Bi bond distances. The difference may be largely semantic but it points out the types of ambiguities that are involved in understanding these E–M cluster compounds. Regardless of the interpretation of bonding, the important point is that appropriate E–M combinations can lead to stable molecules with lower electron counts than would normally

TABLE XIII

"SPIKED" TETRAHEDRAL EM_4 COMPLEXES, **29**

Compound	Ref.
$Co_3(CO)_9PFe(CO)_4$	126
$Co_3(CO)_9PML_n$ (ML_n = $Cp(CO)_2Mn$, $Cr(CO)_5$, $W(CO)_5$)	224
$[Co_3(CO)_8P]_3$	225
$Co_3(CO)_8(PR_2R')PMn(CO)_2Cp$ (R = R′ = ${}^tBu$, cy, Ph; R = CH_2Ph, R′ = Ph; R = Ph, R′ = CH_2Ph)	224
$Co_3(CO)_8\{P(O^iPr)_3\}PMn(CO)_2Cp$	224
$Co_3(CO)_7\{P(OR)_3\}_2PMn(CO)_2Cp$ (R = Me, ${}^iPr$, Ph)	224
$Co_3(CO)_6\{P(OR)_3\}_3PMn(CO)_2Cp$ (R = Me, Ph)	224
$Cp_3(CO)_6M_3PM(CO)_5$ (M = Mo, W)	125
$Re_5(CO)_{14}(\mu\text{-}PMe_2)(\mu_4\text{-}PMe)\{\mu_3\text{-}PRe(CO)_5\}$	85, 226
$[Fe_3(CO)_{10}PFe(CO)_4]^-$	227
$H_2Fe_3(CO)_9PAuPR_3$ (R = Ph, Et)	228
$H_2Fe_3(CO)_9PAu(PR_3)_2$ (R = Ph, Et)	228
$H_2Os_3(CO)_9PR$ (R = Ph, cy)	229, 230
$[(OC)_3Mo(OMe)_3Mo_3(CO)_9AsMo(CO)_5]^{2-}$	231
$Mo_3Cp_3(CO)_6AsML_n$ (ML_n = $Cp(CO)_2Mn$, $(CO)_5M''$(M′ = Cr, Mo, W))	127, 128
$Cp_3Ni_2Mo(CO)_2AsMo_3Cp_3(CO)_5As$	117
$(\mu\text{-}H)Fe_2Mo(CO)_8CpAsMo(CO)_3Cp$	119, 128
$CpMo_2Fe(CO)_7AsMo(CO)_3Cp$	128
$(HC)Fe_3(CO)_9AsML_n$ (ML_n = $M(CO)_5$ (M = Cr, Mo, W))	129
$[(HC)Fe_3(CO)_9As]_2ML_n$ (ML_n = $M(CO)_4$ (M = Cr, Mo))	129
$(HC)Fe_3(CO)_9EML_n$ (E = As, Sb; ML_n = $Cp^{1M}(CO)_2Mn$)	129
$Co_3(CO)_9AsCr(CO)_5$	224
$Co_3(CO)_9AsCo_4(CO)_{11}$	224
$[Co_8(\mu_6\text{-}As)(\mu_4\text{-}As)(\mu_4\text{-}AsPh)_2(CO)_{16}]_2$	232
$[Co_3(CO)_8As]_3$	120, 225
$[Et_4N]_2[HFe_3(CO)_9\{SbFe(CO)_4\}]$	70
$[Et_4N][H_2Fe_3(CO)_9\{SbFe(CO)_4\}]$	70
$[Et_4N][Fe_3(CO)_{10}\{SbFe(CO)_4\}]$	233

have been predicted based on localized bonding pictures. This can be contrasted to the already-mentioned ability of the heavier main group elements to adopt electron-rich or hypervalent bonding modes. Combined, these two capacities give this class of compounds a degree of flexibility that is very important in determining reaction pathways.

$$\overset{\ominus}{M}{\equiv}\overset{\oplus}{E}{:} \qquad M{=}\ddot{E}{-}R \qquad \overset{\ominus}{M}{\equiv}\overset{\oplus}{E}{-}R \qquad M{=}E{=}M$$

30 **31** **32** **33**

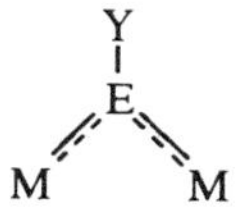

34 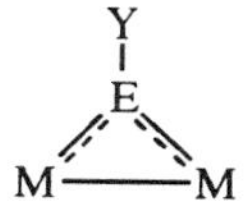 **35** 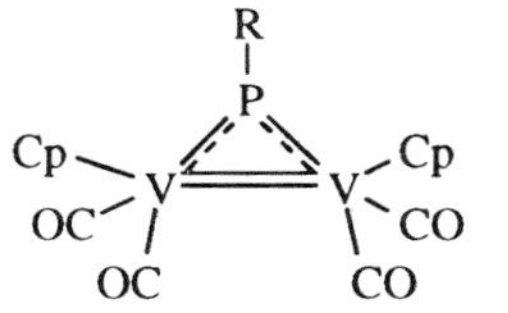 **36**

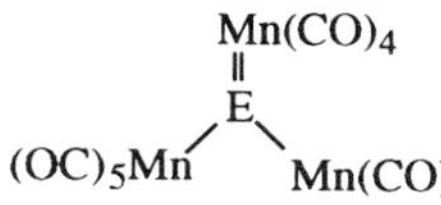

37 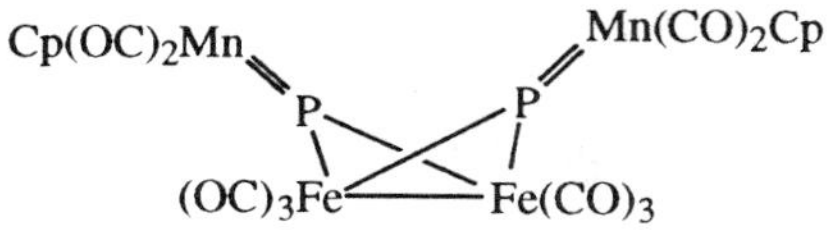 **38**

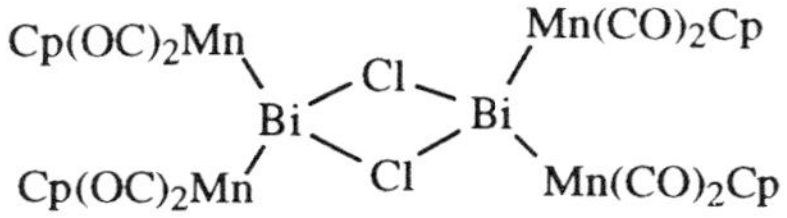

39

Terminally bound main group elements with a double or triple bond to a metal center are rare. Gas-phase studies have given evidence for the existence of $[(CO)_{n-2}MSb]^-$ (M = Cr, Mo, W, n = 6; M = Fe, n = 5).[254] A more useful synthetic strategy is to capture the P–R fragment arising from the cleavage of a phosphaketene. Thus, when $(C_6H_2^tBu_3)P{=}C{=}O$ was treated with $WCl_2(PPh_2Me)_4$, the resulting complex was found to be $Cl_2(CO)(PPh_2Me)_2WP(C_6H_2^tBu_3)$.[234] The W–P distance at 2.370(2) Å is consistent with triple bond formation.

An interesting tautomerism is possible for compounds of type **19** (or **18**). As shown in Eq. (1), the lone pair of electrons could be moved onto the metal fragments, resulting in breakage of the M–M bond, creating a planar EM_3 unit (**40**). This creates an electron-deficient E center, but such molecules have precedent in $[E\{Fe(CO)_4\}_3]^{2-}$ (E = Sn, Pb). Such intermediates could well be stabilized by partial multiple bonding as for the other "inidene" compounds.

(1)

19 **40**

TABLE XIV

ME, M_2E, M_2ER, AND M_3E "INIDENE" COMPLEXES: **30–38**

Compound	Ref.
$W(PMePh_2)_2Cl_2(CO)P$-2,4,6-$^tBu_3C_6H_2$	234
$[Cp^*_2Zr]_2P$	235
$\{Cp_2ClZr\}_2PC_6H_2{}^tBu_3$-2,4,6	236
$\{(Cp)ClZrC_5H_4ZrCp_2\}PSiPh_3$	236
$\{(Cp)ZrC_5H_4\}_3P$	236
$[Cp_2Zr(\mu\text{-}Cl)ZrCp_2[PZr(Cl)Cp_2$	236
$[Cp^*_2U(OMe)]_2PH$	237
$Cp_2V_2(CO)_4P$-2,4,6-$^tBu_3C_6H_2$	238
$\{(CO)_5Cr\}_2PN(SiMe_3)_2$	239
$\{(CO)_5CrPN(SiMe_3)_2Cr(CO)_4$	239
$\{Fe(CO)_3\}_2(POC_6H_2{}^tBu$-2,6-Me-4$\}_2$	240, 241
$Fe_2(CO)_6\{P{=}Mn(CO)_2Cp\}_2$	242, 243
$[Cp(CO)_2Mn]_2PPh$	244
$[\{Cp(CO)_2Mn\}_2As][CF_3SO_3]$	245
$\{Cp^*(CO)_2MN\}_2AsH$	246, 247
$[(CO)_5Cr]_2AsCl$	248, 249
$[(CO)_5Cr]_2AsPh$	250, 251
$\{Cp(CO)_2Mn\}_2AsCl$	252
$\{(CO)_5Mn\}_2As\{{=}Mn(CO)_4\}$	41
$\{(CO)_5Cr\}_2AsN(SiMe_3)_2$	239
$\{(CO)_4Fe\}_2AsN(SiMe_3)_2$	239
$\{Cp(CO)_7CrMo\}As\{Cr(CO)_5\}$	253
$\{CpCO)_7CrW\}As\{Cr(CO)_5\}$	253
$\{Cp(Cp^{1M})(CO)_4MoMn\}As\{Mn(CO)_2Cp^{1M}\}$	253
$(CO)_nMSb$ (M = Cr, Mo, W)	254
$\{L_nM\}_2SbR$ (M = $Cr(CO)_5$, R = ^{t}Bu, Cl, EtS; ML_n = $Mn(CO)_2Cp$, R = Cl; ML_n = $Mn(CO)_2Cp^{1M}$, R = Br)	44, 255
$\{Cp(CO)_2Mn\}_2SbPh$	256
$\{(\eta^6\text{-arene})(CO)_2Cr\}_2SbX$ (X = Cl, Br, I; arene = benzene, MesH, hexamethylbenzene, $C_6H_5CF_3$)	257
$[Bi\{Mn(CO)_2Cp^{1M}\}_2][O_3SCF_3]$	258
$\{Cp^{1M}(CO)_2Mn\}_2BiX$ (X = Cl, I)	258
$[\{Cp(CO)_2Mn\}_2BiCl]_2$	61

E. *Trigonal Bipyramidal E_2M_3 Molecules*

The simplest molecules that have more than one E atom attached to metals but that have no direct E–E bonds are the E_2M_3 trigonal bipyramids. Although E_2M_3 structures could have one of three possible trigonal bipyramidal isomeric forms (**41–43**), two of which would have E–E bonding, in fact only the one with both E groups being axial has been confirmed structurally. It is possible that $Bi_2Os_3(CO)_9$ adopts one of the other geometries, as its infrared spectrum differs from that of the ruthenium analogue, but this has not yet been confirmed.[136] These compounds are listed in Table XV and may be viewed as a fusion of two tetrahedral EM_3 units about a common M_3 triangle. They all possess six pairs of electrons for skeletal bonding as expected from Wade's electron counting rules for a five-vertex cluster. The number of M–M and M–E contacts is nine so that the bonding picture can be viewed as being delocalized. This is in contrast to most of the compounds discussed so far in that those, with the exception of the inidene molecules, can be considered in terms of two center–two electron bonds. A bonding ambiguity persists here as one can view these alternatively as M_3 cluster triangles to which the μ_3-E fragment acts as a simple ligand donating three electrons for E: and E:ML_n fragments (**44**). For these molecules, the metal centers still obey the 18-electron rule. One molecule with only one E atom of the trigonal bipyramid coordinated has been observed. It is $Fe_3(CO)_6\{P(OMe)_3\}_3\{PMn(CO)_2Cp\}\{P\}$ and is noteworthy also because it possesses a naked μ_3-P atom,[259] a rarity owing to the high basicity of the phosphorus lone pair. It is most likely probably stabilized by the steric crowding/protection of the $P(OMe)_3$ ligands. As noted earlier, King[85] has suggested that the edge-localized view is the appropriate one for the trigonal bipyramidal molecule if the four connected vertices are considered to use four internal orbitals for cluster bonding rather than the conventional three. This has some implications for the ability of E to occupy the equatorial sites in E_2M_3 trigonal bipyramids. The characterization of $Bi_2Os_3(CO)_9$ might prove useful in understanding this localized–delocalized bonding picture. To date, no trigonal bipyramidal E_2M_3 molecules with E–R fragments have been reported.

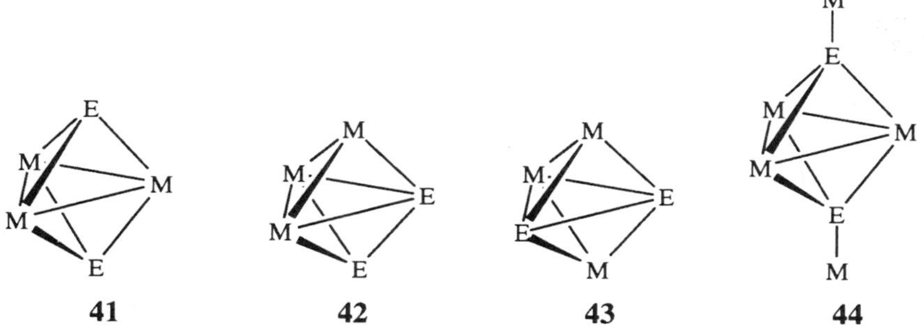

F. *Square Pyramidal E_2M_3 and EM_4 Molecules*

Square pyramidal compounds can be derived formally from the corresponding trigonal bipyramidal molecules by addition of an electron pair.

TABLE XV

TRIGONAL BIPYRAMIDAL M_3E_2 AND $M_3(EML_n)_2$ COMPLEXES (**41** AND **44**)

Compound	Ref.
$Fe_3(CO)_6\{P(OMe)_3\}_3(PMn(CO)_2Cp)P$	259
$Fe_3(CO)_9\{PMn(CO)_2Cp\}_2$	259, 260
$Fe_3(CO)_8(PR_3)\{PMn(CO)_2Cp\}_2$ (R = Me, OMe)	259
$Fe_3(CO)_7(PR_3)_2\{PMn(CO)_2Cp\}_2$ (R = Me, OMe)	259
$As_2Fe_3(CO)_9$	261
$Fe_3(CO)_9(AsML_n)_2$ (ML_n = $CpMn(CO)_2$, $Cr(CO)_5$, $Mo(CO)_5$, $W(CO)_5$)	262
$Fe_3(CO)_9(SbML_n)_2$ (ML_n = $CpMn(CO)_2$, $Cr(CO)_5$, $Mo(CO)_5$, $W(CO)_5$)	262
$Bi_2Fe_3(CO)_9$	263
$Bi_2Ru_3(CO)_9$	137
$Bi_2Os_3(CO)_9$	136

This results in seven skeletal pairs of electrons for bonding in the core structure, and one of the bonds of the trigonal bipyramid is broken. An alternative view of the square pyramid is to remove a vertex from an octahedron. For these molecules the localized view of bonding is inadequate, as there are eight bonding contacts in the square pyramidal array for the seven electron pairs. Following the cluster counting rules, the structures of these compounds are based on the octahedral skeleton ($N + 1$ skeletal pairs, N = number of vertices = 6). The square pyramid is missing one vertex, which makes it a *nido*-octahedral framework.

Structurally, the E_2M_3 square pyramidal framework **45** is a fusion of two EM_3 units (Scheme 1) that share the three metals. These compounds are summarized in Table XVI. Square pyramidal clusters with only one E

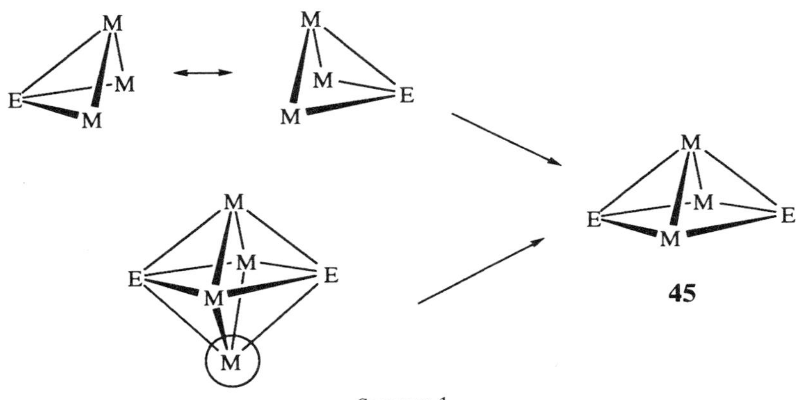

SCHEME 1

46

SCHEME 2

vertex are also known and formally are formed in a similar process by adding a μ_3-metal center to an EM_3 unit (Scheme 2). The parent molecules of type **46** are $Ru_4(CO)_{13}PPh$ and $H_2Ru_4(CO)_{12}PPh$.[110,111,196] They have been listed along with types **23**, **24**, and **25** in Table IX.

An interesting structural feature is observed for the metallated $M_3(EM')_2$ and unmetallated E_2M_3 compounds. For the former the M–E–M angles are considerably more open than those found in the nonmetallated versions, so that it appears in the metallated versions as if there is a weak E···E attraction. The possible cause of this structural distortion may not, in fact, be the formation of an E···E bond, but rather the rehybridization at E upon coordination to an external metal center. For the nonmetallated clusters, the E atoms may well prefer p^3 hybridization with angles tending toward 90°, whereas the addition of the fourth metal atom forces the involvement of the *s* electrons. This would of course induce sp^3 hybridization and an opening of the M–E–M angles.

G. *Edge-Bridged Tetrahedral Molecules*

Following the reasoning used up to this point, addition of one more electron pair to the E_2M_3 core **45** with eight skeletal electron pairs should show breakage of one more bond in the cluster skeleton, although electron counting theories do not predict which bond will be removed. The structural type that results (**47**) is an edge-bridged tetrahedron. An alternative allowed by cluster counting formalisms is that, upon addition of a skeletal electron pair to a five-vertex system, a rearrangement results to a structural type

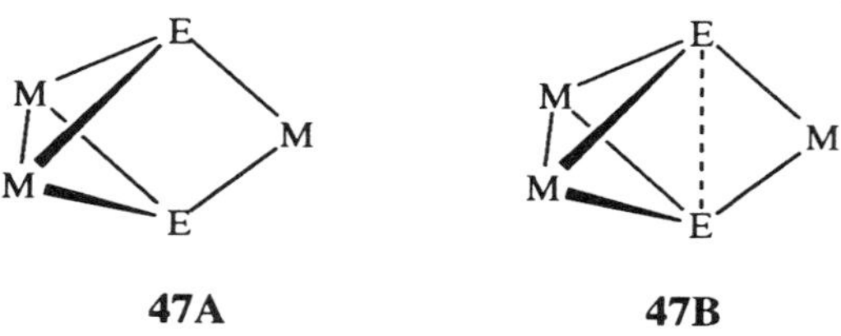

47A **47B**

TABLE XVI

SQUARE PYRAMIDAL M_3E_2, $M_3(ER)_2M_3(EML_n)_2$ COMPLEXES

Compound	Ref.
$\{Cp^{4P}\}_2Ni_2W(CO)_4P_2$	264
$\{Cp^{4P}\}_2Ni_2W(CO)_4(P{=}O)_2$	264
$Fe_3(CO)_9(NMe)_2$	265, 266
$Fe_3(CO)_9(NEt)_2$	267
$Fe_3(CO)_9(NPh)_2$	268–270
$Fe_3(CO)_9(NR)_2$ (R = Ph, Tol, C_6H_4OMe, C_6H_4Cl, $C_6H_4CH_2Cl$)	271
$Fe_3(CO)_9(NN{=}CPh_2)_2$	272
$Fe_3(CO)_9(NH)(NEt)$	151
$Fe_3(CO)_9(NC_6H_4OMe\text{-}p)_2$	270
$Fe_3(CO)_8\{P(OMe)_3\}(NR)_2$ (R = Ph, NC_6H_4OMe-*p*)	270
$Fe_3(CO)_7\{P(OMe)_3\}_2(NPh)_2$	270
$Fe_3(CO)_9(NR)(P^tBu)$ (R = Et, Ph)	273
$Fe_3(CO)_9(S)(NTol)$	274
$Ru_3(CO)_9(NPh)_2$	158, 275, 276
$Ru_3(CO)_9(NC_6H_4Y)_2$ (Y = Cl, H, Me, MeO)	277
$Ru_3(CO)_9(NCF_2C(H)FCF_3)_2$	278
$Ru_3(CO)_8(CNC_6H_3Me_2\text{-}2,6)(NPh)_2$	276
$Ru_3(CO)_7(dppm)(NPh)_2$	276
$Ru_3(CO)_9(NPh)(N^tBu)$	156
$Ru_3(CO)_9(NC_6H_4X)(NC_6H_4X')$ (X = H, X′ = Cl; X = X′ = Cl)	277
$Os_3(CO)_9(NPh)_2$	277
$Fe_3(CO)_9(PH)_2$	279
$[Fe_3(CO)_9(PH)(P)]^-$	279
$[Fe_3(CO)_9P_2]^{2-}$	279
$Fe_3(CO)_9(PMe)_2$	107, 108, 194, 279, 280
$Fe_3(CO)_9(P^iPr)_2$	185, 281
$Fe_3(CO)_9(P^tBu)_2$	194, 196, 280, 281
$Fe_3(CO)_9(PCH_2Ph)_2$	185
$Fe_3(CO)_9(PCH_2SiMe_3)_2$	240
$Fe_3(CO)_9\{PSi(^iPr)_3\}_2$	181, 182
$Fe_3(CO)_9\{PNEt_2\}_2$	195
$Fe_3(CO)_9(PPh)_2$	108, 280, 282–288
$Fe_3(CO)_9(Tol)_2$	108
$Fe_3(CO)_8(NCR)(PPh)_2$ (R = Me, Et)	284
$Fe_3(CO)_7(NCR)_2(PPh)_2$ (R = Me, Et)	284
$[Fe_3(CO)_9(PPh)_2]^-$	286
$Fe_3(CO)_9(PTol)_2$	280
$Fe_3(CO)_9(PC_6H_2Me_3\text{-}2,4,6)_2$	194, 240
$Fe_3(CO)_9(PC_6H_4OMe)(PPh)$	285
$Fe_3(CO)_9(PC_6H_4OMe)(PMe)$	285
$[Fe_3(CO)_9\{PMn(CO)_2Cp\}_2]^{2-}$	259, 260
$Fe_3(CO)_9(PFp)_2$	182
$Fe_3(CO)_9(PFp^*)_2$	182
$Fe_3(CO)_9(PFp)(PFp^*)$	182
$Fe_3(CO)_9(PFp)(PH)$	182

TABLE XVI (*continued*)

Compound	Ref.
$Fe_3(CO)_9(PFp^*)(PH)$	182
$Fe_3(CO)_9(PH)(PMe)$	279
$Fe_3(CO)_9(PCl)(PMe)$	279
$Fe_3(CO)_9(PCl)_2$	279
$Fe_3(CO)_9(PPh)(PR)$ (R = Ph, Tol)	92
$Fe_3(CO)_9(PMe)(PR)$ (R = ^{i}Pr, CH_2Ph, ^{t}Bu)	106
$[\{MePFe_3(CO)_9P\}_2Au]^-$	289
$\{MePFe_3(CO)_9P\}[AuPFe_3(CO)_9P]_n\{PFe_3(CO)_9PMe\}$ (n = 0, 1, 2, ...)	289
$Fe_3(CO)_9(PPh)(AsMe)$	92
$Fe_3(CO)_9(S)(PR)$ (R = ^{t}Bu, C_6H_4OMe-4, cy, Ph, NEt_2, C_6H_4Br-4, Me)	105, 184, 290–294
$Fe_3(CO)_9(E)(PR)$ (E = S, Se, Te; R = ^{i}Pr, ^{t}Bu)	104
$Fe_3(CO)_9\{SM(CO)_5\}(PR)$ (R = alkyl, aryl, M = Cr, W)	291
$Fe_3(CO)_9(E)(PR)$ (E = Se, Te; R = ^{t}Bu, Ph)	295
$Fe_3(CO)_8(CNEt)(E)(PR)$ (E = S, Se; R = ^{i}Pr, ^{t}Bu)	295
$Fe_3(CO)_8\{P(OMe)_3\}(E)(P^tBu)$ (E = Se, Te)	295
$Fe_3(CO)_7\{P(OMe)_3\}_2(E)(P^tBu)$ (E = Se, Te)	295
$Os_3(CO)_9(\mu_3\text{-}S)(N^tBu)$	296
$Fe_3(CO)_9\{AsPh\}_2$	285, 297, 298
$[Fe_3(CO)_9\{AsFe(CO)_4\}_2]^2$	68
$Co_2Cr(CO)_{10}\{AsCr(CO)_5\}_2$	224
$Fe_3(CO)_9(S)(AsR)$ (R = cy, ^{t}Bu, Ph)	291
$Fe_3(CO)_9\{SM(CO)_5\}(AsR)$ (R = alkyl, aryl, M = Cr, W)	291
$[Fe_3(CO)_9\{SbFe(CO)_4\}_2]^{2-}$	233
$[Bi_2Fe_3(CO)_9]^{2-}$	299
$[Et_4N][Fe_3(CO)_9Bi\{BiFe(CO)_4\}]$	300, 301

based on a higher nuclearity parent. Thus, the eight-skeletal-pair system would have as its parent the seven-vertex pentagonal bipyramid (**48**, Scheme 3). What is not accounted for by either of these counting theories is the formation of an interaction between the E–E atoms. For the two bismuth compounds listed, the Bi–Bi distances lie in the range of Bi–Bi single bond values, although for tellurium, where other such structures have been reported, the expected bond order is considerably less. A molecular orbital analysis of these compounds has been reported that shows formation of an extra nonbonding orbital created by the mismatch in symmetry between the E and M fragments.[302] Note that $Rh_6(CO)_{11}(\mu\text{-}As^tBu_2)(\mu_4\text{-}As^tBu)$ **49** and also $[E_2Co_4(CO)_{11}]^{1-/2-}$ (E = Sb, Bi) can be based similarly on the pentagonal bipyramidal geometry and have simliar E–E or E–M contacts through the "pentagonal" ring.[303–305] The orbital analysis of

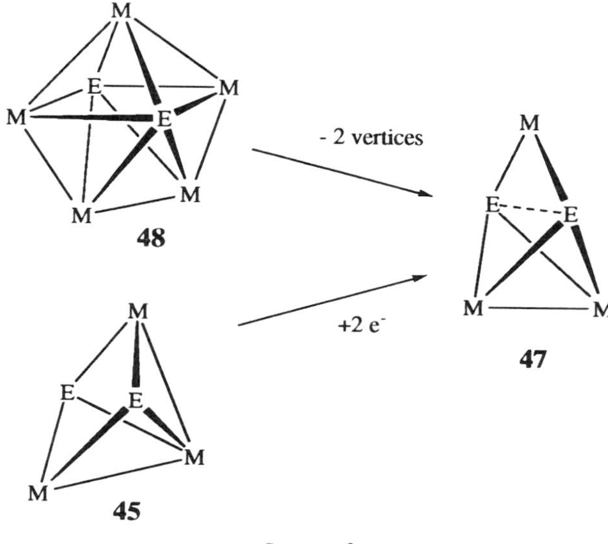

SCHEME 3

$[E_2Co_4(CO)_{11}]^{1-/2-}$, $[Bi_4Fe_4(CO)_{13}]^{2-}$, and the bridged tetrahedral structures are similar in the formation of nonbonding orbitals that can hold more electrons than normally predicted.

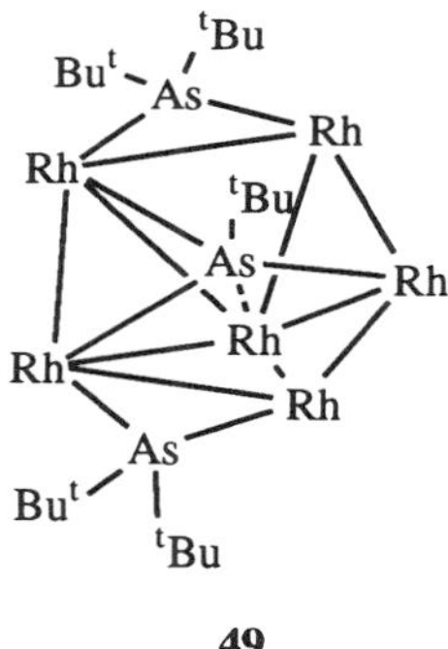

49

Compounds having the bridged tetrahedal structural type are listed in Table XVII. Also listed in the table are EM_4 molecules that have a similar edge-bridged tetrahedral core. The examples known are tetrametal systems with a PR group (**50**). Like the E_2M_3 described previously, compounds **50** have eight skeletal electron pairs. Their electron distributions obey the 18-electron rule at the metals; however, to do this the phosphorus atoms become 10-electron centers and are formally hypervalent.

TABLE XVII

M_3E_2 AND M_4E BRIDGED TETRAHEDRAL COMPLEXES: **47**

Compound	Ref.
$[Fe_2(CO)_6Bi_2Co(CO)_4]^-$	300
$[HFe_2(CO)_6Bi_2Fe(CO)_4]^-$	299
$CpRhCo_2Ru(CO)_{10}PPh$	306
$CpRhRu_2Co(CO)_{10}H(PMe)$	306
$CpRhFeCoRu(CO)_{10}(H)(PMe)$	306

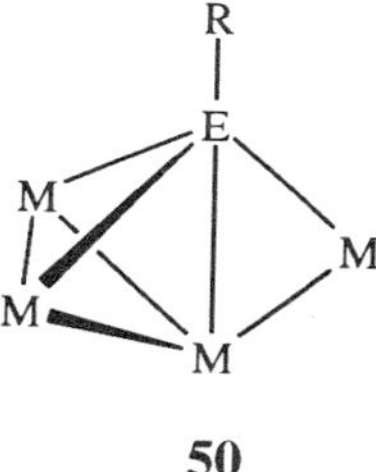

50

H. $(\mu_4\text{-}E)_xM_y$ Compounds

A more common arrangement for EM_4 species is the square pyramidal array with E occupying the apical vertex (**51**), which is related to the octahedral E_2M_4 (**52**) and EM_5 (**53**) arrangements. Note that **51** is the structural isomer of **46** where the E atom adopts the apical rather than the basal position. These compounds are listed in Tables XVIII–XX, whereas Table XXI lists a number of more complicated molecules that have several

TABLE XVIII

SQUARE PYRAMIDAL $(\mu_4\text{-}E)M_4$ AND RELATED COMPLEXES: **51**

Compound	Ref.
$Fe_4(CO)_{10}\{\mu\text{-}P(Me)^iPr\}\{^iPrPCH_2P^iPr\}PH$	313
$Fe_4(CO)_9\{P(Me)R_2\}(\mu_3\text{-}PR)(\mu_4\text{-}PR)$	313
$Ru_4(CO)_{11}(\mu_4\text{-}S)(\mu_4\text{-}PPh)$	314
$Ru_4(CO)_{10}(PEt_3)(\mu_4\text{-}Se)(\mu_4\text{-}PPh)$	314
$Ru_4(CO)_9(PEt_3)_2(\mu_4\text{-}Se)(\mu_4\text{-}PPh)$	314
$Ru_4(CO)_{10}(PEt_3)(\mu_4\text{-}Te)(\mu_4\text{-}PPh)$	314
$HRu_5(CO)_{10}(\mu\text{-}PPh_2)(\mu_3\text{-}PPh)(\mu_4\text{-}PPh)$	206
$Ni_5(CO)_5(\mu_3\text{-}P^tBu)_2(P_3{}^tBu_3)(\mu_4P^tBu)$	116

(μ_4-E)M_4 units in the same molecule. The known compounds of types **51** and **52** possess seven skeletal pairs, as expected for the both *nido-* and the *closo*-octahedral frameworks, respectively. An exception is $Cp_2Ni_2Ru_3(CO)_9(PPh)$ (**54**), which has one more electron pair and one fewer M–M bond. Furthermore, this molecule appears to have a weak interaction between the P atom and the fifth metal vertex that is not anticipated based on the electron counting rules. As has been seen already in Section I,G, secondary interactions of this type are quite common.

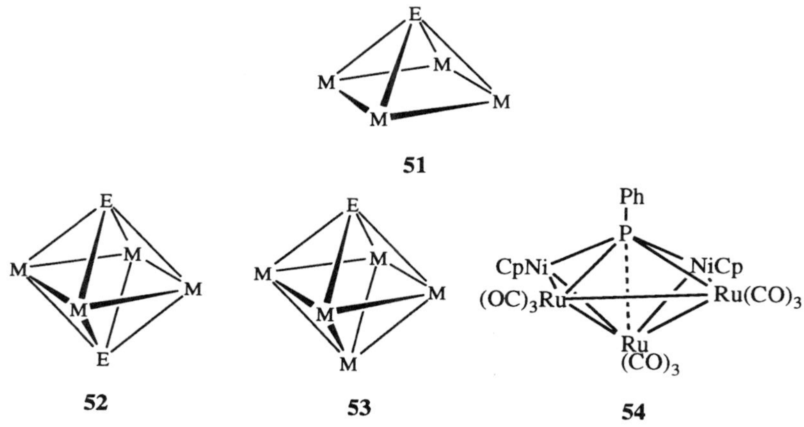

TABLE XIX

OCTAHEDRAL (μ_4-E)M_5 COMPLEXES: **53**

Compound	Ref.
$Ru_5(CO)_{15}PR$ (R = Ph, Me, Et, CH_2Ph, ^tBu)	202, 212, 315
$HRu_5(CO)_{13}\{\mu\text{-}PPh(O^nPr)\}(\mu_4\text{-}PPh)$	207
$Ru_5(CO)_{12}(\mu_3\text{-}CH_2{}^iPr)(\mu\text{-}PPh_2)(\mu_4PPh)$	206
$Ru_6(CO)_{11}(\mu_3\text{-}P^tBu)_3(\mu_4\text{-}P^tBu)$	213
$Ru_7(CO)_{14}(\mu_3\text{-}P^tBu)_2(\mu_4\text{-}P^tBu)_2$	213
$Ru_3Rh_2(CO)_{13}(PEt_3)(PPh)$	214
$Os_5(CO)_{15}POMe$	316
$M_2Os_3(CO)_{15}PR$ (M = Ru, Os; R = Ph, cy)	229
$Cp_2Rh_2FeCo_2(CO)_8(PMe)$	306
$Cp_2Rh_2RuCo_2(CO)_8(PPh)$	306
$Cp_2Rh_2Ru_2Co(CO)_8(H)(PMe)$	306
$Cp_2Rh_2FeRuCo(CO)_8(H)(PMe)$	306
$Cp_2Rh_2FeRuCo(CO)_8(H)(PMe)$	306
$Cp_2Rh_2FeRuCo(CO)_8(H)(P^tBu)$	306
$Re_5(CO)_{14}(\mu\text{-}PMe_2)(\mu_4\text{-}PMe)\{\mu_3\text{-}PRe(CO)_5\}$	226
$Cp_2Ni_2Ru_3(CO)_9(PPh)$	317

TABLE XX

OCTAHEDRAL E_2M_4 COMPLEXES: **52**

Compound	Ref.
$Fe_4(CO)_{11}(NEt)_2$	318
$Ru_4(CO)_{11}(NPh)_2$	318
$FeRu_3(CO)_{11}(NPh)_2$	318
$Fe_4(CO)_{12}(PR)_2$ (R = Me, ^{i}Pr, ^{t}Bu, Ph, Tol)	108, 178, 191, 280, 307, 319
$Fe_4(CO)_{11}(L)(PR)_2$ (L = CO, $P(OMe)_3$, ^{t}BuNC; R = Ph, Tol)	308
$Fe_4(CO)_{11}(PMe)_2$ (R = Me, ^{t}Bu, Ph, Tol)	108, 280, 308
$Fe_4(CO)_{10}(L)(PPh)_2$ (L = CNTol, PMe_3, PMe_3Ph, PPh_3)	319
$Fe_4(CO)_{11}(CNTol)(PPh)_2$	319
$Fe_4(CO)_9(CNTol)(L)(PPh)_2$ (L = CNTol, $P(OMe)_3$, PME_3)	319
$Fe_4(CO)_9\{(MeO)_2PCH_2CH_2P(OMe)_2\}(PPh)_2$	319
$[Fe_4(CO)_{11}(PR)_2]^{n-}$ (R = Ph, ^{t}Bu; n = 0, 1, 2)	320
$[Fe_4(CO)_{10}\{P(OMe)_3\}(PR)_2]^{n-}$ (R = Ph, ^{t}Bu; n = 0, 1, 2)	319, 320
$[Fe_4(CO)_{10}\{CNTol\}(PR)_2]^{n-}$ (R = Ph, ^{t}Bu; n = 0, 1, 2)	319, 320
$Fe_3Ru(CO)_{11}(PPh)_2$	319
$Fe_2Ru_2(CO)_{11}(PPh)_2$	319
$FeRu_3(CO)_{11}(PPh)_2$	319
$[Fe_3Ru(CO)_{11}(PPh)_2]^-$	320
$[Fe_2Ru_2(CO)_{11}(PPh)_2]^-$	320
$[Fe_4(CO)_{10}(PPh)_2(L{-}L)Fe_4(CO)_{10}(PPh)_2]^{n-}$ (L—L = $(MeO)_2PC_6H_4P(OMe)_2$, $(MeO)_2PC_2H_4P)OMe)_2$, CNC_6H_4NC; n = 0, 1, 2)	319, 320
$Fe_4(CO)_9L'(PPh)_2(L{-}L)Fe_4(CO)_9L''(PPh)_2]$ (L—L = $(MeO)_2PC_6H_4P(OMe)_2$, L′ = CO, L″ = $P((OMe)_3$; L′ = L″ = $P(OMe)_3$)	319, 320
$[Fe_4(CO)_{10}(PPh)_2(L{-}L)Fe_4(CO)_9(PPh)_2(L{-}L)]Fe_4(CO)_{10}(PPh)_2]^{n-}$ (L—L = $(MeO)_2PC_6H_4P(OMe)_2$; n = 0, 1, 2)	319, 320
$Fe_4(CO)_{10}(\mu\text{-}COEt)(PR)_2$ (R = ^{t}Bu, Ph)	321
$Ru_4(CO)_8(\mu\text{-}P^tBu)_2(\mu_4P^tBu)_2$	212
$[Cp^*RhFe_3(CO)_8(PPh)_2]^{n-}$ (n = 0, 1, 2)	309
$[Cp^*RhFe_3(CO)_9(PPh)_2]^{n+}$ (n = 0, 1, 2)	309
$Ru_4(CO)_{11}(PPh)_2$	93, 202, 207
$[Ru_4(CO)_{11}(PPh)_2]^{n-}$ (n = 0, 1, 2)	320
$[Ru_4(CO)_{10}\{P(OMe)_3\}(PPh)_2]^{n-}$ (n = 0, 1, 2)	320
$Ru_4(CO)_8\{\mu\text{-}PHPh\}_2(PPh)_2$	207
$Fe_2Co_2(CO)_{11}(PR)_2$ (R = Me, Ph)	108, 322
$Co_4(CO)_{10}(PPh)_2$	224, 323–325
$Co_4(CO)_9(NCR)(PPh)_2$ (R = Me, Et)	284
$Co_4(CO)_8(PPh_3)_2(PPh)_2$	324
$Co_4(CO)_8(diphos)(PPh)_2$	325
$Co_4(CO)_{10-x}L_x(PPh)_2$ (x = 0–4; L = $P(OMe)_3$)	325, 326
$Co_4(CO)_3[(F_2P)_2NMe]_4(PPh)$	310
$Co_4(CO)_{10}(PCH_2SiMe_3)\{PCH(SiMe_3)_2\}$	215
$Co_4(CO)_{10}(P^tBu)_2$	327
$Rh_4(cyclooctadiene)_2(PPh)_2$	328
$[Fe_4(CO)_{12}\{SbFe(CO)_4\}_2]^{2-}$	98
$Ru_4(CO)_{12}Bi_2$	136, 137

TABLE XXI

COMPLEX $M_x(\mu_4\text{-}E)_y$ COMPOUNDS WITH $y \geq 2$

Compound	Ref.
$Re_6(CO)_{18}(PMe)_3$	85, 226
$Ru_6(CO)_{12}(\mu_3\text{-}PPh)_2(\mu_4\text{-}PPh)_2$	207
$Ru_6(CO)_{12}(\mu_3\text{-}PPh)_2(\mu_4\text{-}PPh)_3$	207
$Ni_8(CO)_8(PPh)_6$	85, 329
$Ni_8(PPh_3)_4(PPh)_6$	330
$Ni_9(PPh_3)_4(PPh)_6$	330
$Ni_8(CO)_4(PPh_3)_4(PPh)_6$	78
$Ni_8X_4(PPh_3)_4(PPh)_6$ (X = Cl, Br)	78
$Ni_8(PCMe_3)_2(PMe)_2(CO)_{12}$	331
$[Ni_{10}(CO)_{10}(\mu_4\text{-}PMe)_5(\mu_5\text{-}PMe)_2]^{2-}$	332
$Cu_{12}(PPh)_6(PPh_3)_6$	333
$[Co_8(\mu_6\text{-}As)(\mu_4\text{-}As)(\mu_4\text{-}AsPh)_2(CO)_{16}]_2$	232
$Ni_9(\mu_4\text{-}As)_6(PPh_3)_5Cl_3$	334
$Ni_9(\mu_4\text{-}As)_6(PPh_3)_6Cl_2$	334
$Pd_9(PPh_3)_8As_6$	335
$Pd_9(PPh_3)_8(As_2)_4$	336
$Pd_9(PPh_3)_8Sb_6$	336
$[Bi_4Co_9(CO)_{14}]^{2-}$	337
$[Bi_8Co_{14}(CO)_{20}]^{2-}$	337

The octahedral E_2M_4 molecules have two common electron counts. Many of the compounds do, in fact, obey the cluster formalisms and possess seven skeletal pairs of electrons. Quite a few, however, have eight pairs and still retain all of the bonds between vertex atoms. These compounds have the general formula $M_4(CO)_x(PR)_2$ (M = Fe, Ru; x = 11, 12). In many cases the reversible interconversion of the $x = 11$ and $x = 12$ species has been observed.[307,308] The $x = 11$ complexes will readily add ligands and subsequently eliminate CO in an overall substitution process. Similar chemistry is observed for the heterometallic "unsaturated" cluster $Cp^*RhFe_3(CO)_8(PPh)_2$.[309] It was prepared via the addition of $[Cp^*RhCl_2]_2$ to $[Fe_3(CO)_9(PPh)_2]^{2-}$. Like the other octahedral clusters, it readily adds CO. Substitution of $Co_4(CO)_{10}(PPh)_2$ with $F_2PN(Me)PF_2$ gives the complex $Co_4(CO)_3\{F_2PN(Me)PF_2\}_4(PPh)_2$.[310] The number of skeletal electron pairs is nine, one more than the other octahedral clusters, and consequently one Co–Co bond is found to be broken.

Theoretical analyses of the bonding in M_4E_2 clusters have been made.[85,311,312] The structures show formation of a weak bonding E···E interaction that could explain the high electron count, but it was found that the HOMO is not E···E bonding. Rather, it is M–M antibonding, and the increase in the M–M distances while keeping the E–M distances constant results in a shorter E–E interaction.

The tricapped, trigonal prismatic array of metals observed for $Re_6(CO)_{18}(PMe)_3$ (**55**) is capped on the Re_4 faces with PR groups. This molecule has nine skeletal electron pairs, which is consistent with an edge-localized bonding scheme for the Re_6 core. The trigonal prismatic array of metals has been commonly observed for Re clusters as well as some compounds of Co and Rh. The octahedral arrangement is much more common for the other transition metals. Fourfold bridging of PR groups is also observed for $Ni_8(CO)_{12}(PCMe_3)_2(PMe)_2$ (**56**), which can be viewed as having a cuneane geometry.[331] Two of the PR fragments are isolated from each other and adopt a μ_4-bonding mode, while the other two are connected, giving each P atom a connectivity of 5 within the cluster core. It is interesting that upon addition of two electrons, this molecule rearranges to form a P–P bond between the originally isolated PR groups concomitantly with Ni–Ni bond cleavage. The result is an icosahedral Ni_8P_4 core geometry. The molecule $Cu_{12}(PPh_3)_6(PPh)_6$ (**57**) has a compressed cuboctahedral Cu_{12} core with PPh groups capping each of the fourfold faces. Six of the Cu–Cu bonds of the cuboctahedron have been broken as a result of the presence of PPh bridging groups and the compression of the cuboctahedral core.

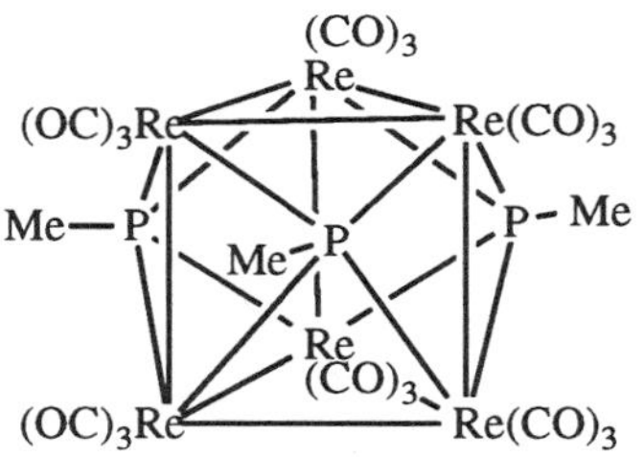

55

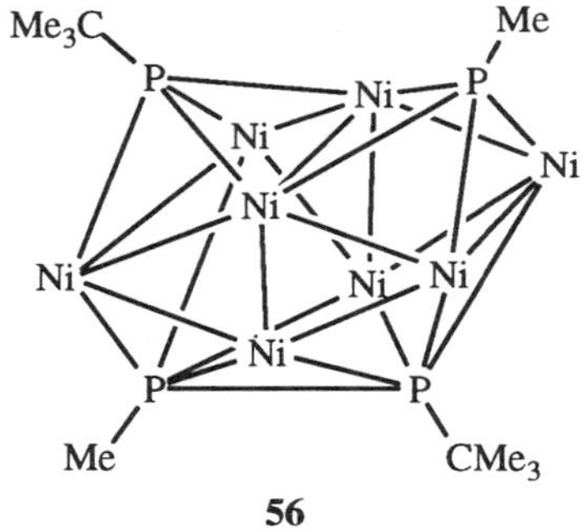

56

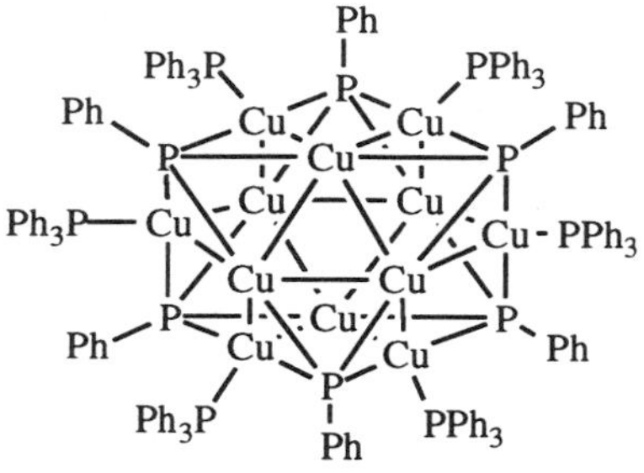

57

Another common structural pattern is that of a cube of metals capped on all faces with μ_4-E fragments. These compounds have been found to occur in both noncentered (**58**) and centered (**59**) varieties.These systems have been examined by molecular orbital calculations.[338–340] A preferred electron count of 120 MVEs (metal valence electrons) for **58**, was calculated but variations in ligands and metal orbitals has allowed compounds in the class to be observed that have as few as 99 MVEs. In fact, most of the known examples are not closed-shell compounds. A similar, although more complicated, picture arises for the centered compounds **59**. For this compound type, more electron-count possibilities are allowed by the match/mismatch of the e_g and t_{2g} orbitals of the interstitial metal atom with the metal cube. Thus, configurations with a significant HOMO–LUMO gap are observed at 120, 124, and 130 MVEs and observed electron counts range from 120 to 130 MVEs. The $Ni_8Cl_4(PPh)_3(PPh)_6$ compound is of interest in that it may be reduced with sodium amalgam to $Ni_8(PPh_3)_4(PPh)_6$.[330] This product readily adds donor ligands to produce $Ni_8(PPh_3)_4L_4(PPh)_6$. When $Ni_8Cl_4(PPh)_3(PPh)_6$ is reduced in the presence of $NiCl_2$, the reduced, centered cluster $Ni_9(PPh_3)_4(PPh)_6$ is obtained. A distortion of the basic cubic geometry is found for $Pd_9(PPh_3)_8(As_2)_4$ (**60**). In this compound, four Pd–Pd bonds of the original Pd-centered Pd_8 cube are broken and four of the faces are bridged by As_2 fragments. One As atom of each of the As_2 units adopts the normal μ_4-position, while the other can be seen to bridge one of the $AsPd_2$ faces so created. An interesting compound is $Ni_5(CO)_5(\mu_3\text{-}P^tBu)_2(P^t_3Bu_3)(\mu_4\text{-}P^tBu)$ (**61**), which looks very much like a fragment of a cubic array with three Ni(CO) groups removed and the dangling valences of the resultant P–R fragments satisfied by forming P–P bonds where appropriate.

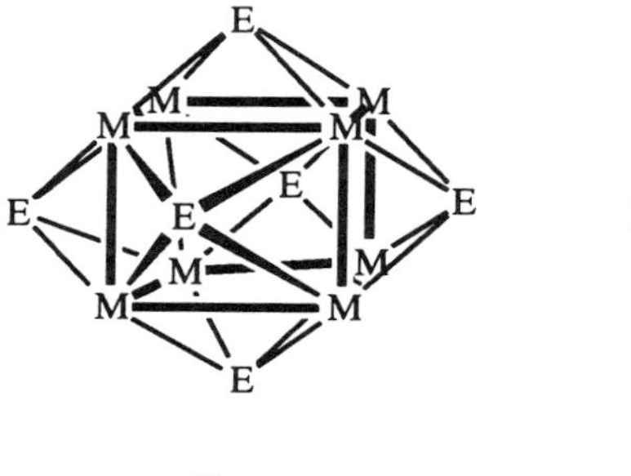

58

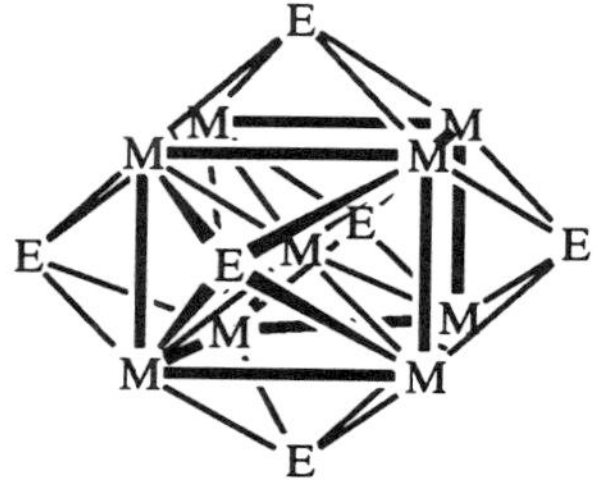

59

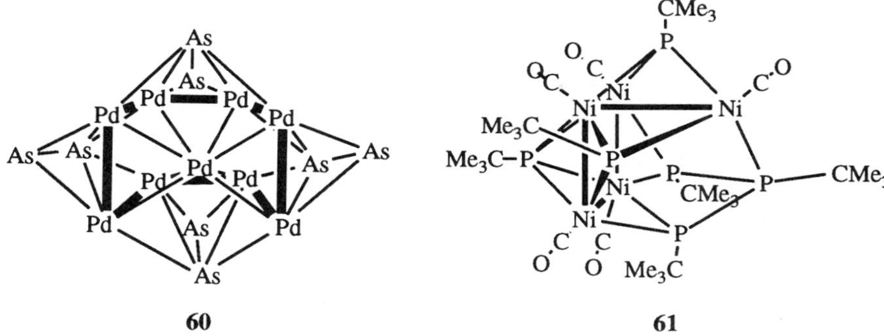

60 **61**

Closely related are $[Bi_4Co_9(CO)_{16}]^{2-}$ and $[Bi_8Co_{14}(CO)_{20}]^{2-}$, the metal cores of which are given as **62A/62B** and **63**. They may be viewed as tetracapped, tetragonally distorted M_8 arrays that also possess interstitial metal atoms. Compound **63** consists of two of the **62** metal frameworks fused about one of the Co_4 faces. Compared to **59**, **62** is seen to be missing two of the μ_4-E groups. In addition, the Co–Co bonds parallel to the C_4 axis of the molecule have been broken and have been replaced with four Bi–$Co_{interstitial}$ interactions. The result is that **61** can be viewed as a cuboctahedron, which is more apparent if the compound is viewed from a different perspective (**62B**). This also emphasizes the close-packed nature of the metal array.

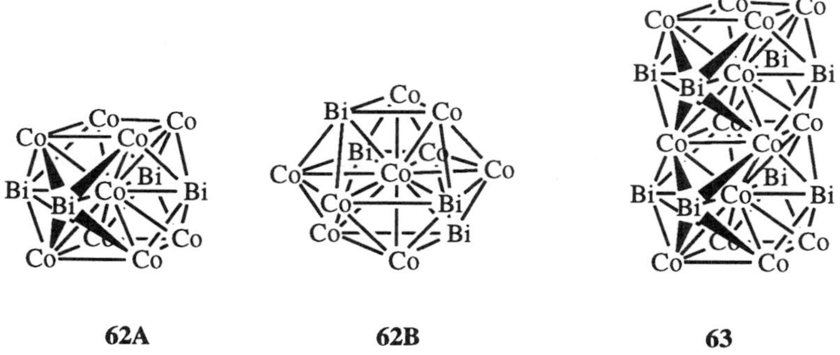

62A **62B** **63**

I. $(\mu_5\text{-}E)_xM_y$ Compounds

In addition to bridging square arrays of metal, E fragments may also adopt a μ_5-configuration. Complexes of this type have been summarized

TABLE XXII

ICOSAHEDRAL AND RELATED COMPLEXES WITH μ_5-E ATOMS AND FRAGMENTS

Compound	Ref.
$[As_3Co_3(PEt_2Ph)_3]_2(As_6)$	343
$[Ni_8(PCMe_3)_2(PMe)_2(CO)_{12}]^{2-}$	331
$[Ni_8(CO)_{12}(PMe)_4]^{2-}$	332
$[Ni_9(CO)_{15}(PMe)_3]^{2-}$	332
$[Ni_{10}(CO)_{18}(PMe)_2]^{2-}$	332
$[Ni_9(CO)_{14}(\mu\text{-}PMe_2)(PMe)_3]^-$	332
$[Ni_{10}(CO)_{10}(\mu_4\text{-}PMe)_5(\mu_5\text{-}PMe)_2]^{2-}$	332
$[Ni_{10}(CO)_{18}(AsMe)_2]^{2-}$	85, 344
$[Ni_9(CO)_{15}(AsPh)_3]^{2-}$	85, 344
$[Ni_{10}(CO)_{18}(SbPh)_2]^{2-}$	345
$[Ni_{10}(CO)_{18}(\mu_{12}\text{-}Ni)(SbMe)_2]^{2-}$	346
$[Ni_{10}(CO)_{18}(\mu_{12}\text{-}Ni)\{SbNi(CO)_3\}_2]^{x-}$ (x = 2, 3 or 4)	341, 346, 347
$[Ni_{10}(CO)_{18}(\mu_{12}\text{-}Ni)Bi_2]^{n-}$ (n = 2, 3, 4)	348
$Pd_8\{\mu\text{-}Pd(PPh_3)_2\}_{12}(\mu_5\text{-}As)_{12}$	335
$Cu_{14}(PPh)_6Cl_2(PMe_3)_8$	333

in Table XXII. The obvious choice for a cluster framework to support a μ_5-ligand is the icosahedron, the most prevalent structural type found in the table. Icosahedral geometries containing two (**64**), three (**65**), four (**66**), and six (**67**) E or ER fragments have been reported.[341] Metal-centered icosahedra (**68**) have also been observed for molecules containing two E fragments. The Co complex **67** has been elaborated by the addition of two As_3 fragments to the central icosahedral core in such a fashion as to form octahedral arrays with the Co atoms. King has discussed the electron counting formalisms for $[Ni_9(CO)_{15}(AsPh)_3]^{2-}$, $[Ni_{10}(CO)_{18}(AsMe)_2]^{2-}$, and $[Ni_{10}(CO)_{18}(Ni)\{SbNi(CO)_3\}_2]^{4-}$.[342] Of note for the latter compounds is the

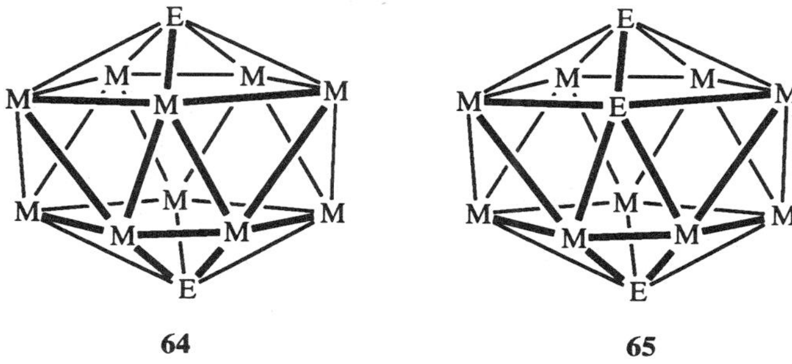

64 **65**

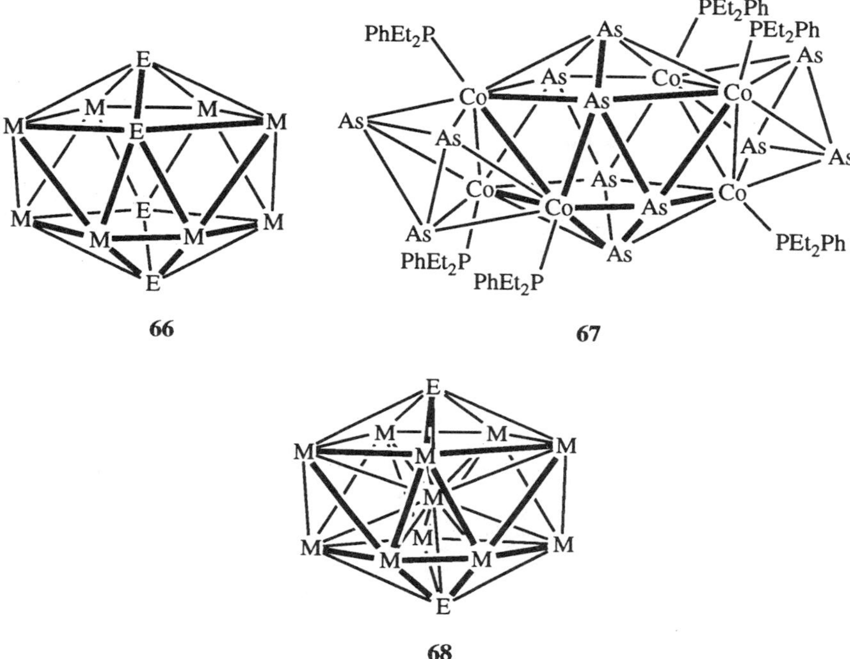

66

67

68

observation that the oxidation to the 3− and 2− ions involves primarily removal of electrons from Ni–Ni nonbonding orbitals, so little change in the cluster framework would be expected, consistent with experimental observations.

J. *Compounds with Interstitial E Atoms*

Higher coordination numbers at E produce complexes in which E may be considered as occupying an interstitial site, nomenclature derived from the solid state where certain types of coordination geometries are created by the packing arrangement of metal spheres in a lattice (Table XXIII). The most common of these are the octahedron and the trigonal prism. The tetrahedron is too small to accommodate an interstitial main group atom, in general, although such an arrangement may be possible if the M–M bonding is weak. Such is the case for the gold cluster complexes where weak Au–Au bonding interactions have been found to be important. This leads to yet further ambiguities in the classification of these molecules according to structural and bonding types. Some of the clusters classified

TABLE XXIII

COMPLEXES WITH ENCAPSULATED AND SEMIENCAPSULATED E ATOMS

Compound	Ref.
$Cp^*_3(CO)_8Mo_3Co_2(\mu_3\text{-NH})(\mu_4\text{-N})$	141
$[Fe_4(CO)_{12}N]^-$	145, 355–357
$HFe_4(CO)_{12}N$	145, 355, 358
$Fe_4(CO)_{11}(NO)N$	359
$[Fe_4(CO)_9(NO)(PPh_3)_2N]$	359
$HFe_4(CO)_{11}(PPh_3)N$	359
$[Fe_4(CO)_{11}(PMe_2Ph)N]^-$	359
$[FeRu_3(CO)_{12}N]^-$	155, 360
$[FeRu_3(CO)_{10}\{P(OMe)_3\}_2N]^-$	155
$HFeRu_3(CO)_{12}N$	145, 155
$HRu_4(CO)_{12}N$	155, 166, 353
$HRu_4(CO)_{11}\{P(OMe)_3\}N$	361
$H_3Ru_4(CO)_{11}N$	362
$[Ru_4(CO)_{12}N]^-$	363
$[Os_4(CO)_{12}N]^-$	362
$Ru_4(CO)_{12}(NCO)N$	353, 364
$Ru_4(CO)_{12}(NO)N$	353, 364
$Cp^*_2Mo_2Co_3(CO)_{10}N$	141
$[Fe_5(CO)_{14}N]^-$	356, 365, 366
$HFe_5(CO)_{14}N$	356
$[Ru_5(CO)_{14}N]^-$	155, 363, 367
$[Ru_6(CO)_{16}N]^-$	363, 367
$[M_6(CO)_{15}N]^-$ (M = Co, Rh)	351, 352, 354
$[Rh_6M(CO)_{15}N]^{2-}$ (M = Co, Rh, Ir)	368
$[N\{AuPMe_3\}_4]^+$	130
$[N\{AuL\}_5]^{2+}$ (L = PMe_3, PPh_3)	349, 350
$[N\{AuPMe_3\}_5]^{2+} \cdot 2Me_3PAuCl$	350
$[PtRh_{10}(CO)_{21}N]^{3-}$	369
$[HRh_{12}(CO)_{23}(N)_2]^{3-}$	370
$HFe_3(CO)_9Au_2(PR_3)_2P$ (R = Ph, Et)	228
$HFe_3(CO)_9Au_2(PR_3)_3P$ (R = Ph, Et)	228
$Fe_3(CO)_9Au_3(PR_3)_3P$ (R = Ph, Et)	228
$[Fe_3(CO)_9Au_4(PPh_3)_4P]^+$	371
$Fe_5(CO)_{16}\{\mu_3\text{-PMn}(CO)_2Cp\}(\mu_5\text{-P})$	262
$Ru_5(CO)_{16}(\mu\text{-PPh}_2)(\mu_5\text{-P})$	372
$[Os_6(CO)_{18}P]^-$	95, 373
$Os_6(CO)_{18}PAuPPh_3$	373
$[Co_6(CO)_{18}P]^-$	374
$H_3Ru_9(CO)_{20}(\mu_3\text{-P}^t\text{Bu})_3(\mu_7\text{-P})$	213
$[Rh_9(CO)_{21}P]^{2-}$	375–377
$[Rh_{10}(CO)_{22}P]^{3-}$	376, 378
$Cu_{96}P_{30}\{P(SiMe_3)_2\}_6(PEt_3)_{18}$	379
$[P(AuP^tBu_3)_4]^+$	380

TABLE XXIII (*continued*)

Compound	Ref.
$[P(AuPPh_3)_5]^{2+}$	381
$[PAu_5(PPh_3)_6]^{2+}$	382
$[P(AuL)_6][BF_4]_3$	383
$[Rh_9(CO)_{21}As]^{2-}$	376, 384
$[Rh_{10}(CO)_{22}As]^{3-}$	376, 384
$[Co_8(\mu_6\text{-As})(\mu_4\text{-As})(\mu_4\text{-AsPh})_2(CO)_{16}]_2$	232
$[Rh_{12}(CO)_{27}Sb]^{3-}$	376, 385

as having the E atom in an interstitial site do not have that atom completely surrounded by M, but may be viewed as derived from a common metal geometry with one or more metal vertices removed. For example, the transformation octahedral (**69**) → square pyramidal (**70**) → butterfly (**71**) is now textbook chemistry (Scheme 4). It should be noted that the interstitial nitrido metal carbonyl clusters have a chemistry that parallels that of the E = C, B compounds. The gold compound $[N\{AuL\}_5]^{2+}$ (**72**) has the nitrogen atom in the center of a trigonal bipyramid of metal atoms, unlike the square pyramidal arrangement **70** normally observed for the NM_5 examples.[349,350] The Au–Au distances lie between 2.9 and 3.1 Å, whereas the $N{-}Au_{equatorial}$ distances are slightly shorter than those to the axial Au atoms. This compound has been found to crystallize with two additional molecules of $ClAuPPh_3$, which interact with the NAu_5 core as shown in **73**.[350] The extra gold atoms lie 3.2 to 3.3 Å from the N-bound Au atoms. The compounds $[M_6(CO)_{15}N]^-$ (M = Co, Rh), which have nine skeletal electron pairs, adopt a trigonal prismatic geometry **74**, as might be expected for edge-localized bonding.[351,352]

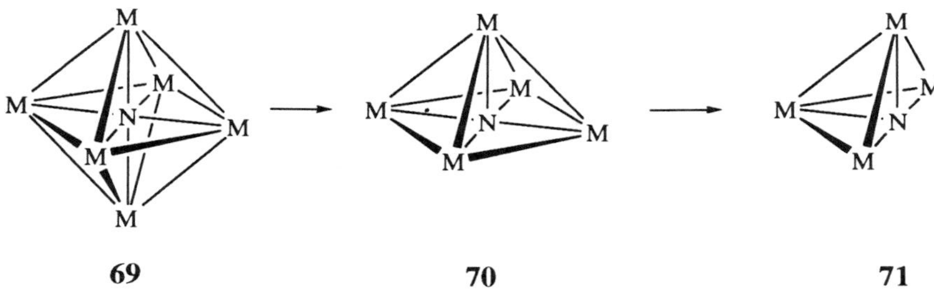

Scheme 4

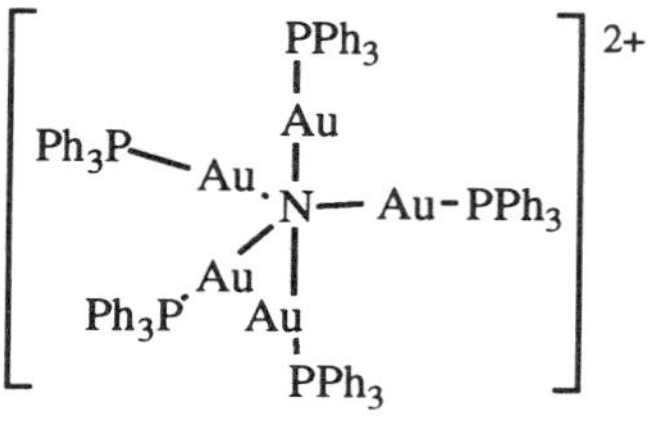

72 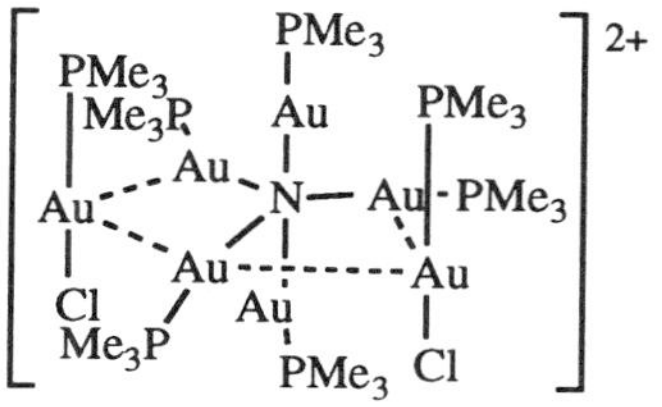 **73**

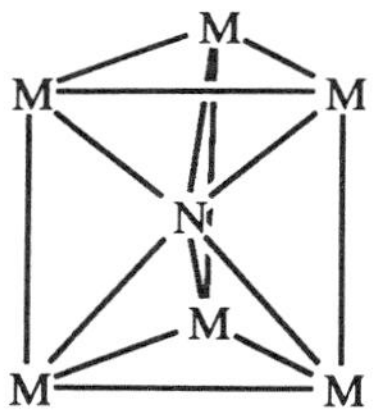

74

Three M–N IR stretches have been found for butterfly clusters with a Ru_4N or Os_4N core. The compounds studied included $HRu_4(CO)_{12}N$, $Ru_4(CO)_{12}(NCO)N$, $Ru_4(CO)_{12}(NO)N$, $H_3Ru_4(CO)_{11}N$, and $HOs_4(CO)_{12}N$. The stretches were found to lie in the region from 590 to 870 cm^{-1}.[353] The vibrational frequencies of the M–N bonds in $[M_6(CO)_{15}N]^-$ (M = Co, Rh) were located in an isotope labeling study.[354] Two bands are observed and were interpreted as having one lying along and one perpendicular to the threefold axis of the cluster. For the M = Co complex these are at 722 and 698 cm^{-1}, and for M = Rh they lie at 645 and 622 cm^{-1}. The axial and equatorial force constants were calculated from this data to be 209 and 261 N m^{-1} for Co and 167 and 207 N m^{-1} for Rh.

For $[PtRh_{10}(CO)_{21}N]^{3-}$ (**75**), the nitrogen atom sits in an irregular cavity and is considered to be bonded to five metal atoms, the platinum and four rhodium atoms, which gives it a coordination geometry similar to that of the well-established square pyramidal M_5N complexes.[369] The Pt atom sits in the center of a distorted six-membered ring of Rh atoms, and this layer is connected to a distorted Rh_4 ring. The large cluster $[HRu_{12}(CO)_{23}(N)_2]^{3-}$ (**76**) possesses two isolated interstitial nitride atoms.[370] The metal framework is very irregular, but may be roughly viewed as a trigonal prismatic Rh_6 unit fused by the face to a square antiprism with a nitride in each cavity. Although the N atom in the trigonal prismatic unit is bonded to all six proximal metal atoms (Rh–N distances ranging from 2.02 to 2.30 Å), the square antiprismatic nitride is bonded only to five of the eight metal

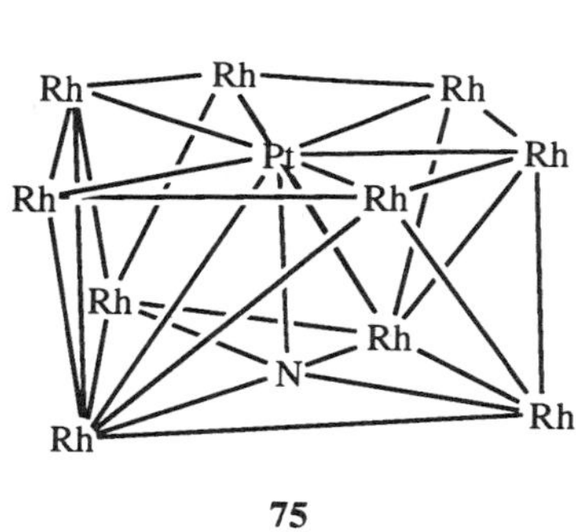

75

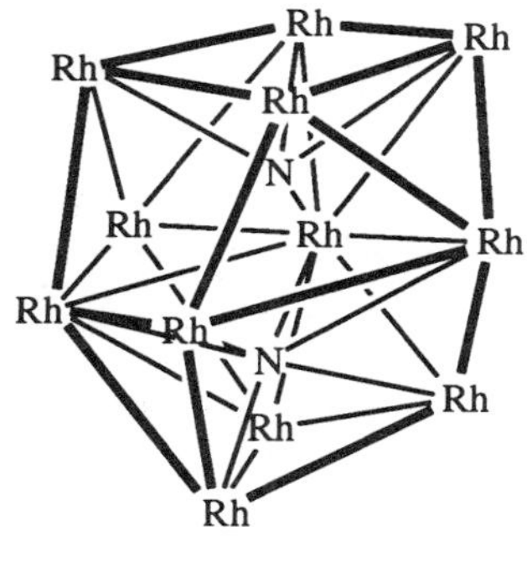

76

atoms (Rh–N ranging from 2.02 to 2.12 Å). In addition, an Rh_2 fragment is situated so that one atom is capping a face of the trigonal prism and one interacts with the square antiprismatic unit.

Interstitial geometries are most common for the lighest elements and decrease in importance with an increase in size. This is to be expected, as the size of the E atom cannot be accommodated in the more common metal packing arrangements. Thus, the interstitial geometry is most commonly observed for N and is well known but not as prevalent for P, but only rare examples exist for As, Sb, with none being yet observed for Bi. Also, as one gets to larger E and higher nuclearity, the geometries are less regular. The P atom in $Ru_5(CO)_{16}(\mu\text{-}PPh_2)(\mu_5\text{-}P)$ (**77**) interacts with all five metals, but much more weakly with one than the others.[372] This can be compared to $Fe_5(CO)_{17}(P)\{PMn(CO)_2Cp\}$ (**78**), which has a similar M_5E core but with one fewer M–E bond. It may be better described as a spiked tetrahedral structure. The hexanuclear cluster $[Co_6(CO)_{16}P]^-$ (**79**) shows a similar partially encapsulated P atom.[374] For this molecule the electron count is consistent with localized Co–Co bonding.

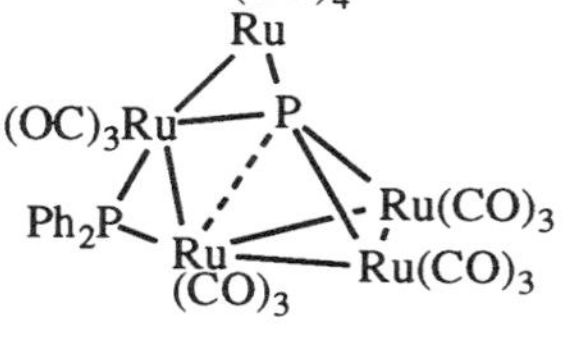

$(CO)_4$
Ru
$(OC)_3Ru$ — P
Ph_2P — Ru $(CO)_3$ — $Ru(CO)_3$ — $Ru(CO)_3$

77

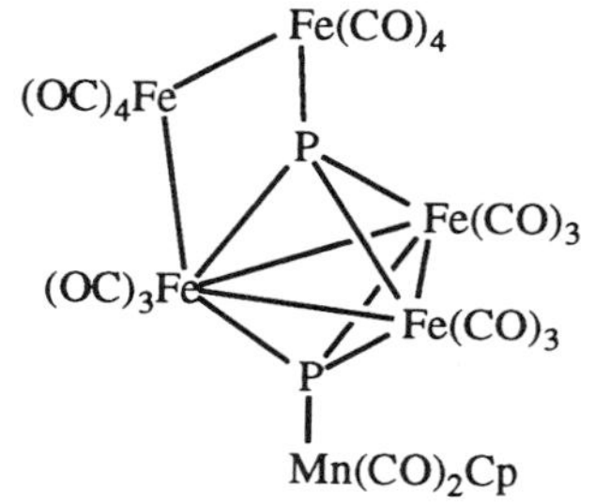

78

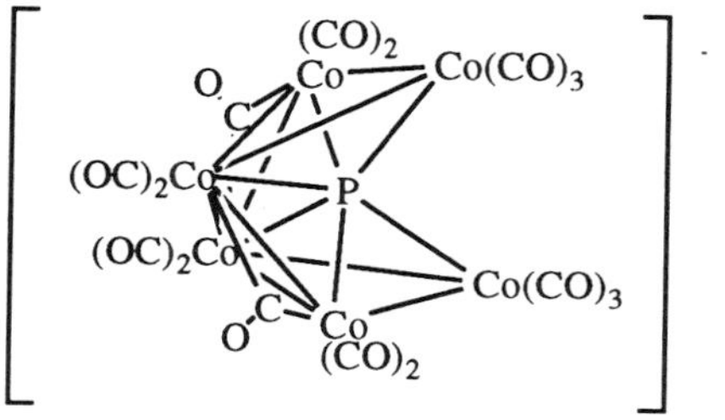

79

Gold(I) readily forms metallated complexes with N and P. The trigonal bipyramidal $[N\{Au(PPh_3)\}_5]^{2+}$ has already been mentioned. Molecules having four, five, and six gold moieties attached to P have been characterized (**80–81**).[380–383] These exhibit Au–Au bonding as associated with homo- and heterometallic gold cluster compounds in general. It is noteworthy that $[P\{AuPPh_3\}_5]^{2+}$ (**81**) adopts a square pyramidal geometry rather than the trigonal bipyramidal configuration seen for the nitrido complex. This PAu_5 framework has also been characterized as a PPh_3 adduct (**82**). The $[P\{AuPR_3\}_6]^{3+}$ compounds are presumably distorted octahedral molecules, but structural characterization has not yet proven possible. The Au–Au bonding interactions in these homometallic species are similar to those observed in the heterometallic Au–Fe system, which starts with the triiron complex $H_2Fe_3(CO)_9PAuPPh_3$ (**83**) and replaces the hydride ligands successively with $AuPPh_3$ units to give penta- (**84**), hexa- (**85**), and heptanuclear (**86**) compounds as shown in Scheme 5.[228,371] The Au–Au distances are on the order of 2.9 to 3.2 Å, lying in the same range as those found for the homometallic gold species. The bonding in all of these compounds involves complex multicenter interactions that are not readily described by conventional rules.

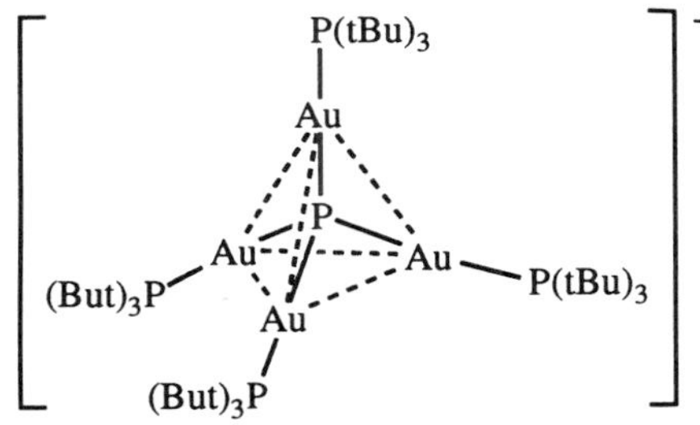

80

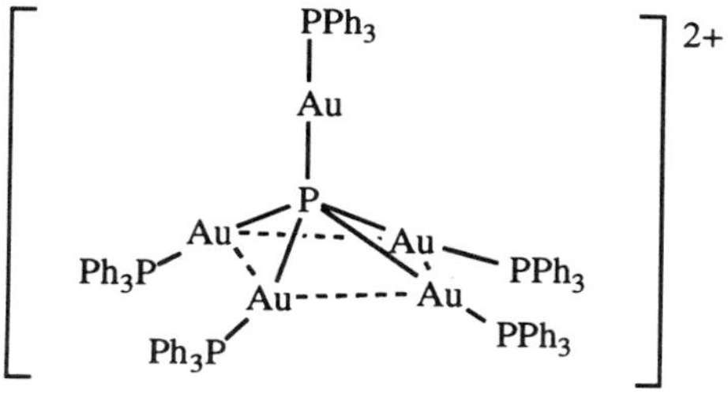

81

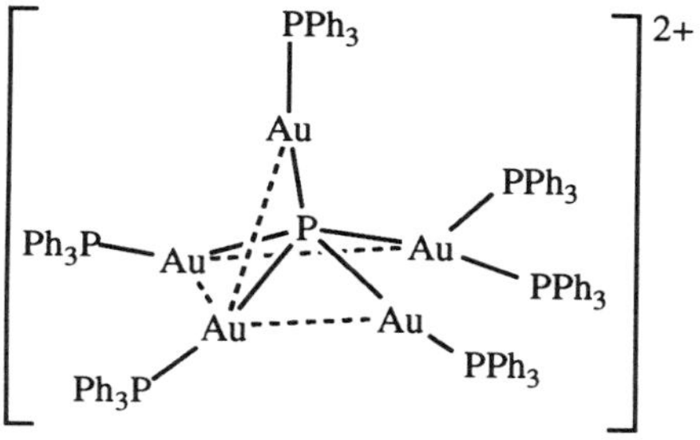

82

For Rh, the size match with P is better and square antiprismatic cavities are seen for $[Rh_9(CO)_{21}P]^{2-}$ and $[Rh_{10}(CO)_{22}P]^{3-}$. Both of these clusters possess 11 skeletal pairs of electrons and obey Wade's rule for *nido*- and *closo*-clusters, respectively. Arsenic analogues of these compounds are known.

Even higher coordination numbers have been reported for the complex high-nuclearity aggregate $Cu_{96}\{P(SiMe_3)_2\}_6(PEt_3)_{18}P_{30}$.[379] The structure is based upon an 11-coordinate P^{3-} ion, lying in the center of a trigonal prism of Cu atoms with all five faces of the trigonal prism capped by additional Cu^+ ions. P atoms on the periphery of the compound adopt μ_6- and μ_7-coordination environments, whereas others closer to the cluster center have μ_9- and μ_{10}-configurations. The compound has been compared to the solid-state phase Cu_3P. Unlike the other complexes discussed in this section, the metal–metal interactions are thought to be relatively weak or nonexistent.

For antimony, the only example of an interstitial geometry is based on an icosahedral array of metal atoms with a μ_{12}-E atom in the center of the cage (**89**). The icosahedron is quite distorted with Rh–Rh bond distances

SCHEME 5

ranging from 2.807(1) to 3.334(1)Å. The longest of these distances are beyond the normal values expected for Rh–Rh single bonds; however, the electron count of 13 skeletal electron pairs is consistent with the *closo*-icosahedral structure and suggests that these interactions are real.

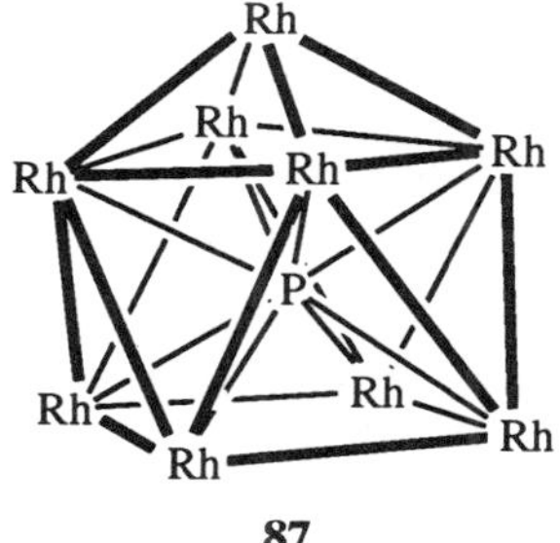

87

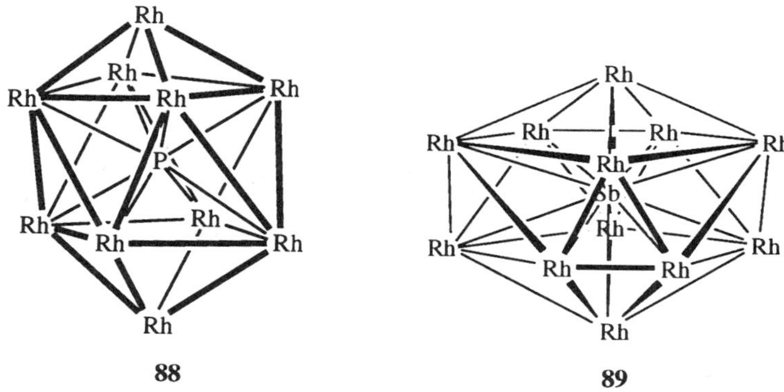

88

89

K. *Compounds with E–E Bonds*

As has already been noted, there is a strong tendency to form E–E interactions within E–M cluster compounds that may have a significant impact on cluster structure and reactivity. As with all molecules, an analysis of the bonding between two atoms A and B is complicated if they are also bridged by other fragments. Be that as it may, there still exists a large number of molecules where direct E–E interactions are clearly present and of considerable importance to the cluster framework. These molecules can be viewed as being derived from E_x ($x \geq 2$) units bonded to one or more metal fragments. The next series of complexes is discussed in terms of increasing x, but it is often a judgment call as to whether a fragment is best described as, for example, an E_5 unit or as E_2 and E_3 fragments. The bonding in these fragments is often asymmetric with some E–E distances being rather long. Some indication of which view is most appropriate is given by structural data where available but, even so, symmetry and properties often argue for more idealized viewpoints.

For the homoatomic main group elements, it has long been known that their reduction with active metals (sodium, potassium, etc.) in liquid ammonia or organic amine solvents leads to cluster anions that have come to be known as the Zintl ions.[26–40] Most often, upon solvent removal these discrete compounds are not stable and solid-state phases are produced. Since many of the organometallic transition metal units are electron rich, their interaction with main group elements can be viewed in one sense as a type of reduction process. Thus, the formation of E–M clusters might be expected to promote E–E bond formation. Structures of a number of E–M cluster compounds have been reported in which the E_x fragment shows high similarity to arrangements present in the discrete Zintl ions or related solid-state phases. Preformed Zintl ions have also been explored for their reactions with transition-metal fragments.

1. *Compounds with E_2 Fragments*

Compounds that contain E_2 and E_2R_2 fragments are summarized in Table XXIV. These fragments may be considered separately from the metal fragments to which they are bound (**90–94**). On "naked" E atoms, a lone

TABLE XXIV

COMPLEXES WITH TETRAHEDRAL E_2M_2 CORE STRUCTURES

Compound	Ref.
$Fe_2(CO)_6(NMe)_2$	386, 388
$Fe_2(CO)_6(NEt)_2$	151
$Fe_2(CO)_6(NPh)_2$	266
$\{Cp(CO)_2Cr\}_2P_2$	389–391
$\{Cp^*(CO)_2Mo\}_2P_2$	392
$Fe_2(CO)_6\{P^tBu\}_2$	393, 394
$Fe_2(CO)_7P_2\{Cr(CO)_5\}_2$	243, 262
$Co_2(CO)_6P_2$	126
$Co_2(CO)_5(PPh_3)P_2$	395
$Co_2(CO)_6\{PCr(CO)_5\}_2$	242, 243
$Co_2(CO)_5(L)P_2\{Ml_n\}_x$ (L = CO, PBu_3, PPh_3; ML_n = $Cr(CO)_5$, $Mo(CO)_5$, $W(CO)_5$, $Fe(CO)_4$; x = 1, 2)	396
$\{Cp^{4P}\}_2Ni_2P_2$	264
$Cp^*_2(CO)_2Mo_2(P_2)_2$	397
$Cp^*_2(CO)_2Mo_2(P_2\{Cr(CO)_5\})_2$	397
$W_2(CO)_7(I)(\mu\text{-}I)As_2$	387
$Cp^*_2(CO)_4Mo_2As_2$	397, 398
$Cp_2(CO)_4M_4As_2$ (M = Mo, W)	128, 387
$Cp_2(CO)_4M_2As\{AsCr(CO)_5\}$ (M = Mo, W)	387
$Cp_2(CO)_4M_2\{AsCr(CO)_5\}_2$ (M = Mo, W)	387
$Cp_2(CO)_4Mo_2\{AsMn(CO)_2Cp\}_2$	128
$Co_2(CO)_6As_2$	399, 400
$Co_2(CO)_{6-x}(PPh_3)_xAs_2$ (x = 1, 2)	399, 400
$Co_2(CO)_{6-x}(L)_xAs_2$ (x = 1, 2, 3, 4; L = PMe_3, $P(OMe)_3$)	401
$Co_2(CO)_6As\{AsM(CO)_5\}$ (M = Cr, Mo, W)	401
$Co_2(CO)_4\{P(OMe)_3\}_2\{AsM(CO)_5\}_2$ (M = Cr, Mo, W)	401
$Co_2(CO)_{6-x}(PMe_3)_x\{AsW(CO)_5\}_2$ (x = 1, 2)	401
$Co_2(CO)_3\{P(OMe)_3\}_3\{AsCr(CO)_5\}_2$	401
$[(Cp^{1M})Mo]_2[As_2]_2$	402
$Co_2(CO)_4(PR_3)_2Sb_2$ (R =Ph, p-Tol)	57
$\{Mo(CO)_2Cp\}_2Sb_2$	403
$Co_2(CO)_4(PPh_3)_2Bi_2$	57
$M_2(CO)_4(Cp^x)_2Bi_2$ (M = Mo, W; Cp^x = Cp, Cp^{1M})	47, 404
$W_2(CO)_8\{Bi(Me)W(CO)_5\}Bi_2$	89

pair is generally retained for bonding external to the metal cluster, whereas the R groups occupy this location in the E_2R_2 derivatives. As is readily apparent, these fragments have a number of possible bonding modes in which there is variety in the number of elecrons that may be used for bonding to metal fragments. As with other systems, when the E_2 unit is bridged by a metal, there may be ambiguity about the direct bonding between the E atoms. The tetrahedral E_2M_2 and related butterfly configurations have been examined theoretically for electron counts of 5, 6, and 7 skeletal electron pairs.[234] The bonding has simliarities to that drawn for an olefin coordinated to a metal [Eq. (2)]: Is it a π-complex (**94**), or is the electron donation so complete that the unit may be better viewed as a metallacyclopropane (**95**)? Consider the case of the simple M_2E_2 complexes such as $Co_2(CO)_6P_2$ [Eq. (3)]. As with the olefin example, the P_2 unit (an acetylene analogue) can be considered to donate in a π-fashion to the $Co_2(CO)_6$ fragment (**97**), or alternatively one can draw full-fledged single bonds between each P atom and each Co atom (**98**). The structural parameters of this class of molecules generally show short E–E distances compared to normal single-bond values, which suggests at least partial retention of the multiple bond, but care must be exercised in any arguments based solely on bond parameters. Other factors such as oxidation state assignment and hybridization can also affect the expected values, and these quantities may be as difficult to assess. Fenske–Hall calculations on a series of $Fe_2(CO)_6X_2$ molecules, including X = NMe, concluded that systematic changes in X had little effect on the Fe–Fe bonding interaction.[386] A comparison of the molecular orbital structure of $Cp_2(CO)_4Mo_2P_2$ and $Cp_2(CO)_4Mo_2Bi_2$ has been made.[47] The molecular orbital picture shows that the lone pairs of electrons available on the P atoms are more available for bonding to external ML_n fragments, whereas for the Bi_2 molecule the most available electrons are those in the Bi–Bi bond, which thus prefers the addition of an unsaturated metal fragment across that bond rather than to a Bi lone pair. An extensive study of the ability of the As_2 fragment in As_2M_2 clusters to function as a four-, six-, or eight-electron donor has been reported.[387]

:E≡E:

90

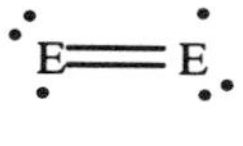

E=E

91

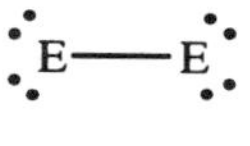

E—E

92

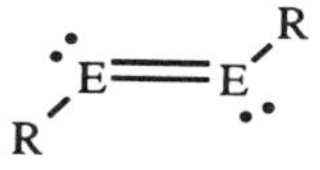

R–E=E–R

93

R–E—E–R

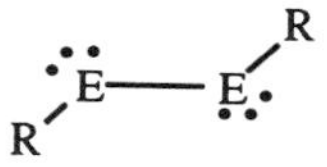

94

$$\text{(2)}$$

95 **96**

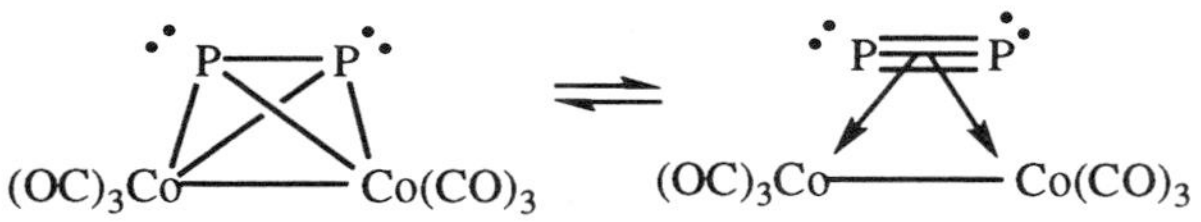

$$\text{(3)}$$

97 **98**

As evident from the table, the tetrahedral class of molecules is one of the largest represented, as is true for the general class for E_xM_{4-x} ($x = 1–4$) compounds (Scheme 6). The E atoms in the E_2M_2 core structure retain an external valence orbital that accommodates a lone pair of electrons, a lone pair donated to an external metal fragment or a bonding pair to an R group (**103–105**). As with metal carbonyl compounds in general, substitution reactions of PR_3 for CO are also possible at the metal centers. For compounds such as E_2M_2, the E_2 unit provides four electrons for donation to the metals. To achieve a closed-shell configuration, the M_2L_n fragments should have 30 electrons, as is the case with $Co_2(CO)_6$ or $Cp_2(CO)_4Mo_2$. An E_2R_2 fragment, however, has 6 electrons available for the cluster core, and so the metal fragment needs only to possess 28 electrons each, as is the case for $Fe_2(CO)_6$. Also of note is the recent report of $[Fe_2(CO)_6AsS]^-$ (**106**) showing that heteronuclear E_2 fragments may behave similarly to the homonuclear species.[405] A unique structure is observed for $Cp^*{}_2Mo_2(P_2)$ (**107**), where two tetrahedral units share the M–M edge.[397]

The E_2R_2 as a six-electron donor fragment can satisfy the electron demand of other M_x fragments where $x > 2$. The 18-electron rule predicts that M_3 triangles should prefer having 48 electrons total. Thus, the $M_3(CO)_9$ fragments (M = Fe, Ru) have 42 electrons and can form closed-shell configurations with an E_2R_2 group (**108**; Table XXV). For the tetrametal system $M_4E_2R_2$ (Table XXVI), the commonly observed geometry is that given in

99 **27** **100** **101** **102**

SCHEME 6

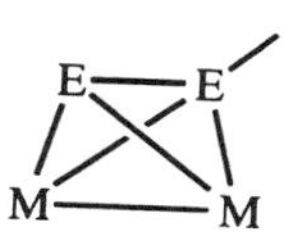

103 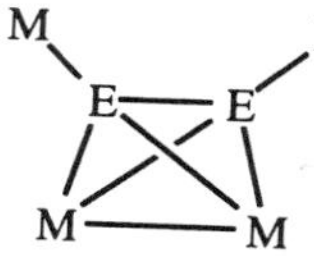 **104** 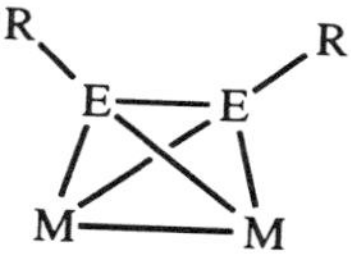 **105**

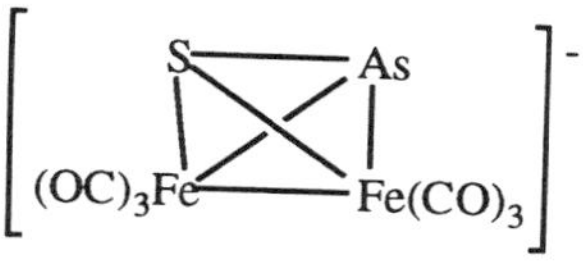

106 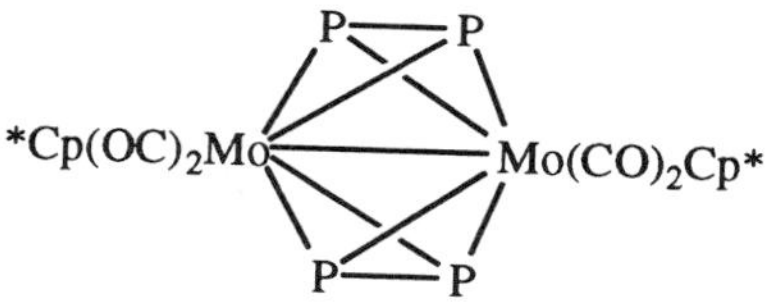 **107**

109. According to the 18-electron rule an M_4 fragment with five M–M bonds would require a total of 62 electrons. This is observed for $M_4(CO)_{12}N_2R_2$ (M = Fe or Ru). It is noteworthy that the clusters **109** have the N_2R_2 fragment bonded in an η^2-μ_2-μ_2 configuration rather than in the more symmetrical η^2-μ_3-μ_3 arrangement observed for alkyne coordination (**110**). The bonding framework of the N_2R_2 molecule has one more electron pair than predicted for the alkyne derivative. The latter fits the cluster counting rules, having seven skeletal electron pairs with an octahedral framework. An octahedral framework is also observed for the isoelectronic $Os_4(CO)_{12}Bi_2$ (**111**).[136] It is interesting to note that this is a structural isomer of $Ru_4(CO)_{12}Bi_2$ (**112**) where the two bismuth atoms occupy *trans* vertices of the octahedron.[136,137] One should keep in mind that the cluster electron counting rules will not predict which of a set of possible isomers having the same electron count and same overall cluster framework will be favored.

TABLE XXV

Molecules with $M_3(\eta^2, \mu_3\text{-}E_2)$ Core Structures

Compound	Ref.
$Fe_3(CO)_9(EtNNEt)$	267, 318
$Ru_3(CO)_9(RNNR)$ (R = Me, Et)	318
$Fe_2Ru(CO)_9(EtNNEt)$	318
$Cr_3(CO)_{10}(P^tBu)(P_2{}^tBu_2)$	175

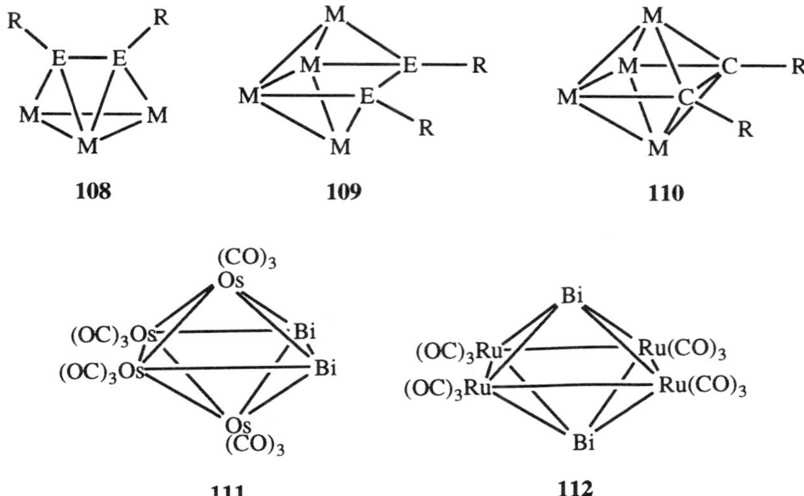

The P_2R_2 fragment can bond to metallocene fragments, as has been found for both $Cp_2^*ZrP_2H_2$ and $Cp_2MoP_2H_2$ (**113**). Neither of these molecules has been structurally characterized, however, so that the extent of involvement of the lone pairs on P with the metal center is not clear. The bonding mode with the metal fragment bridging the E–E bond (**114**) has been established for other compounds (Tables XXVII and XXVIII). The E_2R_2 fragment may also bind one or two metal fragments as a simple two-electron donor ligand (**115**, **116**; Table XXIX). The ability to donate the lone pairs is observed even in compounds that also contain a μ-M fragment (**117**, **118**).

TABLE XXVI

MOLECULES WITH $M_4(\eta^2, \mu_4\text{-}E_2)$ CORE STRUCTURES

Compound	Ref.
$Ru_4(CO)_{12}(RNNR)$ (R = Me, Et)	318, 406
$FeRu_3(CO)_{12}(RNNR)$ (R = Me, Et)	318
$Fe_2(CO)_6\{P_2R_2\}_2Fe_2(CO)_6$ (R = Me, Ph, Tol)	107
$Fe_2(CO)_6(MePCH_2CH_2Me)(MePPMe)Fe_2(CO)_6$	107
$[Fe_2(CO)_6(P(H)Tol)_2]_2(MeC_6H_4PPTol)$	107
$\{Fe_2(CO)_6\}_2\{Fe(CO)_4\}_2(P_2)_2$	407
$Ph_2P_2[Ir_2(C_8H_{12})_2(Cl)]_2(PPh)_2$	96
$Ir_4(CO)_6(P_2{}^tBu_2)(P_4{}^tBu_4)$	327
$[(\mu\text{-}{}^tBuS)Fe_2(CO)_6]_2(\eta^2, \mu_4\text{-}MeAsAsMe)$	408

TABLE XXVII

Complexes with $E_2(\mu\text{-}M)$ Fragments

Compound	Ref.
$Cp_2^*Zr(PH)_2$	235
$Cp_2^*ZrP_2(C_6H_2Me_3)_2$	409
$[Cp_2^*ZrP_2]^{2-}$	235
$Cp_2Mo(HPPH)$	410
$Ph_2P_2\{\mu\text{-}Cr(CO)_5\}\{Cr(CO)_5\}_2$	411
$R_2P_2\{\mu\text{-}Fe(CO)_4\}\{Fe(CO)_4\}$ (R = $OC_6H_2{}^tBu_3$-2,4,6, Ph, C_6H_2Me-2,4,6, CH_2SiMe_3, $N(SiMe_3)_2$, $CH(SiMe_3)_2$)	239, 240, 412–415
$Ph_2P_2Ni[MeNC_4O_2(PPh_2)_2]$	416
$R_2P_2Ni(R_2PCH_2CH_2PR_2)$ (R = Et, cy, Ph; R′ = Me, ^{t}Bu, Ph)	417
$Ph_2P_2Pd(dppe)$	418, 419
$Ph_2P_2\{W(CO)_5\}_2(\mu\text{-}Pd(dppe)$	418
$Ph_2As_2[Cr(CO)_5]_2[\mu\text{-}Cr(CO)_5]$	420
$(C_6F_5)_2As_2\{\mu\text{-}Fe(CO)_4\}$	421
$Ph_2As_2Ni[MeNC_4O_2(PPh_2)_2]$	416
$R_2Sb_2\{W(CO)_5\}_3$ (R = Me, Ph, ^{t}Bu)	44, 422
$\{CH(SiMe_3)_2\}_2Sb_2\{\mu\text{-}Fe(CO)_4\}$	87, 88

As with previous examples, the structurally characterized molecules have short E–E contacts indicating at least partial retention of the multiple bond.

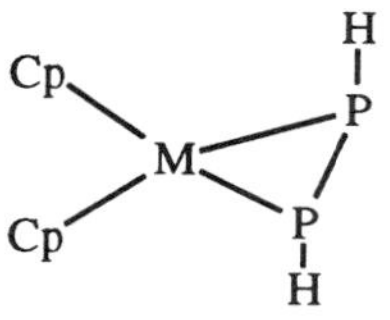

113 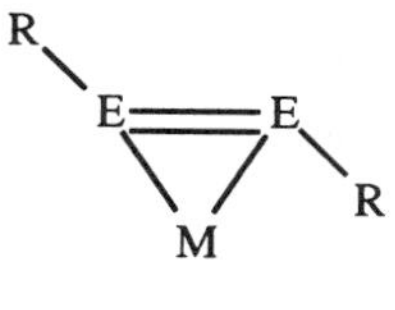 **114** 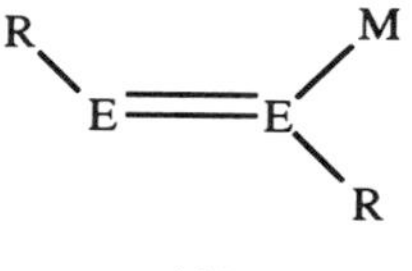 **115**

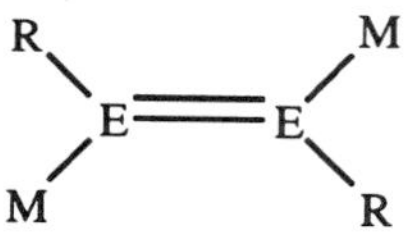

116 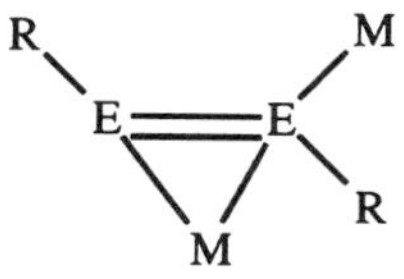 **117** 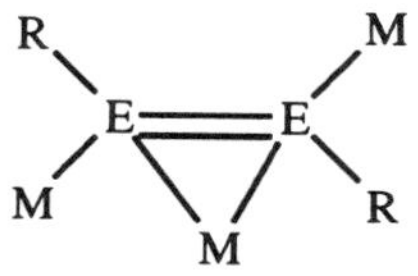 **118**

Dinitrogen is the classic example of a naked E_2 fragment. The triple-bonded analogues for the heavier congeners are observed only in the gas phase, but have been stabilized by bonding to metal fragments. Compounds

TABLE XXVIII

COMPLEXES WITH $E_2(\mu\text{-}M)_2$ FRAGMENTS

Compound	Ref.
$[Cp_2^*Zr]_2P_2$	235
$\{Cp(CO)_2Mo\}_2\{PhPPh\}$	423
$[Cp^*Co]_3(P_2)_3$	343
$[(R_2PCH_2CH_2PR_2)Ni]_2P_2$ (R = Et, cy, Ph)	417
$\{Cp^*(CO)_2Mn\}_2As_2$	246, 247
$[(Ph_3P)_2Pd]_2As_2$	335
$[Cp^*Co]_3\{As_2\}\{As_4\}$	424
$\{Cp_2^*Sm\}_2Bi_2$	425

with two metal fragments bridging an E_2 unit are also known, as exemplified by $[Cp_2^*Zr]_2P_2$, $[(Ph_3P)_2Pd]_2As_2$ (**119**), and $[Cp^*Sm]_2Bi_2$ (**120**) (see Table XXVIII). The molecules may be planar or butterfly-shaped. Three metal centers bridge the E_2 unit in $E_2[\mu\text{-}W(CO)_5]_3$ (E = As, Sb, Bi) and $Bi_2[\mu\text{-}Mn(CO)_2(Cp^{1M})]_3$ (**121**). Structures for all these have been reported (although they have only been produced in low yields) and all have very short E–E distances showing retention of the E–E multiple bond character (Table XXX). There is no bonding between the transition-metal centers. We have already mentioned the formation of E–E interactions in the E_2M_3 cluster series upon successive addition of two electrons to the *closo*-cluster core, and the examples known for E = Group 15 elements are $[Fe_2(CO)_6Bi_2\{\mu\text{-}Co(CO)_4\}]^-$ and $[(\mu\text{-}H)Fe_2(CO)_6Bi_2\{\mu\text{-}Fe(CO)_4\}]^-$.[300] The Bi–Bi distances are consistent with single-bond values, which was not anticipated based on conventional electron counting rules. A similar situation obtains in the case of $[E_2Co_4(CO)_{11}]^{-/2-}$ (**122**; E = Sb, Bi).[65,303,428] The mono- and

TABLE XXIX

COMPLEXES WITH E_2R_2 FRAGMENTS BONDED η^1 TO A TRANSITION-METAL FRAGMENT

Compound	Ref.
$RP{=}PR\{Cr(CO)_5\}$ (R = $CH(SiMe_3)_2$)	412
$\{(Me_3Si)_2NP{=}PN(SiMe_3)_2\}Cr(CO)_5$	239
$Ph\{Cr(CO)_5\}P{=}P\{Cr(CO)_5\}Ph$	411
$\{Cp^*(CO)_2M\}P{=}P(C_6H_2{}^tBu_3\text{-}2,4,6)$ (M = Fe, Ru, Os)	426
$\{Cp^*(CO)_2M\}\{Fe(CO)_4\}P{=}P(C_6H_2{}^tBu_3\text{-}2,4,6)$ (M = Fe, Ru, Os)	426
$\{Fe(CO)_4\}_2PhP{=}P(H)Ph\{W(CO)_5\}$	413, 414
$\{Fe(CO)_4\}_2RP{=}P(H)R\{Fe(CO)_4\}$ (R = Ph, Me)	414
$[Cp(CO)(L)Fe][(CO)_5Cr]As{=}AsC_6H_2{}^tBu_3$ (L = CO, PPh_3)	427
$\{Cp(CO)_2Mn\}_4As_2$	246, 247, 387

TABLE XXX

Complexes with $E_2(\mu\text{-}M)_3$ and $E_2(\mu\text{-}M)_4$ Fragments

Compound	Ref.
$As_2\{Mo(CO)_5\}_3$	387
$As_2\{W(CO)_5\}_3$	429
$Sb_2\{W(CO)_5\}_3$	422
$Bi_2\{W(CO)_5\}_3$	89, 430
$Bi_2\{Mn(CO)_2Cp^{1M}\}_3$	258, 431
$[Fe_2(CO)_6Bi_2\{\mu\text{-}Co(CO)_4\}]^-$	300, 301
$[HFe_2(CO)_6Bi_2\{\mu\text{-}Fe(CO)_4\}]^-$	299
$[E_2Co_4(CO)_{11}]^{x-}$ (E = Sb, Bi; x = 1, 2)	65, 303, 428
$Bi_2Os_4(CO)_{12}$	136

dianionic forms of the Sb molecules have been structurally characterized, as has the monoanion for Bi. They all show essentially the same structure. Formally the molecules are electron rich, having 8.5 skeletal electron pairs for the monoanions and 9 pairs for the dianions, whereas a bicapped tetrahedral structure (or monocapped trigonal bipyramid) would be satisfied with six pairs. An alternative view is to consider the molecules as *nido*-pentagonal bipyramids that should have eight pairs, but this view also does not predict an E–E interaction. The orbital interaction between the main group fragment and the transition metals is such that a set of nonbonding orbitals is created that can hold as many as six skeletal electrons more than expected. The primary structural change upon going from the Sb monoanion to the Sb dianion is an increase in the CO-bridged Co–Co bond by ca. 0.136 Å, suggesting that the HOMO of the monoanion is antibonding with respect to this Co–Co bond.

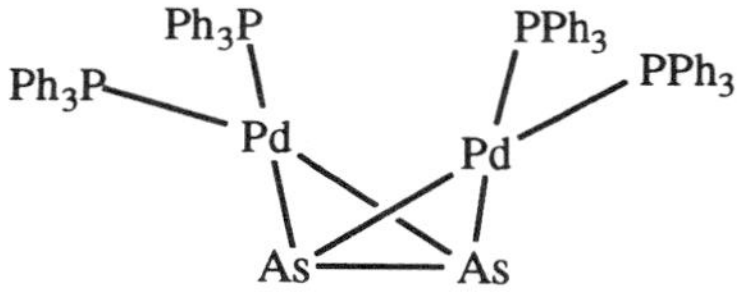

119

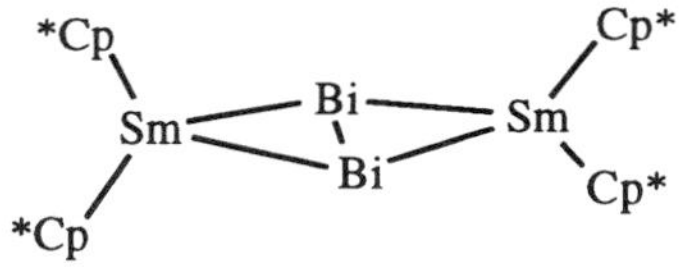

120

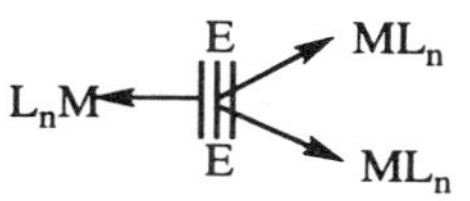

121

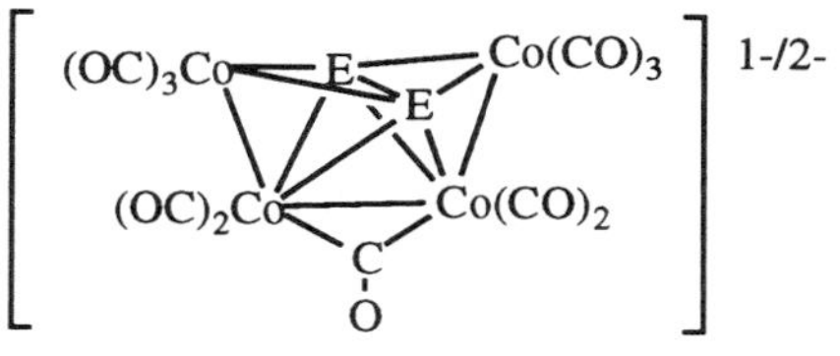

122

Two E_2 units may connect two metal centers (**123**; Table XXXI). These compounds resemble *trans*-E_4M_2 octahedra with two E–E bonds broken (or weakened) in the final product. To date the observed molecules are all made of metal fragments that have either 13 or 14 valence electrons. Unfortunately, the structure of $\{Cp^*Fe\}_2(P_2)_2$ was disordered, making its detailed comparison with the other examples difficult. The Nb compound has also not been structurally characterized, but the $Cp^*Nb(CO)_2$ fragment is isoelectronic to the CpCo and CpRh units so that it would be expected to adopt a similar arrangement.

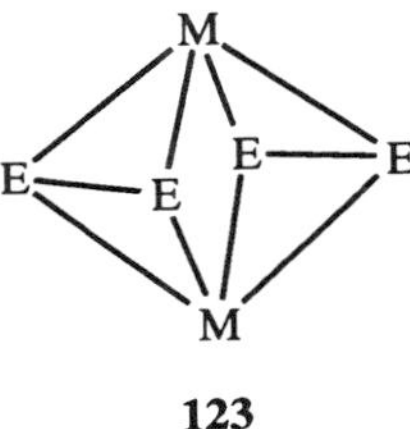

123

2. *Compounds with E_3 Units*

Numerous molecules containing E_3 units with and without organic substituents have been reported (Tables XXXII–XXXV). The most prevalent is a triangular E_3 unit with P being the most represented, followed by As. No Sb examples are known, but one with a Bi_3 triangle has been observed. The neutral triangular unit can serve as a three-electron source (**124**), but it is often commonly considered as an E_3^{3-} ion (**125A**), which can then donate six electrons, each E atom retaining an external lone pair. This ion has six π electrons and may thus be considered a pseudo benzene analogue (**125B**). In this view it can form π complexes to a single metal ion (**126**) in the same fashion as an aromatic ring (**127**). This type of analysis will also be made for E_5^- as an analogue to Cp^- and for E_6 as a benzene analogue.

TABLE XXXI

MOLECULES WITH $M_2(\mu\text{-}E_2)_2$ CORE STRUCTURES

Compound	Ref.
$\{Cp^*Fe\}_2(P_2)_2$	432
$\{Cp^*Co\}_2(P_2)_2$	432
$[(Cp^{M4E})Rh]_2(P_2)_2$	433
$[Cp^*Nb(CO)]_2(As_2)_2$	434
$\{(Cp^x)Co\}_2(As_2)_2$ (Cp^x = Cp^*, Cp^{M4E})	424

A theoretical study of the E_3 unit has been reported.[435] It was found that metal-bound P_3 rings were more strongly bound than corresponding C_3 units, in part because of a better orbital interaction between the P_3 unit and the metal center.

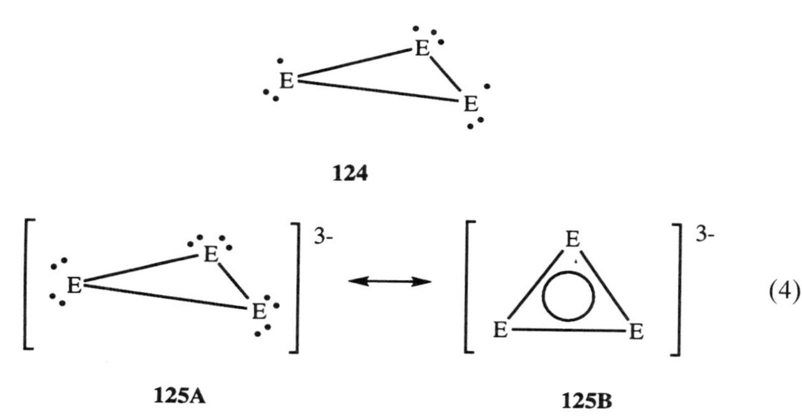

TABLE XXXII

Complexes with Tetrahedral E_3M Core Structures

Compound	Ref.
$Cp(CO)_2CrP_3$	389–391
$Cp^x(CO)_2MoP_3$ (Cp^x = Cp, Cp^{1B}, Cp^{2B})	392, 439
$[Cp^*Fe]_3\{P_3Fe\}P_6$	343
$[\{Cp^xRu\}_3Ru(\eta^3\text{-}As_3)(\mu_3\text{-}\eta^3\text{-}As_3)(\mu_3\text{-}As)_3$ (Cp^x = Cp*, Cp^{M4E})	437
$(CO)_3CoP_3$	126, 435
"$C_3H_3CoP_3$"	435
$N(CH_2CH_2PPh_2)_3CoP_3$	440
$[(triphos)CoP_3]_2(CoBr)_6$	441
$(triphos)MP_3$ (M = Co, Rh, Ir)	442–445
$(triphos)CoP_3\{Cr(CO)_5\}_x$ ($x = 1, 2$)	224, 442, 446
$[(triphos)MP_3Me]^+$ (M = Co, Rh)	447
$\{(Cp^{2B})Ni\}P_3$	79
$CpNiP_3$	435
"$C_3H_3NiP_3$"	435
$[(triphos)NiP_3]^+$	444
$[(CO)_3Ni]_2\{P_3{}^iPr_3\}[Ni(CO)_2]_2\{P_3{}^iPr_3\}[Ni(CO)_3]_2$	327
$[(CO)_3Ni]\{P_3{}^iPr_3\}[Ni_3(CO)_4]\{P_3{}^iPr_3\}[Ni(CO)_3]$	96
$[(triphos)MP_3]^+$ (M = Pd, Pt)	444, 448
$Co(CO)_3As_3$	449
$Cp^*(CO)_2MoAs_3$	397, 450
$Cp^x(CO)_2CrAs_3$ (Cp^x = Cp*, Cp^{M4E})	451
$\{Cp^{4P}Ni\}As_3$	452

TABLE XXXIII

COMPLEXES WITH TRIGONAL BIPYRAMIDAL E_3M_2 CORE STRUCTURES

Compound	Ref.
$\{Cp^{2B}\}_2ThP_3Th(Cl)\{Cp^{2B}\}_2$	453
$[\{(triphos)Ni\}_2P_3][BR_4]_2$ (R = F, Ph)	443
$[(triphos)CoP_3Co(triphos)]^{n+}$ (n = 1, 2)	436
$[(triphos)ME_2SM'(triphos)]^{2+}$ (E = P, As; M, M′ = Co, Rh)	454
$[NiP_3\{Ni(triphos)\}_2]^{n+}$ (n = 1, 2)	436
$[(triphos)NiP_3Co(triphos)]^{n+}$ (n = 1, 2)	436
$[LPdP_3PdL]^+$ (L = triphos, $N(CH_2CH_2PPh_2)_3$)	448
$[As_3\{M(triphos)\}_2][BX_4]_2$ (M = Co, Ni; X = F, Ph)	436, 455

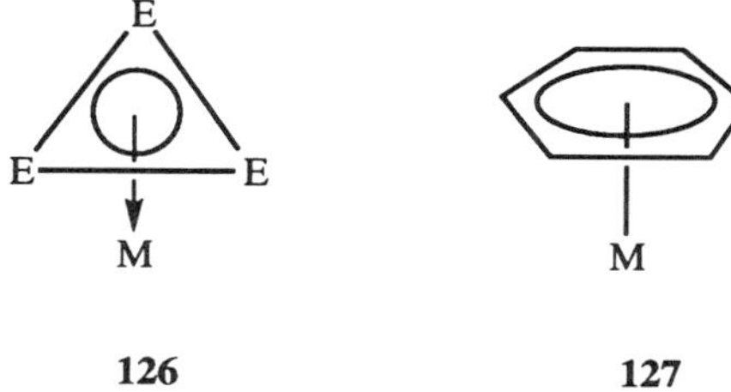

126 **127**

As with the E_2M_2 molecules, the tetrahedral E_3M unit occurring in E_3M (**101**) and trigonal bipyramidal E_3M_2 cores (**128**) are the most commonly observed structural types. A number of the trigonal bipyramidal E_3M_2 compounds have been prepared by reaction of simple M^{2+} salts with a tridentate ligand and elemental E.[436] The products have electron counts ranging from 31 to 34. E_3M_3 units are also known. This has already been seen in **67**, where the icosahedral central Co_6As_6 core has two As_3 units fused onto it so that two As_3Co_3 octahedral cavities result.[343] An octahedral cavity was also found in $Co_4(PPh_3)_4(\mu_3\text{-}As)_3(\mu_3, \eta^3\text{-}As_3)$ (**129**),[223] which

TABLE XXXIV

COMPLEXES WITH $(\eta^3, \mu_3\text{-}E_3)M_3$ UNITS

Compound	Ref.
$[As_3Co_3(PEt_2Ph)_3]_2(As_6)$	343
$(PPh_3)_4Co_4(\mu_3\text{-}As)_3(\mu_3, \eta^3\text{-}As_3)$	223
$[\{(Cp^x)Ru\}_3Ru(\eta^3\text{-}As_3)(\mu_3\text{-}\eta^3\text{-}As_3)(\mu_3\text{-}As)_3$ (Cp^x = Cp*, Cp^{M4E})	437
$[Et_4N]_2[Bi_3\{Fe(CO)_3\}_3BiFe(CO_4)]$	304, 438
$[Cp^*Fe]_3\{P_3Fe\}P_6$	343

TABLE XXXV

COMPLEXES WITH NONTRIANGULAR E_3 FRAGMENTS

Compound	Ref.
$[K(THF)_x][Cp_2^*ZrP_3]$	235
$Cp^*ZrP_3Ph_3$	409
$M_2(CO)_7(P_3{}^tBu_3)$ (M = Mn, Re)	327
$(Et_2N)_2PP\{Fe(CO)_4\}P\{Fe(CO)_4\}OC(NEt_2)Fe(CO)_3$	456
$(Et_2N)_2PP\{Fe(CO)_4\}POC(NEt_2)Fe_2(CO)_6$	456
$Fe_2(CO)_6\{PPh\}_3$	282
$[Fe(CO)_3]_2\{P_3{}^tBu_3\}_2$	96
$[Fe(CO)_3]_2\{P_3{}^iPr_3\}_2$	96
$\{P(H)Ph\}Fe_2(CO)_6(PPh)_3Fe_2(CO)_6\{P(H)Ph\}$	282
$[\{MeC(CH_2PPh_2)_3Co\}_2P_3Et]^{2+}$	457
$Ni_5(CO)_5(\mu_3\text{-}P^tBu)_2(P_3{}^tBu_3)(\mu_4\text{-}P^tBu)$	116
$[\{(triphos)Co\}_2As_3Ph]^{2+}$	457
$\{CpMo\}_2(As_3)(AsS)$	458

can be viewed as a Co_4 tetrahedron capped on three faces by μ_3-As atoms and on the fourth by the As_3 unit. For $\{Cp^xRu\}_3Ru(\eta^3\text{-}As_3)(\mu_3\text{-}\eta^3\text{-}As_3)(\mu_3\text{-}As)_3$ (**130**; Cp^x = Cp*, Cp^{M4E}),[437] $Cp_3^*Fe_3(P_3)(\mu_4\text{-}P)_3Fe(\eta^3,\mu_3\text{-}P_3)$ (**131**), and $[Et_4N]_2[Bi_3\{Fe(CO)_3\}_3BiFe(CO)_4]$ (**132**),[304,438] there are no M–M bonds joining the three metals. The metal frameworks of **130** and **131** are essentially the same, but subject to interpretation of the bonding between the basal E_3 unit and the other bridging E atoms. For **130** the As–As distances between the basal triangle and the other bridging As atoms are on the order of 2.71 to 2.75 Å, whereas in **131** the corresponding P–P distances are ca. 2.41 Å. These values are about 7% (**130**) and 14% (**131**) longer than the E–E distances within the triangle. A careful examination reveals that the core structures of **130**, **131**, and **132** are also similar. This is most obvious if one removes the E atoms that are connected to three metals in **130** and **131**. A molecular orbital analysis of **132** is similar to that of $[E_2Co_4(CO)_{11}]^{n-}$ in that extra nonbonding orbitals are created that accommodate more than the expected number of electrons.[303,304]

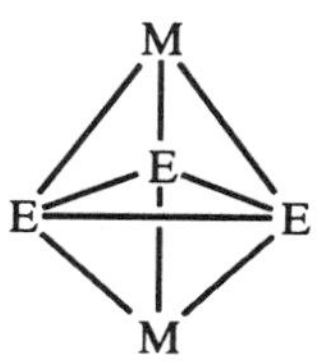

128

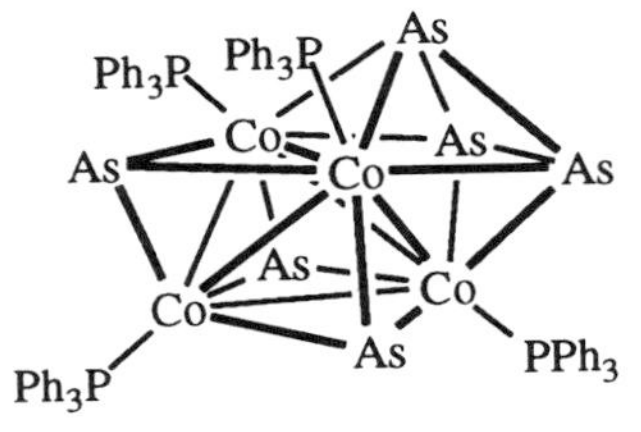

129

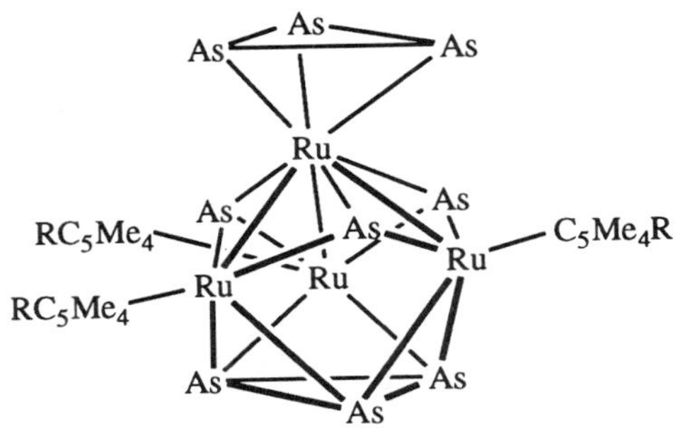

130

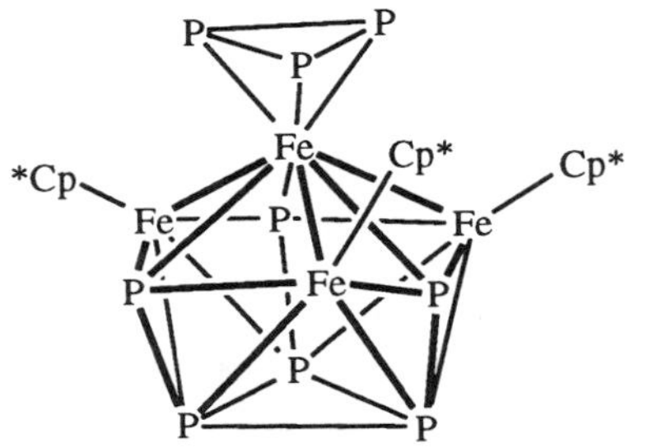

131

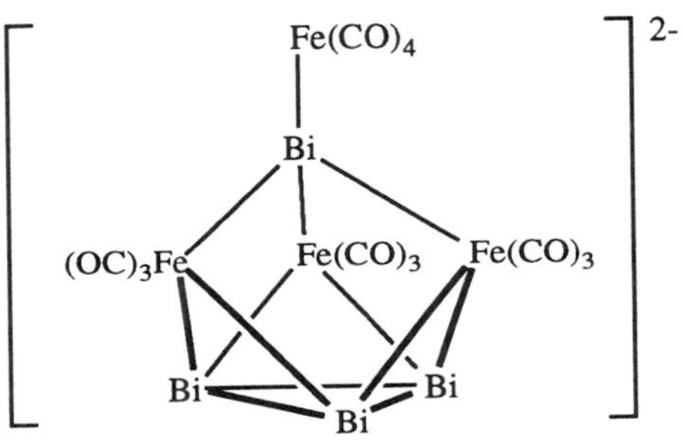

132

In addition to the naked E_3 unit, *cyclo*-E_3R and -E_3R_3 groups have also appeared coordinated to metals. The cyclic nature of E_3R_3 gives rise to two isomers (**133A** and **133B**), of which only the second type has been characterized. Examples include **134** (M = Co, Rh), **135**, and **136**.

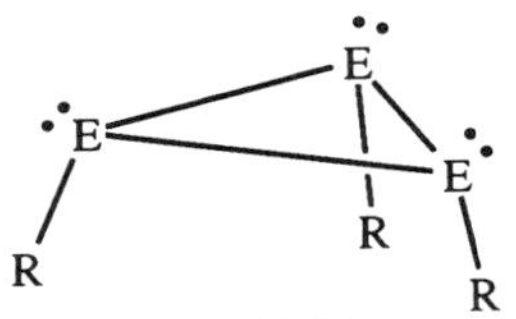

133A

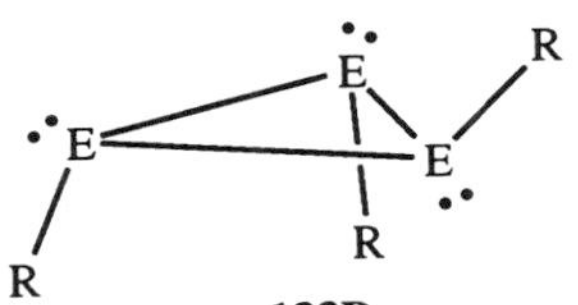

133B

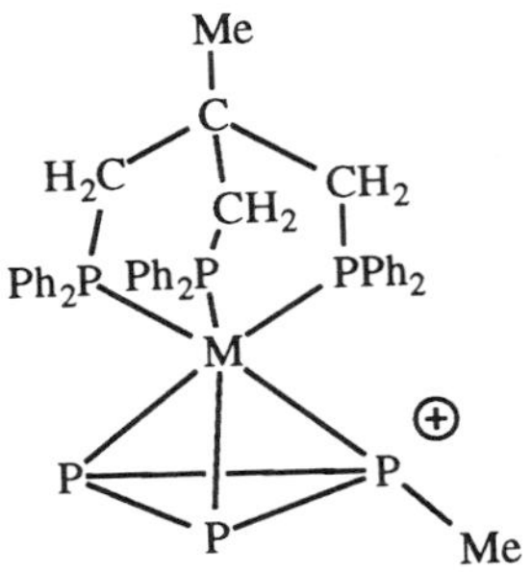

134

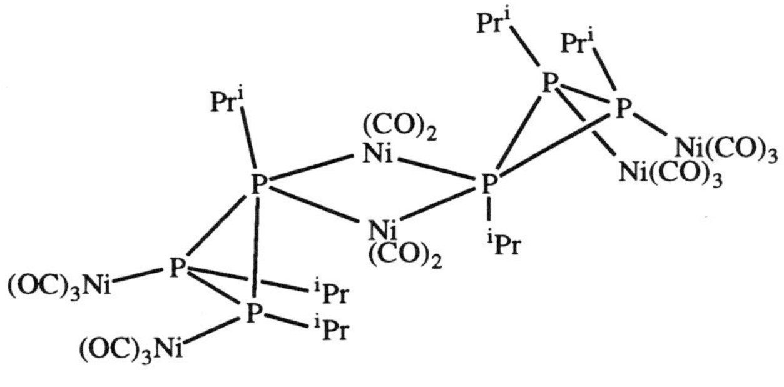

135

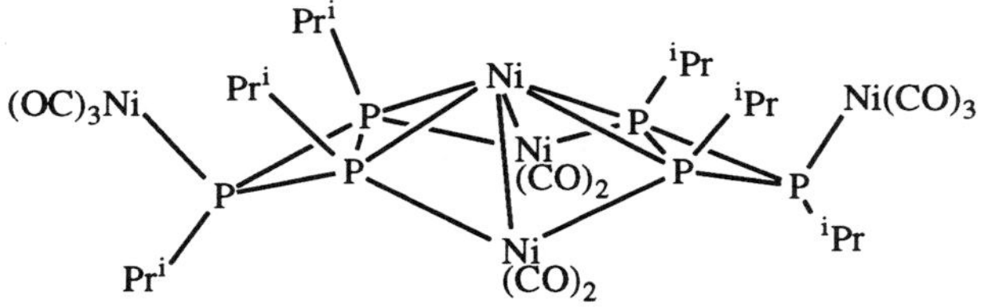

136

A variety of coordination geometries have been observed for nontriangular E_3 units. The neutral fragments in question are **137–139**, and all have been observed coordinated to metals (**140–146**). The most variety is seen for the E_3R_3 unit. It may be bonded to a single metal or to two metals. The terminal E atoms show a preference for bridging between two metal centers, and the examples show that these end units may bridge the same metals, as in **143** and **144**, or different metals, as in **146**. A bond between the two metals may be present, as in **143** and **144**, or absent, as in **145**. Two related molecules based on E_3R_4 fragments deserve mention in this context. They are **147** and **148**, and their general structural features are consistent with the observations made earlier.

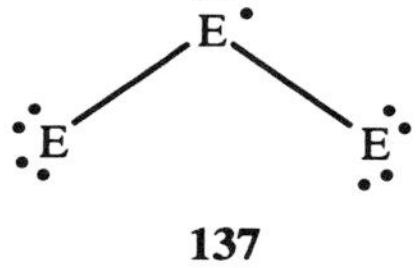

137 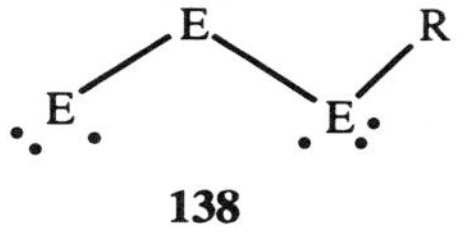 **138** 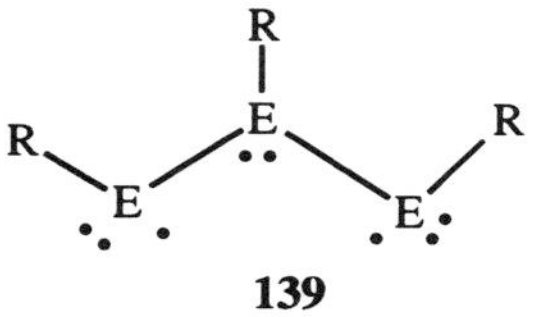 **139**

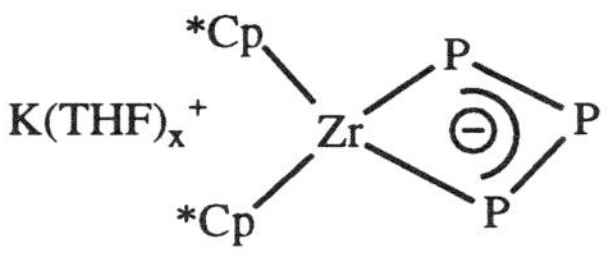

140

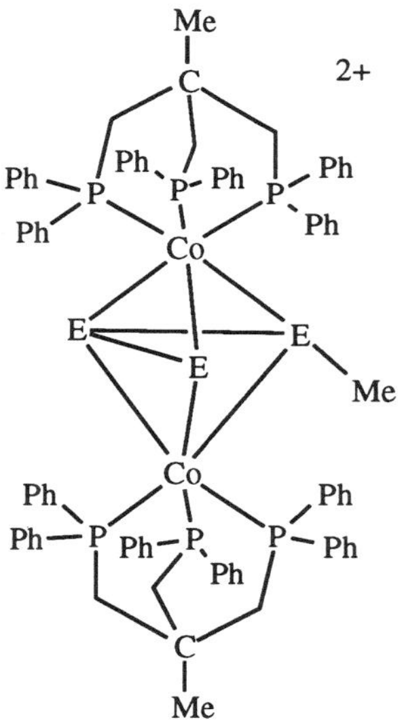

141

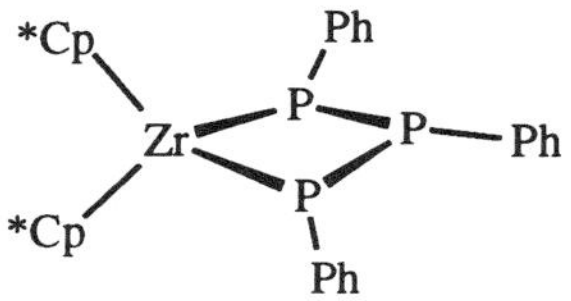

142

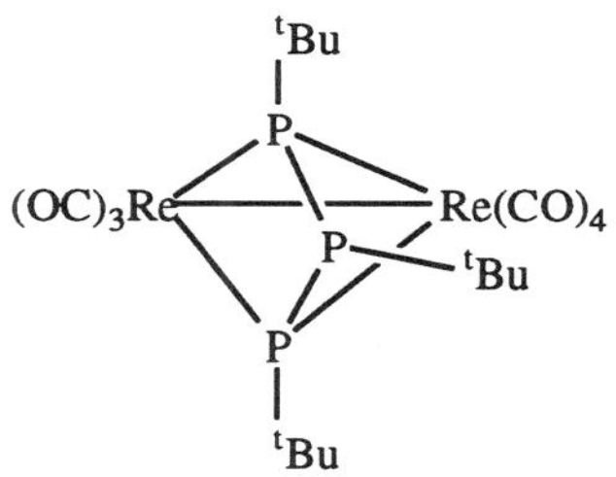

143

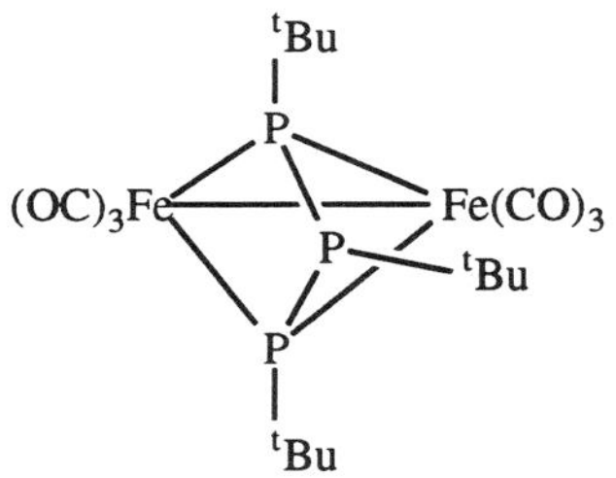

144

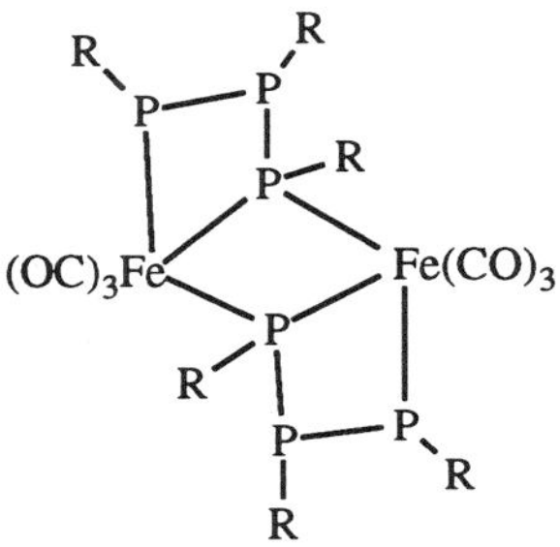

145

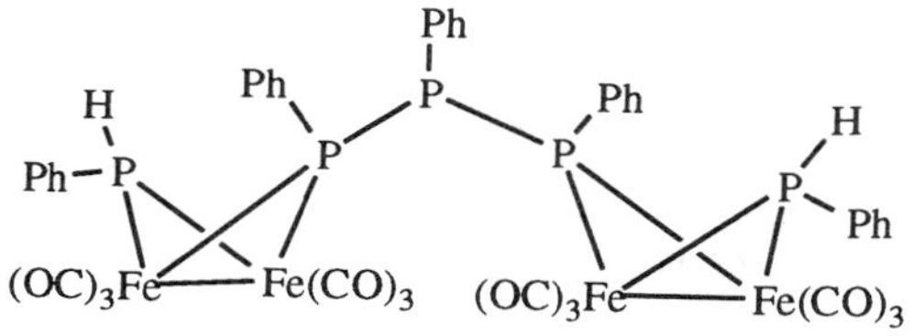

146

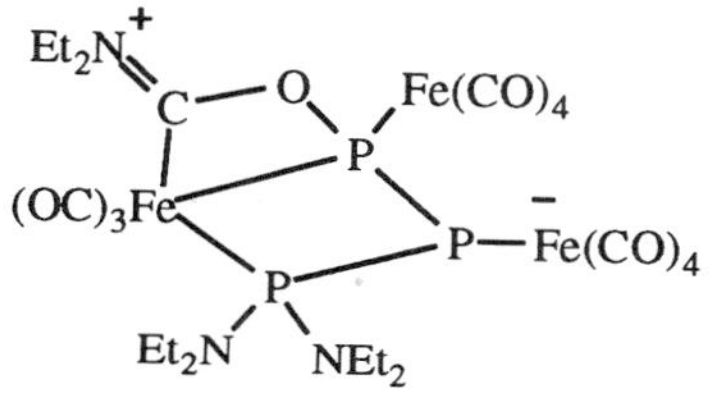

147 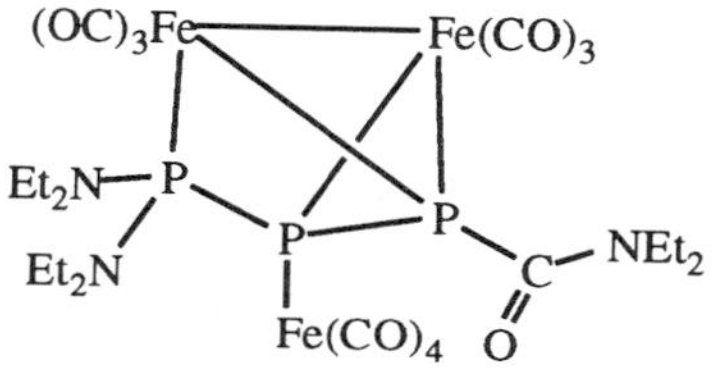 **148**

It may be argued as to whether $[CpMo]_2(As_3)(AsS)$ (**149**) belongs in this class or in the M_2E_5 category. As can be seen, it is very similar to a number of triple-decker, stacked sandwich complexes M_2E_5 to be discussed shortly in which the metals form the apical vertices of pentagonal bipyramidal arrangements. The E–E distances are asymmetric for **149** and also for $[CpMo]_2As_5$ but less so for the latter. The $[CpCr]_2P_5$ molecule is undistorted. The As_3 unit in **149** is considered allylic-like and the central As atom is considerably farther from the Mo atoms than the two terminal atoms.

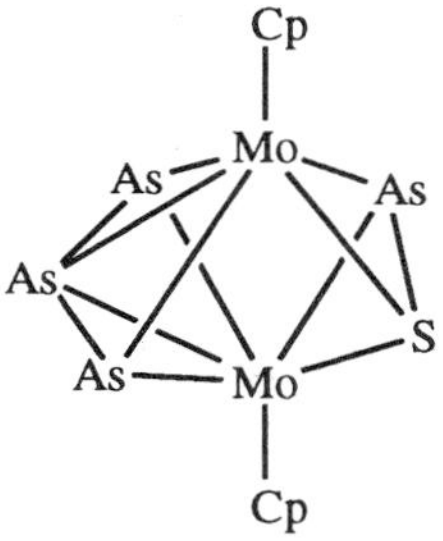

149

3. *Compounds with E_4 Fragments*

The E_4 unit shows capabilities similar to those of E_3 (Table XXXVI). The *cyclo*-E_4 unit is well represented by P and As examples and may be

TABLE XXXVI
COMPOUNDS WITH E_4 FRAGMENTS

Compound	Ref.
$(Cp^{2B})MP_4$ (M = Zr, Hf)	459
$Cp^*(CO)_2NbP_4$	460
$(CO)_4M\{PW(CO)_5\}_4$ (M = Cr, W)	407, 461, 462
$(CO)_2(MeCN)_2(Cl)WP_4(Cl)\{W(CO)_5\}_3$	407
$(CO)_2(MeCN)_2(Cl)WP_4(OH)\{W(CO)_5\}_3$	407
$P_4[Fe(CO)_4]_3$	463
$Fe_2(CO)_6\{PMe\}_4$	464
$Fe_2(CO)_6\{PPh\}_4$	282, 464
$(Cp^{2B})CoP_4\{Cr(CO)_5\}_3$	465
$Cp^*(CO)CoP_4$	466
$[Cp^*(CO)Co]_2P_4$	466
$(Cp^{M4E})RhP_4Rh(Cp^{M4E})(CO)$	459
$(Cp^{1B})(CO)RhP_4\{Cr(CO)_5\}_4$	407, 467
$(Cp^{1B})RhP_4\{Cr(CO)_5\}_4$	407, 467, 468
$(Cp^{2B})(CO)RhP_4\{Cr(CO)_5\}_4$	407, 468
$(Cp^{1B})RhP_4\{Cr(CO)_5\}_3$	407, 468
$(Cp^{1B})RhP_4\{Cr(CO)_5\}$	468
P_4RhL_2Cl (L = PPh_3, $P(Tol\text{-}p)_3$ $P(Tol\text{-}m)_3$, $AsPh_3$)	469, 470
$\{(Cp^{4P})Ni\}_2E_4$ (E = P, As)	452
$\{(Cp^{4P})Ni\}_2P_4\{W(CO)_5\}_2$	452
$(PPh_3)_2PtP_4\{Cr(CO)_5\}_2$	471
$\{Pt(PPh_2R)\}_2P_4$ (R = Me, Ph)	407
$Cp^*(CO)_2NbAs_4$	434
$[Cp^*(CO)_2Fe]_2As_4(C_6H_2{}^tBu_3)_2$	427
$Cp^*(CO)CoAs_4$	472
$\{Cp^*(CO)Co\}_2As_4$	472
$\{Cp^*(CO)Co\}\{Cp^*Co\}As_4$	472
$[Cp^*Co]_3[As_4][As_2]$	424
$Fe_2(CO)_6\{AsMe\}_4$	473, 474

thought of as a cyclobutadiene analogue (**150**). Metallated examples having an intact *cyclo*-E_4 ring are shown in **151–155**. Note that the lone pairs on the E atoms may also donate to external metal fragments. In **156**, a second metal fragment has inserted into one of the P–P bonds, whereas for **157** there are some secondary As–As interactions that make it debatable as to whether this is a molecule with an As_4 and an As_2 unit or one based on As_6. Likewise, incipient P–P bond formation in **158** suggests that it contains a *cyclo*-P_4 moiety. For comparison, the structure of the nonmetallated As derivative (**159**) has been reported, and it appears that there are only three As–As bonds. For both **158** and **159**, there are nine skeletal pairs of electrons, consistent with edge-localized bonding in a trigonal prismatic structure. One exception to the similarity between E_4 and E_3 units is that the former have no examples that fit in the multidecker classification. For E_3 units, there are the trigonal bipyramidal compounds **128**, but corresponding examples of the E_4 unit giving octahedral compounds **160** are not yet known. The known examples of this empirical formula fit better as $M_2(E_2)_2$ complexes **123** where there may also be an M–M interaction as in **107**.

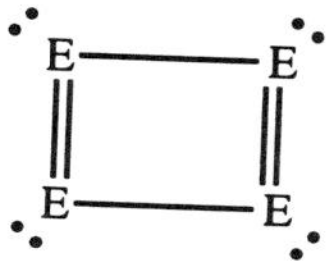

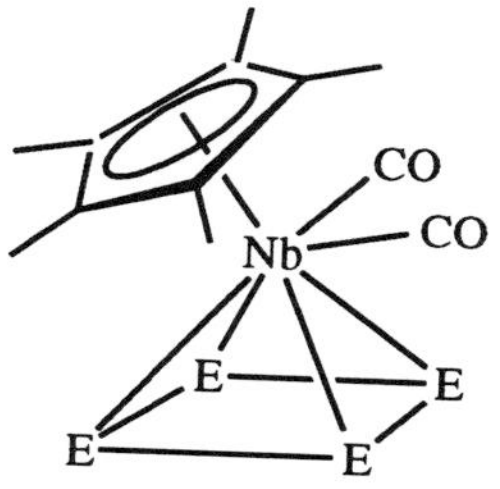

E = P, As

150 **151**

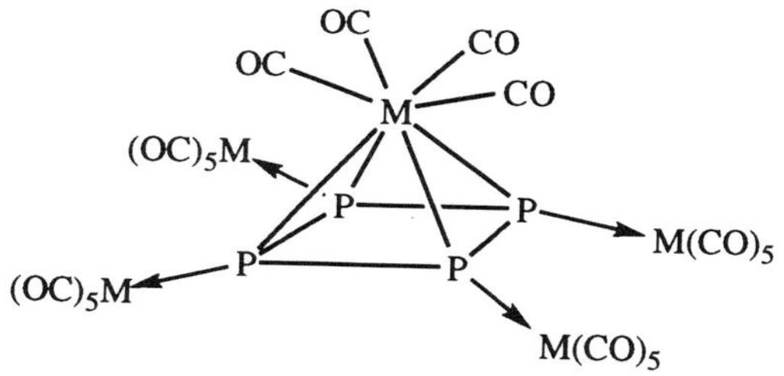

152

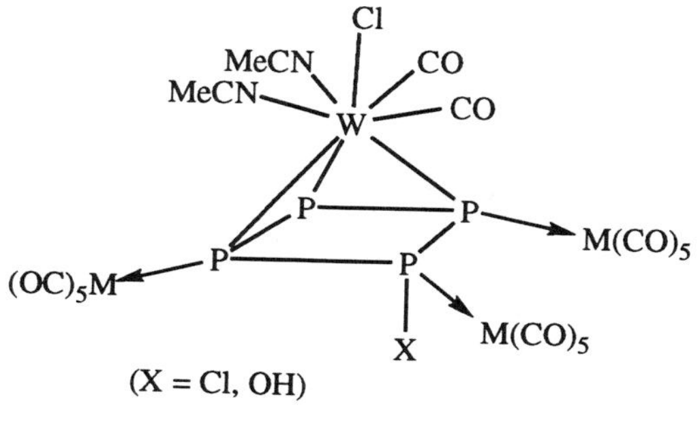

(X = Cl, OH)

153

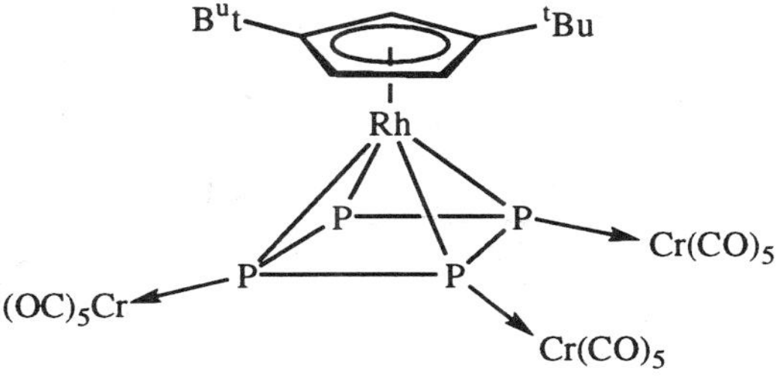

154

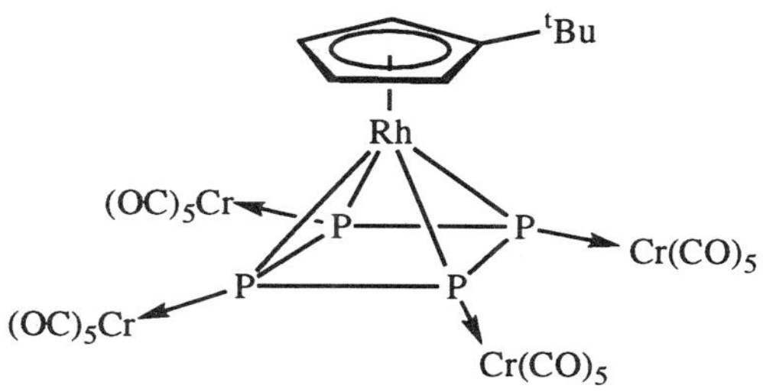

155

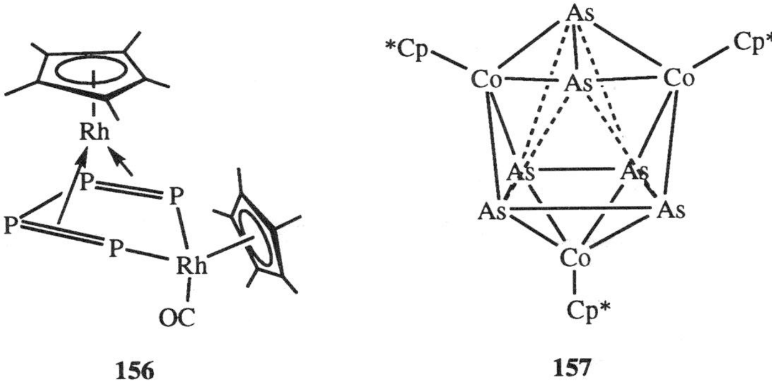

156 **157**

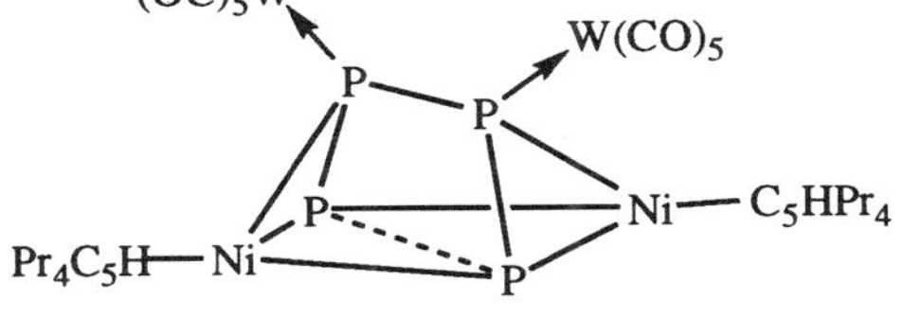

$(OC)_5W$ $W(CO)_5$

P P

P Ni — C_5HPr_4

Pr_4C_5H — Ni P

158

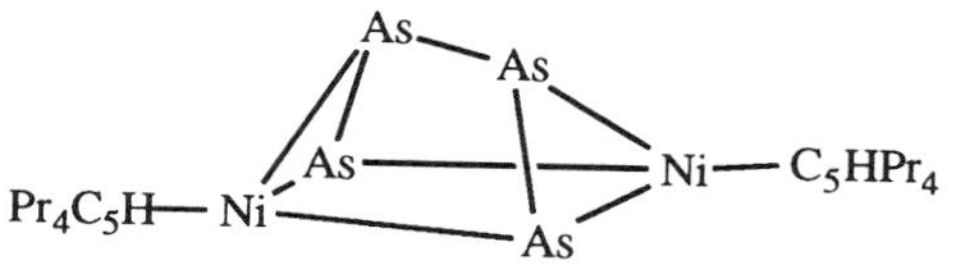

159

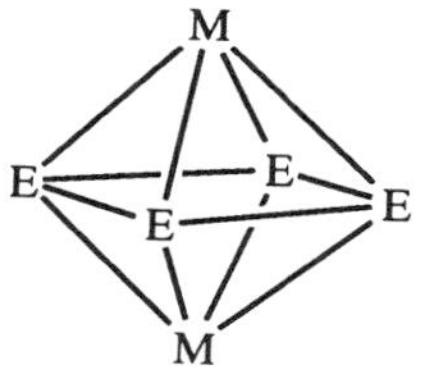

160

An alternative geometry to the cyclic structure is the tetrahedral E_4 unit, which is the geometry adopted by white phosphorus, P_4 (**102**). Metallated structures have been observed that may be derived from this structure by insertion of metal fragments into E–E bonds of the tetrahedron (**161–166**). As with the cyclic E_4 structures shown earlier, the lone pairs may also serve as donors. Note that the addition of the second metal fragment in **165** gives a cyclic E_4 structure, but here the ring is qualitatively different from the other cyclic structures in that it is puckered, rather than having a planar cyclobutadiene-like arrangement.

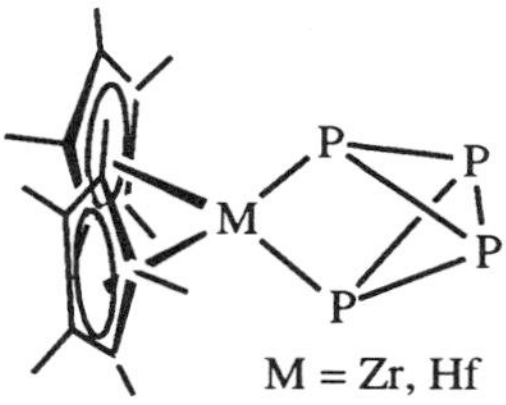

M = Zr, Hf

161

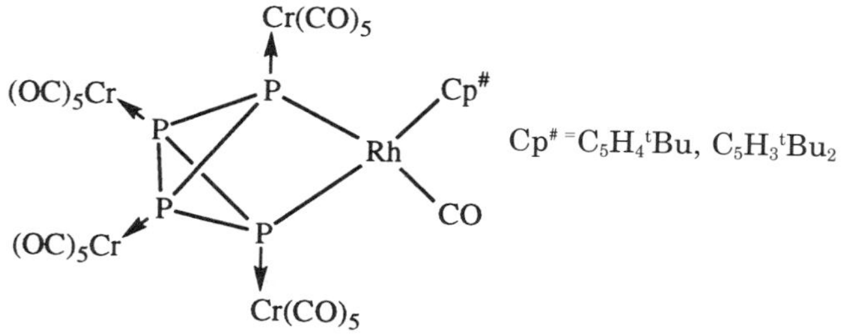

$Cp^{\#} = C_5H_4{}^tBu$, $C_5H_3{}^tBu_2$

162

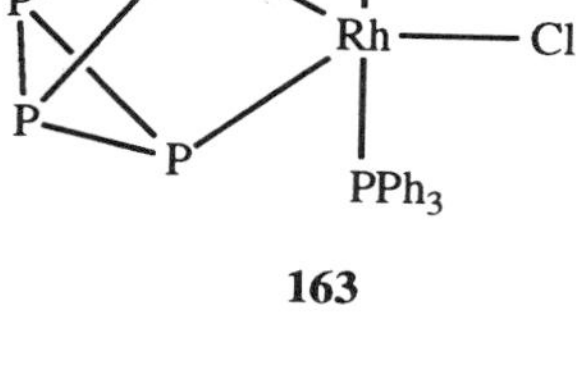

163

E = P, As

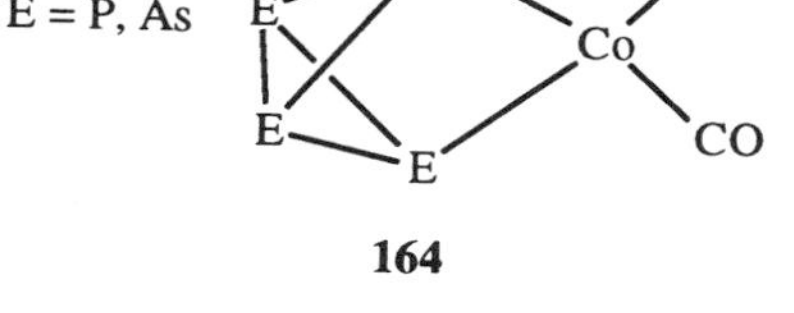

164

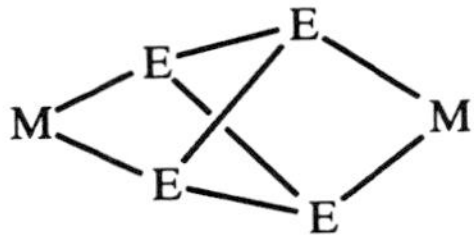

E = P, As, M = Cp*(CO)Co,
E = P, M = $Pt(PR_3)_2$

165

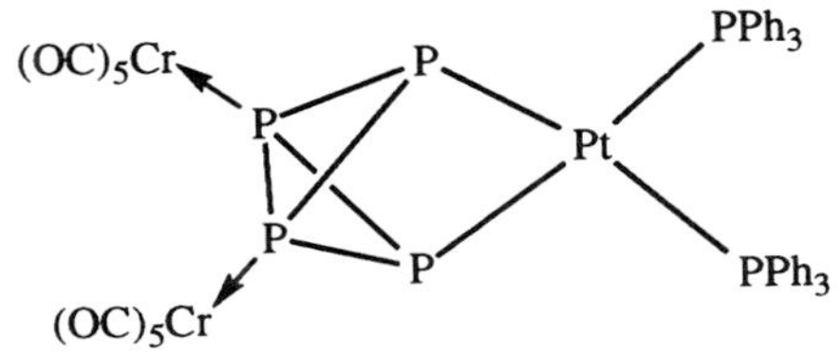

166

Only two types of organo-E_4 derivatives have been reported. These include $[Cp^*Fe]_2As_4(C_6H_2{}^tBu_3)_2$ (**167**) and $Fe_2(CO)_6E_4R_4$ (**168**; E = P, R = Me, Ph; E = As, R = Me).

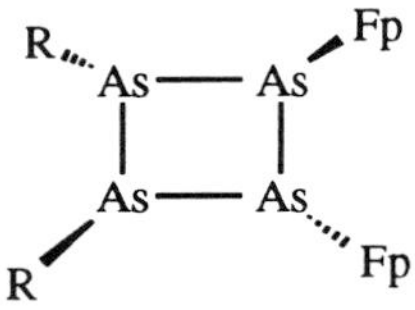

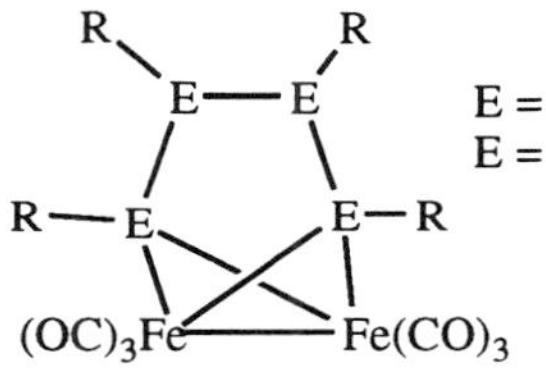

E = P, R = Me, Ph;
E = As, R = Me

167 **168**

4. *Compounds with E_5 Fragments*

As for E_4, the E_5 group is well represented by P and As examples (Table XXXVII). The planar $[E_5]^-$ ring is isoelectronic to Cp^- and forms similar complexes with transition metals (**169–172**). Thus, one Cp^- ring of metallocene complexes can be replaced with the $E_5{}^-$ unit, and analogues to the $CpM(CO)_3$ molecules are known. Triple-decker sandwich complexes such as **170** are well established. This geometry has been studied by extended Hückel calculations,[475–477] and electron counts of 26 to 34 electrons were found to be stable, depending on the degree of interaction between the metal fragments and the central ring. The M–M interaction through the rings was found to be partly responsible for stabilizing the electron-deficient electron counts.[477] The $[CpMo]_2As_5$ compound, which shows two long As–As distances, may be better written as $[CpMo]_2(As_3)(As_2)$. The asymmetry in the As–As bonding was attributed to a Jahn–Teller-type distortion. Additional metal fragments can add to the E–E bonds as in **173**, or insert into them as in **174–176**.

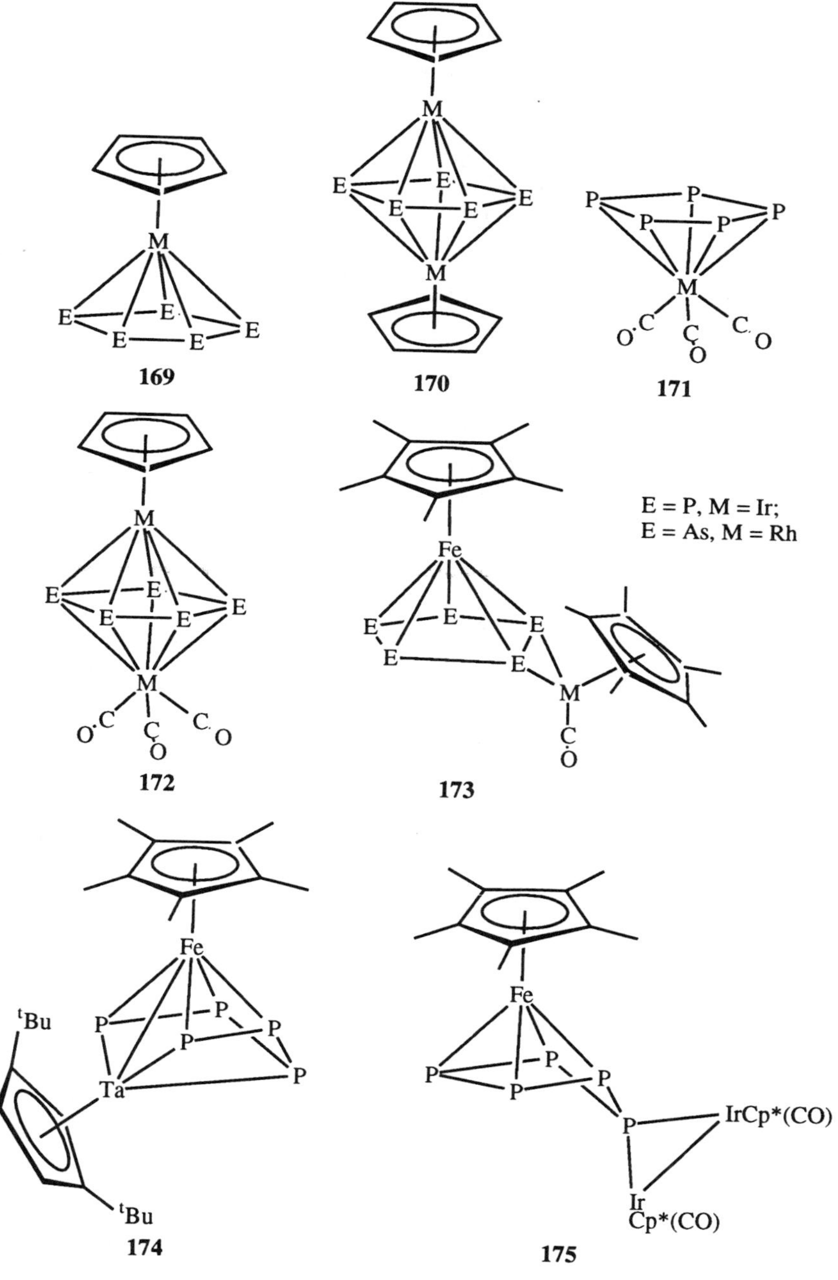
M
E
M
E
P
M
C
O
169
170
171
172
173
Fe
E = P, M = Ir;
E = As, M = Rh
174
175
Ta
tBu
IrCp*(CO)
Ir
Cp*(CO)

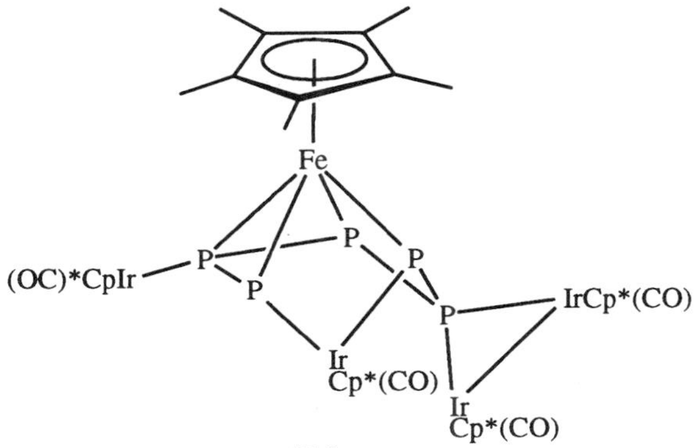

176

Two examples of metallated E_5R_5 compounds are known and have similar coordination geometries (**177**, **178**). The major difference between the arrangements in **177** and **178** is the additional coordination of the middle PR unit in **177**. As with E_3R_3 and E_4R_4 units, there is a preference for the terminal E atoms to bridge between metal centers.

TABLE XXXVII

COMPOUNDS WITH E_5 FRAGMENTS

Compound	Ref.
$[(CO)_3MP_5]^-$ (M = Cr, Mo, W)	478
$(CO)_3WP_5SiMe_3$	478
$[CpM]_2E_5$ (E = P, As; M = Cr, Mo)	476, 477
$(CO)_3MnP_5$	478
$\{Cp^{1B}\}FeP_5$	439
Cp^*FeP_5	475, 479, 480
$Cp^*FeP_5M(CO)_3$ (M = Cr, Mo)	479
Cp^*FeP_5TaR (R = Cp*, Cp^{1B})	481
$Cp^*FeP_5Ir(CO)Cp^*$	482
$Cp^*FeP_5Ir_2(CO)_2Cp^*_2$	481
$Cp^*FeP_5\{Ir_2(CO)_2Cp^*_2\}\{Ir(CO)Cp^*\}\{\mu\text{-}Ir(CO)Cp^*\}$	481
Cp^*RuP_5	483
$(Cp^x)OsP_5$ (Cp^x = Cp*, Cp^{M4E})	483
$[Fe(CO)_3][FeCl_2]\{P_5{}^tBu_5\}$	96
$\{CpMo(CO)_2\}_2\{AsMe\}_5$	484
Cp^*MAs_5 (M = Fe, Ru)	437, 485
$Cp^*FeAs_5Rh(CO)Cp^*$	482
$\{Cp^xCr\}_2As_5$ (Cp^x = Cp*, Cp^{M4E})	451
$Cp^*FeAs_5M(CO)_3$ (M = Cr, Mo, W)	479
$Cp^*RuAs_5Mo(CO)_3$	479
$(Cp^x)RuAs_5$ (Cp^x = Cp*, Cp^{M4E})	437

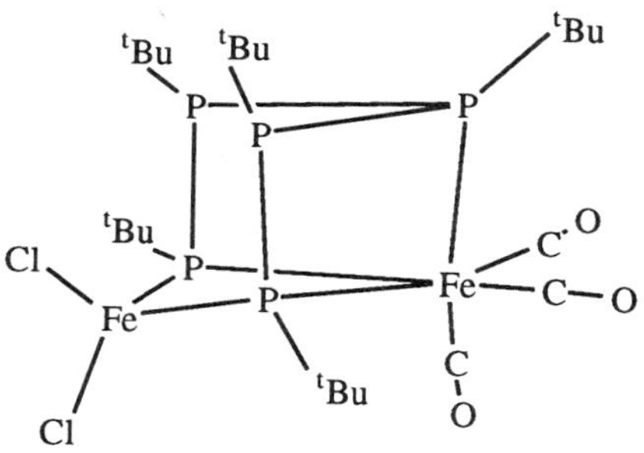

177 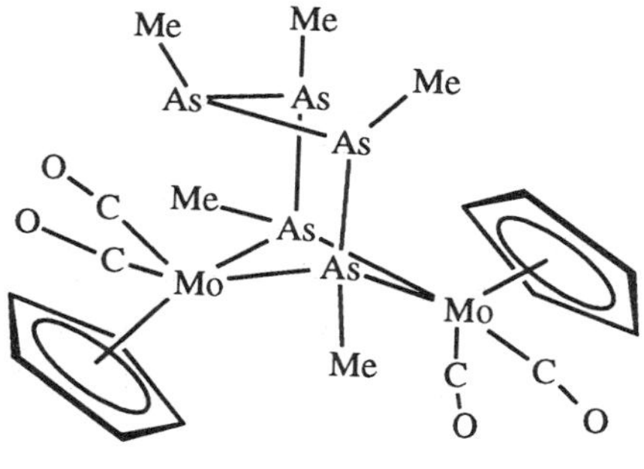 **178**

5. *Compounds with E_6 Fragments*

Planar six-membered rings from triple-decker complexes (**179**) for P and As (Table XXXVIII) with metals including V, Nb, Cr, Mo, and W. These complexes have been examined by extended Hückel calculations, as have the E_5 analogues.[476] Unlike the E_5 system, no compounds with only one metal moiety attached to a planar E_6 group have been described. The compound $[CpTi]_2P_6$ has titanium atoms on opposite sides of a P_6 ring, but each Ti atom is only bonded to three of the P atoms and the result is a cubane-like structure (**180**). Various isomeric forms of P_6 have been examined by SCF MO calculations.[216]

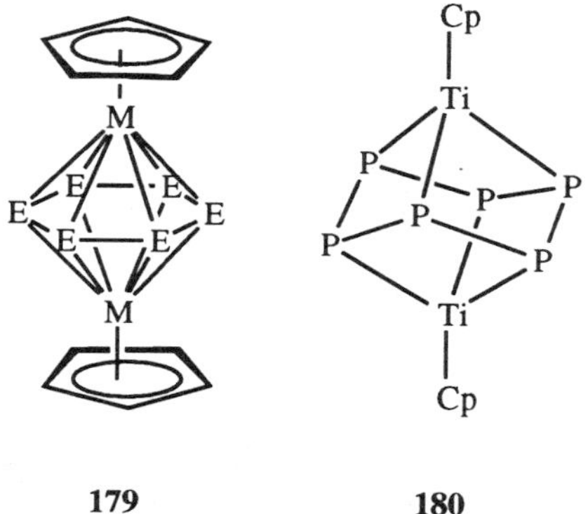

179 **180**

Complexes **181–183** can be derived from As_6 trigonal prisms in which varying degrees of bond breaking between the As atoms have occurred. These complex molecules can also be viewed as derived from other cluster frameworks. For example, the framework for **182** can also be derived from

TABLE XXXVIII

COMPOUNDS WITH P_6 FRAGMENTS

Compound	Ref.
$[Cp_2^{2B}Th]_2P_6$	453
$[Cp^*Ti]_2P_6$	486
$[Cp^xV]_2P_6$ (Cp^x = Cp*, Cp^{M4E})	487
$[(Cp^{1E})Nb]_2P_6$	460
$[\{Cp^{2B}\}Nb]_2P_6$	488
$Cp_2Zr\{(Me_3Si)_2P_6(SiMe_3)_2\}$	489
$CpME_6MCp$ (E = P, As; M = Cr, Mo)	476, 477
$\{Cp^*Mo\}_2P_6$	392
$[Cp^*W]_2P_6$	487
$[Cp^*Fe]_3P_6$	343
$[Cp^*Fe]_3\{P_3Fe\}P_6$	343
$Ni(P^tBu)_6$	116, 490, 491
$Fe_2(CO)_4\{P_6{}^iPr_6\}$	96
$[Co\{P_6(CH_2PPh_2)_2\}]^+$	492
$[Co\{P_6(CH_2PPh_2)_2\}\{W(CO)_5\}]^+$	492
$Mo(CO)_4As_6Ph_6$	493
$\{CpMo\}_2As_6$	82
$\{Cp^*Co\}_2(As_6)$	424, 494
$\{Cp^*Co\}_2\{Mo(CO)_3\}\{As_6\}$	494
$\{Cp^*Co\}_2\{Cp^*CoMo(CO)_3\}\{As_6\}$	494

an As_4MoCo trigonal prism capped on its faces by two As atoms and one Co fragment. It has been compared to Bi_9^{5+}, which has a similar framework. Both have 22 skeletal electrons.[494] Compound **183** has been described as a distorted sphenocorona structure with some similarity to Au_{11} cluster compounds containing one Au atom in an interstitial site. Quite different geometries of the P_6 unit have been noted for $[(Cp^{1B})_2Th]_2P_6$ (**184**), which is based on a tricyclic main group fragment, whereas in **185** the compound described previously as **131** is redrawn to emphasize the P_6 fragment, which can be viewed as arising from the P_7^{3-} ion with removal of the apical P atom.

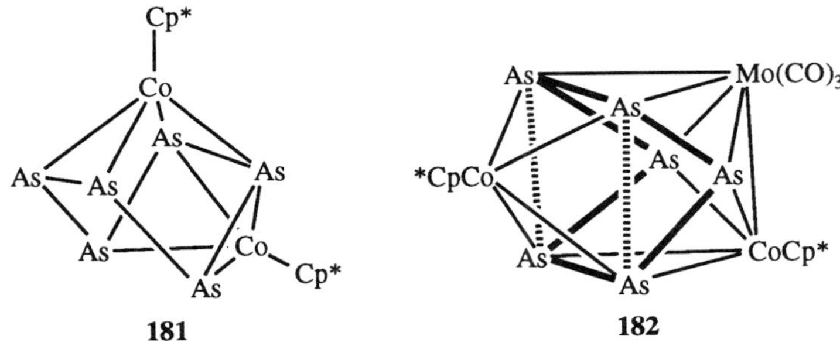

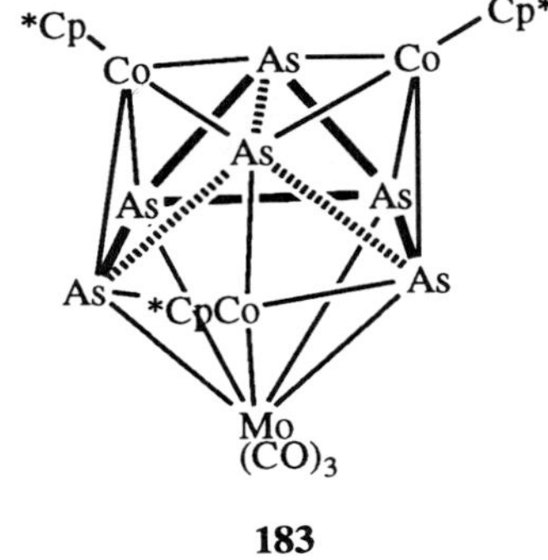

183

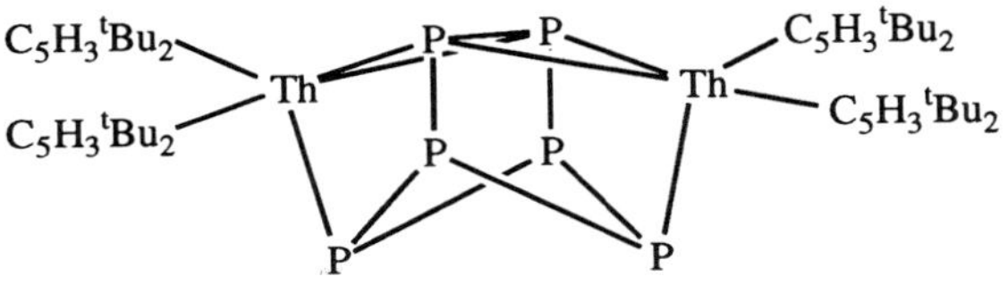

184

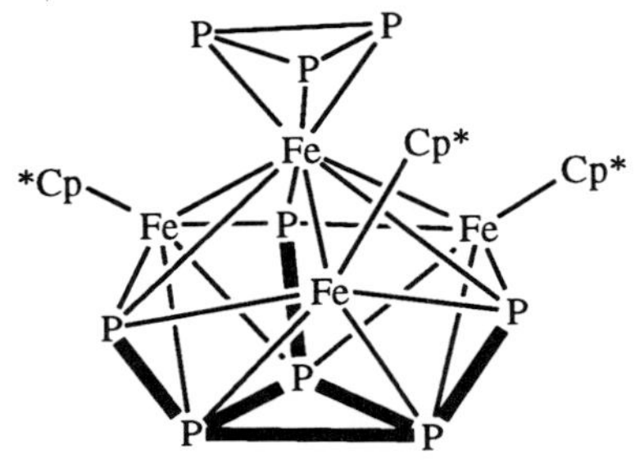

185

Metallated organic derivatives of both P_6 and As_6 units have been reported. A very intriguing molecule is based on a nearly planar $P_6^tBu_6$ ring with a nickel atom located at its center (**186**). This arrangement suggested an unusually high electron count at Ni,[491] but calculations later concluded that the compound is effectively a 16-electron species at Ni.[490] Compound **186** has a similar orbital structure to $[In\{Mn(CO)_4\}_5]^-$, in which an In atom sits at the center of a nearly planar pentagon of Mn atoms. The $P_6^iPr_6$ chain has been observed bound to $Fe_2(CO)_4$ (**187**), which is similar to complexes of the other P_xR_x (x = 3–5) units. Again, the terminal P atoms show the tendency toward bridging two metal centers. The two centermost P atoms are also coordinated. The more unusual $P_3\{P(SiMe_3)_2\}\{PP(SiMe_3)_2\}$ ligand has been isolated bound to a Cp_2Zr fragment (**188**). The cyclic As_6Ph_6 ring coordinates in a chair conformation to $Mo(CO)_4$ as shown in **189**.

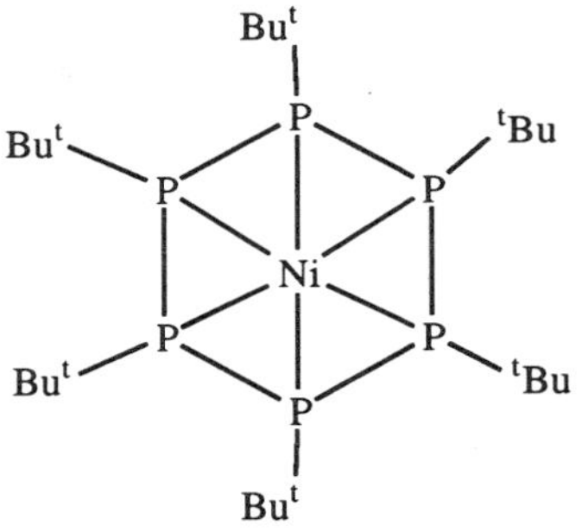

186

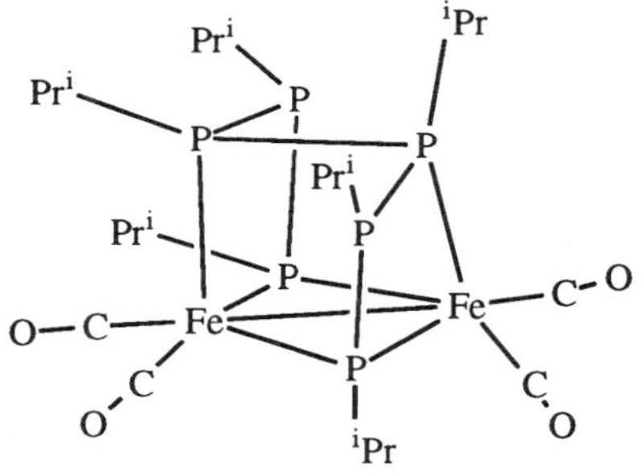

187

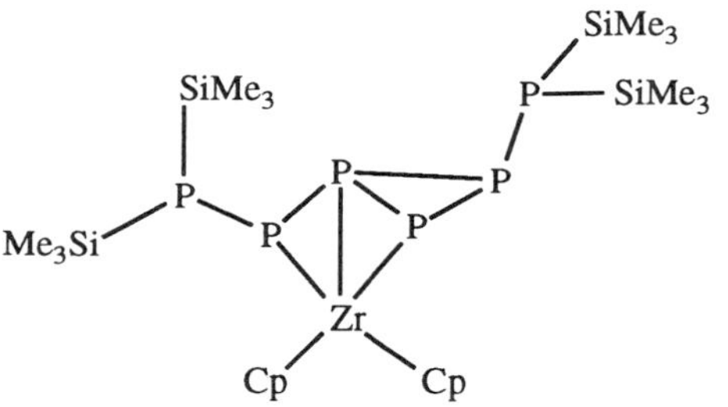

188

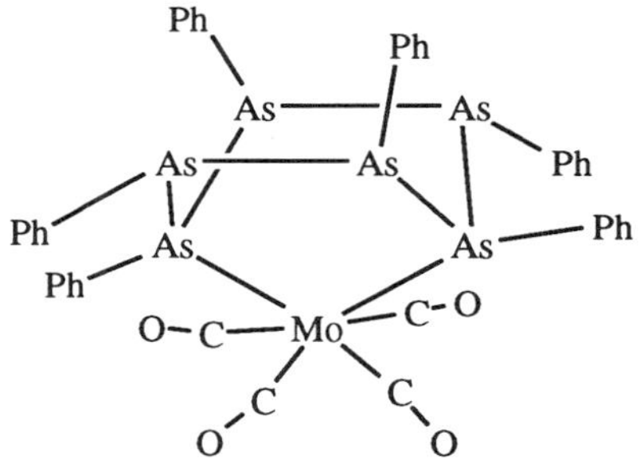

189

6. *Compounds with E_7 Fragments*

The E_7^{3-} ions (**190**) are known for P, As, and Sb (but not Bi), and they form the basis for a number of metallated derivatives (Table XXXIX). For ease of discussion of the **190** structure, we will refer to the E atoms in the triangular unit as *basal*, those that are only two-coordinate as *bridging*, and the remaining P atom as *apical*. The bridging E atoms can be readily seen from the diagram to be relatively highly charged. Functionalization by cationic (R^+, R_2P^+) species occurs here: P_7R_3 (**191**), $P_7(PR_2)_3$ (**192**), $P_7(PR_2)(P_2R_2)$ (**193**). These derivatives are mentioned because they have also been metallated. Similar to the alkylation reaction, P_7^{3-} forms the metallated derivative $P_7(Fp)_3$ (**194**) with the metals attached to the bridging

TABLE XXXIX

COMPOUNDS WITH E_7 FRAGMENTS

Compound	Ref.
$P_7Et_3\{Cr(CO)_5\}$	495
$P_7R_3\{Cr(CO)_5\}_2$ (R = Et, ^{i}Pr)	495
$P_7{}^iPr_3\{Cr(CO)_5\}\{Cr(CO)_4\}$	495
$P_7Et_3\{Cr(CO)_5\}_3\{Cr(CO)_4\}$	495
$P_7(P^tBu_2)_3\{Cr(CO)_5\}$	496
$P_7(P^tBu_2)_3\{Cr(CO)_5\}\{Cr(CO)_4\}$	496
$P_7(P^tBu_2)_3\{Cr(CO)_5\}_2$	496
$P_7(P^tBu_2)_3\{Cr(CO)_4\}$	496
$P_7(P^tBu_2)_3\{Cr(CO)_4\}_2$	496
$P_7(P^tBu)_2(^tBu)\{Fe(CO)_4\}$	497
$P_7(P^tBu)_2(^tBu)\{Fe(CO)_4\}_2$	497
$P_7(P^tBu)_2(^tBu)\{Cr(CO)_5\}$	498
$P_7(P^tBu)_2(^tBu)\{Cr(CO)_5\}_2$	498
$P_7(P^tBu_2)_3\{Cr(CO)_4\}_3$	496
$P_7(^iPr)_2(SiMe_3)\{Cr(CO)_5\}$	499
$P_7(^iPr)_2(H)\{Cr(CO)_5\}$	499
$P_7(^iPr)_2(H)\{Cr(CO)_5\}_2$	499
$P_7(^iPr)_2(SiMe_3)\{Cr(CO)_5\}\{Cr(CO)_4\}$	499
$P_7(^iPr)_2(H)\{Fe(CO)_4\}$	499
$[P_7M(CO)_3]^{3-}$ (M = Cr, Mo, W)	500, 501
$[HP_7M(CO)_4]^{2-}$ (M = Mo, W)	500
$[E_7M(CO)_3]^{3-}$ (E = P, As, Sb; M = Cr, Mo, W)	500–503
$[HE_7M(CO)_3]^{2-}$ (E = P, As, Sb; M = Cr, Mo, W)	500, 501
$[(en)W(CO)_3P_7]^{3-}$	501
$[(en)(CO)_3WP_7W(CO)_3]^{3-}$	501
$[PhCH_2P_7M(CO)_4]^{2-}$ (M = Mo, W)	500
$P_7\{Fp\}_3$	343
$P_7H(^iPr)_2\{Cr(CO)_5\}$	499
$P_7H(^iPr)_2\{Cr(CO)_5\}_2$	499
$P_7(SiMe_3)(^iPr)_2\{Cr(CO)_5\}$	499
$P_7(SiMe_3)(^iPr)_2\{Cr(CO)_4\}\{Cr(CO)_5\}$	499
$P_7(Et)_3\{Fe(CO)_4\}$	497
$P_7\{Fe(CO)_4\}\{P^tBu_2\}_3$	497
$[Sb_7Ni_3(CO)_3]^{3-}$	504

P atoms. The compounds **190–193** may serve as simple donor ligands to 16-electron metal fragments. This donation may occur via the basal atoms, the bridging atoms, or both (**195**, **196**). **191** can also function as a chelating ligand with two bridging atoms attached to a single metal center (**197**). The PR_2-derived compound also may act as a donor ligand at the basal sites and/or at the PR_2 groups as illustrated in **198**, or it may chelate using one PR_2 function and one bridging P atom (**199**). Molecules having one, two, or three metal atoms chelated as in **199** have been reported.

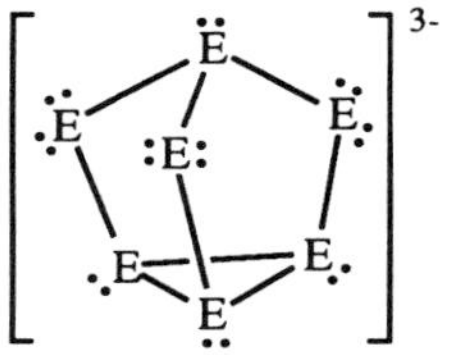

190 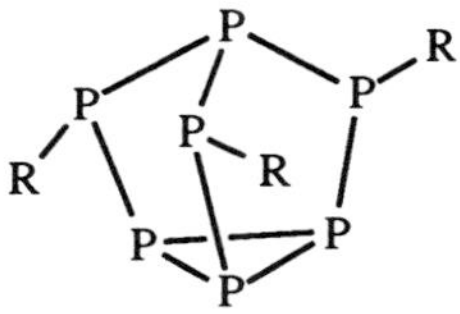 **191** 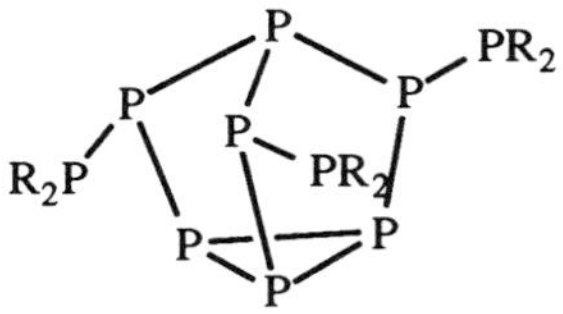 **192**

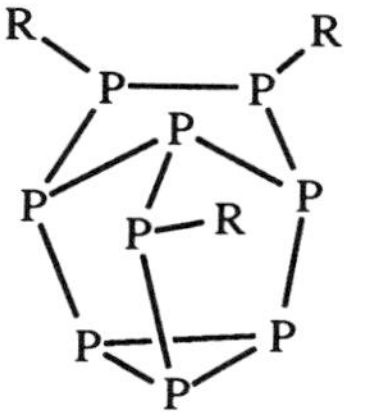

193 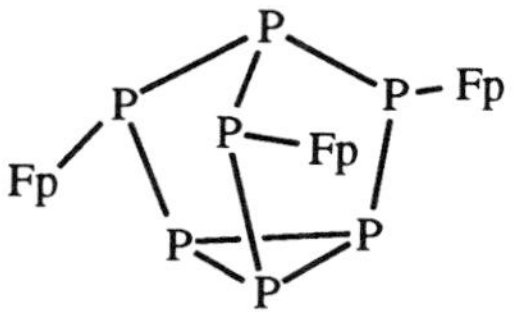 **194** 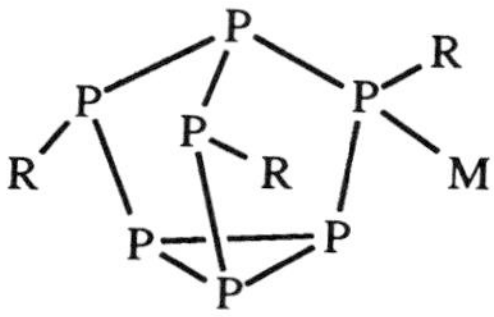 **195**

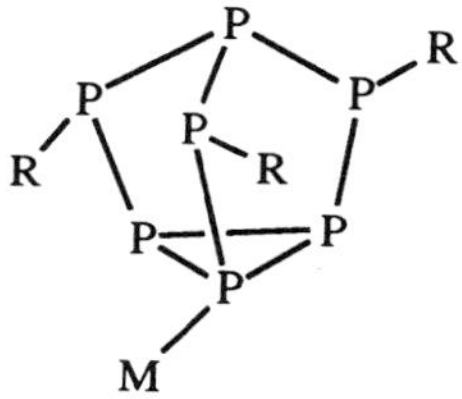

196 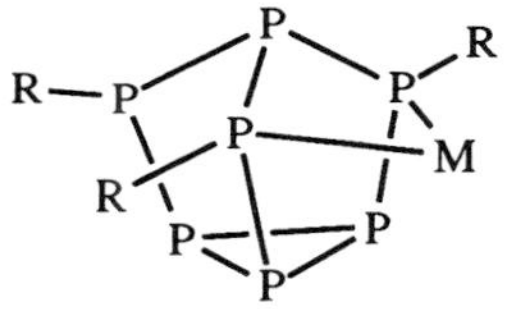 **197**

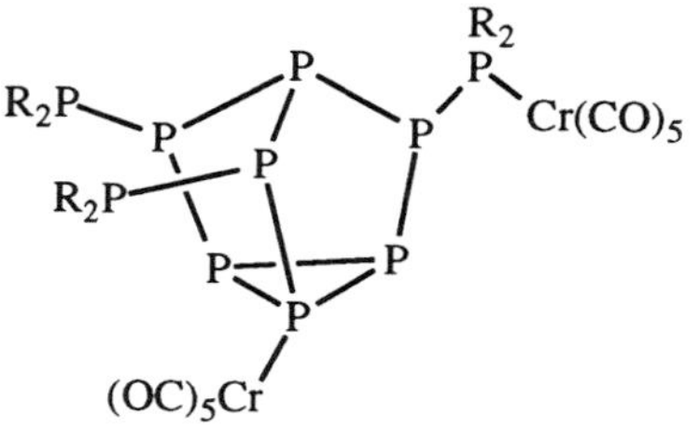

198 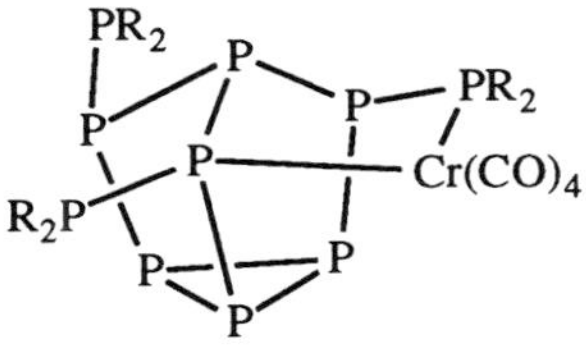 **199**

The bonding situation becomes more cluster-like for **200** and its derivatives. Here a μ_4-$M(CO)_3$ fragment bridges two basal and two bridging E atoms. The structure distorts so that the basal E–E bond that is bridged by the metal unit is greatly weakened or perhaps broken completely. Such compounds are known for E = P, As, and Sb. Alkylation or protonation occurs at the remote bridging E atom (**201**), and that E atom may also donate to a metal fragment (**202**).

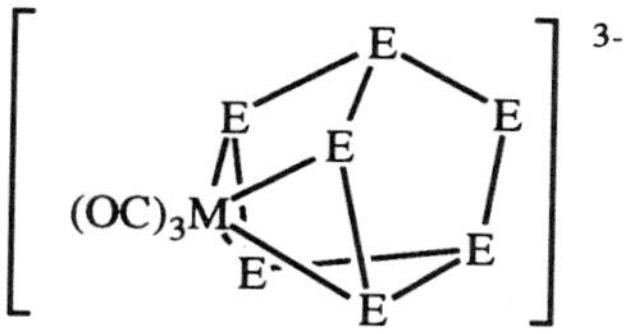

200

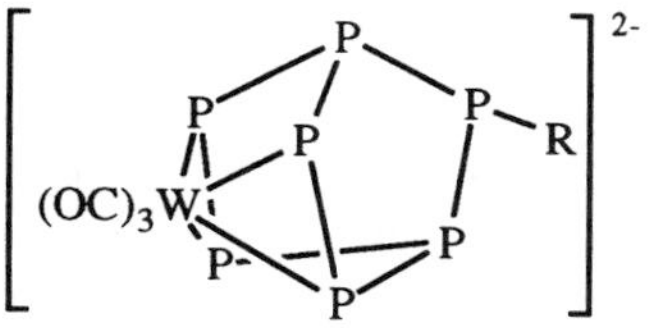

R = alkyl, H

201

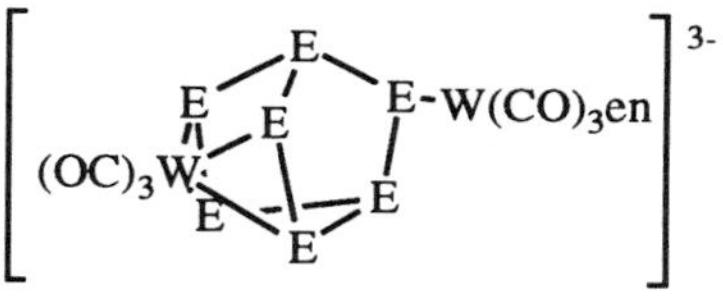

202

Of the compounds listed, only $[Sb_7Ni_3(CO)_3]^{3-}$ (**203**) does not retain the original framework of the E_7^{3-} ion. The molecule has been interpreted as having a 24-electron cuneane core with 12 2-center–2-electron bonds.

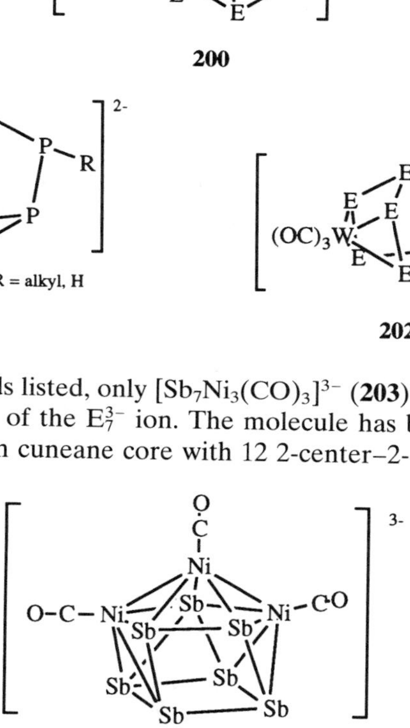

203

7. *Compounds with E_x Fragments ($x \geq 8$)*

Even higher nuclearity E_x fragments have been observed coordinated to metal fragments. These include P_9, P_{10}, P_{14}, As_8, As_9, and As_{10} examples as found in Table XL. The P_9 and some of the P_{10} derivatives have already been examined, as they are structurally based on the P_7 framework (see **192**, **193**), and they have been included in Table XXXIX. The P_{10} unit in

TABLE XL

COMPOUNDS WITH Ex ($x \geq 8$) FRAGMENTS

Compound	Ref.
$\{Cp(CO)_2Cr\}_5P_{10}$	389, 505
$[(Cp^{M4E})Rh]_4P_{10}$	506
$[Ni\{P^tBu_3\}_2]_4P_{14}$	343
$(CO)_6Mo_2As_8Ph_8$	507
$[\{Cp^{2B}\}Nb]_2As_8$	508
$[\{Cp^{2B}\}Nb]_2As_8[Cr(CO)_5]$	508
$[(CO)_4Re][(CO)_3Re]_2[As_8Me_7]$	509
$[(CO)_3Mo]_2As_9Ph_9$	507
$Mo_2(CO)_6As_{10}Me_{10}$	507

$[Cp(CO)_2Cr]_5P_{10}$ (**204**) is based on a norbornane framework with additional P and P_2 units attached, whereas in $[(Cp^{M4E})Rh]_4P_{10}$ (**205**) it is structurally related to dihydrofulvalene. The norbornane structure also appears in $[Ni\{P^tBu_3\}_2]_4P_{14}$ (**206**), where two P_7 units are connected.

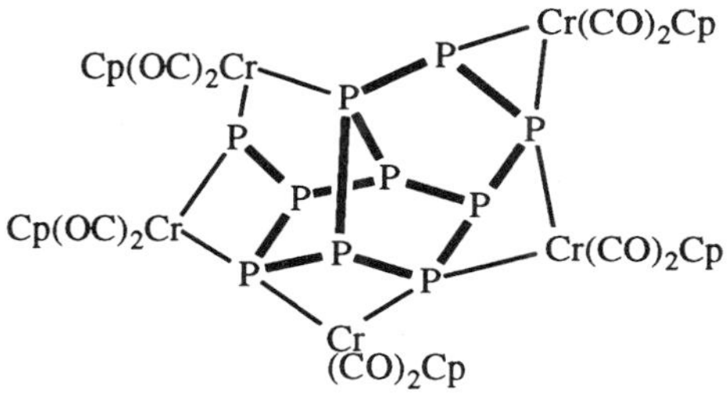

204

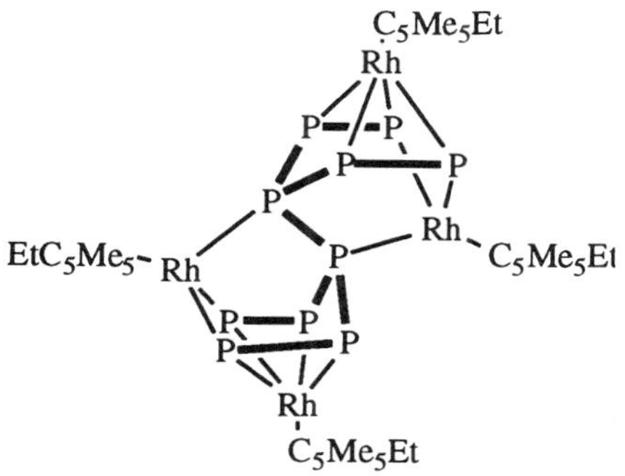

205

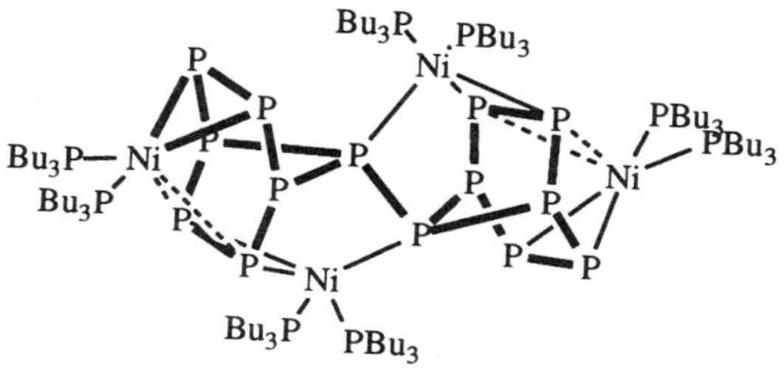

206

The naked eight-membered ring structure is found for As in $[Cp^{2B}Nb]_2As_8$ (**207**) and has been derivatized by further coordination to a $Cr(CO)_5$ fragment (**208**). The other compounds with large arsenic arrays are organic derivatives having cyclic or chain structures. As_7 and As_8 chains are found in $[(CO)_4Re][(CO)_3Re]_2[As_8Me_7]$ (**209**) and $Mo_2(CO)_6As_8Ph_8$ (**210**). As with the other chain structures, the terminal atoms bridge between two metal centers, but for **209** there is no supporting M–M bond, whereas one is present in **210**. Cyclic As_9 and As_{10} units are also found ligated to

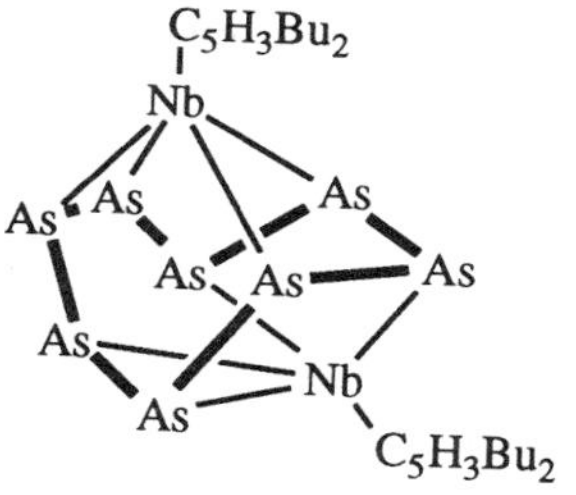

207

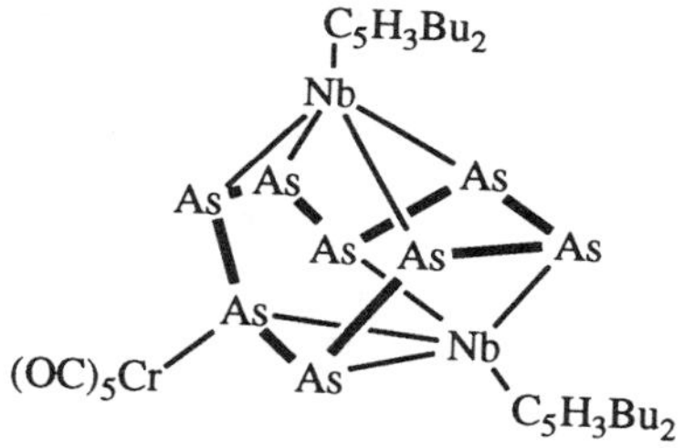

208

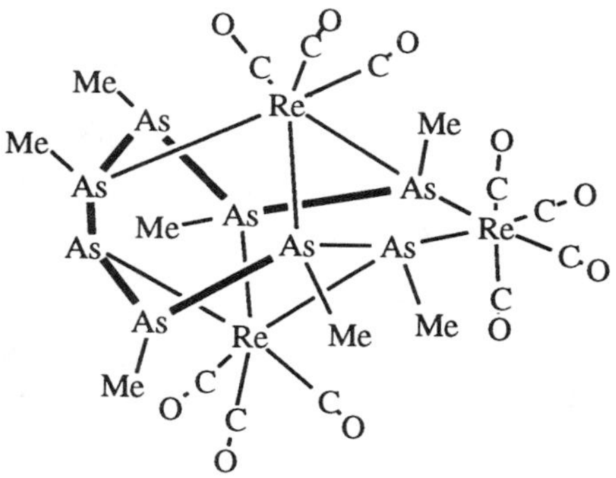

209

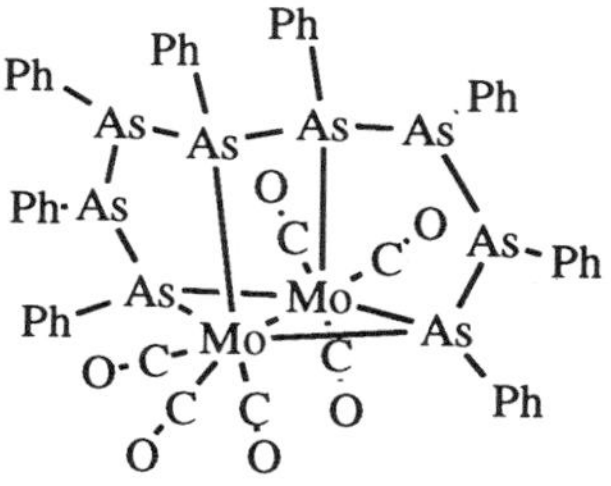

210

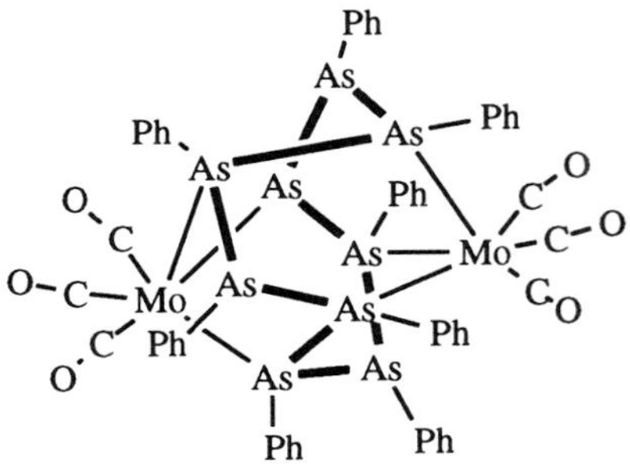

211 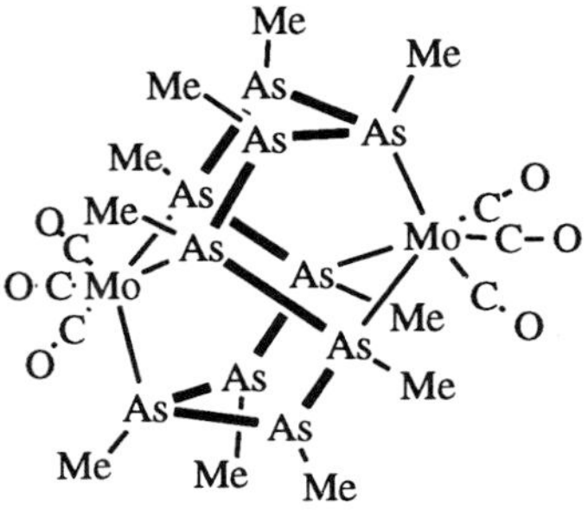 **212**

$Mo(CO)_3$ groups in **211** and **212**. The Mo and Re compounds have been discussed as arising by insertion of bridging groups into a cubane-framework.[507,509]

III

SYNTHETIC METHODOLOGIES AND REACTIVITY PATTERNS

A. *Ion Metathesis or Salt Elimination Reactions*

Perhaps the most widely used methodology for creating E–M bonds is the simple displacement of halide from a Group 15 halide complex by a metal carbonylate anion. It is a general reaction applicable to a large variety of different reactants. In the simplest cases, there is direct replacement of halide by an ML_n fragment [Eq. (5)],[41,42,46,48,49,57,58,60,125] but E–E and/or M–M bond formation may also be observed. A wide variety of metal carbonylate fragments have been used, including $[Cp^x(CO)_3Mo]^-$ (Cp^x = Cp, Cp^{1B}, Cp^{2B}),[439] $[Fp^{1B}]^-$,[439] $[Fe(CO)_2(NO)(PPh_3)]^-$,[59] $[Fe(CO)_4]^{2-}$,[70,123,456] $[Co(CO)_4]^-$,[126,215,225,303,374,428] $[Co(CO)_3(PPh_3)]^-$,[59] and $[Ir(CO)_4]^-$.[140] In a few cases, formation of M–M bonds upon ligand loss has been observed, but for the reaction of $BiCl_3$ with $[Ir(CO)_4]^-$ [Eq. (6)][140] no intermediates are seen and the *closo*-compound is obtained directly even at −20°C. When treated with PCl_3, $[Co(CO)_4]^-$ produces, as one of several products, $PCo_3(CO)_9$ [Eq. (7)].[126,225,374] This molecule, like the Ir case, shows no evidence of an intermediate open $P\{Co(CO)_4\}_3$ species. Furthermore, the high basicity of P causes the molecule spontaneously to undergo a self-substitution reaction to give $[PCo_3(CO)_8]_3$ (**213**). Reaction of $[Co(CO)_4]^-$, $SbCl_3$ produces a very short-lived intermediate that spontaneously converts into a mixture of the mono- and dianions of $[Sb_2Co_4$

$(CO)_{11}]^{n-}$.[303,428] Examples which show coupling of E–E fragments are $Co_2(CO)_6E_2$ (E = P, As),[126,225] $Co_2(CO)_4(PPh_3)_2Bi_2$.[57] The corresponding Sb tetrahedral species is similarly obtained, but appears to proceed via the formation of $Sb[Co(CO)_3(PPh_3)]_3$.[57] The reaction with Bi is highly dependent on stoichiometry: 2:1 and 3:1 reactions lead to the simple salt displacement products, whereas the 1:1 reaction gives the tetrahedral $Bi_2Co_2(CO)_4(PPh_3)_2$. In most instances, the metal carbonylate anions are used as ionic alkali metal salts (Na^+ or K^+ being most common), but the more covalently bound $Hg[Fe(CO)_2(NO)(L)]_2$ (L = CO, PPh_3)[59] and $Zn[Co(CO)_4]_2$[323,324] have also been used. A variation on this reaction is to employ a metal-coordinated ligand $L_nM \leftarrow EX_3$. The complex $(CO)_5CrPCl_3$ has been used in several reactions.[439] With $K[Mo(CO)_3Cp^x]$ (Cp^x = Cp, Cp^{1B}, Cp^{2B}) it gives $Cp^x(CO)_2MoP_3$, with $K[Fp^{1B}]$ $Cp^{1B}FeP_5$ being obtained, whereas with $Na_2Cr_2(CO)_{10}$ the product is $\{(CO)_5Cr\}Cl_2PPCl_2\{Cr(CO)_5\}$.

$$EX_3 + x\,[ML_n]^- \rightarrow EX_{3-x}[ML_n]_x + x\,X^- \quad (x = 1\text{-}3) \tag{5}$$

$$BiCl_3 + 3\,[Ir(CO)_4]^- \rightarrow BiIr_3(CO)_9 + 3\,CO + 3\,Cl^- \tag{6}$$

$$PCl_3 + 3\,[Co(CO)_4]^- \Rightarrow \mathbf{213} + Co_2(CO)_6P_2 + (CO)_3CoP_3 \tag{7}$$

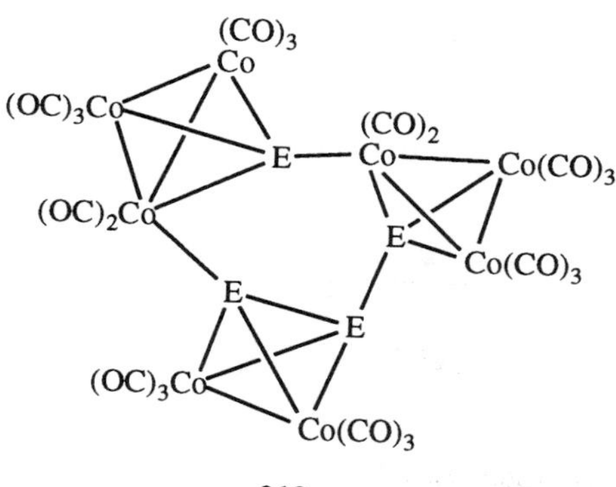

213

Metal cluster anions can be employed in place of the mononuclear metal carbonylate anions. Complexes that have been used include $[Fe_2(CO)_8]^{2-}$,[107] $[Fe_3(CO)_{11}]^{2-}$,[92,132] $[Fe_4(CO)_{13}]^{2-}$,[132,227,233] $[HRu_3(CO)_{11}]^-$,[124,136,137] $(H_3Ru_4(CO)_{11}]^-$,[136,137] and $[Ni_6(CO)_{12}]^{2-}$.[329,331,332,344,345,347,348] The products from these reactions are quite varied, so it is hard to generalize. Most of those obtained from the iron and ruthenium reagents are conventional small clusters of the type EM_3, EM_4, or E_2M_4. The reactions with

$[Ni_6(CO)_{12}]^{2-}$ by and large result in icosahedral compounds of P, As, Sb, or Bi (see Table XXII), except in one case[329] where the cubic compound $Ni_8(CO)_8(PPh)_6$ was obtained.

Hetero-main group element or heterometallic complexes can be prepared from anionic metallates that already contain an E or ER fragment. A common reaction type is the addition of main group element halides or transition-metal halide complexes to metal cluster anions, which already contain E. Examples of the hetero-main group syntheses are given in Eqs. (8),[105,293] (9),[92] and (10).[92] This approach was also successfully employed starting from clusters containing a main group fragment other than from Group 15 as shown in Eqs. (11)[408] and (12).[184] Use of R_3PMX (M = Cu, Ag, Au) has been particularly popular for homometallic transition metal clusters as well as those containing E from other groups than 15. A series of $Fe_3P(Au)_x$ clusters has been prepared via the successive deprotonation/metallation of $[H_2Fe_3(CO)_9P]^-$ using this methodology (see Section II,I; Scheme 5).[92,228] Similarly, two $[MePFe_3(CO)_9P]^-$ may be linked via a gold atom bridge, yielding $[MePFe_3(CO)_9P{-}Au{-}PFe_3(CO)_9PMe]^-$.[289] Based on this methodology, the difunctional compound $[Fe_3(CO)_9P_2]^{2-}$ can be used to create polymeric systems in which a number of the Fe_3P_2 units are linked via metal bridges into small oligomers (**214**).[289] $[MePFe_3(CO)_9P]^-$ units were used in that system as capping groups for the polymer chains.

$$Na_2Fe(CO)_4 + RP(S)Cl_2 \Rightarrow Fe_3(CO)_9(S)(PR) \quad (R = Me, Ph, MeOC_6H_4) \tag{8}$$

$$[Fe_3(CO)_9PR]^{2-} + R'EX_2 \rightarrow Fe_3(CO)_9(PR)(ER) + 2\,X^- \quad (R = Ph, Tol;\ R'E = PhP, MeC_6H_4P, MeAs) \tag{9}$$

$$[Fe_3(CO)_9PR]^{2-} + Me_2SnBr_2 \rightarrow Fe_3(CO)_9(PR)(SnMe_2) + 2\,Br^- \tag{10}$$

$$[Fe_2(CO)_7(S^tBu)]^- + MeAsI_2 \Rightarrow {}^tBuSFe_2(CO)_6As(Me)As(Me)Fe_2(CO)_6(S^tBu) \tag{11}$$

$$[Fe_3(CO)_9(S^tBu)]^- + REX_2 \Rightarrow Fe_3(CO)_9(\mu_3\text{-}S)(\mu_3\text{-}PR) + {}^tBuX + X^- \quad (E = P, As) \tag{12}$$

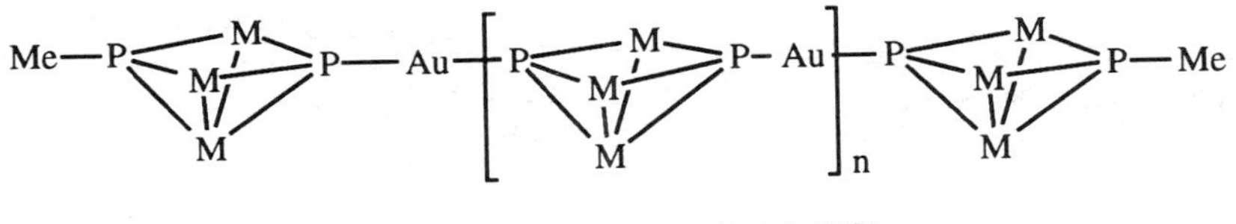

Me—P(M₃)P—Au—[P(M₃)P—Au]$_n$—P(M₃)P—Me

M = $Fe(CO)_3$

214

In some reactions, halide ions may be retained at E.[512] This is potentially useful as the residual halide becomes a functionality that may allow for further reactions. Examples are given in Eqs. (13)[53,71] and (14).[72]

$$\underset{(E = Sb, Bi)}{ECl_3} + [Fe_2(CO)_8]^{2-} \text{ or } [HFe(CO)_4]^- \rightarrow [ClE\{Fe(CO)_4\}_3]^{2-} \quad (13)$$

$$SbCl_3 + 3\ Na[Fp] \Rightarrow [Fp_3SbCl]_2[FeCl_4] \quad (14)$$

Unsaturated systems may be obtained via halide displacement reactions. In some instances these may have E–E unsaturation, whereas others show E–M unsaturation. Some reactions produce products of both types. The variety of metal reagents that will support this type of chemistry is remarkable. The general conceptual approach is the addition of metal carbonyl dianionic species to REX_2 (E = P, As, Sb, Bi). The initial expectation is the formation of $L_nM{=}ER$ with the elimination of 2 X^-. These $L_nM{=}ER$ would then be expected to oligomerize, a process that may be controlled somewhat by the size and electronic properties of R. This has been the predominant method for the production of the inidene complexes listed in Table XIV, especially when the metal fragment was $[Fe(CO)_4]^{2-}$, $[M(CO)_5]^{2-}$ (M = Cr, Mo, W), or $[M_2(CO)_{10}]^{2-}$ (M = Cr, Mo, W). The outcomes are very sensitive to reagent employed, and not always for obvious reasons. For example, the related reaction of $PhAsCl_2$ with $Na_2Fe(CO)_4$ does not give inidene complexes analogous to the phosphorus system, but rather the cluster $Fe_3(CO)_9(AsPh)_2$.[298] Unsaturated systems that contain halide ions may be further reacted with metal carbonylates or other anions to displace X^-. In some cases the product retains the unsaturation as in Eqs. (15)[253] and (16),[255] but in others the unsaturation is lost as in Eqs. (17)[224] and (18).[262]

$$[ML_n]_2AsCl + [Cp(CO)_3M']^- \rightarrow [ML_n]_2AsM'Cp(CO)_3 + Cl^- \quad (15)$$
$$(ML_n = Cr(CO)_5,\ (Cp^{1M})Mn(CO)_2;\ M' = Mo,\ W)$$

$$[ML_n]_2SbCl + X^- \rightarrow [L_nM]_2SbX + Cl^- \quad (16)$$
$$(ML_n = Cr(CO)_5,\ Cp^{1M}Mn(CO)_2;\ X^- = Br^-,\ I^-,\ SEt^-)$$

$$[(CO)_5Cr]_2AsCl + Na[Co(CO)_4] \Rightarrow Co_3(CO)_9AsCr(CO)_5 \quad (17)$$

$$(ML_n)_2EX + Na_2Fe(CO)_4 \Rightarrow Fe_3(CO)_9\{EML_n\}_2 \quad (18)$$
$$(E = P,\ As,\ Sb;\ ML_n = Cr(CO)_5,\ Mo(CO)_5,\ W(CO)_5,\ CpMn(CO)_2)$$

The polarity of the displacement reaction can be reversed where suitable reagents are available. Thus, metal halides can be treated with anionic main group fragments. This has been an important method for extending the E–M chemistry to the earlier transition metals for which fewer readily accessible metal–carbonyl complexes are known. Examples for production of Zr complexes are given in Eqs. (19)–(22).[235,409] The method also works for late transition-metal–halide complexes that may or may not contain CO [Eqs. (24)–(26)].[305,328,418,419] The Zintl ions have become a popular choice for the anionic E group complex and constitute a special class that is discussed in Section III,D.

$$Cp_2ZrHCl + K[PH(C_6H_2{}^tBu_3)]/KH \Rightarrow Cp_2ZrH\{P(C_6H_2{}^tBu_3)\}K(THF)_2 \quad (19)$$

$$Cp_2ZrCl_2 + PH_2(C_6H_2{}^tBu_3)/xs\ KH \Rightarrow [Cp_2Zr]_2P_2 + [Cp_2Zr]_2P \quad (20)$$

$$[K(THF)_x]_2[Cp^*{}_2ZrP_2] + Cp^*{}_2ZrCl_2 \rightarrow 2\ KCl + Cp^*{}_2ZrP_2ZrCp^*{}_2 \quad (21)$$

$$Cp^*{}_2ZrCl_2 + 2\ LiPHPh \Rightarrow Cp^*{}_2Zr\{PHPh)_2 \Rightarrow Cp^*{}_2Zr(PPh_2) \Rightarrow Cp^*{}_2Zr\{PPh)_3 \quad (22)$$

$$Cp^*{}_2ZrCl_2 + 2\ LiPH(C_6H_2Me_3) \Rightarrow \Rightarrow Cp^*{}_2Zr\{P(C_6H_2Me_3)\}_2 \quad (23)$$

$$cis\text{-}[MCl_2L_2] + Li_2[PhPPPh] \Rightarrow L_2M(PhP{=}PPh) \quad (24)$$
$(M = Pd, L_2 = dppe; M = Pt, L_2 = dppe\ or\ (PPh_3)_2)$

$$Rh_2Cl_2(cyclooctadiene)_2 + Li_2PPh \rightarrow Rh_4(cyclooctadiene)_4(PPh)_2 \quad (25)$$

$$[Rh(CO)_2Cl]_2 \xRightarrow{Li[As^tBu_2]} Rh_6(CO)_{11}(As^tBu_2)_2(\mu_4\text{-}As^tBu) + Rh_2(CO)_4(\mu\text{-}As^tBu_2)_2 \quad (26)$$

Metal complexes of E fragments that contain E–H bonds may be deprotonated by strong bases. The E atoms in these complexes are often highly negatively charged and consequently very basic. When these complexes are subjected to reactions involving the ionic displacement strategy, the site of reaction is often at E as exemplified in Eqs. (27)[282] and (28).[282]

$$Fe_2(CO)_6(PHPh)_2 \xRightarrow{1.\ NEt_3 \quad 2.\ PhPCl_2} [(PHPh)Fe_2(CO)_6]_2\{P_3Ph_3\} \quad (27)$$

$$Fe_2(CO)_6(PHPh)_2 \xRightarrow{1.\ NEt_3 \quad 2.\ Ph(I)PP(Ph)I} [(PHPh)Fe_2(CO)_6]_2\{P_4Ph_4\} \quad (28)$$

B. *Other Elimination Reactions*

In the cases where the fragments A and B form a stable, strongly bonded compound A–B of reasonable volatility, the reaction of E–A and M–B species can lead to elimination of A–B and the formation of an E–M bond. A notable example is the reaction of E–C bonds with metal hydrides, which can lead to the thermodynamically favored elimination of R–H while forming the desired E–M bond [Eq. (29)[45]]. In some cases this strategy occurs more cleanly than the salt elimination process.

$$BiMe_3 + Cp(CO)_3MH \rightarrow Cp(CO)_3MBiMe_2 + CH_4 \quad (29)$$

A widely used elimination reaction is that of Me_3SiX (X = Cl, F, $OSiMe_3$), which has proven to be applicable to a wide variety of conditions and reagents. It has been extensively used for E = P and As, but less so for E = N and Sb. Successful use of trimethylsilyl bismuth complexes for such reactions has not yet been reported. This methodology has been found by Schmidbaur to be generally useful for the production of EAu_x (E = N, P) clusters.[131,350,380,381,383] The diphosphine $P_2(SiMe_3)_4$ has proven useful, as well as $P(SiMe_3)_3$[381] compounds, although the chemistry of many of these complexes has not been developed owing to their low solubility. The generality of this reaction is illustrated by Eqs. (30)–(44).[69,78,116,223,333,335,336,379]

$$As(SiMe_3)_3 + 4\ FpCl \rightarrow 3\ Me_3SiCl + [As\{Fp\}_4]^+Cl^- \quad (30)$$

$$CoCl_2{\cdot}2PPh_3 + (Me_3Si)_2PPh \Rightarrow Co_4(PPh_3)_4(\mu_3\text{-}PPh)_4 \quad (31)$$

$$NiCl_2 \xRightarrow{PPh_3,\ (Me_3Si)_2PPh,\ MeC(=O)Cl} Ni_8Cl_4(PPh_3)_4(PPh)_6 + Ni_8(CO)_4(PPh_3)_4(PPh)_6 \quad (32)$$

$$CuCl + PHPh_2 + PhP(SiMe_3)_2 + Ph_2PSiMe_3 \Rightarrow Cu_{18}(PHPh_2)_3(PPh_2)_{10}(PPh)_4 \quad (33)$$

$$CuCl + PhP(SiMe_3)_2 + PPh_3 \Rightarrow Cu_{12}(PPh_3)_6(PPh)_6 \quad (34)$$

$$CuCl + PhP(SiMe_3)_2 + PMe_3 \Rightarrow Cu_{14}(PMe_3)_8Cl_2(PPh)_6 \quad (35)$$

$$CuCl + P(SiMe_3)_3 + PEt_3 \Rightarrow Cu_{96}\{P(SiMe_2)_6\}(PEt_3)_3P_{30} \quad (36)$$

$$CoCl_2(PPh_3)_2 + PhAs(SiMe_3)_2 \Rightarrow Co_4(\mu_3\text{-}As)_3(\mu_3,\eta^3\text{-}As_3)(PPh_3)_4 \quad (37)$$

$$NiCl_2 + P(SiMe_3)_3 + P_3{}^tBu_3 \Rightarrow NiP_6{}^tBu_6 \quad (38)$$

$$NiCl_2 + PPh_3 + PhP(SiMe_3)_2 \Rightarrow Ni_8(PPh_3)_4(PPh)_6 \quad (39)$$

$$Pd_3S_2Cl_2(PPh_3)_4 + As(SiMe_3)_3 \Rightarrow Pd_9(PPh_3)_8(As)_6 \quad (40)$$

$$NiCl_2(PPh_3)_2 + PhAs(SiMe_3)_2 \Rightarrow Ni_9(PPh_3)_5Cl_3(As)_6 + Ni_9(PPh_3)_6Cl_2(As)_6 \quad (41)$$

$$PdCl_2(PPh_3)_2 \xRightarrow{As(SiMe_3)_3,\ \text{THF or toluene}} [(Ph_3P)_2Pd]_2As_2 + Pd_{20}(PPh_3)_{12}As_{12} + Pd_9(PPh_3)_8(As)_6 \quad (42)$$

$$PdCl_2(PPh_3)_2 + As(SiMe_3)_3 \xRightarrow{MeCN} Pd_9(PPh_3)_8(As_2)_4 \tag{43}$$

$$PdCl_2(PPh_3)_2 + Sb(SiMe_3)_3 \xRightarrow[(E = P, As)]{PhE(SiMe_3)_2} Pd_9(PPh_3)_8Sb_6 \tag{44}$$

When chelating ligands are attached to the transition metal, the aggregation into cluster compounds is limited and mono- or dinuclear complexes are observed as in Eqs. (45)–(47).[416,417]

$$\{MeNC_4O_2(PPh_2)_2NiCl_2 \Rightarrow \{MeNC_4(OSiMe_3)_2(PPh_2)_2NiP_2Ph_2 \tag{45}$$

$$(R_2PCH_2CH_2PR_2)NiCl_2 \xRightarrow[(R = Et, cy, Ph; R' = Ph, {}^tBu)]{R'P(SiMe_3)_2 \text{ or } R'(SiMe_3)PPR(SiMe_3)} (R_2PCH_2CH_2PR_2)NiP_2R'_2 \tag{46}$$

$$(R_2PCH_2CH_2PR_2)NiCl_2 \xRightarrow[(R = Et, cy, Ph; X = H, SiMe_3, Li, P(SiMe_3)_2)]{XP(SiMe_3)_2} [(R_2PCH_2CH_2PR_2)Ni]_2P_2 \tag{47}$$

Besides the synthetic procedures for making E–M clusters, the elimination strategy may be employed to functionalize clusters to further reaction. The conversion of $Fe_3(CO)_9\{PSi^iPr_3\}_2$ into $Fe_3(CO)_9\{PH\}_2$ upon hydrolysis is one example.[279] Another is the dehalogenation of $ClE[Mn(CO)_2R]_2$ using $Me_3SiO_3SCF_3$ to give $[E\{Mn(CO)_2R\}_2]CF_3SO_3$ (E = As, R = Cp; E = Bi, R = Cp′) and Me_3SiCl.[245,258]

In the case of the formation of gold complexes from gold oxonomium reagents, elimination of H_2O drives the reaction as in Eqs. (48)–(52).[130,131,350,382] In Eq. (50), both silyl and water elimination are active.

$$[(Me_3PAu)_3O]^+ \xRightarrow{NH_3} [(Me_3PAu)_4N]^+ \tag{48}$$

$$[(Me_3PAu)_3O]^+ \xRightarrow{RNH_2\ (R = {}^tBu, CH_2Ph)} [(Me_3PAu)_3NR]^+ \tag{49}$$

$$[(Me_3PAu)_3O]^+ \xRightarrow{(Me_3Si)_2NH} [(Me_3PAu)_3NSiMe_3]^+ + [(Me_3PAu)_4N]^+ \tag{50}$$

$$[(Ph_3PAu)_3O]^+ \xrightarrow{RPH_2} [RP(AuPPh_3)_3][BF_4] \tag{51}$$

$$[(Ph_3PAu)_3O][BF_4] \xrightarrow[NaBF_4]{PH_3} [PAu_5(PPh_3)_6][BF_4]_3 \quad (52)$$

Normally, one expects organic compounds with azide functional groups to readily eliminate N_2. This is observed as per Eqs. (53)–(56).[147,148,172,176] In the case of $H_2Os_3(CO)_{10}$ reacting with N_3Ph, an intermediate is observed but under thermolysis is converted into the expected nitrene species [Eqs. (55)–(56)]. If an electron-withdrawing group is present on N, the products are $HOs_3(CO)_{10}(\mu\text{-}NHR)$ instead of the nitrene species.[172]

$$Fe_2(CO)_9 + Me_3SiN_3 \Rightarrow Fe_3(CO)_{10}NSiMe_3 \quad (53)$$

$$CpWFe_2(CO)_9(\mu_3\text{-}CTol) + N_3Tol \rightarrow CpWFe_2(CO)_7(\mu_3\text{-}CTol)(\mu_3N\text{-}Tol) \quad (54)$$

$$H_2Os_3(CO)_{11} + PhN_3 \rightarrow HOs_3(CO)_{10}(\mu\text{-}HN_3Ph) \quad (55)$$

$$HOs_3(CO)_{10}(\mu\text{-}HN_3Ph) \xrightarrow{85^\circ C} H_2Os_3(CO)_9(\mu_3\text{-}NPh) + N_2 + CO \quad (56)$$

Another oft observed pathway is benzene elimination from E–Ph complexes attached to clusters that have hydride ligands [Eqs. (57)[107]and (58)[328]].[209,217] Pyrolysis of $HM_3(CO)_{10}(PPh_2)$ (M = Ru, Os) under CO in toluene leads to formation of $M_4(CO)_{13}(PPh)$.[209] The osmium compound's conversion required more forcing conditions. Heterometallic derivatives may be obtained by a similar pathway as exemplified by the addition of $Cp_2Ni_2(CO)_2$ to $HRu_3(CO)_9(PPh_2)$, which formed $Cp_2Ni_2Ru_3(CO)_9PPh$ along with benzene.[317] In the presence of H_2, benzene may also be eliminated from E–Ph groups attached to metals, as found when $HRu_3(CO)_9(PPh_2)$ is treated with H_2 to give $H_2Ru_3(CO)_9PPh$.[110]

$$Fe_2(CO)_6\{P(Ph)H\}_2 \xrightarrow{h\nu} Fe_3(CO)_9(PPh\}_2 + Fe_4(CO)_{11}(PPh)_2 + Fe_3(CO)_8(\mu\text{-}PHPh)_2(PPh) + Fe_2(CO)_4(C_6H_8)(PHPh)_2 \quad (57)$$

$$2\ Rh_2(PPh_2)(cyclooctadiene)Cl \xrightarrow{+2\,LiHBEt_3,\ -2\,C_6H_6,\ -2\,LiCl} Rh_4(cyclooctadiene)_4PPh \quad (58)$$

C. *Metal Insertion into E–H Bonds*

A widely developed methodology to create E–M compounds has been the insertion of metal fragments into E–H bonds. In some cases simple substitution products of the type $L_nM \leftarrow ER_{3-x}H_x$ are first prepared and then reacted further with metal fragments.[200] The stepwise approach allows the possibility of creating heterometallic systems, although redistribution reactions make prediction of the exact final products uncertain. For the iron-group metals, the initial reaction of three metal carbonyl units with H_2PR leads in a general fashion to $H_2M_3(CO)_9(PR)$ [M = Fe, Ru, Os; Eq. (59)].[177,181,201,202,205,230,283] Subsequent reactions give rise to higher nuclearity products.[202,209] Commonly observed products include $M_3(CO)_9(PR)_2$ and $M_4(CO)_x(PR)_2$ (x = 11, 12). Since Ru–Ru and Os–Os bonds are stronger, they tend to form stable higher nuclearity complexes quite readily. In addition to the products already mentioned, $Ru_4(CO)_8(PPhH)_2(PPh)_2$, $Ru_5(CO)_{15}(PPh)$, $Ru_6(CO)_n(PPh)_2$ (n = 14, 15), $Ru_6(CO)_{12}(\mu_3\text{-}PPh)_2(\mu_4\text{-}PPh)_2$, and $HRu_5(CO)_{13}(\mu\text{-}PPhO^{n}Pr)(\mu_4\text{-}PPh)$ have been isolated from the reaction of $Ru_3(CO)_{12}$ with $PhPH_2$.[207] For Ru, the reaction was found to be promoted by the presence of halide ions [Eq. (60)[203]], which has been demonstrated for the reaction of ruthenium carbonyls with organic nitrogen compounds (see Section III,T).

$$Fe_3(CO)_{12} + RPH_2 \rightarrow H_2Fe_3(CO)_9(PR) + 3\,CO \qquad (59)$$

$$[Ru_3(CO)_{11}Cl]^- + H_2PR \rightarrow [HRu_3(CO)_{10}(\mu_3\text{-}PR)]^- + HCl + 2\,CO \qquad (60)$$

Derivatized compounds may also be used as starting materials. For example, upon treatment with $PhPH_2$ the acetylide complex $Ru_5(CO)_{13}(C_2^iPr)(\mu\text{-}PPh_2)$ is converted first into $Ru_5(CO)_{12}(CCH_2^iPr)(PPh_2)$, then $Ru_5(CO)_{11}(CH_2CH_2^iPr)(PPh_2)(PPh)_2$, and finally $HRu_5(CO)_{10}(PPh_2)(PPh)_3$.[206] In this process the oxidative addition of the P–H bonds to the metal cluster leads to the hydrogenation of the acetylide moiety and its ultimate elimination as $CH_3CH_2^iPr$.

Selectivity can be enhanced by using activated metal complexes, i.e., those with more easily displaceable ligands such as THF in $Cp^*Mn(CO)_2(THF)$ or MeCN in $Os_3(CO)_{11}(NCMe)$ and $Os_3(CO)_{10}(NCMe)_2$. This avoids the use of more forcing reaction conditions so that initial reaction products can be isolated before further reactions to more intimate bonding modes or condensation to higher nuclearity species occur. The methodology was extended to the production of multiple clusters linked by an organic spacer when $X_2P{-}R{-}PX_2$ (X = H, Cl, NEt_2) were employed.[183]

When treated with arsine, $Cp^*Mn(CO)_2(THF)$ gives $[Cp^*Mn(CO)_2]_2AsH$, which converts into $[\mu\text{-}(Cp^*)Mn(CO)_2]_2As_2$ with elimination of H_2.[246,247] This contrasts with the reaction of the analogous nonsubstituted cyclopentadienyl complex, which under similar circumstances yields $[CpMn(CO)_2]_4As_2$.[246,247] To achieve selectivity in the synthesis of the sensitive $H_2Fe_3(CO)_9(PSi^iPr_3)$, the starting material $Fe(CO)_3$(cyclooctene)$_2$ was chosen. With $^iPr_3SiPH_2$, it yields $H_2Fe_3(CO)_9(PSi^iPr_3)$.[181] If $Fe(CO)_3$(cyclooctene)$_2$ is used instead, the reaction gives $HFe_3(CO)_9\{P(H)(Si^iPr_3)\}(PSi^iPr_3)$.[182]

Precoordinated complexes with E–H bonds have also been used. Heterometallic complexes have been prepared by treating $Cr(CO)_5PH_3$ with $Co_2(CO)_8$ where the product was found to be $CrCo_2(CO)_{11}PH$,[103] while reaction of $Cp(CO)_2MnPH_2Ph$ and $Fe_3(CO)_{12}$ yielded $CpMnFe_2(CO)_8(PPh)$.[177] Multinuclear starting materials with E–H bonds have also been used. Reaction of $Fe_2(CO)_6(PRH)_2$ and $Co_2(CO)_8$ affords a mixture of $Fe_2Co(CO)_9(\mu\text{-}PHR)(PR)$ and $Fe_2Co_2(CO)_{11}(PR)_2$ (R = Me, ^{t}Bu, Ph, Tol).[108,322] The P—H bond in $Fe_3(CO)_9\{PFp\}\{PH\}$ is displaced with FpCl with production of HCl and $Fe_3(CO)_9\{PFp\}_2$.[182] The compound $HOs_3(CO)_{10}(PH_2)$ [**215**, Eqs. (61)[95] and (62)[229], Scheme 7] has been examined in a number of cluster expansion reactions.[94,95] The first addition inserts another Os_3 fragment into one of the P—H bonds of the μ-PH_2 group to give a μ_3-PH ligand bridging two nonbonded Os_3 triangles **216**. Complex **216** may in turn be derivatized in the presence of carbonate to the metalloester **217**.

$$Os_3(CO)_{11}(NCMe) \xrightarrow{+PH_3,\ -MeCN} Os_3(CO)_{11}PH_3 \xrightarrow{1.\ Na_2CO_3;\ 2.\ H^+} HOs_3(CO)_{10}(PH_2) \quad (61)$$

$$HOs_3(CO)_{10}P(R)H + M_3(CO)_{12} \Rightarrow Os_3M_3(CO)_{17}PR \Rightarrow Os_3M_3(CO)_{15}PR \quad (62)$$

The use of R = SiR_3 has the advantage that the SiR_3 group in the product can be displaced by addition of halide ions. This allows the synthesis of P—H derivatives that are unobtainable by other methods. It is interesting that addition of $[NBu_4]F$ to $H_2Fe_3(CO)_9PSi(^iPr)_3$ in CH_2Cl_2 gives $[HFe_3(CO)_9PH]^-$, whereas the same reagent in THF affords $[HFe_3(CO)_9PSi(^iPr)_3]^-$. Other metallated derivatives can be obtained using suitable metal halides and corresponding R_3SiX elimination reactions as exemplified by Eq. (63).[181,182]

$$HFe_3(CO)_9\{P(H)Si^iPr_3)\}(PSi^iPr_3) \xrightarrow{FpCl} Fe_3(CO)_9\{PFp\}\{PH\} \quad (63)$$

The ruthenium sulfoximido complex $HRu_3(CO)_9\{NS(O)MePh\}$ is of note because it formally has two electrons fewer than the normal tetrahedral structure [Eq. (64)].[170] Upon addition of CO it is converted into an electron-precise μ-imido complex [Eq. (65)[170]].

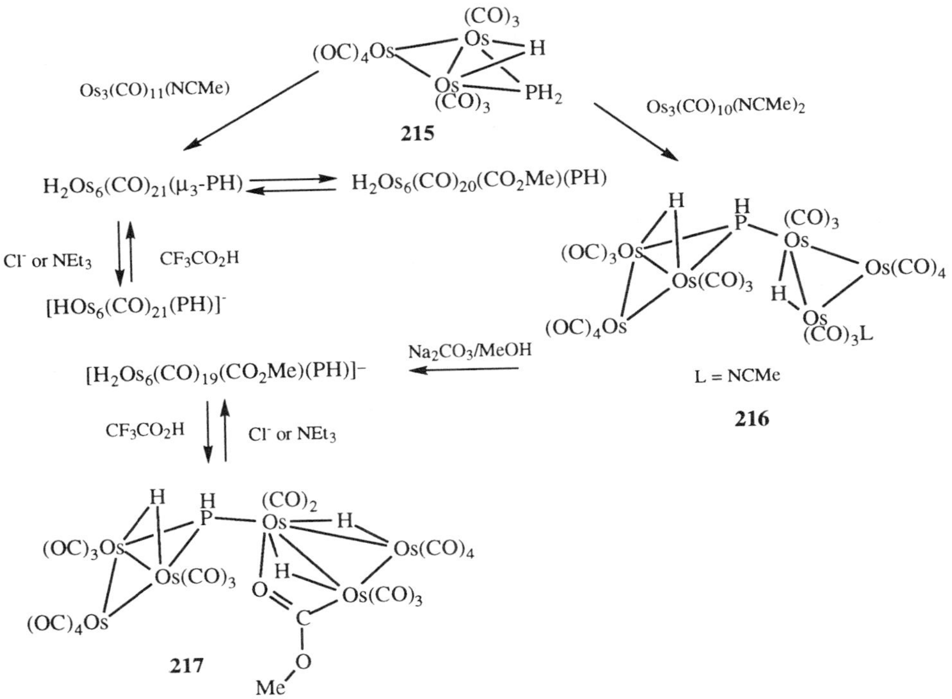

$(OC)_4Os(\mu\text{-}H)(\mu\text{-}PH_2)Os(CO)_3Os(CO)_3$ (**215**)

$Os_3(CO)_{11}(NCMe)$ → $H_2Os_6(CO)_{21}(\mu_3\text{-}PH)$ ⇌ $H_2Os_6(CO)_{20}(CO_2Me)(PH)$

$H_2Os_6(CO)_{21}(\mu_3\text{-}PH)$ ⇌ $[HOs_6(CO)_{21}(PH)]^-$ (Cl^- or NEt_3 / CF_3CO_2H)

$Os_3(CO)_{10}(NCMe)_2$ → **216** (L = NCMe)

216 → $[H_2Os_6(CO)_{19}(CO_2Me)(PH)]^-$ (Na_2CO_3/MeOH)

$[H_2Os_6(CO)_{19}(CO_2Me)(PH)]^-$ ⇌ **217** (CF_3CO_2H / Cl^- or NEt_3)

Scheme 7

$$Ru_3(CO)_{12} + MePhS(O)NH \xrightarrow{20^\circ C} HRu_3(CO)_9(\mu_3\text{-}NS(O)MePh) + 3\,CO \quad (64)$$

$$HRu_3(CO)_9(\mu_3\text{-}NC(O)MePh) + CO \rightarrow HRu_3(CO)_{10}(\mu\text{-}N{=}S(O)MePh) \quad (65)$$

Arsenic–hydrogen bonds also easily undergo metal insertion. In this way $Fe_4(CO)_{12}(AsMe)_4$ was produced from $Fe_2(CO)_9$ and $AsMeH_2$[80,81] and $FeCo_2(CO)_9AsMe$ from $Fe(CO)_4AsMeH_2$ and $Co_2(CO)_8$ or $C_3H_5Co(CO)_3$.[81] Reaction of arsine with $Fe(CO)_5$ was temperature dependent giving $[Fe_2(CO)_8]As[Fe_2(CO)_6]As[Fe_2(CO)_8]$ at 70°C but $Fe_3(CO)_9As_2$ at 110°C.[120] Room-temperature treatment of $Co_2(CO)_8$ with AsH_3 afforded $[Co_3(CO)_8As]_3$ having the structure **213** like the phosphorus analogue.[120]

Multidecker M_2E_3 complexes are most often obtained by treating M^{2+} ions and tripodal ligands with elemental phosphorus or arsenic. If $PhAsH_2$

is used instead, the resulting product is $[(triphos)Co(As_3Ph)Co(triphos)]^{2+}$, which retains an organic group attached to arsenic.[457]

D. *Syntheses from Zintl Ions and Related Compounds*

The use of the Zintl ions E_x^{n-} to form metallated complexes is now a well-established subfield of the E–M cluster area; reactions are illustrated by Eqs. (66)–(78).[343,478,480,500–504] Phosphorus has been the most studied, but As and Sb compounds are also known. The reaction patterns are similar to those already encountered. In some cases, neutral metal complexes with displaceable ligands form simple substitution complexes with the Zintl ion as a ligand. The most commonly studied have been the Cr, Mo, and W carbonyls. With $[E_7]^{3-}$ these give tetracarbonyl complexes in which the $M(CO)_4$ unit bridges between two μ-E atoms of the starting cage structure. Subsequent CO loss gives a complex that has the metal fragment bridging four of the E atoms (**200**). The metallated species can accommodate protonation or alkylation at E [Eqs. (70)–(71)]. In most cases, the Zintl ion retains its basic integrity upon metallation, but in one case a coupling has been observed [Eq. (76)], and for Sb a cluster rearrangement occurs even though the number of E atoms has remained the same [Eq. (77)]. As Eq. (78)[343] illustrates, the Zintl ions may be generated *in situ* before treatment with metal complexes. The Zintl ions also undergo salt metathesis reactions with metal halides such as FpCl [Eq. (75); **194**] or $NiCl_2L_2$ [Eq. (76); **206**].[343] Similar reactions are observed with R, PR_2, and P_2R_2 derivatized P_7 compounds (see Section II,K,6).[343,496,497,499] In most of those cases, the main group element cluster also remains intact, but when R = $SiMe_3$ and metal halides are employed [Eqs. (79) and (80)], fragmentation occurs.

$$P_5^- + M(CO)_3L_3 \rightarrow [(CO)_3MP_5]^- + 3\,L \quad (66)$$
(M = Cr, Mo, W; L = CO, NCMe)

$$P_5^- + Mn(CO)_5Br \rightarrow (OC)_3MnP_5 + Br^- + 2\,CO \quad (67)$$

$$P_5^- + FeCl_2 + Cp^{*-} \rightarrow Cp^*FeP_5 + 2\,Cl^- \quad (68)$$

$$E_7^{3-} + M(CO)_3L_3 \rightarrow [E_7\{M(CO)_3\}]^{3-} \quad (69)$$
(E = P, As, Sb; M = Cr, W, L_3 = MesH; M = Mo, L_3 = cycloheptatriene, (CO)(bipy))

$$[E_7\{M(CO)_x\}]^{3-} + H^+ \rightarrow [HP_7\{M(CO)_x\}]^{2-} \quad (70)$$
E = P, As, Sb; M = Cr, Mo, W; x = 3,4)

$$[P_7\{M(CO)_4\}]^{3-} + R_4N^+ \rightarrow [RP_7\{M(CO)_4\}]^{2-} \quad (71)$$
(M = Mo, W; R = alkyl)

$$[P_7\{W(CO)_3\}]^{3-} + W(CO)_3(arene) \xrightarrow{en} [E_7\{W(CO)_3\}\{W(CO)_3en\}]^{3-} \quad (72)$$

$$[P_7\{M(CO)_3\}]^{3-} + CO \rightleftarrows [P_7\{M(CO)_4\}]^{3-} \quad (73)$$
(M = Mo, W)

$$[RP_7\{M(CO)_3\}]^{2-} + CO \rightleftarrows [RP_7\{M(CO)_4\}]^{2-} \quad (74)$$
(R = H, alkyl; M = Mo, W)

$$P_7^{3-} + 3\ FpBr \rightarrow P_7\{Fp\}_3 \quad (75)$$

$$P_7^{3-} + NiCl_2(PBu_3)_2 \Rightarrow P_{14}[Ni(PBu_3)_3]_4 \quad (76)$$

1. K/NH_3
2. $CoCl_2(PEt_2Ph)_2$

$$Sb_7^{3-} + Ni(CO)_2(PPh_3)_2 \Rightarrow [Sb_7Ni_3(CO)_3]^{3-} \quad (77)$$

$$As_7^{3-} + CoCl_2(PEt_2Ph)_2 \Rightarrow Co_6(PEt_2Ph)_6(As_3)_2As_6 \quad (78)$$

$$P_7(SiMe_3)_3 + LiCp^* + CoCl_2 \Rightarrow (Cp^*Co)_3(P_2)_3 \quad (79)$$

$$P_7(SiMe_3)_3 + LiCp^* + FeCl_2 \Rightarrow [(Cp^*Fe)_3(P_6)]^+ \text{ or } (Cp^*Fe)_3P_3FeP_6 \quad (80)$$

Zintl-ion-like species can arise in reactions other than those that start with the preformed cluster anions. The reaction of $[BiFe_3(CO)_{10}]^-$ with CO in CH_2Cl_2 affords $[Bi_4Fe_4(CO)_{13}]^{2-}$, which can be viewed as having a Bi_3^{3-} ligand attached to a $BiFe_4^+$ fragment.[304,438] It has some relationship to the E_7^{3-} structures if one views the μ-E atoms as being replaced by μ_3-$Fe(CO)_3$ groups. Similarly, other species arising from the reaction of the main group elements (Section III,L) may be viewed as Zintl ions that have been formed by reduction by a metal fragment followed by trapping of that main group element cluster by the metal fragments.

The heteronuclear compound P_4S_3, which has the same structure as P_7^{3-} but with the μ_2-P groups replaced by S, has also been examined. It reacts with $Cp_2Cr_2(CO)_6$ to give a mixture of $CpCr(CO)_2H$, $Cp_4Cr_4(CO)_9P_4S_3$, $[CpCr(CO)_2]_2S$, and $CpCr(CO)_2P_3$.[391] The E_7 framework is not retained for the heteroatomic species.

E. *Metal Addition and Elimination*

Metal fragments may add to some existing E–M compounds to give higher nuclearity complexes. If a lone pair of electrons is present on E, especially for the lighter E atoms, which are more basic, the metal fragment may simply coordinate to that electron pair [Eqs. (81)–(91)]. This process may be reversible as in Eq. (83),[125] where addition of a competing ligand is seen to displace the ML_n fragment from the lone pair. As Eqs. (81)–(91)[125,127–129,264,396,397,401,418,426,442,446,452,492] illustrate, a wide variety of metal-bound E fragments—both isolated E and E_x units, saturated, or unsaturated—are able to undergo this metal addition reaction.

$$Cp_3(CO)_6M_3E + M'L_n(THF) \rightarrow Cp_3(CO)_6M_3EML_n \quad (81)$$
(M = Mo, W; E = P, As; $M'L_n$ = $Cr(CO)_5$, $Mo(CO)_5$, $W(CO)_5$, $Cp^{1M}CO)_2Mn$)

$$HCFe_3(CO)_9As + ML_n(THF) \rightarrow HCFe_3(CO)_9AsM(CO)_5 \quad (82)$$
(ML_n = $Cr(CO)_5$, $Mo(CO)_5$, $W(CO)_5$, $(Cp^{1M})Mn(CO)_2$)

$$Cp_3(CO)_6M_3P\ M(CO)_5 + PPh_3 \rightarrow Cp_3(CO)_6M_3P + Ph_3PM(CO)_5 \quad (83)$$
(M = Mo, W)

$$Cp^*{}_2Mo_2(CO)_2(P_2)_2 + 2\ Cr(CO)_5(THF) \rightarrow Cp^*{}_2Mo_2(CO)_2(P_2\{Cr(CO)_5\})_2 + 2\ THF \quad (84)$$

$$Co_2(CO)_5(L)E_2 \xrightarrow{\text{"}ML_n\text{"}} Co_2(CO)_5(L)\{E_2ML_n\} \xrightarrow{\text{"}ML_n\text{"}} Co_2(CO)_5(L)\{EML_n\}_2 \quad (85)$$
(E = P, As; L = CO, PBu_3, PPh_3)

$$[Co[Ph_2PCH_2P(PH)_2P_4P(Ph)_2CH_2PPH_2\}]^+ + M(CO)_6 \Rightarrow [Co[Ph_2PCH_2P(PH)_2P_4P(Ph)_2CH_2PPH_2\}\{M(CO)_5\}_x]^+ \quad (86)$$
(M = Cr, x = 2; M = W, x = 1)

$$(dppe)Pd(PhP{=}PPh) + 2\ W(CO)_5(THF) \rightarrow (dppe)Pd(Ph\{W(CO)_5\}P{=}P\{W(CO)_5\}Ph) \quad (87)$$

$$(triphos)CoP_3 + n\ Cr(CO)_6 \rightarrow (triphos)CoP_3\{Cr(CO)_5\}_n \quad (88)$$
(n = 1, 2)

$$[(Cp^{4P})Ni]_2P_2 + W(CO)_5(THF) \Rightarrow [(Cp^{4P})Ni]_2W(CO)_4P_2 \quad (89)$$

$$Cp^*MP{=}PR + Fe_2(CO)_9 \Rightarrow Cp^*M[Fe(CO)_4]P{=}PR + Fe_2(CO)_6\{RPC(O)PM(CO)_2Cp^*\} \quad (90)$$

$$[(Cp^x)Ni]_2P_4 + 2\ W(CO)_5(THF) \rightarrow [(Cp^x)Ni]_2P_4\{W(CO)_5\}_2 \quad (91)$$

The clusters $Co_3(CO)_9E$ (E = P, As) are an interesting example showing the basicity of the main group element. For these molecules the discrete trimetallic compound is only observed for E = As, and then it is only short-lived. These molecules undergo a spontaneous trimerization to give $[Co_3(CO)_8E]_3$ in which three clusters each have an E atom donating to a

Co of an adjacent metal triangle (**213**).[126] The tetrahedral complex can be trapped by ML_n fragments as shown in Eq. (92).[126]

$$PI_3 + Co_2(CO)_8 \xrightarrow[2.\ Fe_2(CO)_9]{1.\ CO,\ thf} Co_3(CO)_9PFe(CO)_4 \quad (92)$$

In complexes containing a planar E_x unit, it is common for the addition of metal fragments to give a "multidecker" complex [Eqs. (93)–(96)[441,454,479,485]]. A metal fragment may attach to several E atoms to give a more complicated cluster core as in Eqs. (97) and (98),[494] or may insert into an E–E bond in the E_x unit [Eqs. (99)–(103)[481,482]].

$$2\,(triphos)CoP_3 + 6CuBr \rightarrow (triphos)CoP_3[Cu_6Br_6]P_3Co(triphos) \quad (93)$$

$$[(triphos)CoE_2S]^+ + M^{n+} + triphos \Rightarrow ([triphos)ME_2SM(triphos)]^{2+} \quad (94)$$
($E = P, As; M^{n+} = Co^{2+}, Rh^+$)

$$Cp^*ME_5 + M'(CO)_3(NCMe)_3 \rightarrow Cp^*ME_5M'(CO)_3 + 3\,MeCN \quad (95)$$
(M = Fe, Ru; M' = Cr, Mo, W)

$$(Cp^x)FeAs_5 + [CpFe(C_6H_6)]PF_6 \rightarrow [(Cp^x)FeAs_5FeCp]PF_6 + C_6H_6 \quad (96)$$
($Cp^x = Cp^*, Cp^{M4E}$)

$$[Cp^*Co]_2As_6 + Mo(CO)_5(THF) \Rightarrow [Cp^*Co]_2[Mo(CO)_3]As_6 \quad (97)$$

$$[Cp^*Co]_2[As_2][As_4] + Mo(CO)_5(THF) \Rightarrow [Cp^*Co]_2[Mo(CO)_3][As_2][As_4] \quad (98)$$

$$Cp^*FeAs_5 + [Cp^*Rh(CO)]_2 \Rightarrow Cp^*FeAs_5Rh(CO)Cp^* \quad (99)$$

$$Cp^*FeP_5 + (Cp^{2B})Ta(CO)_4 \Rightarrow Cp^*FeP_5Ta(Cp^{2B}) \quad (100)$$

$$Cp^*FeP_5 + [Cp^*Ir(CO)]_2 \Rightarrow Cp^*FeP_5Ir_2(CO)_2Cp^*_2 \quad (101)$$

$$Cp^*FeP_5 + 2\,[Cp^*Ir(CO)]_2 \Rightarrow Cp^*FeP_5[Ir_2(CO)_2Cp^*_2][Ir(CO)Cp^*]_2 \quad (102)$$

$$Cp^*FeP_5 + Cp^*Ir(CO)_2 \Rightarrow Cp^*FeP_5Ir(CO)Cp^* \quad (103)$$

Addition reactions may occur at the metal framework of the cluster [examples are given in Eqs. (104)–(107)[117,208,406]]. A variety of trinuclear PR-capped clusters $M_3(CO)_9PR$ were expanded by the addition of one or two "CpRh" fragments as $Cp(CO)_2Rh$.[306] Addition of one such fragment occurred to bridge an M–P bond, whereas the second addition resulted in octahedral M_5P clusters. An additional gold fragment will add to tetrahedral $[\{Ph_3PAu\}_4N]^+$ in the form of $[Ph_3PAu]BF_4$ to give trigonal bipyramidal $[\{Ph_3PAu\}_5N]^{2+}$ as determined by X-ray diffraction.[349] Cluster expansion of $[Rh_6(CO)_{15}N]^-$ occurs when it is treated with $[M(CO)_4]^-$ (M = Co, Rh, Ir) or $[PtRh_4(CO)_{14}]^{2-}$.[352,368,369] The products were $[Rh_6M(CO)_{15}N]^{2-}$ or $[PtRh_{10}(CO)_{21}N]^{3-}$, respectively. For $[Rh_6M(CO)_{15}N]^{2-}$ the added metal fragment caps a rectangular face of the starting trigonal prismatic compound, whereas in the latter the interstitial nitride is bonded to five metal atoms in an irregular geometry. An interesting feature of $Ru_3(CO)_8\{C_6H_4Cr(CO)_3\}(P^tBu)$ [Eq. (107); **218**] is the formation of a bonding interaction between the Cr atom and one of the Ru atoms of the metal triangle.[208] The $Cr(CO)_3$ fragment is π-bonded to the C_6 ring in the same fashion as with $(C_6H_6)Cr(CO)_3$, except that two adjacent H atoms of the aromatic ring have been removed so that the two carbon atoms are bonded in a μ_3, η^2 fashion to the Ru_3 unit.

$$Cp_3(CO)_6Mo_3As + Cp_2Ni_2(CO)_2 \Rightarrow Cp_3Mo_2Ni(CO)_4As + Cp_6Mo_4Ni_2(CO)_7As_2 \quad (104)$$

$$Cp_3(CO)_6Mo_3As + Pt(C_2H_4)(PPh_3)_2 \Rightarrow CpMoPt_2(CO)_4(PPh_3)_2As \quad (105)$$

$$M_3(CO)_9(N_2Et_2) + ML_n \rightarrow M_4(CO)_{12}(N_2Et_2) \quad (106)$$
(M = Fe; ML_n = $Fe_2(CO)_9$; M = Ru, ML_n = $Ru(CO)_5$)

$Ru_3(CO)_{12} + P[C_6H_5Cr(CO)_3]_2{}^tBu \Rightarrow$

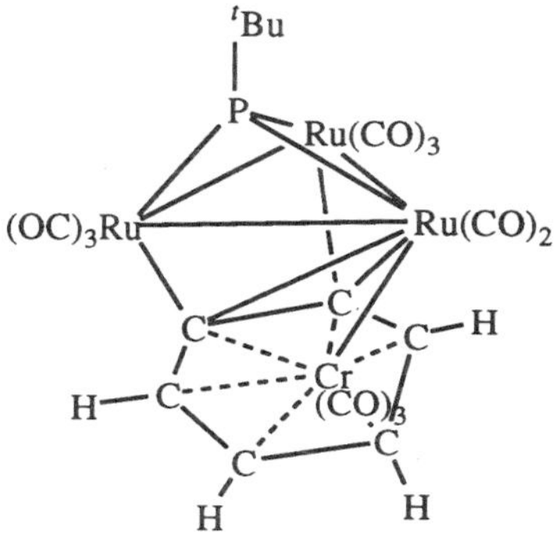

$Ru_3(CO)_8\{C_6H_4Cr(CO)_3\}(P^tBu)$, **218** (107)

Another common strategy for addition of metal fragments to a cluster is the combination of a cluster anion source with an organometallic metal cation source. Numerous examples exist, several of which have already been mentioned in Section III,A. The addition of $[R_3PMX]_x$ (M = Cu, Ag, Au) is perhaps one of the most common. Further examples are given in Eqs. (108)[350] and (109).[373] Equation (108) is interesting in that the reaction is a simple addition to form a weak adduct based on the Au···Au interactions.

Deprotonation of $H_2Ru_3(CO)_9PPh$ followed by addition of $[Au(PR_3)_2]PF_6$ gave $HRu_3(CO)_9(\mu\text{-}AuPR_3)(PPh)$, whereas use of $[Rh(CO)_3(PEt_3)_2]BF_4$ gave $Ru_3Rh_2(CO)_{13}(PEt_3)(\mu_4\text{-}PPh)$.[214] The latter molecule shows insertion of the Rh fragments into the Ru–P bonds as the μ_4-PPh ligand is capping a square Rh_2Ru_2 face in the product.

$$[(Me_3PAu)_4N]^{2+} + 2\ Me_3PAuCl \rightarrow [(Me_3PAu)_4N]^{2+}\cdot 2\ Me_3PAuCl \quad (108)$$

$$[Os_6(CO)_{18}(\mu_6\text{-}P)]^- + Ph_3PAuX \Rightarrow Os_6(CO)_{18}(\mu\text{-}AuPPh_3)(\mu_6\text{-}P) \quad (109)$$

Thermal treatment of some tetrairon and tetraruthenium clusters results in vertex elimination [Eqs. (110)[108] and (111)[111]]. Vertex elimination can be promoted by the addition of ligands, especially CO [Eqs. (212)[363] and (113)[378,384]] and may result in fragments that then aggregate to larger clusters. Thus, treatment of $[BiFe_3(CO)_{10}]^-$ with CO in CH_2Cl_2 leads to the cluster $[Bi_4Fe_4(CO)_{13}]^{2-}$. The overall charge balance suggests that the solvent CH_2Cl_2 is important to this reaction, possibly with the elimination of some Cl^-.[304,438]

$$Fe_4(CO)_{11}(PR)_2 \Rightarrow Fe_3(CO)_9(PR)_2 \quad (110)$$
$$(R = Me, Ph, {}^tBu, Tol)$$

$$H_2Ru_4(CO)_{12}(PPh) \Rightarrow H_2Ru_3(CO)_9(PPh) + Ru_3(CO)_{12} \quad (111)$$

$$[Ru_6(CO)_{16}N]^- + CO \Rightarrow [Ru_5(CO)_{14}N]^- \quad (112)$$

$$[Rh_{10}(CO)_{22}E]^{3-} + 3\ CO \rightleftarrows [Rh_9(CO)_{21}P]^{2-} + [Rh(CO)_4]^- \quad (113)$$
$$(E = P, As)$$

Added unsaturated metal fragments have seen to function as CO scavengers rather than addition to the cluster [Eq. (114)].[407]

$$(Cp^{1B})(CO)RhP_4\{Cr(CO)_5\}_4 \xrightarrow[\substack{-\ Cr(CO)_6 \\ -\ THF \\ x\,=\,3,\,4}]{+\ Cr(CO)_5(THF)} (Cp^{1B})RhP_4\{Cr(CO)_5\}_x \quad (114)$$

F. *Metal Exchange Reactions*

The previous section deals with addition and elimination of metal vertices. The next logical extension of these processes is to have both proceed in the same reaction. This results in an exchange of the metal atoms in a cluster, which is now a well-established reaction, especially for tetrahedral EM_3 species [Eqs. (115)–(122)]. The original intent of many reactions of

E_xM_y compounds with additional ML_n fragments was to obtain cluster growth and higher nuclearity species. Instead, upon addition of the new ML_n species, one or more of the other metal vertices are displaced by the incoming fragments. Vahrenkamp has argued that reactivity and outcome of these processes is largely controlled by steric factors pertaining to the ligands at the various metal fragments.[97] The reaction type is very sensitive to metal, so that replacing Mo with W was not successful in Eqs. (115)–(117).[97]

$$FeCo_2(CO)_9AsMe + Na[Cp(CO)_3Mo] \Rightarrow CpMoFeCo(CO)_8AsMe \quad (115)$$

$$FeCo_2(CO)_9AsMe + Cp(CO)_3MoAsMe_2 \Rightarrow CpMoFeCo(CO)_8AsMe \quad (116)$$

$$FeCo_2(CO)_9AsMe + [Cp(CO)_3Mo]_2 \Rightarrow CpMoFeCo(CO)_8AsMe \quad (117)$$

$$Cp_3(CO)_6Mo_3As + Fe(CO)_5 \Rightarrow Cp(H)MoFe_2(CO)_8AsMo(CO)_3Cp \quad (118)$$

$$Cp_3(CO)_6Mo_3As + Fe_2(CO)_9 \Rightarrow Cp_2Mo_2Fe(CO)_7AsMo(CO)_3Cp \quad (119)$$

$$Fe_3(CO)_9Bi_2 + [Co(CO)_4]^- \Rightarrow [Fe_2(CO)_6Bi_2\{Co(CO)_4\}]^- \quad (120)$$

$$(CO)_4M\{PM(CO)_5\}_4 + 2\,Pt(PR_3)_4 \Rightarrow P_4(PtPR_3)_2 \quad (121)$$
(M = Cr, W)

$$Cp^*FeE_5 + [Cp^*M(CO)_2]_2 \Rightarrow Cp^*ME_5 \quad (122)$$
(M = Ru, E = P, As; M = Os, E = P)

The addition reactions of $[Fe(CO)_4]^{2-}$ and $[Co(CO)_4]^-$ to $Fe_3(CO)_9Bi_2$ make an interesting comparison. In the former case, addition occurs at one Bi atom to form the square pyramidal cluster $[Fe_3(CO)_9(Bi)\{BiFe(CO)_4\}]^{2-}$, whereas in the case of the cobalt anion one of the iron atoms is extruded and a bridged-tetrahedral structure is obtained instead [Eq. (120)].[300,301] This difference has been attributed to the different bonding capabilities of the bismuth atoms due to the charge differences in the resulting products. Upon addition of $[CpMo(CO)_3]_2$ to $[(Me_3SiO)_2(Me)NC_4(PPh_2)_2]NiP_2Ph_2$, the only characterized product was one in which the P_2Ph_2 group was completely transferred to Mo.[423] The product was $[CpMo(CO)_2]_2P_2Ph_2$, in which the two Mo atoms bridge the P–P bond. Likewise, the reaction of $(CO)_4M\{PM(CO)_5\}_4$ with $Pt(PR_3)_4$ transfers the P_4 unit to different metal centers [Eq. (121)].[407] An exchange of the E_5 group from one metal to another has been observed for the Cp^*FeE_5 system [Eq. (122)].[483] This is in contrast to related reactions in which the metal fragment added to give a multidecker complex [see Eqs. (93)–(96), Section III,E].

G. *Reduction of Main Group Oxides via Metal Carbonyls and Carbonylate Anions*

Main group element oxo compounds can react with neutral metal carbonyl complexes, often with the generation of CO_2 and the formation of E–M bonds. This reaction type has been documented for all group 15 E except P. Iron pentacarbonyl reacts readily with NO_2^- to give $[Fe(CO)_3(NO)]^-$, which can be used for building of higher nuclearity clusters containing the naked nitride ligand, via elimination of another molecule of CO_2.[145,357,360] In this way $[Fe_4(CO)_{12}N]^-$ was prepared from $Fe_3(CO)_{12}$[45,357] and $Cp'MoFe_2(CO)_7(NO)(\mu_3\text{-}NH)$ from $Cp'Mo(CO)_3Cl$.[142] The neutral complexes $Fe(CO)_2(NO)_2$ and $Co(CO)_3(NO)$ have been used as well. The former reacts with $[Fe_3(CO)_{11}]^{2-}$ to give $[Fe_3(CO)_8(NO)(NH)]^-$.[146] Upon photolysis, $Cp^*_2(CO)_4Mo_2$ and $Co(CO)_3(NO)$ yield $Cp^*_3Mo_3Co_2(CO)_{10}(\mu_5\text{-}N)$ or **219**.[141]

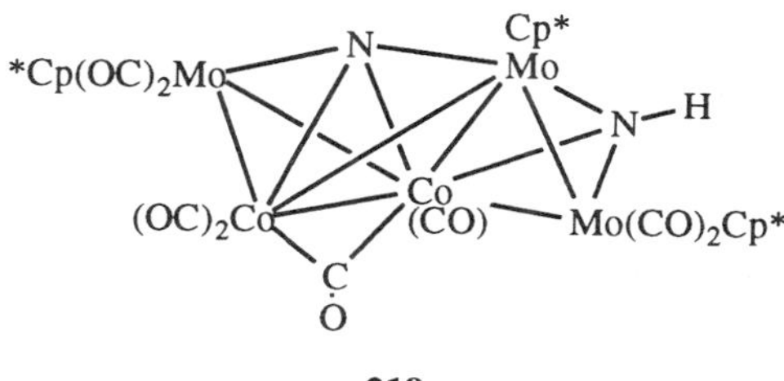

219

Cluster-bound nitrosyl ligands are also converted into amine, amide, and nitride ligands. Treatment of $HRu_3(CO)_{10}(NO)$ with 50 atm of H_2(g) yields a mixture of $HRu_3(CO)_9NH_2$, $HRu_4(CO)_{12}N$, $H_2Ru_3(CO)_9NH$, and $H_4Ru_4(CO)_{12}$.[166] Carbonylation of $Ru_3(CO)_{10}(NO)_2$ produces the isocyanate compound $Ru_3(CO)_{10}(NO)(NCO)$ along with $Ru_4(CO)_{12}(NO)N$, $Ru_4(CO)_{12}(NCO)N$, and $Ru_3(CO)_{12}$.[364]

In a simplification of the procedure, the intermediate metal-bound nitrosyl complex is not isolated, but rather addition of NO^+ to metal complexes, especially anionic ones, is taken directly to the metal amido or nitrido species. Mixtures of $[Fe_2(CO)_8]^{2-}$, $Fe(CO)_5$, and NO^+ give $HFe_4(CO)_{12}N$ upon heating to ca. 130°C, but above that temperature the product is $HFe_5(CO)_{14}N$.[356] Addition of NO^+ to ruthenium carbonyl anions has given $HRu_3(CO)_{10}(NO)$, $HRu_3(CO)_9NH_2$, and $HRu_4(CO)_{12}N$ starting from $[HRu_3(CO)_{11}]^-$[166] and $HRu_4(CO)_{12}N$ or $H_3Ru_4(CO)_{11}N$ when using $[H_3Ru_4(CO)_{12}]^-$.[361] The heterometallic compounds $FeCo_2(CO)_9NH$ and $CoRu_3(CO)_{12}N$ were obtained starting with $[FeCo_3(CO)_{12}]^-$ or $[CoRu_3(CO)_{13}]^-$, respectively,[167] but use of $[H_2CoRu_3(CO)_{12}]^-$ gave the homometallic compound $H_2Ru_3(CO)_9NH$.[167] The osmium cluster $[H_3Os_4(CO)_{12}]^-$ behaves similarly to ruthenium, yielding $HOs_4(CO)_{12}N$ upon treatment with NO^+. This compound is deprotonated by $[PPN]NO_2$.[362] The hexanuclear cobalt and rhodium nitrido-clusters $[M_6(CO)_{15}N]^-$ are obtained from $[M_6(CO)_{15}]^{2-}$.[351]

In some cases, the reduced nitrogen atom from a nitrosyl ligand (or indirectly from NO^+) is observed to be protonated, giving a μ_3-NH ligand. Naked μ_3-nitrido ligands, as noted previously for the phosphorus analogues, would be very basic and expected to pick up a proton quite readily. Among the species isolated to date are $[Fe_3(CO)_8(NO)(NH)]^-$,[146] $H_2Ru_3(CO)_9NH$,[166,167] $FeCo_2(CO)_9NH$,[167] $Cp^{1Me}MoFe_2(CO)_7(NO)NH$,[142] and $Cp^*_3Mo_3Co_2(CO)_8NH$.[141]

Metal-bound isocyanates give similar results to nitrosyl compounds.[363,367] Starting with $[Ru_4(NCO)(CO)_{13}]^-$, one can obtain $[Ru_4(CO)_{12}N]^-$ in 78% yield upon thermolysis. Treatment of $[Ru_3(CO)_x(NCO)]^-$ ($x = 10, 11$) with $Ru_3(CO)_{12}$ leads to the hexanuclear nitride $[Ru_6(CO)_{16}N]^-$ in 82% yield. Organic nitroso and nitro can be similarly reduced by metal carbonyls or carbonylate anions. These reactions are discussed in the later section on reactions with organic substrates.

The deoxygenation reactions of As, Sb, and Bi oxides are known as well. A number of examples are given in Eqs. (123)–(131).[73,133,136,231,233,405] Equation (131) is noteworthy in that it produces a heteromain group ligand—AsS—in a coreduction process.

$$NaAsO_2 + Mo(CO)_6 \xrightarrow{MeOH} [(OC)_5MoAsMo_3(CO)_9(\mu_3\text{-}OMe)Mo(CO)_3]^{2-} \quad (123)$$

$$NaBiO_3 + 3\,Fe(CO)_5 \rightarrow [BiFe_3(CO)_{10}]^- + 3\,CO_2 + 2\,CO \quad (124)$$

$$NaBiO_3 + Mo(CO)_6 \xrightarrow{MeOH} [BiMo_3(CO)_9(\mu_3\text{-}OMe)Mo(CO)_3]^{2-} \quad (125)$$

$$NaSbO_3 + Fe(CO)_5 \Rightarrow [Fe_3(CO)_9\{SbFe(CO)_4\}_2]^{2-} \quad (126)$$

$$NaBiO_3 + Os_3(CO)_{12} \Rightarrow H_3Os_3(CO)_9Bi + Os_3(CO)_9Bi_2 + Os_4(CO)_{12}Bi_2 \quad (127)$$

$$NaBiO_3 + Os_5(CO)_{16} \Rightarrow Os_3(CO)_9Bi_2 \quad (128)$$

$$NaBiO_3 + Os_6(CO)_{18} \Rightarrow Os_4(CO)_{12}Bi_2 \quad (129)$$

$$NaBiO_3 + H_2Os_6(CO)_{18} \Rightarrow Os_5(CO)_{14}Bi_2 \quad (130)$$

$$NaAsO_2 + Na_2S_2O_4 + Fe_3(CO)_{12} \Rightarrow [Fe_2(CO)_6AsS]^- \quad (131)$$

Metal carbonylate anions can often act as strong reducing agents. In a number of reactions oxides of the group 15 elements (and also the other main groups) are efficiently reduced by such reagents [Eqs. (132)–(134)[66,67,73,74,136]]. The most prominent example of this type of reactivity employs $Fe(CO)_5/KOH$ mixtures in which the active metal carbonyl species is the hydride $[HFe(CO)_4]^-$, which gives particularly clean reactions with yields often in the neighborhood of 70 to 80%. The mechanism of this reaction has not been elucidated, but may involve elimination of CO_2 as with the neutral metal carbonyls described previously, or could involve elimination of OH^-.

$$Fe(CO)_5/KOH + NaBiO_3 \Rightarrow [Bi\{Fe(CO)_4\}_4]^{3-} \quad (132)$$

$$\underset{(E = As, Sb)}{Fe(CO)_5/KOH} + NaAsO_2 \text{ or } E_2O_3 \Rightarrow [HE\{Fe(CO)_4\}_3]^{2-} \quad (133)$$

$$NaBiO_3 + [HOs_3(CO)_{11}]^- \Rightarrow H_3Os_3(CO)_9Bi \quad (134)$$

The basicity of the oxygen atom of a CO ligand increases as one goes from terminal to μ_2 to μ_3 configurations. This transformation is accompanied by a weakening of the C–O bond order. Metal-bound nitrosyl ligands behave similarly. Like CO, the oxygen atom of μ_3-CO may be protonated or alkylated. This simple procedure is an effective reduction of the NO bond order and the resulting ligand can be thought of just as any other NR species. Note that in $[Ru_3(CO)_{10}(NO)]^-$ the NO ligand begins in a μ_2 geometry but adopts the μ_3 arrangement upon alkylation or protonation. The addition of $[PPN]CF_3SO_3$ causes the conversion of $Ru_3(CO)_{10}(NOH)$ into $HRu_3(CO)_{10}(NO)$.[163] Examples of this type of chemistry are outlined in Eqs. (135)–(140).[143,144,162,163,176,510] In Eq. (137),[143] the presence of excess acid promotes the conversion of the NOH ligand to NH, chemistry similar to that seen for CO reduction on cluster compounds.

$$[Ru_3(CO)_{10}(NO)]^- + \underset{(R = H, Me)}{RO_3SCF_3} \rightarrow Ru_3(CO)_{10}(NOR) \quad (135)$$

$$Cp'_3Mn_3(NO)_4 + H^+ \rightleftarrows [Cp'_3Mn_3(NO)_3(NOH)]^+ \quad (136)$$

$$Cp'_3Mn_3(NO)_4 + \text{xs } H^+ \rightarrow [Cp'_3Mn_3(NO)_3(NH)]^+ \quad (137)$$

$$[CpWFe_2(CO)_7(NO)(\mu_3\text{-CTol})]^- + H^+ \rightarrow CpWFe_2(CO)_7(\mu_3\text{-CTol})(\mu_3\text{-NOH}) \quad (138)$$

$$\underset{(R = H, Me)}{Cp'MnFe_2Cp_2(CO)_2((NO)(\mu_3\text{-NO})} \xrightarrow{+ CF_3SO_3R} [Cp'MnFe_2Cp_2(CO)_2(NO)(\mu_3\text{-NOR})]^+ \quad (139)$$

$$Cp'MnFe_2Cp_2(CO)_2((NO)(\mu_3\text{-}NO) + Ag[PF_6] \Rightarrow [Cp'MnFe_2Cp_2(CO)_2((NO)(\mu_3\text{-}NH)]^+ \quad (140)$$

H. *Ligand Elimination/Addition Processes*

Addition of an electron pair (reduction) or ligand can result in M–M bond breakage, whereas removal of an electron pair (oxidation) or ligand can result in M–M bond formation. The most common form of this reactivity pattern is the elimination of CO induced either thermally or photochemically. In the case of $E[ML_n]_3$ complexes, intermediates with only one or two M–M bonds are often not observed, and the resultant complexes are the *closo*-tetrahedral complexes as exemplified in Eqs. (141)[65,139] and (142).[50,51,135] The reverse of these reactions may be observed upon addition of L with the overall process being effectively a substitution reaction [Eq. (143)].[511] The direct substitution reaction of $Bi[Co(CO)_4]_3$ has been observed [Eqs. (144) and (145)].[60] As indicated in Eq. (145), only two $SbPh_3$ substitute the Bi–Co compounds, whereas phosphines and arsines gave trisubstituted species.

$$Bi[Co(CO)_4]_3 \rightarrow BiCo_3(CO)_9 + 3\ CO \quad (141)$$

$$Bi[Fe(CO)_2Cp^x]_3 \rightarrow BiFe_3(CO)_3(Cp^x)_3 + 3\ CO \quad (142)$$
$(Cp^x = Cp, Cp^{1M})$

$$BiCo_3(CO)_9 + 3\ PR_3 \rightarrow Bi[Co(CO)_3PR_3]_3 \quad (143)$$

$$BiCo_3(CO)_{12} + 3\ EPh_3 \rightarrow Bi[Co(CO)_3(EPh_3)]_3 \quad (144)$$
$(E = P, As)$

$$BiCo_3(CO)_{12} + 2\ SbPh_3 \rightarrow Bi[Co(CO)_3(EPh_3)]_2[Co(CO)_4] \quad (145)$$

The $Fe_3(CO)_{10-x}L_xPR$ (**28**; x = 1–3) system is especially flexible in its ability to add and subtract ligands. Stepwise addition occurs to give complexes with two and one M–M bond(s). These reactions are reversible so that substituted derivatives of each of the compound types can be obtained for L = CO, $P(OMe)_3$, or CN^tBu. This is illustrated for the all-CO example in Eq. (146).[91] The addition of two CO ligands to $Ru_4(CO)_{11}(EPh)_2$ (**220**; E = P, As) causes breakage of an E–M and an M–M bond to give **221** [Eq. (147)]. Related heterometallic systems have also been examined [Eqs. (148)–(150)].

$$\underset{\mathbf{28}}{Fe_3(CO)_{10}PR} + CO \rightleftarrows \underset{\mathbf{24}}{Fe_3(CO)_{11}PR} + CO \rightleftarrows \underset{\mathbf{22}}{Fe_3(CO)_{12}PR} \quad (146)$$

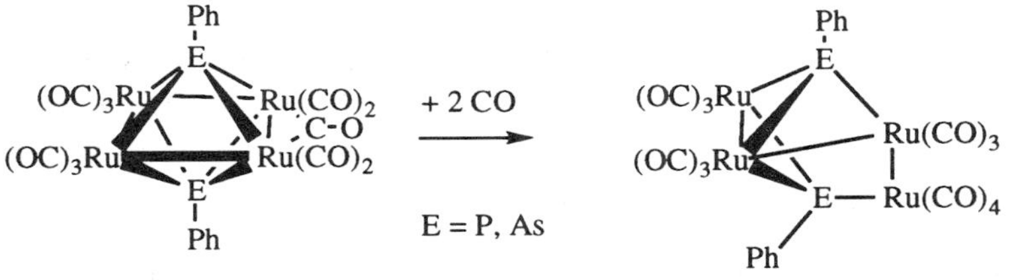

$$\xrightarrow[E = P, As]{+2\,CO} \qquad (147)$$

220 **221**

$$FeCo_2(CO)_9AsMe + 2\,L \rightarrow \{FeCo(CO)_6L\}\{Co(CO)_3L\}AsMe \quad (L = PMe_2Ph) \qquad (148)$$

$$CpMnFe_2(CO)_8PPh + L \rightleftarrows \{Cp(CO)_2Mn\}\{Fe_2(CO)_6L\}PPh \quad (L = CO, PPh_3) \qquad (149)$$

$$FeCo_2(CO)_8\{AsMe_2Mo(CO)_3Cp\}(\mu_3\text{-}PR) \xrightarrow{-3\,CO} CpMoCo_2Fe(CO)_8\{\mu\text{-}AsMe_2)(\mu_3\text{-}PR) \qquad (150)$$

Halide ions can add to E–M clusters by simple displacement of two-electron donor ligands such as CO. In contrast to CO, they have the added capability of functioning as a bridging ligand with donation of an extra electron pair. This is seen in Scheme 8 for $Os_3(CO)_{10}NPh$ (**222**).[101] Similar chemistry is observed for the Ru analogues.[156,161,164,168] Both bridging halide **223** and terminal halide isomers **224** are observed, the ratio depending on which halide is employed. For Cl^- the major isomer is the bridging form, whereas the trend is reversed for I^-. About equal amounts of the terminal and bridging forms are seen when $X^- = Br^-$. Protonation of the μ-X isomer occurs at an Os–Os edge to give **225**. The halide substitution/addition activates the cluster to carbonylation of the NPh group, a reaction that does not occur for the parent all-CO molecule. As can be seen in products **226–228**, insertion of a CO into an N–Os bond occurs.

Chemistry related to the addition of halides to an E–M cluster is that of XCN^- (X = S, Se, Te) addition.[104,295] Initial addition of XCN^- to the cluster framework is followed by removal of an isocyanide upon alkylation [Eq. (151)[104]]. The reaction pathway when X = O is different [Eq. (152)[178]], resulting in coupling of the oxygen with the PR unit.[178] A minor by-product in that reaction is $Fe_4(CO)_{12}(\mu_4\text{-}PR)$, which has a trigonal bipyramidal Fe_4P core that is unusual in that the P vertex is in one of the equatorial sites.[191] This cluster possesses six skeletal electron pairs, which is consistent with the trigonal bipyramidal arrangement. In a few documented cases, ligand addition/subtraction does not affect the cluster core noticeably. This situa-

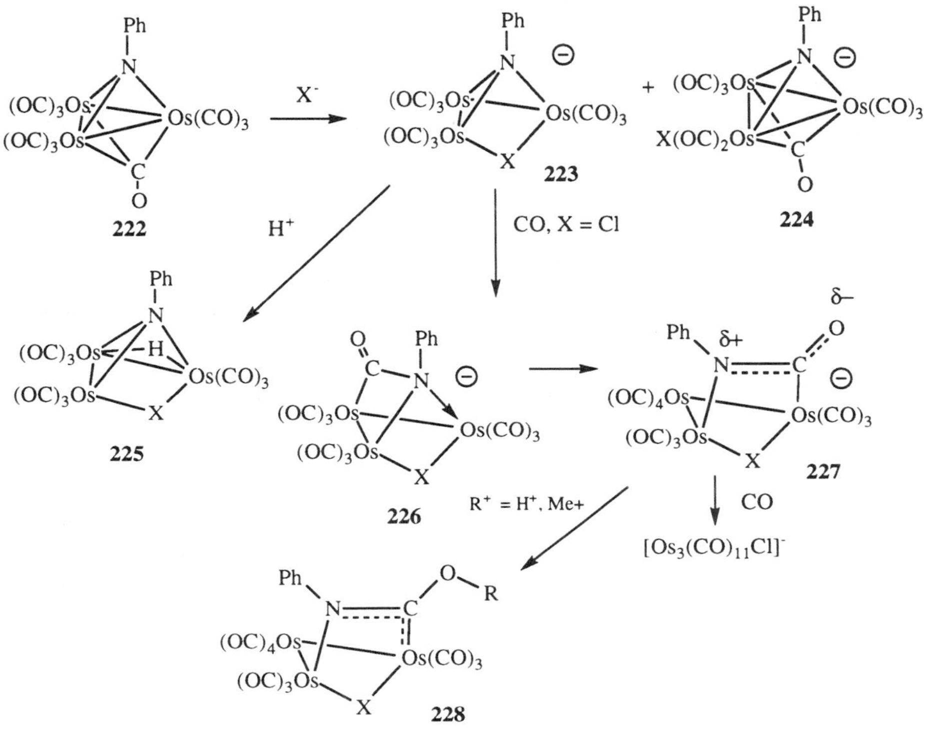

SCHEME 8

tion has been particularly well studied for the $Fe_4(CO)_x(PR)_2$ system [x = 11 or 12, Eq. (153)[108]]. We also note another type of ligand addition reaction in which an organic-based ligand is converted into a nitrene-type compound [Eq. (154)[150]]. Essentially this reduces the N=C bond by forcing the metal-bound hydride ligands onto the organic fragment.

$$Fe_3(CO)_{10}(PR) \xrightarrow[]{\text{1. } XCN^-\text{; 2. } Et_3O^+\text{, - EtNC, -CO}} Fe_3(CO)_9(X)(PR) \quad (151)$$

$$Fe_3(CO)_{10}(PR) \xrightarrow[]{\text{1. } OCN^-\text{; 2. } Et_3O^+} HFe_3(CO)_9(\mu_3,\eta^2\text{- }RP{=}OEt) + H_2Fe^3(CO)_9(PR) \quad (152)$$

$$Fe_4(CO)_{11}(PR)_2 \underset{-\,CO}{\overset{+\,CO}{\rightleftarrows}} Fe_4(CO)_{12}(PR)_2 \quad (153)$$

$$HFe_3(CO)_9(\mu_3, \eta^2\text{-}N{=}C(H)R) + CO \rightarrow HFe_3(CO)_9(\mu_3\text{-}NCH_2R) \quad (154)$$

I. *Alkylation and Related Reactions*

Halogens attached to E can be replaced by alkyl groups using AlR_3 reagents. This has been documented for $[Cp(CO)_2Fe]_2BiCl$, which reacts with $AlEt_3$ to give $[Cp(CO)_2Fe]_2BiEt$ and $AlEt_2Cl$.[50] Addition of alkyl halides or other strong carbocation reagents such as methyl triflate to E–M anionic complexes may result in alkylation at lone pairs on E. In some cases the structure of the product may be retained but, especially for the heavier elements, may vary considerably from that of the parent complex. For example, $[Fe_3(CO)_9(PH)P]^-$ forms the expected $Fe_3(CO)_9(PH)(PMe)$ upon addition of MeO_3SCF_3.[279] The formation of the dimethyl complex can be achieved by addition of $LiCuMe_2$ to $Fe_3(CO)_9(PH)(PCl)$. Complexes with E_x ligands can be alkylated with retention of the cluster core [Eqs. (155)[447,457] and (156)[478]]. However, the $[RBi\{Fe(CO)_4\}_3]^{2-}$ compounds appear to be prone to fragmentation leading ultimately to cyclic compounds of the formula $[RBiFe(CO)_4]_2$ [Eq. (157)–(160)[53,55,56]]. This fragmentation, depending on the nature of R and X, may happen spontaneously or may be induced via addition of acid. The molecule $[BiFe_3(CO)_9(\mu_3\text{-}CO)]^-$ has two potential sites of alkylation, the lone pair on Bi and the O atom of the μ_3-CO ligand. The reaction with MeO_3SCF_3 showed that the μ_3-CO ligand is the more basic site and the methoxy alkylidyne complex $BiFe_3(CO)_9(\mu_3\text{-}COMe)$ was produced.[134] Alkylation of the dianions $[Fe_3(CO)_9PR]^{2-}$ with CH_2I_2 results in C–P double bond formation, giving $Fe_3(CO)_{10}(\mu_3, \eta^2\text{-}H_2C{=}PR)$.[179] In the case of R = anisyl, cluster fragmentation was observed, yielding $Fe_2(CO)_6(\mu\text{-}I)[P(anisyl)Me]$.[179] It is important to run the reaction under conditions that prevent further reaction with NaI or else cluster degradation is also observed.

$$MeC(CH_2PPh_2)_3CoP_3 + R_3O^+ \rightarrow [MeC(CH_2PPh_2)_3Co(P_3R)]^+ \quad (155)$$
(R = Me, Et)

$$[(CO)_3WP_5]^- + Me_3SiCl \rightarrow (CO)_3WP_5SiMe_3 \quad (156)$$

$$[ClBi\{Fe(CO)_4\}_3]^{2-} + MeI \rightarrow [MeBiFe(CO)_4]_2 \quad (157)$$

$$[Bi\{Fe(CO)_4\}_4]^{3-} + RBr \rightarrow [RBiFe(CO)_4]_2 \quad (158)$$
(R = Me, Et, CH_2Ph)

$$[Bi\{Fe(CO)_4\}_4]^{3-} + {}^iBuBr \rightarrow [{}^iBuBi\{Fe(CO)_4\}_3]^{2-} + [{}^iBuBiFe(CO)_4]_2 \quad (159)$$

$$[RBi\{Fe(CO)_4\}_3]^{2-} + HOAc \rightarrow [RBiFe(CO)_4]_2 \quad (160)$$
(R = ^{i}Pr, ^{n}Bu, ^{i}Bu, $PhCH_2$, $Cl(CH_2)_2$, $Br(CH_2)_4$)

J. *Reactions of Inidene Complexes*

Inidene complexes are formally electron deficient and thus would be expected to function as Lewis acid sites [Eqs. (161)[249] and (162)[44]]. The ability to undergo a valence tautomerism in some cases, however, also gives them the capability of serving as Lewis bases [see Eq. (1)].[44] Note in the latter case the formation of the M–M bond that occurs in the valence tautomerism. It has proven possible in one case to convert the Lewis acid adduct form into the Lewis basic form directly [Eq. (163)[44]]. The distilbene complex $[Ph\{W(CO)_5\}Sb{=}Sb\{W(CO)_5Ph\}]\{\mu\text{-}W(CO)_5\}$ reacts with PPh_3 to cleave the Sb–Sb bond with formation of the inidene base adduct $Ph\{W(CO)_5\}_2Sb(PPh_3)$.[44] The molecules $[XE\{Fe(CO)_4\}_3]^{2-}$ (E = Sb, Bi; X = halide) are related to the inidene complexes in that removal of X^- should result in formation the isoelectronic inidene species $[E\{Fe(CO)_3\}_3]^-$. Although this species has never been directly observed, addition of metal fragments results in addition to what would be the valence tautomer of $[E\{Fe(CO)_3\}_3]^-$ [E = Sb, Bi; Eq. (165)[53]]. Reactions of alcohols that contain an appended Lewis basic functionality (HO–R–base) with $ClAs[Cr(CO)_5]_2$ eliminate HCl and generate chelating alkoxide complexes where the internal Lewis base is also coordinated to As.[248] Similar chemistry is observed with the bismuth inidene complexes and chelating bases.[258] As discussed in Section III,A, inidene complexes with halide ions of the form $[L_nM]_2EX$ can react with metal carbonylate anions to displace X^-.

$$ClAs[Cr(CO)_5]_2 + X^- \rightarrow [\{(CO)_5Cr\}_2As(Cl)X]^- \quad (161)$$
$(X^- = Cl^-, Br^-, CN^-, N_3^-, SCN^-)$

$$\{(CO)_5W\}_2SbR + PPh_3 \rightarrow \{(CO)_5W\}_2Sb(PPh_3)R \quad (162)$$

$$\{(CO)_5W\}_2SbR + W(CO)_5(THF) \rightarrow \{(CO)_{10}W_2\}\{(CO)_5W\}SbR \quad (163)$$

$$\{(CO)_{10}W_2\}\{(CO)_5W\}SbCl + 2\,PPh_3 \rightarrow (CO)_5W\}_2Sb(PPh_3)Cl + W(CO)_5PPh_3 \quad (164)$$

$$[ClE\{Fe(CO)_4\}_3]^{2-} + Cr(CO)_5(THF) \rightarrow [E\{Fe_2(CO)_8\}\{Fe(CO)_4\}\{Cr(CO)_5\}]^- \quad (165)$$
(E = Sb, Bi)

Thermolysis of ${}^tBuP\{Cr(CO)_5\}_2$ at 100°C leads to P–P coupling and formation of a trinuclear cluster complex $Cr_3(CO)_{10}(\mu_3\text{-}P^tBu)(P^t_2Bu_2)$.[175] This contrasts with thermolysis of similar complexes with other R groups, which leads instead to trimetallated diphosphene compounds $({}^tBu\{Cr(CO)_5\}P{=}P\{Cr(CO)_5\}{}^tBu)\{\mu\text{-}Cr(CO)_5\}$.[175]

K. *Redistribution Reactions*

A complication to the formation of a single compound from halide displacement reactions is the ability of the main group element center to undergo redistribution reactions. Examples are given in Eqs. (166)–(169).[41,45] These reactions probably account for why the complexes $[Co(CO)_4]_2ECl$ and $[Co(CO)_4]ECl_2$ (E = P, As, Sb) have not been isolated.[57]

$$2\ L_nM\text{-}EX_2 \rightleftarrows [L_nM]_2EX + EX_3 \quad (166)$$

$$2\ [L_nM]_2EX \rightleftarrows [L_nM]_3E + [L_nM]EX_2 \quad (167)$$

$$2\ [L_nM]_3E + EX_3 \rightleftarrows 3\ [L_nM]_2EX \quad (168)$$

$$Me_2BiBr + [Cp(CO)_3M]^- \rightarrow [Cp(CO)_3M]BiMe_2 + [Cp(CO)_3M]_2BiMe + BiMe_3 + \ldots \quad (169)$$

L. *Reactions of the Main Group Elements with Transition-Metal Complexes*

The reaction of the transition-metal fragments with main group 15 elements directly has proven a very fruitful field for exploration. The methodology has been successful for a wide range of metal complexes. These fall generally into three basic types: (1) reactions with cyclopentadienyl metal carbonyls, (2) reactions with homoleptic metal carbonyls and substituted derivatives, and (3) reactions with metal cations in the presence of a multidentate chelating ligand.

The cyclopentadienyl reactions have been illustrated for early as well as late transition metals and also the actinides. It is remarkable that the reactions are very similar across the entire group. This methodology results often in sandwich-type complexes of the E_x units and has been found successful for both phosphorus and arsenic, but only one example has been reported for antimony. There are no reports of bismuth metal reacting with cyclopentadienyl metal fragments. Neither Sb nor Bi readily form the soluble E_4 species as do P and As, and so their reactivities may be limited by solubility factors (Table XLI).

The reactions with metal carbonyls are less regular but also show retention of E–E bonding [Eqs. (170)–(177)].[126,129,407,461–463,469–471] One exception is the reaction of Bi metal with $Co_2(CO)_8$, which gives $Bi[Co(CO)_4]_3$.[60] As noted previously, Bi is unlike P and As in that it does not form the discrete E_4 tetrahedral species and, perhaps therefore, is less likely to maintain the E–E bonding in the products. Equation (177)[129] represents a very unusual reaction that produces interesting products that contain both E and a methyne fragment. These are produced in only very low yields, which is unfortunate, as it would be interesting to probe the effect of the E atom on the reactivity of the methyne ligand.

$$M(CO)_6 + P_4 \xrightarrow{h\nu} (CO)_4M\{PM(CO)_5\}_4 \tag{170}$$

$$M(CO)_5(THF) + P_4 \Rightarrow (CO)_4M\{PM(CO)_5\}_4 \tag{171}$$
(M = Cr, W)

$$Fe_2(CO)_9 + P_4 \Rightarrow \{Fe_2(CO)_6\}_2\{Fe(CO)_4\}_2(P_2)_2 \tag{172}$$

$$Fe_2(CO)_9 + P_4 \Rightarrow [Fe(CO)_4]_3P_4 \tag{173}$$

$$Co_2(CO)_8 + P_4 \Rightarrow Co_3(CO)_9P + Co_2(CO)_6P_2 + (CO)_3CoP_3 \tag{174}$$

$$L_3MX + P_4 \rightarrow L_2RhClP_4 + L \tag{175}$$
(M = Rh, Ir; L = PPh_3, $P(p\text{-}Tol)_3$, $P(m\text{-}Tol)_3$, $AsPh_3$; X = Cl, Br, I)

$$Pt(PPh_3)_2(C_2H_4) + P_4/Cr(CO)_5(THF) \Rightarrow (PPh_3)_2PtP_4\{Cr(CO)_5\}_2 \tag{176}$$

$$Fe_5(CO)_{15}C + E \xrightarrow{H_2SO_4 \text{ or } H_2O} EFe_3(CO)_9CH \tag{177}$$
(E = As, Sb, Bi)

The reactions of metal cations with P_4 and As_4 in the presence of chelating ligands (most often tridentate compounds, of which triphos is the most used) has been extensively studied (Table XLII). This reaction type has been reviewed.[21] The predominant outcome is the formation of ME_3 tetrahedral and M_2E_3 trigonal bipyramidal compounds. Some of these have flexibility in the observed oxidation states. For comparison purposes, a reaction of P_4S_3 is included in Table XLII as the products from the reaction are similar to those obtained using only P_4.

In two reactions insertion of P_4 is observed to occur into a M– linkage. In the first, the covalent bonds of $Cp_2Zr\{P(SiMe_3)_2\}_2$ react with P_4 to give $Cp_2Zr\{P_4(P(SiMe_3)_2)\}$ (**188**) showing insertion of P_4 into the $ZrP(SiMe_3)_2$ bonds.[489] In the second, P_4 reacts with Co^{2+} in the presence of dppm to give the product $[Co\{Ph_2PCH_2P(Ph)_2P_4P(Ph_2)CH_2PPh_2\}]^{2+}$ in which the P_4 unit has inserted into two P–Co dative bonds, one on each of the dppm ligands.[492]

M. *Reactions of Metal Complexes with Polyphosphines and Polyarsines*

Phosphorus and arsenic form cyclic compounds of the general formula $(ER)_x$. They have been extensively examined in reactions with neutral organometallic fragments (Table XLIII). In many of these reactions the

TABLE XLI

PRODUCTS FROM THE REACTION OF ELEMENTAL E WITH CYCLOPENTADIENYL METAL FRAGMENTS

Metal complex	E	Product(s)	Ref.
$Cp^*_2Ti(CO)_2$	P_4	$[Cp^*Ti]_2P_6$	486
$(Cp^{2B})M(CO)_2$ (M = Zr, Hf)	P_4	$(Cp^{2B})ZrP_4$	459
$(Cp^x)V(CO)_4$ ($Cp^x = Cp^*, Cp^{M4E}$)	P_4	$[(Cp^x)V]_2P_6$	487
$Cp^*Nb(CO)_4$	P_4	$Cp^*(CO)_2NbP_4$	460, 476
		$[Cp^*(CO)_2Nb]_2P_4$	
		$[Cp^*Nb]_2P_6$	
$Cp^*Nb(CO)_4$	As_4	$Cp^*Nb(CO)_2As_4$	434
		$[Cp^*Nb(CO)]_2(As_2)_2$	
$(Cp^{2B})Nb(CO)_4/Cr(CO)_5(THF)$	As_4	$[(Cp^{2B})Nb]_2As_8$	508
		$[(Cp^{2B})Nb]_2As\{Cr(CO)_5\}$	
$[CpCr(CO)_3]_2$	P_4	$Cp_2Cr_2(CO)_4P_2$	389, 390, 505
		$CpCr(CO)_2P_3$	
		$Cp_2Cr_2(CO)_4$	
		$[CpCr(CO)_2]_5P_{10}$	
$[(Cp^x)Cr(CO)_2]_2$ ($Cp = Cp^*, Cp^{M4E}$)	As_4	$(Cp^x)Cr(CO)_2As_3$	451
		$[(Cp^x)Cr(CO)_2]_2As_5$	
Cp_2MoH_2	P_4	$Cp_2Mo(P_2H_2)$	410
$[Cp^*Mo(CO)_2]_2$	P_4	$[Cp^*Mo]_2P_6$	392
		$Cp^*Mo(CO)_2P_3$	
		$Cp^*_2Mo_2(CO)_4P_2$	
		$Cp^*_2Mo_2(CO)_2(P_2)_2$	
$Cp_2(CO)_6Mo_2$	As_4	$Cp_2(CO)_4Mo_2As_2$	128
		$Cp_3(CO)_6Mo_3As$	
$[Cp^*Mo(CO)_2]_2$	As_4	$Cp^*Mo(CO)_2As_3$	397
		$Cp^*_2Mo_2(CO)_4As_2$	
		$Cp^*_2Mo_2(CO)_2(As_2)_2$	
$Cp^*_2Mo_2(CO)_4$	As_4S_4	$[Cp^*Mo(CO)_2]_2As_2$	450
		$Cp^*Mo(CO)_2As_3$	
		$[Cp^*Mo]_2(S_2)(As_2S)$	
$[CpMo(CO)_3]_2$	Sb	$Cp_2Mo_2(CO)_2Sb_2$	403
$[Cp^*(CO)_nW]_2$ ($n = 2, 3$)	P_4	$[Cp^*W]_2P_6$	487
$Cp^*_2Fe_2(CO)_4$	P_4	$[Cp^*Fe]_2(P_2)_2$	432
$Cp^*_2Fe_2(CO)_4$	P_4	Cp^*FeP_5	475
$[(Cp^x)Fe(CO)_2]_2$ ($Cp^x = Cp^*, Cp^{M4E}$)	As_4	$(C_p{}^x)FeAs_5$	485
$[(Cp^x)Ru(CO)_2]_2$ ($Cp^x = Cp^*, Cp^{M4E}$)	As_4	$(Cp^x)RuAs_5$	437
		$[(Cp^x)Ru]_3As_3Ru(As_3)_2$	
$[Cp^*Co(CO)]_2$	P_4	$Cp^*(CO)CoP_4$	466
		$[Cp^*(CO)Co]_2P_4$	
$CpCo(CO)_2$	P_4	$Cp_4Co_4P_4$	77
$Cp^*Co(CO)_2$	P_4	$Cp^*_2Co_2(P_2)_2$	32
$[(Cp^{2B})Co(CO)]_2/Cr(CO)_5(THF)$	P_4	$(Cp^{2B})(CO)CoP_4\{Cr(CO)_5\}x$ ($x = 3, 4$)	465
$[Cp^*Co(CO)]_2$	As_4	$Cp^*(CO)CoAs_4$	472
		$[Cp^*(CO)Co]_2As_4$	
		$[Cp^*Co]As_4[Co(CO)Cp^*]$	
		$[Cp^*Co]_2(As_2)_2$	

TABLE XLI (*continued*)

Metal complex	E	Product(s)	Ref.
$(Cp^x)Co(CO)_2$ (Cp^x = Cp*, Cp^{M4E})	As_4	$[\{(Cp^x)Co\}(As_2)]_2$ $[(Cp^x)Co]_2As_6$ $[(Cp^x)Co]_3(As_2)(As_4)$	424
$(Cp^{M4E})Rh(CO)_2$	P_4	$[Cp^{M4E}Rh][Cp^{M4E}(CO)Rh]P_4$ $[Cp^{M4E}Rh]_2P_4$	459
$(Cp^{1B})Rh(CO)_2/Cr(CO)_5(THF)$	P_4	$(Cp^{1B})(CO)RhP_4\{Cr(CO)_5\}_4$ $(Cp^{1B})RhP_4\{Cr(CO)_5\}_4$	467, 468
$(Cp^{2B})Rh(CO)_2/Cr(CO)_5(THF)$	P_4	$(Cp^{2B})(CO)RhP_4\}Cr(CO)_5\}_4$ $(Cp^{2B})RhP_4\{Cr(CO)_5\}_x$ (x = 1, 3)	468
$(Cp^{2B})Rh(CO)_2$	P_4 or As_4	$[(Cp^{2B})Rh]_2E_{10}$ $[(Cp^{2B})Rh]_2(E_2)_2$	506
$[(Cp^{1M})Ni(CO)]_2$	P_4	$[(Cp^{1M})Ni]_4P_4$	79
$[Cp^*Ni(CO)]_2$	P_4	$[Cp^*Ni]_3P_5$	79
$[(Cp^{2B})Ni(CO)]_2$	P_4	$(C_p{}^{2B})NiP_3$	79
$Cp^*Rh(CO)_2$	P_4	$[Cp^*Rh]_2(P_2)_2$	433
$Cp^xRh(CO)_2/Cr(CO)_5(THF)$ (Cp^x = Cp^{1B}, Cp^{2B})	P_4	$Cp^x(CO)RhP_4\{Cr(CO)_5\}_4$	407
$(Cp^{2B})_2Th(C_4H_6)$	P_4	$[(Cp^{2B})_2Th]_2(\mu,\eta^3,\eta^3\text{-}P_6)$	453
$(Cp^{2B})_2Th(C_4H_6)$	P_4	$[(Cp^{2B})_2Th]_2P_3$	453
$[(Cp^{4P})Ni(CO)]_2$	P_4 or As_4	$[(Cp^{4P})Ni]_2E_4$ $(Cp^{4P})NiE_3$	452

TABLE XLII

PRODUCTS FROM THE REACTION OF E_4 WITH METAL CATIONS IN THE PRESENCE OF CHELATING LIGANDS

Metal cation	Ligand	E	Product	Ref.
Co^{2+}	np_3	P	$[np_3CoP_3]^{2+}$	440
Co^{2+}	triphos	P	$(triphos)CoP_3$	442, 443
Co^{2+}	dppm	P	$[CoP_4\{P(Ph_2)CH_2PPh_2\}_2]^{2+}$	492
$[RhL_2Cl]_2$ (L = C_2H_4, CO)	triphos	P	$(triphos)RhP_3$	445
$Ir(PPh_3)_2(CO)Cl$	triphos	P	$(triphos)RhP_3$	445
Ni^{2+}	triphos	P	$[(triphos)NiP_3Ni(Triphos)]^{2+}$	442, 443
Ni^{2+}	triphos	P_2S_3	$[(triphos)NiP_3]^+$	444
M^{2+} (M = Co, Ni; E = P, As; n = 1, 2)	triphos	E	$[(triphos)ME_3M(triphos)]^{n+}$	436, 455
$MCl_2(PBu_3)_2$ (M = Pt, Pd)	triphos	P	$[(triphos)MP_3]^+$	448
(triphos)Pd		P	$(triphos)PdP_4$	448
$PdCl_2(PBu_3)_2$	triphos np_3	P	$[L_3PdP_3PdL_3]^+$	448

TABLE XLIII

PRODUCTS FROM REACTIONS OF POLYPHOSPHINES AND POLYARSINES PRESERVING E–R AND E–E BONDING

Metal complex	$(ER)_x$	Products	Ref.
$Mo(CO)_6$	P_4Et_4	$(CO)_3Mo(PEt)_5$	514, 515
	(PR_x)	$(CO)_{6-y}M(PR)_5$	
$M(CO)_6$			
(R = Et, Ph, x = 4; E = Me, Ph, x = 5; y = 1–3)			
$Mo(CO)_6$	$(AsPh)_6$	$Mo(CO)_4(AsPh)_6$	493
$Mo(CO)_3L_3$	$(AsMe)_5$	$Mo_2(CO)_6(AsMe)_{10}$	507
(L_3 = MesH, $(CO)_3$)			
$Mo(CO)_3L_3$	$(AsPh)_6$	$Mo_2(CO)_6(AsPh)_9$	507
(L_3 = MesH, $(CO)_3$)			
$[CpMo(CO)_3]_2$	As_5Me_5	$[CpMo(CO)_2]_2(AsMe)_5$	82
		$Cp_4Mo_4As_4$	
$[(Cp^{1M})Mo(CO)_3]_2$	$(AsMe)_5$	$[(Cp^{1M})Mo(CO)]_2(As_2)_2$	402
$[CpMo(CO)_3]_2$	$(MeAsS)_n$	$Cp_2(CO)_4Mo_2As_2$	458
		$[Cp(CO)_2Mo]_2(As_2)_2$	
		$[Cp(CO)_2Mo]_2(As_3)(AsS)$	
$[CpMo(CO)_3]_2$	$(AsPh)_6$	$[CpMo(CO)_2]_2As_2$	398
$[CpMo(CO)_3]_2$	$(AsMe)_5$	$[CpMo(CO)_2]_2As_5$	82
$W(CO)_3(NCMe)_3$	$(PEt)_3$	$(CO)_4W(PEt)_4$	515
$CpW(CO)_3H$	$(AsPh)_6$	$[CpW(CO)_2]_2As_2$	398
$M_2(CO)_{10}$ (M = Mn, Re)	$(P^tBu)_3$	$M_2(CO)_7(P^tBu)_3$	327
$Mn_2(CO)_{10}$	As_5Me_5	$Mn_2(CO)_8(AsMe)_5$	474
$Mn_2(CO)_{10}$	As_5Me_5	$Mn_2(CO)_6[(AsMe)_4]_2$	474
$[Cp(CO)_2Mn]_3(PPh_3)$		$[Cp(CO)_2Mn]_2PPh$	244
$Re_2(CO)_{10}$	As_5Me_5	$\{Re(CO)_3\}_2As_8Me_7$	509
$Fe(CO)_5$	$(PR)_x$	$Fe_2(CO)_6(PR)_4$	464, 513, 516
(R = Me, Ph, x = 5; R = Et, Ph, x = 4; R = Ph, x = 6)			
$Fe(CO)_5$	$(AsR)_x$	$Fe_2(CO)_6(AsR)_4$	473, 474
(R = Me, Et, x = 5; R = Ph, x = 6)			
$Fe(CO)_5$	$(As(C_6F_5))_4$	$Fe(CO)_4[As(C_6F_5)]_2$	421
$Fe_2(CO)_9$	$(P^iPr)_3$	$Fe_2(CO)_6[P^iPr)_3]_2$	
		$Fe_2(CO)_4(P^iPr)_6$	96
$Fe_2(CO)_9/{}^iPrPCl_2$	$(P^tBu)_3$	$Fe_2(CO)_3Cl_2(P^tBu)_5$	96
$Ru_3(CO)_{12}$ or $H_4Ru_4(CO)_{12}$	$P_4{}^tBu_4$	$Ru_4(CO)_8(\mu\text{-}P^tBu)_2(\mu_4\text{-}P^tBu)_2$	212, 213
		$Ru_5(CO)_{15}(\mu_4\text{-}P^tBu)$	
		$H_2Ru_6(CO)_{12}(\mu_3\text{-}P^tBu)_3$	
		$Ru_6(CO)_{11}(P^tBu)_4$	
		$Ru_7(CO)_{14}(P^tBu)_4$	
		$H_3Ru_9(CO)_{20}(P^tBu)_3(\mu_7\text{-}P)$	
$Cp_2Ru_2(CO)_3(C_2Ph_2)$	$(AsR)_x$	$Cp_2Ru_2(CO)_3(AsR)$	90
(R = Me, x = 5; R = Ph, x = 6)			
$Co_2(CO)_8$	$(P^tBu)_3$	$Co_4(CO)_{10}(P^tBu)_2$	327
$Co_2(CO)_8$	$(P^iPr)_3$	$[Co_2(CO)_5]_2[P^iPr]_3$	96

TABLE XLIII (*continued*)

Metal complex	$(ER)_x$	Products	Ref.
$Co_2(CO)_8$	$(AsPh)_6$	$[Co_8(CO)_{16}(\mu_6\text{-As})(\mu_4As)(\mu_4\text{-AsPh})_2]_2$	232
$Co_2(CO)_8$	$(AsMe)_5$	$(CO)_3CoAs_3$	449
$[Ir(C_8H_{12})Cl]_2$	$K_2[P_4Ph_4]$	$[Ir(C_8H_{12})Cl]_2(PPh)_2(P_2Ph_2)$	96
$Ni(CO)_4$	$(P^iPr)_3$	$Ni_2(CO)_4\{P_3{}^iPr_3Ni(CO)_3\}_2$	327
$Ni(CO)_4$	$(PPh)_4$	$(CO)_3Ni(PPh)_4$	516
$Ni(CO)_4$	$(P^tBu)_3$	$Ni_3(CO)_5(\mu_3\text{-}P^tBu)_2(\mu_4\text{-}P^tBu)(P_3{}^tBu_3)$	16

$(ER)_x$ unit is retained but coupling may occur to give higher order $(ER)_x$ species. In some cases this is a simple coupling of two of the original E_x units, but in others there is clearly a more substantial fragmentation and reorganization process involved. For example, reaction of $(PEt)_4$ with $Mo(CO)_6$ affords a product with a $(PEt)_5$ fragment attached to Mo. Other examples can be readily gleaned from the table. Under forcing conditions the R groups may also be lost, giving products with naked E units in addition to showing E–E bond cleavage.

N. *Halogenation and Dehalogenation Reactions*

Neutral metal carbonyls are effective at cleaving E–X bonds and have been extensively encountered. In cases where the X group is completely removed from the molecule, it may leave as a metal halide, although this has not been confirmed. In some cases, a halide ion becomes incorporated in the final product bonded to the metal centers, often in a bridging mode. The dehalogenation of EX_3 (or, in one case, EX_5) has been reported in a number of instances; examples of these dehalogenation reactions are given in Eqs. (178)–(186). The products belong to a variety of structural types, so it is not possible to rationalize a single reaction pathway. In a couple of cases where activated metal complexes are employed [Eqs. (184)–(186)[61,118,121,126,129,252,257,261,429]], inidene-type products resulted. It is possible that such unsaturated molecules might be intermediates in some of the other reactions because, as we have seen in Section III,J, inidene species will react further with metal fragments to generate saturated cluster species.

$$Fe_2(CO)_9 + \underset{(E = P, As)}{ECl_3} \Rightarrow [Fe_2(CO)_8]E[Fe_2(CO_6Cl] \qquad (178)$$

$$CpCo(CO)_2 + AsF_3 \Rightarrow [\{Cp_2(CO)_2Co_2\}_2As][BF_4] \qquad (179)$$

$$AsF_3 + Fe(CO)_5 \xRightarrow{120°C} As_2Fe_3(CO)_9 \quad (180)$$

$$PI_3 + Co_2(CO)_8 \Rightarrow Co_3(CO)_9P + P_3Co(CO)_3 + Co_2(CO)_6P_2 \quad (181)$$

$$SbCl_5 + Fe_5(CO)_{15}C \Rightarrow SbFe_3(CO)_9CH \quad (182)$$

$$W(CO)_5(THF) + AsCl_3 \Rightarrow As_2[W(CO)_5]_3 \quad (183)$$

$$Cp(CO)_2Mn(THF) + ECl_3 \Rightarrow [\{Cp(CO)_2Mn\}_2ECl]_x \quad (184)$$
$(E = As, x = 1; E = Bi, x = 2)$

$$(arene)Cr(CO)_2(THF) + SbX_3 \Rightarrow [(arene)Cr(CO)_2]SbX \quad (185)$$
$(X = Cl, Br, I)$

$$(Cp^{1M})(CO)_2Mn(THF) + BiCl_3 \Rightarrow [(Cp^{1M})(CO)_2Mn]_2BiCl + [(Cp^{1M})(CO)_2Mn]_3Bi_2 \quad (186)$$

When REX_2 fragments are used, the E–X bonds are again cleaved, but generally the R group is retained in the products, as seen in Eqs. (187) and (188).[194] These reactions, to date, have most often yielded a mixture of saturated di- and trinuclear products.

$$Fe_2(CO)_9 + RPX_2 \Rightarrow Fe_2(CO)_6(\mu\text{-}PRX)(\mu\text{-}X) + Fe_3(CO)_{10}PR + Fe_3(CO)_9(PR)_2 \quad (187)$$
$(R = Mes, X = Cl, Br)$

$$Fe_2(CO)_9 + (Me_3Si)_2CHPCl_2 \Rightarrow Fe_3(CO)_{10}PCH(SiMe_3)_2 \quad (188)$$

Preformed metal complexes containing E–X bonds may be employed. Both $L_nM–EX_3$ and $L_nM–EX_2R$ species have been employed. Examples are found in Eqs. (189)–(197).[195,199,224,242,256,262,297,315] The original metal fragment M may, but need not necessarily, be retained in the product. Hetero-main group element complexes may also be prepared by this methodology. When treated with $^tBuPCl_2$, $H_2Fe_3(CO)_9(NR)$ (R = Et, Ph) gave $Fe_3(CO)_9(NR)(P^tBu)$.[273] An "assisted" dehalogenation process is observed in the reaction of $RP(S)Cl_2$ with $Fe_3(CO)_{12}$ in the presence of magnesium to yield the hetero-main group element complexes $Fe_3(CO)_9(S)(PR)$.[290]

$$L_nMPX_3 + Fe_2(CO)_9 \Rightarrow Fe_3(CO)_9\{PML_n\}_2 \quad (189)$$
$(ML_n = Cp(CO)_2Mn, Cr(CO)_5, Mo(CO)_5, W(CO)_5)$

$$(CO)_5CrPBr_3 + Fe_2(CO)_9 \Rightarrow Fe_2(CO)_7\{PCr(CO)_5\}_2 \quad (190)$$

$$CpMn(CO)_2PBr_3 + Fe_2(CO)_9 \Rightarrow Fe_2(CO)_6\{P{=}Mn(CO)_2Cp\}_2 \quad (191)$$

$$L_nMPX_3 + Co_2(CO)_8 \Rightarrow Co_3(CO)_9PML_n \quad (ML_n = Cp(CO)_2Mn,\ Cr(CO)_5,\ W(CO)_5) \tag{192}$$

$$Cp(CO)_2MnPCl_2Ph + Fe_2(CO)_9 \Rightarrow CpMnFe_2(CO)_8(PPh) \tag{193}$$

$$Cp(CO)_2MnPCl_2R + Ru_3(CO)_{12} \Rightarrow Ru_5(CO)_{15}PR \tag{194}$$

$$Cp(CO)_2MnAsCl_2Ph + Fe_2(CO)_9 \Rightarrow Fe_3(CO)_9(AsPh)_2 \tag{195}$$

$$Cp(CO)_2MnSbI_2Ph + Cp(CO)_2Mn(THF) \Rightarrow [Cp(CO)_2Mn]_2SbPh \tag{196}$$

$$Fe(CO)_4(PCl_2NEt_2) + Fe_2(CO)_9 \Rightarrow Fe_2(CO)_6(\mu\text{-}Cl)(\mu\text{-}PClNEt_2) + Fe_3(CO)_{10}PR + Fe_3(CO)_9(PR)_2 \tag{197}$$

Halogenation of E–M clusters is much less studied, but a few examples are known. C_2Cl_6 has been reported to halogenate $(CO)_4W\{PW(CO)_5\}_4$.[407] The product was found to be $[(CO)_2(NCMe)_2(Cl)W]P_4(Cl)\{W(CO)_5\}_3$ (**229**) when the reaction was conducted in MeCN. This product reacted with water to give the hydroxide species $[(CO)_2(NCMe)_2(Cl)W]P_4(OH)\{W(CO)_5\}_3$. The compound $[\{HFe_2(CO)_6Bi_2Cl_2\}^-]_\infty$ (**230**) arising in the reaction of $[Bi_4Fe_4(CO)_{13}]^{2-}$ with $MePCl_2$ [Eq. (198)[512]] is interesting in that in the solid state it adopts a polymeric framework connected via Bi–Cl–Bi bridges. In that reaction the P atom is not retained and polymeric phosphorus compounds are the by-product.

$$[Bi_4Fe_4(CO)_{13}]^{2-} + MePCl_2 \Rightarrow [Fe(CO)_3Bi_3Cl_8]^{3-} + [Fe(CO)_4\{Bi_2Cl_6\}]^{2-} + [HFe_2(CO)_6Bi_2Fe(CO)_4]^- + [\{HFe_2(CO)_6Bi_2Cl_2\}^-]_\infty \tag{198}$$

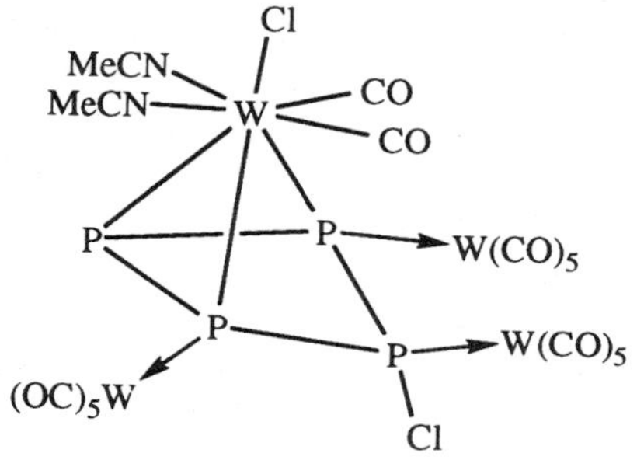

229

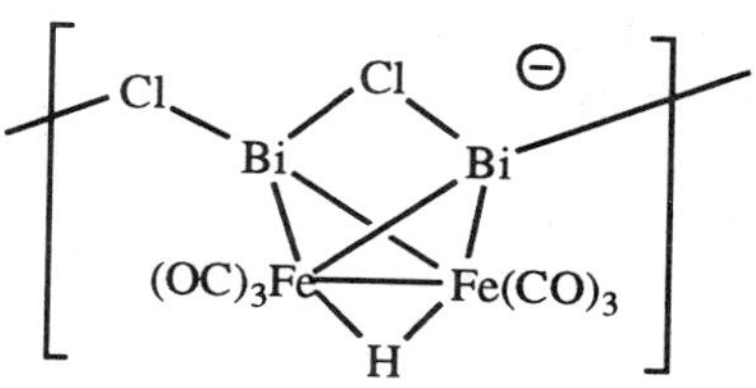

230

O. "Insertion" of Main Group Element Fragments into M–M Bonds

The effective insertion of E into the M–M bond of dimer has been observed in three cases as per Eqs. (199)–(201),[43,59] although it is probably simplistic to believe that the actual mechanism is a simple insertion process. In these particular reactions the counteranions that were isolated were actually EX_4^- or $Sb_2Cl_7^-$ owing to the Lewis acidic nature of EX_3 toward free halide ions. The reaction in Eq. (199) was not completely clean and the neutral compounds $Cp(CO)_2FeEX_2$ (E = As, Sb, Bi) and $[Cp(CO)_2Fe]_2SbX$ were also obtained.[43]

$$[Cp(CO)_2Fe]_2 + \underset{(E = As, Sb)}{EX_3} \rightarrow [\{Cp(CO)_2Fe\}_2EX_2]^+ X^- \quad (199)$$

$$(CF_3)_2SbI + [Cp(CO)_2Fe]_2 \rightarrow [\{Cp(CO)_2Fe\}_2Sb(CF_3)_2]^+ [(CF_3)_2SbI_2]^- \quad (200)$$

$$SbX_3 + Co_2(CO)_6(PPh_3)_2 \rightarrow [Sb\{Co(CO)_3(PPh_3)\}_4]_3[Sb_2Cl_9] \quad (201)$$

P. Miscellaneous Thermolysis and Photolysis Reactions

Thermolysis reactions are often unpredictable and can lead to a large variety of outcomes. Already discussed are cluster-closing reactions in which simple ligand loss occurs with the formation of M–M bonds. Most reactions are much more complex, however, and give a mixture of products. The mechanisms by and large remain unstudied. Often E–M, M–M, or E–C bonds are broken.

Pyrolysis of some E–M compounds containing naked E atoms has been examined. For these it is clear that a point is reached at which the compounds fragment into smaller E–M units, which then reassemble to form the final products. This is most clear for the pyrolysis of $Bi[Co(CO)_4]_3$. Under mild conditions, it has already been noted that it undergoes simple closure to yield the tetrahedral $BiCo_3(CO)_9$, but under prolonged heating $Bi[Co(CO)_4]_3$ yields a mixture of $[Bi_2(Co_4(CO)_{11}]^-$ and $[Co(CO)_4]^-$,[65] whereas heating the closed cluster $BiCo_3(CO)_9$ yields the cubane $Co_4(CO)_{12}Bi_4$.[84] Pyrolysis of $[Cp^x(CO)_3M]_3Bi$ likewise gives the multiply bonded $[Cp^x(CO)_2M]_2Bi_2$ (Cp^x = Cp, Cp^{1M})[47] showing a coupling of BiX_x fragments, whereas heating (or photolyzing) $[Fe(CO)_4]_3P_4$ causes elimination of CO and $Fe(CO)_5$ to produce $[Fe(CO)_3P_2]_n$.[463]

An empirical observation is that there is an alternative pathway available for some of these pyrolysis reactions that contrasts with simple loss of CO to form M–M bonds. In general, compounds $[L_nM]_3E$ with a lone pair undergo ligand loss to give M–M bond formation, whereas $[L_nM]_3EY$ or

$[L_nM]_3EML_n$ prefer to undergo more complicated coupling reactions in which two E atoms end up in the products [Eqs. (202)–(205)].[53,68,98] This coupling was induced photochemically for $[Sb\{Fe_2(CO)_8\}\{Fe(CO)_4\}_2]^-$. It appears that the presence of a fourth substituent on E may be the directing influence in the different pathways taken. The postulate[68] is that the presence of the fourth group forces the E atom to become sp^3 hybridized, whereas the compounds in which E retains a lone pair of electrons would have primarily p^3 hybridization (cf. inert pair effect). The effect is twofold. The first is geometric in that p^3 hybridization would force appended metal atoms closer together to achieve angles closer to 90°, promoting M–M bond formation where E has a lone pair. The second effect would introduce the differences in bond energies that result on going from p^3 to sp^3 hybridization.

$$[ClSb\{Fe(CO)_4\}_3]^{2-} \xrightarrow{\Delta} [Fe_3(CO)_9\{SbFe(CO)_4\}_2]^{2-} \quad (202)$$

$$[HAs\{Fe(CO)_4\}_3]^{2-} \xrightarrow{\Delta} [Fe_3(CO)_9\{AsFe(CO)_4\}_2]^{2-} \quad (203)$$

$$[Sb\{Fe_2(CO)_8\}\{Fe(CO)_4\}_2]^{-} \overset{h\nu}{\Rightarrow} [Fe_4(CO)_{12}\{SbFe(CO)_4\}_2]^{2-} \quad (204)$$

$$[ClBi\{Fe(CO)_4\}_3]^{2-} \xrightarrow{\Delta} [Bi_2Fe_4(CO)_{13}]^{2-} \quad (205)$$

Organo-E compounds and metal complexes of organo-E compounds are a good starting source of E and Er fragments for incorporation into E–M compounds [Eqs. (206)–(228)].[54,95,106,113,115,174,175,185,218–222,226,292,296,316,372,373,425,517,518] One can start with a metal complex and a free organo-E compound or with the organo-E species already attached to the metal(s). The products are so diverse that it is difficult to draw many conclusions about reaction pathways, except that E–Y bonds are easily cleaved and that higher nuclearity complexes are often formed. The yields are generally low and the products are often one of a number that have to be separated chromatographically. There is some indication that metal-complexed E–R units undergo E–C bond cleavage more readily than the parent organo E compound and many of the reactions involve further metal insertion into E–C or E–E bonds of already coordinated fragments. Pyrolysis of $[Rh_6(CO)_{15}N]^-$ under basic conditions yields the higher nuclearity $[HRh_{12}(CO)_{23}(N)_2]^{3-}$ cluster, which contains two isolated, interstitial nitrido atoms.[370] Thermolysis of cluster-bound phosphines and arsines has been a particularly well-examined area. Poly(phosphines) and poly(arsines) have been employed as discussed in Section III,O, where we saw that the E–E framework is reasonably robust and often persists in the products found. It is important to keep in mind that the thermolysis yields fragmentation that then results in reorganization and recombination. E–E coupling starting from complexes that do not already possess E–E bonds is known [Eqs. (206),[425] (226),[296] and (227)[175]].

$$Cp^*{}_2Sm + BiPh_3 \Rightarrow [Cp^*{}_2Sm]_2Bi_2 \quad (206)$$

$$Fe(CO)_4\{BiPh_2\}_2 \xrightarrow{\Delta} [PhBiFe(CO)_4]_2 \quad (207)$$

$$Re_2(CO)_{10} + Re_4Cl_2(CO)_{15}(P_3Me_3) \xRightarrow{h\nu \text{ or } \Delta} Re_6(CO)_{18}(\mu_4\text{-}PMe)_3 + Re_5(CO)_{14}(\mu_4\text{-}PMe)(\mu\text{-}PMe_2)(\mu_3\text{-}PRe(CO)_5) \quad (208)$$

$$Fe_2(CO)_9 + R(H)PCH_2P(R)H \Rightarrow HFe_3(CO)_9(H(R)PCH_2PR) + HFe_3(CO)_9(PRMe)(PR) \quad (209)$$

$(R = {}^iPr, CH_2Ph, Me, {}^tBu)$

$$HFe_3(CO)_9(PRMe)(PR) \xRightarrow{\Delta} Fe_3(CO)_9(PR)(PMe) \quad (210)$$

$$Ru_3(CO)_8(dppm)_2 \xRightarrow{\Delta} Ru_3(CO)_7((\mu\text{-}CHPPh_2)(dppm)(PPh) \quad (211)$$

$$Ru_3(CO)_{12} + PhP[C_6H_5Cr(CO)_3]_2 \xRightarrow{\Delta} Ru_3(CO)_9(C_6H_4Cr(CO)_3)\{PC_6H_5Cr(CO)_3\} \quad (212)$$

$$HRu_3(CO)_9(PPh_2) \xRightarrow{\Delta} Ru_4(CO)_{13}PPh + Ru_5(CO)_{16}(PPh_2)P \quad (213)$$

$$Os_3(CO)_{10}\{Fe(C_5H_4P^iPr_2)_2\} \xRightarrow{\Delta} HOs_3(CO)_7\{Fe(C_5H_4P^iPr_2)_2(C_5H_4)P\}({}^iPr) \quad (214)$$

$$Os_3(CO)_{11}\{PPh(C_5H_4FeCp)_2\} \xRightarrow{\Delta} Os_3(CO)_9(R)(PC_5H_4FeCp) \quad (215)$$

$(R = C_6H_4, C_5H_3FeCp)$

$$Os_3(CO)_{11}\{P(OMe)_3\} \xRightarrow{\Delta} Os_5(CO)_{15}(POMe) \quad (216)$$

$$[HOs_3(CO)_{11}][HOs_3(CO)_{10}]PH \xRightarrow{\Delta} HOs_6(CO)_{18}(\mu_6\text{-}P) \quad (217)$$

$$[\{Os_3(CO)_{11}\}\{HOs_3(CO)_{10}\}PH]^- \xRightarrow{\Delta} [Os_6(CO)_{18}(\mu_6\text{-}P)]^- \quad (218)$$

$$Ir_4(CO)_{12} + PPh_3 \Rightarrow Ir_4(CO)_6(PPh_3)_4(PPh) \quad (219)$$

$$Os_3(CO)_{11}(NCMe) + As(Tol)_3 \Rightarrow Os_3(CO)_9(C_6H_3Me)(As\text{-}Tol) \quad (220)$$

$$Fe_3(CO)_{12} + RP(S)S\text{-}SP(S)R \Rightarrow Fe_3(CO)_9(S)(PR) \quad (221)$$

$$Cp_2Ni + S(N^tBu)_2 \Rightarrow Cp_3Ni_3(\mu_3\text{-}N^tBu) \quad (222)$$

$$Ru_3(CO)_{12} + {}^tBu_2AsNSNAs^tBu_2 \Rightarrow H_2Ru_3(CO)_8(NSNAs^tBu_2)(\mu_3\text{-}As^tBu) \quad (223)$$

$$Fe_3(CO)_{12} + {}^tBu_2AsNSNAs^tBu_2 \Rightarrow Fe_3(CO)_9(S)(\mu_3\text{-}As^tBu) \quad (224)$$

$$Os_3(CO)_{12} + {}^tBu_2AsNSNAs^tBu_2 \Rightarrow HOs_3(CO)_8(NSNAs^tBu_2)(\mu_3\text{-}As^tBu) \quad (225)$$

$$Os_3(CO)_{12} + S(NSiMe_3)_2 \Rightarrow Os_3(CO)_9(S)(\mu_3\text{-}NSiMe_3) \quad (226)$$

$${}^tBuP\{Cr(CO)_5\}_2 \xRightarrow{100°C} Cr_3(CO)_{10}(\mu_3\text{-}P^tBu)(P_2{}^tBu_2) \quad (227)$$

$$RP\{Cr(CO)_5\}_2 \Rightarrow ({}^tBu\{Cr(CO)_5\}P{=}P\{Cr(CO)_5\}{}^tBu)\{\mu\text{-}Cr(CO)_5\} \quad (228)$$

Some use has been made of severe reaction conditions in which a metal salt or complex is reduced in the presence of a main group element compound [Eqs. (229)–(233)][83,375,378,384,385] to give extremely robust compounds. Unfortunately, little of the subsequent chemistry of these high-nuclearity species has been reported, except for spectroscopic investigations of the rhodium compounds that have shown the metal frameworks to be fluxional as well as the ligand complement.

$$Co(OAc)_2 \xRightarrow{150°C,\ MeOH;\ SbCl_3} Co_4(CO)_{12}Sb_4 \quad (229)$$

$$Rh(CO)_2(\text{acetylacetonate}) \xRightarrow{CO/H_2\ (400\ atm);\ 140-160°C;\ [Cs]\text{benzoate};\ PPh_3} [Rh_9(CO)_{21}P]^{2-} \quad (230)$$

$$Rh(CO)_2(\text{acetylacetonate}) \xRightarrow{CO/H_2\ (400\ atm);\ 140-160°C;\ [Cs]\text{benzoate};\ PPh_3;\ NaBH_4} [Rh_{10}(CO)_{22}P]^{3-} \quad (231)$$

$$Rh(CO)_2(\text{acetylacetonate}) \xRightarrow{[M]\text{benzoate};\ (M = \text{alkali metal});\ AsPh_3} [Rh_{10}(CO)_{21}As]^{2-} \quad (232)$$

$$Rh(CO)_2(\text{acetylacetonate}) \xRightarrow[]{\text{[Cs]benzoate, } SbPh_3} [Rh_{12}(CO)_{27}Sb]^{3-} \qquad (233)$$

Photolysis reactions proceed via creation of an unsaturated metal fragment via either ligand loss (often CO) or M–M bond cleavage. Thus, photolysis of square pyramidal $Ru_4(CO)_{13}PPh$ (**231**) with SCH_2CH_2, $SePEt_3$, or $TePEt_3$ resulted in addition of a chalcogen atom to the cluster core to give octahedral *trans*-$Ru_4(E)(PPh)$ [**232**; Eq. (234)].[314] It is interesting that the chalcogen atom does not end up capping the existing open face. This would have resulted in the formation of a P–S bond. Photolysis of the open compound $Bi[Mo(CO)_2Cp]_3$ might be expected to result in CO elimination and the formation of a *closo*-tetrahedral cluster as has been observed for a number of other $E\{ML_n\}_3$, but the major product from that reaction, in addition to $Cp_2Mo_2(CO)_6$, was $Cp_2Mo_2(CO)_4Bi_2$.[404]

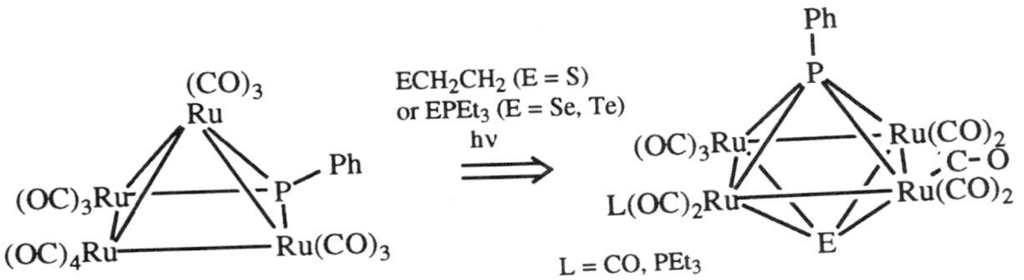

(234)

231 **232**

Q. *Substitution Reactions*

In many cases the metal centers in E–M complexes function in the same way as any other transition-metal center in that the ligands can be readily substituted. Substitution can be achieved directly and, for E–M cluster compounds, as noted in Section III,H, may well occur via an associative mechanism in which M–M or E–M bonds are broken and then reformed upon loss of the ligand being displaced. Most often the substituting ligands are phosphines or phosphites, but H_2/CO and NO/CO exchanges have also been examined. In a few cases the substitution process has been examined in some detail. The tetrahedral complex $H_3Ru_3(CO)_9Bi$ undergoes a number of sequential substitution reactions as per Eq. (235), as do the tetrahedral E_2M_2 compounds [Eq. (236)].[138,387,399,400]

$$H_3Ru_3(CO)_9Bi + L \Rightarrow H_3Ru_3(CO)_{9-x}L_xBi \qquad (235)$$

$(L = PPh_3, As(Tol)_3, SbPh_3, P(OMe)_3; x = 1 - 3))$

$$Co_2(CO)_6As_2 \xrightarrow{+PPh_3,\ -CO} Co_2(CO)_5(PPh_3)As_2 \xrightarrow{+L,\ -CO} Co_2(CO)_4(PPh_3)_2As_2 \quad (236)$$

Substitution reactions for the $Fe_3(CO)_9\{PX\}_2$ system have been particularly well studied. Nitrile derivatives were prepared chemically and photochemically and used to prepare a number of products substituted with PF_3, $P(OR)_3$, PR_3, and AsR_3.[284] The mechanism has been examined (X = Ph, $Mn(CO)_2Cp$); an electron-transfer catalyzed pathway was found, which is common to many of these E–M systems.[259,270,286–288] The substitution rates were found to be highly influenced by the nature of the capping group as determined by a comparison with the structurally related compounds $Fe_3(CO)_9E_2$ (E = S, Se, NPh).[270] For the latter examples, the rate-determining step was found to be dissociation of CO, whereas for E = PPh the rate-determining step was breakage of an Fe–P bond. An interesting feature of that substitution chemistry occurs using X = $Mn(CO)_2Cp$; L = $P(OMe)_3$; the product is $Fe_3(CO)_6\{P(OMe)_3\}_3\{PMn(CO)_2Cp\}\{P\}$, where one of the $Mn(CO)_2Cp$ groups has been eliminated, presumably due to steric crowding.[259] Compounds with a naked μ_3-P ligand are rare due to the high basicity of the P atom. The analogous $Ru_3(CO)_9(NPh)_2$ also undergo ETC catalyzed substitutions.[276]

Substitution of the hetero-main group element compounds $Fe_3(CO)_9(E)(PR)$ has been examined for isonitriles, phosphines, and phosphites. In the case of $P(OMe)_3$, in the initial product the ligand occupies a site on the apical Fe atom but this slowly rearranges to an equatorial location.[295] The related compound $Fe_3(CO)_9(\mu_3\text{-}CO)(NSiMe_3)$ undergoes three successive substitution reactions with PR_3 (R = ^{n}Bu, OMe) upon irradiation.[149] Each substitution occurs at a different iron center. Similarly, $H_2Fe_3(CO)_9(NSiMe_3)$ undergoes photoassisted substitution by PR_3 (R_3 = $(OMe)_3$, $(Omenthyl_3$, Ph_2menthyl, Et_2menthyl) giving mono- and disubstituted products.

The $M_4(CO)_x(PR)_2$ (M = Fe, Ru; x = 11, 12) is an interesting case for substitution in that the unsaturated x = 11 compounds will readily add a variety of ligands to give the "saturated" compounds $M_4(CO)_{11}L(PR)_2$, which then loose CO under vacuum to generate substituted "unsaturated" compounds.[308] The substitution reaction using bifunctional donor ligands has been used to create linked cluster units.[319] Substitution reactions of the related cluster system $Co_4(CO)_{10}(PPh)_2$ have also been reported.[325,326] Like several other clusters noted previously, substitution reactions of $Co_4(CO)_{10}(PPh)_2$ are catalyzed by electron-transfer reactions.

The nitrido clusters $[Fe_4(CO)_{12}N]^-$ and $Fe_4(CO)_{11}(NO)N$ undergo substitution reactions yielding mono- and disubstituted products.[359] The phosphines used (PPh_3 and PMe_2Ph) preferred to bond to the wingtip irons in the metal butterfly.

"Substituted" products may be obtained via routes other than direct substitution reactions. Presubstituted metal fragments may be employed in an appropriate synthetic scheme [Eqs. (237)–(239)]. In addition to normal two-electron donor ligands, H_2 can replace CO, sometimes with little change in the metal framework [Eqs. (240) and (241)].

$$BiCl_3 + 2\ Na[Mo(CO)_2(CN^tBu)Cp] \rightarrow [Cp(CO)_2(CN^tBu)Mo]_2BiCl + 2\ Cl^- \quad (237)$$

$$BiCl_3 + 3\ [Co(CO)_3PPh_3]^- \rightarrow Bi[Co(CO)_3PPh_3]_3 + 3\ Cl^- \quad (238)$$

$$[HRu_3(CO)_9(PPh)]^- + [Rh(CO)_3(PPh_3)_2]^+ \Rightarrow H_2Ru_3(CO)_8(PPh_3)(PPh) \quad (239)$$

$$Fe_3(CO)_{10}(NSiMe_3) + H_2 \xrightarrow{h\nu} H_2Fe_3(CO)_9(NSiMe_3) + CO \quad (240)$$

$$H_2 + Ru_4(CO)_{13}(PPh) \rightarrow H_2Ru_4(CO)_{12}(PPh) + CO \quad (241)$$

In some cases the ensuing reactivity of the complex is changed upon substitution. For example, the $Bi[Co(CO)_3L]_3$ L = phosphine species have exhibited no tendency to form *closo*-tetrahedral molecules like the parent L = CO complex. Presumably, the steric crowding imposed by the phosphines prevents the formation of $BiCo_3(CO)_6L_3$ species.

NO is most often a three-electron donor ligand. It can be used as the neutral molecule and in simple coordination chemistry can substitute for CO readily. Usually two NO ligands will displace three CO ligands as in the $Fe(CO)_5$ and $Fe(CO)_2(NO)_2$ pair. A common method of producing NO-substituted compounds, however, is not the direct reaction of NO with metal fragments, but rather the use of NO^+ salts (BF_4^- or PF_6^- are commonly employed) as in Eq. (242).[357] For metal carbonylate anions this provides a convenient methodology. Notice that CO is still displaced in these reactions.

$$[Fe_4(CO)_{12}N]^- + NO^+ \rightarrow Fe_4(CO)_{11}(NO)N + CO \quad (242)$$

R. *Brønsted Acid/Base Properties*

Protonation reactions may be simply H^+ addition in which the E–M framework of a cluster complex remains essentially unaltered. Where simple protonation is observed, conventional acid–base chemistry is involved and the site of protonation is almost always the electron-rich metal center, though protonation can sometimes be observed at E. Likewise, addition of bases to metal cluster hydrides results in the deprotonation [e.q., Eq. (243)[179]]. Sometimes even halide ions in nonaqueous media are strong enough to deprotonate the metal centers [Eqs. (244)–(246)[95,203,373]]. Osmium phosphinidene complexes such as those outlined in Scheme 7 (Section III,C) are readily deprotonated with bases such as NEt_3 or even with Cl^- in large excess using an appropriate solvent.[94] Norton has determined pK_a values for a number of metal systems.[180,184] Of relevance here are the values for $H_2Fe_3(CO)_9(PR)$ (R = ^{t}Bu, C_6H_5OMe), which were found to be 11.4 and 9.0, respectively. The pK_a values were noted to be extremely sensitive to the nature of the capping group. These values were considerably more acidic than those found for several mononuclear hydrides examined in the same study.

$$H_2Fe_3(CO)_9PR \xrightarrow[-HOAc]{+AcO^-} [HFe_3(CO)_9PR]^- \xrightarrow[-NH_3]{+NaNH_2/60°C} [Fe_3(CO)_9PR]^{2-} \quad (243)$$

$$\underset{(R = Ph,\ ^tBu)}{Ru_3(CO)_{10}(NCMe)_2} + H_2PPh \rightarrow HRu_3(CO)_{10}(PPhH) + [PPN]Cl$$
$$\Rightarrow [PPN][HRu_3(CO)_9PPh] + HCl + CO \quad (244)$$

$$\underset{(X = Cl^-, Br^-)}{[HOs_3(CO)_{11}][HOs_3(CO)_{10}]PH} + X^- \rightleftarrows [\{Os_3(CO)_{11}\}\{HOs_3(CO)_{10}\}PH]^- + HX \quad (245)$$

$$\underset{(X = Cl^-, Br^-)}{HOs_6(CO)_{18}(\mu_6\text{-}P)} + X^- \rightleftarrows [Os_6(CO)_{18}(\mu_6\text{-}P)]^- \quad (246)$$

The major site of protonation of $[M_4(CO)_{12}N]^-$ occurs at the M atoms, but a minor products shows vertex elimination and protonation at N [Eqs. (247)–(249)].[145,155,357] Protonation of $[FeRu_3(CO)_{12}(NO)]^-$ induces the reduction of NO with formation of a nitrido cluster. The site of protonation in the final product is again at the metal centers.

$$[Fe_4(CO)_{12}N]^- + H^+ \rightarrow HFe_4(CO)_{12}N + Fe_3(CO)_{10}NH \quad (247)$$

$$[FeRu_3(CO)_{12}(NO)]^- + H^+ \Rightarrow HFeRu_3(CO)_{12}N \quad (248)$$

$$[FeRu_3(CO)_{12}N]^- \xRightarrow{CO,\ H^+} FeRu_2(CO)_{10}(NH) + [FeRu_4(CO)_{14}N]^- + HFeRu_3(CO)_{12}N \quad (249)$$

$$[Ru_4(CO)_{12}N]^- \xRightarrow{CO,\ H^+} Ru_3(CO)_{10}(NH) + [Ru_5(CO)_{14}N]^- + HRu_4(CO)_{12}N \quad (250)$$

In many cases, however, protonation reactions, especially of the anionic clusters, result in oxidation and/or concomitant fragmentation and reorganization of the cluster. The pathways are complicated and not understood. In the protonation of $[E\{Fe(CO)_4\}_4]^{3-}$(E = Sb, Bi) there appear to be two available pathways. One pathway is simple protonation, which is also observed to promote closure and M–M bond formation. The other is an oxidation "dimerization" pathway such that the final products end up with two main group element groups. The protonation pathway dominates for E = Sb, whereas for Bi the oxidation process is apparently more favorable.[70,263] Upon careful protonation of $[Sb\{Fe(CO)_4\}_4]^{3-}$, M–M bond formation occurs to yield first $[HFe_3(CO)_9SbFe(CO)_4]^{2-}$ and subsequently $[H_2Fe_3(CO)_9SbFe(CO)_4]^{-}$.[70] If the protonation occurs too rapidly, however, the product is $[Fe_3(CO)_9\{SbFe(CO)_4\}]^{2-}$, representing the oxidation pathway.[70] Similarly, protonation of $[Bi\{Fe(CO)_4\}_4]^{3-}$ gives a mixture of $H_3Fe_3(CO)_9Bi$ and $Bi_2Fe_3(CO)_9$.[134,263] The latter dominates with the hydride being obtained in only small yields. When protonated, $[HAs\{Fe(CO)_4\}_3]^{2-}$ gives $[Fe_2(CO)_6][\mu\text{-}AsFe_2(CO)_8]_2$, made up of two spirocyclic As groups.[68] It is interesting to compare these oxidative protonations to the reaction of $[Cu(NCMe)_4][BF_4]$ with $[ClSb\{Fe(CO)_4\}_3]^{2-}$, which yields $Fe_2(CO)_6[\mu\text{-}Sb\{Fe_2(CO)_8\}]_2$.[122]

For the lighter elements, which are more basic, protonation/deprotonation may take place at the main group element [Eqs. (251) and (252)].[235,279] The H atoms attached to phosphorus in $Fe_3(CO)_9\{PH\}_2$ and $[Fe_3(CO)_9(PH)(PMe)]^-$ are displaced by chloride when the compound is treated with CCl_4.[279]

$$Fe_3(CO)_9\{PH\}_2 \xrightarrow{NEt_3} [Fe_3(CO)_9\{PH\}P]^- \xrightarrow{LiBu} [Fe_3(CO)_9P_2]^- \quad (251)$$

$$Cp^*_2Zr(P_2H_2) \xrightarrow{KH} [K(THF)_x]_2[Cp^*_2ZrP_2] \quad (252)$$

The protonation of $[As\{Fp\}_4]^+$ by water accomplishes the elimination of FpH, giving $[HOAs\{Fp\}_3]^+$.[69] The ammonolysis of $Cp^*Ti(NMe_2)_3$ yields $Cp^*_4Ti_4N_4$ with the elimination of $HNMe_2$.[76]

S. *Oxidation/Reduction Processes*

Many of the E–M cluster complexes are versatile at undergoing oxidation and/or reduction reactions. As noted in the previous section, protonation of many E–M complexes, especially the anionic ones, may result in oxidation of the compound. A common outcome is simple E–M or M–M bond cleavage. This is illustrated in Eq. (253). It is this process that is responsible for the production of $[Bi\{Co(CO)_4\}_4]^-$ upon treatment of $Bi[Co(CO)_4]_3$ with Cp_2Co as in Eq. (254).[64,75] In a few cases, oxidation or reduction proceeds with internal bond breakage or formation as anticipated by the change in electron count. For example, oxidation of $[Sb\{Fe(CO)_4\}_4]^{3-}$ by two electrons forms one Fe–Fe bond [Eq. (255)]. Similarly, for $Fe_3(CO)_9\{PMn(CO)_2Cp\}_2$, reduction breaks an Fe–Fe bond to go from a trigonal bipyramidal to a square pyramidal core [Eq. (256)].[259,260] The redox potentials of mono- and disubstituted derivatives with L = $P(OMe)_3$ and PMe_3 were reported. The concentration of the radical anions in solution is controlled by their disproportionation constants. Consistent with the ligand addition/elimination chemistry, $M_4(CO)_{11}(PR)_2$ (M = Fe, Ru) are formally unsaturated (seven skeletal electron pairs) and add electrons cleanly in place of L. Thus, the mono- and dianionic compounds have been prepared that retain the basic octahedral core structure.[320] The radical monoanion has been trapped by reaction with $[Et_3O]BF_4$.[321] The product is $Fe_4(CO)_{10}(COEt)(PR)_2$, in which the oxygen of bridging carbonyl was found to be alkylated. Similar chemistry is found for the heterometallic clusters $Cp^*RhFe_3(CO)_x(PR)_2$ ($x = 8, 9$).[309] The $x = 8$ molecules readily undergo reversible one- and two-electron reductions, whereas the $x = 9$ species can be reversibly oxidized by one or two electrons. The $Co_4(CO)_{10}(PR)_2$ clusters possess eight skeletal pairs and, as expected, undergo oxidation reactions that become successively easier with phosphine substitution. A radical anion has also been observed in this system, and it is substitutionally labile.[325,326]

$$Bi[Co(CO)_4]_3 + 3\,Cp_2Co \rightarrow Bi(s) + 3\,[Cp_2Co][Co(CO)_4] \qquad (253)$$

$$[Co(CO)_4]^- + Bi[Co(CO)_4]_3 \rightarrow [Bi\{Co(CO)_4\}_4]^- \qquad (254)$$

$$[Sb\{Fe(CO)_4\}_4]^{3-} \xrightarrow{+\ 2\,[Cu(NCMe)_4][BF_4]} [Sb\{Fe_2(CO)_8\}\{Fe(CO)_4\}_2]^- + 2\,Cu^0 \qquad (255)$$

$$Fe_3(CO)_9\{PMn(CO)_2Cp\}_2 + 2\,e^- \rightarrow [Fe_3(CO)_9\{PMn(CO)_2Cp\}_2]^{2-} \qquad (256)$$

Oxidation or reduction of EM compounds may promote fragmentation/reorganization reactions. This is observed in the reduction of $BiCo_3(CO)_9$, which gives $[Bi_2Co_4(CO)_{11}]^-$ as the major product along with $[Co(CO)_4]^-$.[303] The compound contains a central Bi_2Co_2 core capped on the $BiCo_2$ faces by $Co(CO)_3$ groups. The compound $[Bi_2Co_4(CO)_{11}]^-$ undergoes a further one-electron reduction to give a dianion with presumably the same core structure. For comparison, the attempted production of $Sb[Co(CO)_4]_3$ leads directly to a mixture of $[Sb_2Co_4(CO)_{11}]^{1-/2-}$ via what must be a disproportionation reaction. The structure determinations of both mono- and dianions have been accomplished. The metal core structure is retained, with the major perturbation being an elongation of the Co–Co bond that is part of the central Co_2Sb_2 tetrahedron. Extensive fragmentation and reorganization is also observed when a reaction solution made by the addition of

$Mo(CO)_3$(toluene) to $[Bi_2Co_4(CO)_{11}]^-$ oxidizes slowly.[337] Here the products are the large cubic $[Bi_4Co_9(CO)_{16}]^{2-}$ and $[Bi_8Co_{14}(CO)_{20}]^{2-}$.

Some reactions that appear otherwise may proceed via redox processes. For example, the reaction of $[E\{Fe(CO)_4\}_4]^{3-}$ with EX_3 leads to the formation of $[XE\{Fe(CO)_4\}_3]^{2-}$ [Eq. (257)].[70] At face value this would appear to be a simple substitution of $[Fe(CO)_4]^{2-}$ by X^-, but the reaction does not proceed if the halide source cannot undergo redox chemistry. Conventional $[NR_4]X$ and [PPN]X do not promote the exchange reaction while $TlCl_3$ will.

$$[E\{Fe(CO)_4\}_4]^{3-} + EX_3 \Rightarrow [XE\{Fe(CO)_4\}_3]^{2-} \quad (E = Sb, Bi) \tag{257}$$

An electrochemical analysis has been carried out on the pseudo-cubane cluster $[Cp^*Ti]_2P_6$.[486] It was found to undergo an electrochemically irreversible oxidation at $E_{pa} = 0.70$ V and a reversible reduction at -1.31 V. The unpaired electron in the radical anion appears largely localized on the metal centers. The difference in oxidation and reduction potentials was found to be considerably larger than for the triple-decker complexes $[Cp^*M]_2P_6$ (M = Mo, W, V), which also contained an E_6 group but one that is planar.[487]

Oxidation can lead to E–E bond formation. Deprotonation of the bridging phosphido ligands in $Fe_2(CO)_6(PHR)_2$ gives dianions $[Fe_2(CO)_6(PR)_2]^{2-}$, which upon treatment with the mild oxidants undergo coupling of the P atoms [Eqs. (258) and (259)].[107,393,394] Similar coupling of phosphide ligands occurs upon mild oxidation of $Li_2[Cp(CO)_2MnPPh]$ by $cyNCl_2$, where the product has been found to be $[Cp(CO)_2Mn]_3P_3Ph_3$.[244] Similar oxidation of the arsenide complex $Li_2[(CO)_5CrAsPh]$ with $cyNCl_2$ yields instead the arsinidene species $[(CO)_5Cr]_2AsPh$.[250,251] Another possible outcome is site-specific oxidation at the main group element center as exemplified by the conversion of the phosphido ligands in $[(Cp^{4P})Ni]_2W(CO)_4P_2$ into P=O units by treatment with bis(trimethylsilyl)peroxide [Eq. (260)].[264]

$$Fe_2(CO)_6\{P(R)H\}_2 \xrightarrow{1.\ MeLi;\ 2.\ C_2H_4Br_2} [Fe_2(CO)_6]_2\{P_2R_2\}_2 \quad (R = Me, {}^tBu, Ph, Tol) \tag{258}$$

$$Fe_2(CO)_6\{P(Ph)H\}_2 \xrightarrow{1.\ MeLi;\ 2.\ SOCl_2} [Fe_2(CO)_6]_2\{P_2Ph_2\}_2 + Fe_3(CO)_9(PPh)(S) \tag{259}$$

$$[(Cp^{4P})Ni]_2W(CO)_4P_2 + Me_3SiO_2SiMe_3 \Rightarrow [(Cp^{4P})Ni]_2W(CO)_4(PO)_2 \tag{260}$$

Higher nuclearity systems such as the cubic M_8 and icosahedral $M_{12-x}E_x$ undergo reduction/oxidation reactions owing to their open shell nature and resulting flexibility in electron count [Eqs. (261)–(263)].[330,331] The reduction of $Ni_8(CO)_{12}(PMe)_2(P^tBu)_2$ induces a cluster rearrangement from a cuneane to an icosahedral geometry.

$$Ni_8Cl_4(PPh_3)_4(PPh)_6 + 4\ Na(Hg) \rightarrow Ni_8(PPh_3)_4(PPh)_6 + 4NaCl \qquad (261)$$

$$Ni_8Cl_4(PPh_3)_4(PPh)_6 + 4\ Na(Hg) + NiCl_2 \rightarrow Ni_9(PPh_3)_4(PPh)_6 + 4NaCl \qquad (262)$$

$$Ni_8(CO)_{12}(PMe)_2(P^tBu)_2 + 2\ e^- \rightarrow [Ni_8(CO)_{12}(PMe)_2(P^tBu)_2]^{2-} \qquad (263)$$

T. *Reactions with Organic Substrates*

Reactions of metal carbonyls with a variety of nitrogen-containing organic compounds—amines, azides, azoalkanes, diazo compounds, nitriles, and nitro compounds—produce either as intermediates or final products metal-bound NR fragments [some examples are given in Eqs. (264)–(267)[99,171,265,266]]. Nitroethane reacts with $Fe_3(CO)_{12}$ to yield a mixture of products, among which are $Fe_3(CO)_9(NEt)(NH)$ and $Fe_3(CO)_{10}(NEt)$.[151] Triiron dodecacarbonyl will reduce a variety of nitrobenzenes to the corresponding anilines via the intermediacy of $[HFe_3(CO)_{11}]^-$.[268] The reaction proceeds without destroying a number of functional groups. The methodology works when the $Fe_3(CO)_{12}$ is supported on silica, where it is also converted to $[HFe_3(CO)_{11}]^-$,[152] or under phase-transfer conditions.[271] Metal-bound nitrene intermediates were proposed for this reaction. Later work showed that $[HFe_3(CO)_{11}]^-$ reacts with ArNO and $ArNO_2$ complexes via an electron transfer process to give $[HFe_3(CO)_9NAr]^-$.[153] In much of this work, an attempt has been made to develop a catalytic cycle for converting the nitrene ligands into isocyanates, but the usefulness has been limited by the high stability of the cluster nitrene complexes.[164,168] Azomethane and azobenzene react with $Fe_2(CO)_9$ giving $Fe_2(CO)_6(NR)_2$ (R = Me, Ph), among other products.[388,519,520] Diazo compounds react similarly, giving a mixture of cluster complexes, among which was $Fe_3(CO)_9(N{-}N{=}CPh_2)_2$.[272] The reduction of nitriles by iron carbonyls on triangular clusters can also produce nitrene ligands. For example, hydrogenation of $HFe_3(CO)_9(N{=}CHR)$ (R = Me, Ph) yields $H_2Fe_3(CO)_9NCH_2R$.[154] Ethyl diazoacetate reacts with $[Cp(CO)_2Mo]_2$, retaining an N atom as a metal-bound nitrido ligand [Eq. (267)[99]].

$$Fe_2(CO)_9 + MeNO_2 \Rightarrow Fe_3(CO)_9(NMe)_2 \qquad (264)$$

$$Fe_2(CO)_9 + \underset{(R = Me,\ Ph)}{RN_3} \Rightarrow Fe_3(CO)_9(NR)_2 \qquad (265)$$

$$Fe_3(CO)_{12} + PhNO_2 \xrightarrow[Co_2(CO)_8]{benzene} FeCo_2(CO)_9(NPh) \quad (266)$$

$$[Cp(CO)_2Mo]_2 + N_2CHCO_2Et \xrightarrow{\Delta} Cp_3Mo_3(CO)_4(O)(N) \quad (267)$$

Similar chemistry to the iron system is observed for $M_3(CO)_{12}$ (M = Ru, Os) [Eqs. (268)–(271)].[156,158,159,165,171,173,269,275–278,318] The metal carbonyl species are dominated by the presence of the $M_3(\mu_3\text{-NR})$ groups. The ruthenium reactions were found to be promoted by addition of halides and cyanide, which is thought to arise by increasing the lability of CO.[156,164] The initially formed species were $[Ru_3(CO)_xX]^-$ ($x = 10, 11$). Use of acetonitrile-substituted clusters also rapidly resulted in formation of μ_3-NR compounds via deoxygenation reactions. This was expected, as the NCMe ligands are weakly bound and labile. Using the acetonitrile activation approach, a number of heterometallic nitrene complexes were prepared. Catalytic carbonylation of $PhNO_2$ to yield PhNCO has been examined using a variety of Ru_xNR species.[165] The species examined included $Ru_3(CO)_{10}NPh$, $HRu_3(CO)_{10}(NPhH)$, $H_2Ru_3(CO)_9NPh$, $[HRu_3(CO)_9(PhNCO)]^-$, $[HRu_3(CO)_9NPh]^-$, and $[Ru_3(CO)_9(PhNCO)]^-$. Heating $HOs_3(CO)_{10}(NHPh)$ promotes the elimination of CO and the subsequent insertion of Os into the N—H bond, giving $H_2Os_3(CO)_9NPh$.[173] Benzonitrile reacts with $Ru_3(CO)_{12}$ and H_2 at 130°C to give $HRu_3(CO)_{10}\{N{=}C(H)Ph\}$.[169] Prolonged hydrogenation of this product converts it in 49% yield into the nitrene complex $H_2Ru_3(CO)_{10}NCH_2Ph$.[169] A nitrile ligand in $[H_3Os_4(CO)_{12}(NCMe)_2]^+$ is converted into a μ_3-amido group upon reaction with $[PPN]NO_2$.[102] The product was found to be $[HOs_4(CO)_{12}(\mu_3\text{-NC(O)Me})]^-$, which reacts with Ph_3PMX to give $HOs_4(CO)_{12}(\mu_3\text{-MPPh}_3)(\mu_3\text{-NC(O)Me})$ (M = Cu, Au).

$$Ru_3(CO)_{12} + PhNH_2 \Rightarrow H_2Ru_3(CO)_9NPh + Ru_3(CO)_9(NPh)_2 + Ru_3(CO)_{10}NPh + HRu_3(CO)_9(NHPh) \quad (268)$$

$$Ru_3(CO)_{12} + PhNO_2 \Rightarrow H_2Ru_3(CO)_9NPh + Ru_3(CO)_9(NPh)_2 + Ru_3(CO)_{10}NPh + HRu_3(CO)_9(NHPh) \quad (269)$$

$$Ru_3(CO)_{12} + PhNO_2 \xrightarrow[Co_2(CO)_8]{benzene} Ru_3(CO)_7(C_6H_6)(NPh) \quad (270)$$

$$Ru_3(CO)_{11}(NCMe) + PhNO \rightarrow Ru_3(CO)_{10}(NPh) + CO_2 + NCMe \quad (271)$$

Fischer carbene formation can be carried out with $Fe_3(CO)_9(N_2Et)_2$ or $Fe_3(CO)_9(NEt)_2$ by the conventional methods of reaction first with LiR followed by alkylation.[267] The products are **233** and **234**, respectively. The $Fe_3(CO)_{10}(PR)$ system may also be derivatized similarly.[197] Two types of product are observed. One has the carbene bonded in a terminal fashion (**235**) to one Fe center, whereas in the other the carbene ligand adopts a μ_3-η^2-configuration (**236**). Ruthenium nitrene complexes may also be treated with LiR to give a cluster-bound acyl ligands (Scheme 9).[156] Protonation of the metal framework occurs to generate the hydrido derivative, which under CO pressure eliminates PhNHC(=O)R.

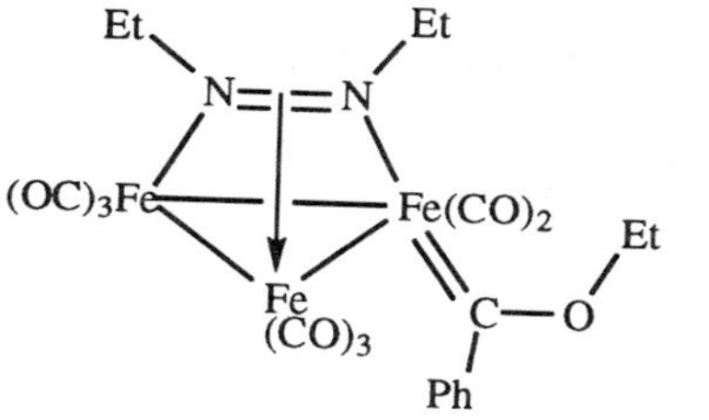

233 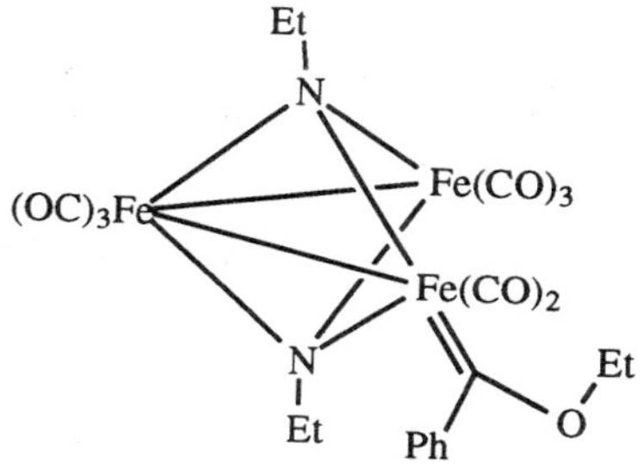 **234**

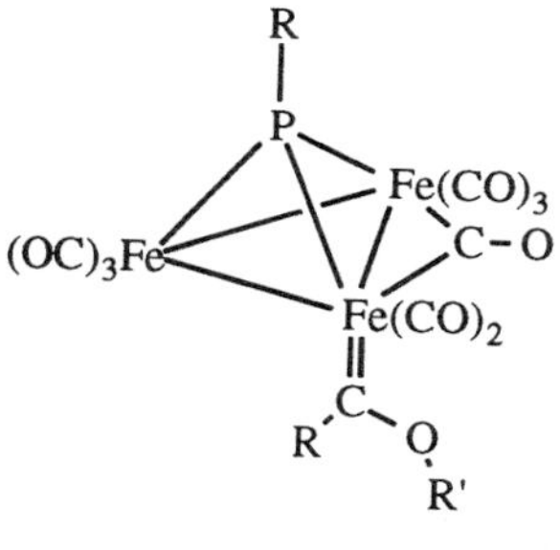

235 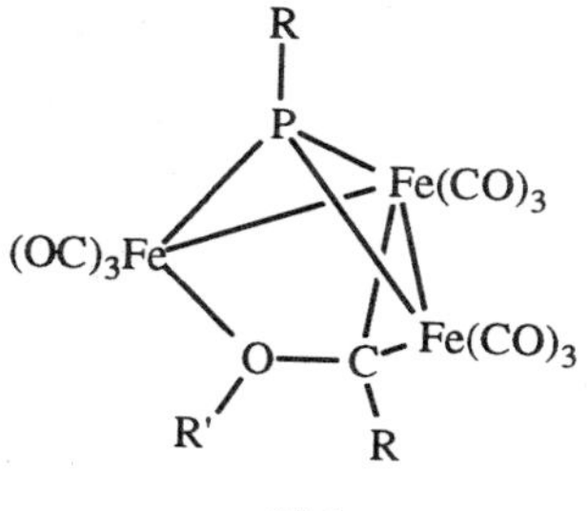 **236**

Carbon monoxide and alkynes can be coupled with nitrene ligands.[160] The presence of halide ligands promotes the carbonylation of μ_3-NR ligands to form isocyanates [Eq. (272)].[156] The major product of addition of alkynes PhCCR to $Ru_3(CO)_{10}(NPh)$ is a dinuclear metallapyrrolidone complex $Ru_2(CO)_6(\mu_2,\eta^3PhC{=}CRC(O)NPh)$ (R = Me, Ph), along with other minor coupling products.[160] Azoarenes ArN=NAr are cleaved using $H_2Ru_3(CO)_9(NAr)$ to give aryl isocyanates and $Ru_3(CO)_9(NAr)_2$.[161] With phenylacetylene, $Fe_3(CO)_9(N_2Et)_2$ can be induced to add one equivalent upon activation by Me_3NO to give **237**. A second equivalent will add at 110°C, and two isomeric products **238a** and **238b** result. It was found that

$$Ru_3(CO)_{10}NPh \; + \; LiR \; \rightarrow \; Li^+[Ru_3(CO)_9\{C(=O)R\}NPh]^-$$

$$\Downarrow H^+$$

$$HM_3(CO)_9\{C(=O)R\}(NPh) + Li^+$$

$$\Downarrow CO$$

$$PhNHC(=O)R$$

SCHEME 9

the tetramethyl analogue to **238b** could be obtained by irradiation of the tetramethyl pyrrole complex $Fe_3(CO)_8(C_4Me_4)$ with azoethane.

$$[Ru_3(CO)_9(X)NR]^- \overset{CO}{\Rightarrow} O{=}C{=}N{-}R \tag{272}$$

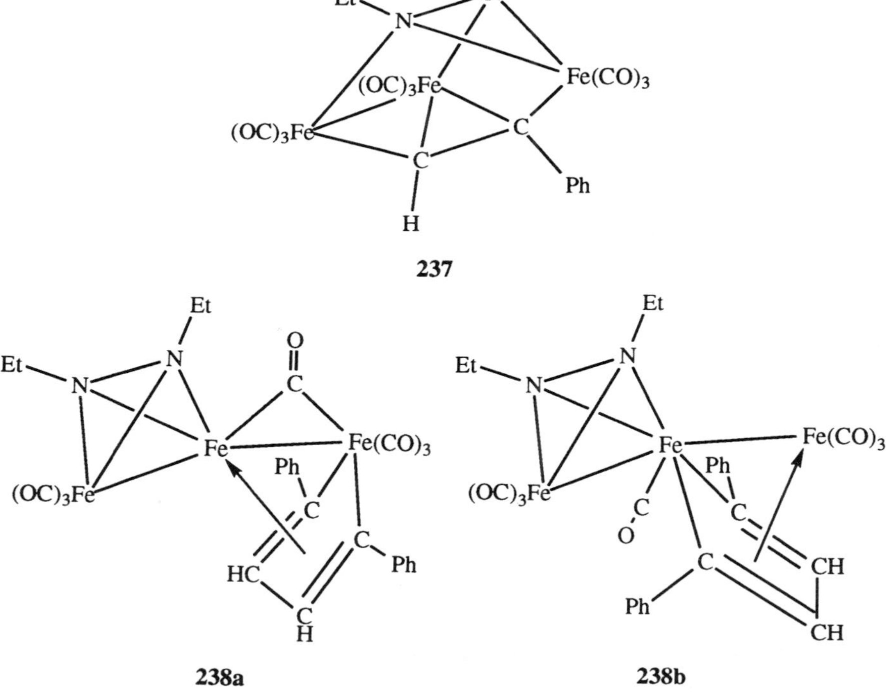

A series of papers has appeared examining the reactions of $Fe_3(CO)_{10}(PR)$ and $Fe_3(CO)_9(PR)_2$ with various alkynes.[186–190,193,195,281,285]

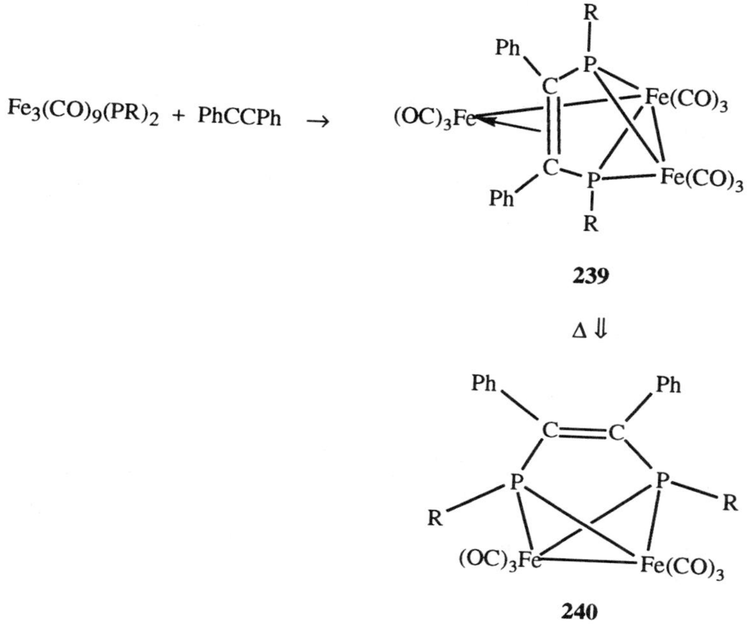

$Fe_3(CO)_9(PR)_2$ + PhCCPh $\rightarrow$

239

$\Delta \Downarrow$

240

SCHEME 10

Generally, insertion of the alkyne into a metal–P bond is observed (Scheme 10).[188,190] When aminoalkynes are used, the formation of a C=N double bond inhibits the interaction of that carbon with the metal centers of the cluster.[186,187] When two PR groups are present, the alkyne has been observed to bridge between them as seen in Scheme 10.[195,285] A second equivalent of diphenylacetylene can substitute for two carbonyl groups on the iron triangle.[195] The hetero-main group element species $Fe_3(CO)_9(NPh)(P^tBu)$ and $Fe_3(CO)_9(NPh)_2$ have been reacted with diphenylacetylene.[273] Some of the products involved in the acetylene addition reaction are shown here (**241–243**).

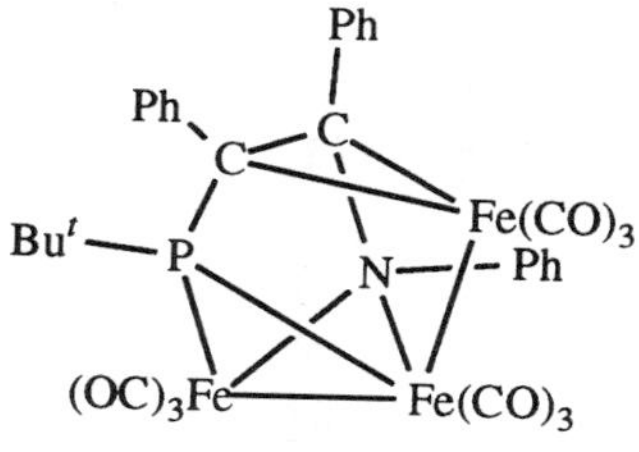

241

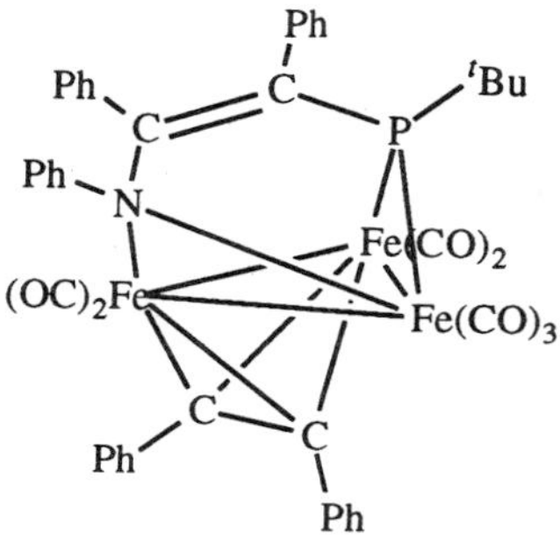

242

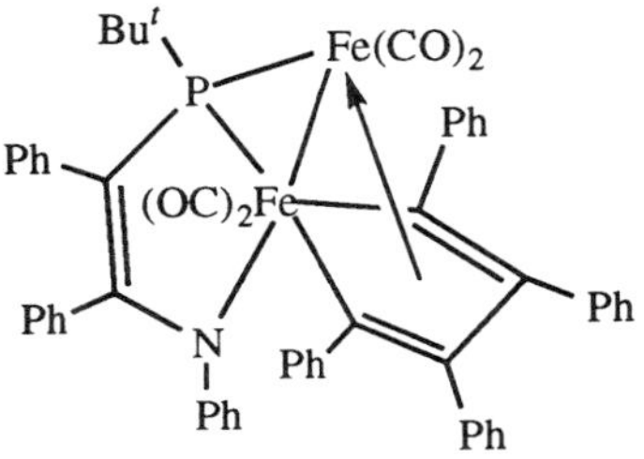

243

Reaction of diynes and $Ru_4(CO)_{10}(PPh)$ have been examined.[210] In the first stages of this reaction only one triple bond is involved (**244**). Upon thermolysis, a cluster rearrangement occurs to give a pentagonal bipyramidal structure (**245**), whereas prolonged reaction causes involvement of the second triple bond (**246**). The compound contains a square pyramidal Ru_4P core with the C_4 fragment bonded in a μ_4-$\eta^1,\eta^1,\eta^3,\eta^3$ fashion to the square face.

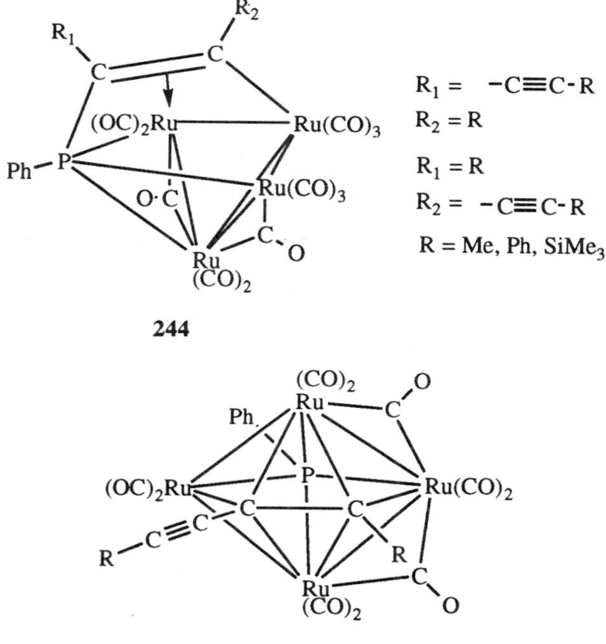

244

245

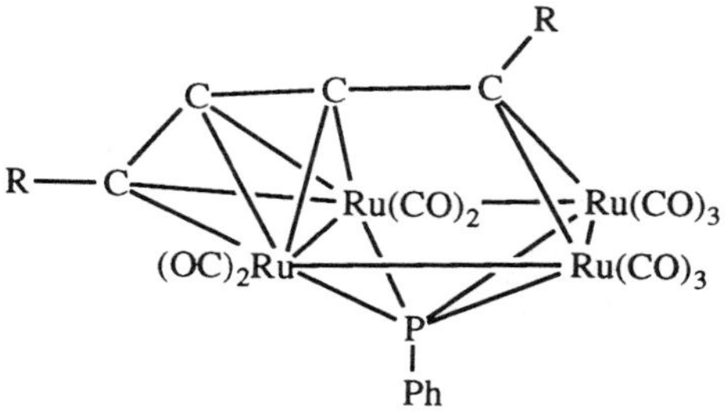

246

Some E–M cluster compounds have been observed to facilitate hydrogenation of organic molecules [e.g., Eq. (273)[111]]. The compound $H_2Fe_3(CO)_9(NSiMe_3)$ catalyzes the hydrogenation of olefins and dienes.[149] A study on the hydrogenation of diphenylacetylene by $Fe_3(CO)_{10}(PR)$ (R = ^{i}Pr, NEt_2), $H_2Fe_3(CO)_9(PC_6H_4OMe)$, $Fe_3(CO)_9(P^tBu)_2$, and $Ru_4(CO)_{13}(PPh)$ gave mixtures of the *cis* and *trans* stilbenes.[196] These same complexes isomerize the stilbenes, which accounts for the mixture of products obtained in the hydrogenation reactions.

$$H_2Ru_4(CO)_{12}(PPh) + \text{cyclohexene} + CO \rightarrow Ru_4(CO)_{13}(PPh) + \text{cyclohexane} \quad (273)$$

IV

CHARACTERIZATION

Conventional analytical techniques may be applied to the characterization of E–M compounds, but they have some unique problems.

Mass spectrometry must be used with caution, as the weakness of the E–M bonding framework in some cases leads to fragmentation of the parent ion, which is thus not observed. For the smaller, more volatile molecules, conventional electron-impact techniques can be valuable,[42] but softer ionization techniques may be required for these as well. Chemical ionization (CI) has been successfully employed for the characterization of $Fe_3(CO)_{10}(PR)$.[192] Electrospray techniques have proven useful for characterizing even nonvolatile, ionic complexes.[231] Positive ion–laser desorption methods were employed successfully to characterize $(CO)_4W\{PW(CO)_5\}_4$.[461]

Nuclear magnetic resonance is potentially useful for analyzing E–M compounds. The other E atoms, however, are quadrupolar and their observation is more problematic. Some use of them has been made, particularly for nitrogen. The ^{15}N signal in $[Rh_6(CO)_{15}N]^-$ using an enriched compound showed the five lines (septet expected) attributed to coupling to Rh. The nuclei most often studied are the C and H atoms of the other ligands attached to M and/or E. Studies of ^{13}CO ligands in metal carbonyls may give some clue as to structure, but such studies are generally complicated by the high mobility of CO about a cluster surface. Under ideal conditions, variable-temperature studies can reveal information about the nature of these dynamic processes. Two exchange processes were noted for scrambling of CO on $Ru_3(CO)_9(\mu_3\text{-}CO)NPh$.[157] One process involved the exchange of the axial and equatorial CO ligands at each Ru atom (turnstile mechanism), whereas the other exchanged the μ_3-CO ligand with the equatorial CO groups. These exchange mechanisms are conventional for metal carbonyl clusters. At room temperature there is total scrambling of CO on the cluster surface of $Ru_4(CO)_{13}(PPh)$ that could be resolved into four independent CO scrambling processes.[209] A variable-temperature study showed that $Fe_3(CO)_8L(PPh)_2$ clusters (L = $P(OMe)_3$, PPh_3, $AsPh_3$) were fluxional and existed as a mixture of axial and equatorial isomers.[284] In contrast, the L = NCMe compound appeared to prefer a single isomer with L in an equatorial site on a basal Fe atom.

^{31}P NMR is a particularly powerful tool for characterizing complexes with more than one phosphorus atom owing to the 100% natural abundance of the ^{31}P nucleus. It has been extensively employed for compounds containing P_x units, where the coupling constants give useful structural information. It has also proven an effective means for examining the degree of oligomerization in the $[MPFe_3(CO)_9P]\text{–}[Au{-}PFe_3(CO)_9P]_x\text{–}[PFe_3(CO)_9PMe]^{(x+1)-}$ system.[289] The ^{31}P NMR spectrum of $Fe_2(CO)_6(PPh)_4$ shows a large coupling constant between the two terminal P atoms ($^2J_{P \cdots P}$ = 125.4 Hz) suggestive of an attractive P···P interaction.[464]

Although it is not surprising that the CO ligands in $[Rh_9(CO)_{21}P]^{2-}$ would be fluxional as confirmed by ^{13}C NMR spectroscopy, what is more unusual is that the ^{31}P and ^{103}Rh NMR spectra indicate that the Rh atoms in the cluster are also fluxional.[375–377] The ^{31}P signal is a complex 10-line multiplet at 35° that splits into more lines upon cooling. The motion is proposed to occur via Rh–Rh bond formation across square faces with concomitant Rh–Rh bond cleavage across Rh_4 rhombs. Similarly, $[Rh_{10}(CO)_{22}P]^{3-}$ also shows fluxional behavior, as does $[Rh_{12}(CO)_{27}Sb]^{3-}$.[376,378]

X-ray diffraction has been an extremely powerful tool for characterizing E–M complexes, and most reported here have had single-crystal determinations reported. Even more exciting results will come in the future as area detectors become more widely used. These allow for the collection of data on larger unit cells and consequently larger cluster molecules. Extended

X-ray absorption fine structure (EXAFS) can be a potentially useful technique that can give information about bonded atoms and coordination environments. EXAFS have been applied very rarely to E–M compounds, but some work has been reported. For example, the EXAFS data of solid samples of $[Cp(CO)_3Mo]_2BiCl$ are consistent with the crystal structure data that indicate that the trimerization via intermolecular Bi–Cl interactions is broken in solution and replaced via coordination of solvent thf to the bismuth center.[46]

Mössbauer spectroscopy may be applied to compounds containing iron. Some limited data have been reported for the iron nitride and related iron carbide clusters.[355,358,365]

V
FUTURE DIRECTIONS

In examining the wealth of knowledge that is known about E–M cluster compounds it is clear that, in spite of the fact that a lot of work has already been done, there are many unanswered synthetic and mechanistic questions. Most of the work to date has been heavily structural in nature. The field has been forging into a new area where just being able to define the players involved has required significant effort. Much of the work in the past decade has helped to systematize the synthetic procedures and develop high-yield procedures so that the chemistry of these complexes can be examined more fully. In some cases, however, yields are still low and better synthetic methods are needed. Other questions that still remain to be answered include why some structures and reactivities differ so dramatically for compounds that one would expect to be similar.

Take as an illustration the case of antimony, which is by far the least-represented element in this review. From the initial indications of reactivity and structural patterns, it often behaves differently from both arsenic and bismuth. Why, for example, does $Sb[Co(CO)_4]_3$ seem to have only a fleeting existence while $Bi[Co(CO)_4]_3$ is stable? Furthermore, it has not been possible to isolate $SbCo_3(CO)_9$ when both the arsenic (as the cyclic trimer) and bismuth compounds are known. And even though $Sb[Co(CO)_4]_3$ may have a fleeting existence, it spontaneously disproportionates to give $[Sb_2Co_4(CO)_{11}]^{1-/2-}$, whereas $Bi[Co(CO)_4]_3$ is a stable, isolable molecule. There are no molecules with the Sb_3^{3-} ion present, whereas examples exist for P, As, and Bi.

A significant challenge is the controlled creation of higher nuclearity species. The best method to date is the addition of main group element silyl complexes to metal halides, but even that method, which has been successful at producing a large number of very esthetically pleasing compounds, is difficult to control. Not all elements have proven amenable to this technique. Bismuth in particular has yet to yield stable metal complexes starting from bismuth silyl compounds. The chemist is still at the mercy of thermodynamics. What crystallizes out is what one gets and, once out of solution, many of the higher nuclearity compounds will not redissolve. This seriously handicaps the ability to elucidate the chemical and physical properties of these very interesting compounds. Nevertheless, newer methods of characterization, including the advent of area detectors for single-crystal X-ray diffraction, are beginning to move this field in the direction of larger and larger cluster sizes that take advantage of the methods developed by protein crystallographers for macromolecule characterization. Synthetic chemists are also beginning to examine their product mixtures for clusters on the nanometer scale by electron microscopy, and very exciting results are in the works using this technique for homometallic, heterometallic, and E–M cluster compounds.

Although some reports have appeared that discuss how E–M cluster compounds interact with organic substrates, there is much to be done and to be understood on this topic. Clearly, the evidence to date indicates that the main group element is integral in determining the structure and reactivity patterns of E–M clusters, and that the E atoms can become directly bonded to the entering group. The chemistry is not simply the same as that of homometallic transition metal clusters and promises to hold surprises for years to come. Some of these compounds have remarkable thermal stability. An effort should be made to survey a number of the higher yield, stable compounds for potential use in catalytic cycles, especially those involving carbonylation and/or hydrogenation.

Acknowledgments

The author acknowledges the support of the National Science Foundation and the Robert A. Welch Foundation, which have made possible the efforts that have been undertaken at Rice University to understand the E–M cluster system. Also acknowledged are the dedicated efforts of the numerous co-workers who have contributed to the development of this area and who are listed in the references.

References

(1) Zanello, P. Stereochemical aspects associated with the redox behavior of heterometal carbonyl clusters, *Struc. Bonding* **1992**, *79*, 101.

(2) Whitmire, K. H. A comparison of bismuth- and antimony-containing transition metal cluster complexes, *J. Cluster Sci.* **1991**, *2*, 231.

(3) Scheer, M.; Herrmann, E. Koordinationschemische Stabilisierung "nackter" Elemente der V. Hauptgruppe (außer Stickstoff)—Synthese, Struktur und Bindung, *Z. Chem.* **1990**, *30*, 41.
(4) Compton, N. A.; Errington, J.; Norman, N. C. Transition metal complexes incorporating atoms of the heavier main-group elements, *Adv. Organomet. Chem.* **1990**, *31*, 91.
(5) Scherer, O. J. Complexes with substituent-free acyclic and cyclic phosphorus, arsenic, antimony and bismuth ligands, *Angew. Chem., Int. Ed. Engl.* **1990**, *29*, 1104; *Angew. Chem.* **1990**, *102*, 1137.
(6) Whitmire, K. H. Clusters of metals and nonmetals, *Clusters of Metals and Nonmetals*; Elsevier: New York, 1989, pp. 503.
(7) Fehlner, T. P. Inorganometallic chemistry, *Comments Inorg. Chem.* **1988**, *7*, 307.
(8) Whitmire, K. H. The interface of main group and transition metal cluster chemistry, *J. Coord. Chem. B* **1988**, *17*, 95.
(9) Norman, N. C. Organotransition metal complexes incorporating bismuth, *Chem. Soc. Rev.* **1988**, *17*, 269.
(10) Fenske, D.; Ohmer, J.; Hachgenei, J.; Merzweiler, K. New transition metal clusters with ligands from main groups five and six, *Angew. Chem., Int. Ed. Engl.* **1988**, *27*, 1277; *Angew. Chem.* **1988**, *100*, 1300.
(11) Huttner, G.; Knoll, K. RP-bridged metal carbonyl clusters: Synthesis, properties, and reactions, *Angew. Chem., Int. Ed. Engl.* **1987**, *26*, 743; *Angew. Chem.* **1987**, *99*, 765.
(12) Scherer, O. J. P_x units as complex ligands, *Comments Inorg. Chem.* **1987**, *6*, 1.
(13) Huttner, G. Unconventional multiple bonds: Coordination compounds of unstable Vth main group ligands, *Pure Appl. Chem.* **1986**, *58*, 585.
(14) Herrmann, W. A. Multiple bonds between transition metals and "bare" main group elements: Links between inorganic solid state chemistry and organometallic chemistry, *Angew. Chem., Int. Ed. Engl.* **1986**, *25*, 56; *Angew. Chem.* **1986**, *98*, 57.
(15) Huttner, G.; Evertz, K. Phosphinidene complexes and their higher homologues, *Acc. Chem. Res.* **1986**, *19*, 406.
(16) Scherer, O. J. *Angew. Chem., Int. Ed. Engl.* **1985**, *24*, 924; *Angew. Chem.* **1985**, *97*, 905.
(17) Gladfelter, W. L. Organometallic metal clusters containing nitrosyl and nitrido ligands, *Adv. Organomet. Chem.* **1985**, *24*, 41.
(18) Schmid, G. Developments in transition metal cluster chemistry—the way to large clusters, *Struct. Bond.* **1985**, *62*, 51.
(19) Nicholls, J. N. Main group heteroatoms in transition metal clusters, *Polyhedron* **1984**, *3*, 1307.
(20) Vahrenkamp, H. Basic metal cluster reactions, *Adv. Organomet. Chem.* **1983**, *22*, 169.
(21) Di Vaira, M.; Sacconi, L. Transition metal complexes with cyclo-triphosphorus (η^3-P_3) and *tetrahedro*-tetraphosphorus (η^1-P_4) ligands, *Angew. Chem., Int. Ed. Engl.* **1982**, *21*, 330; *Angew. Chem.* **1982**, *94*, 338.
(22) Vahrenkamp, H. *The Synthesis of Clusters—Why and How?* Verlag Chemie: Weinheim, 1981, p. 35.
(23) Schmid, G. Tetrahedral carbonyl cobalt clusters, *Angew. Chem., Int. Ed. Engl.* **1978**, *17*, 392; *Angew. Chem.* **1978**, *90*, 417.
(24) Vahrenkamp, H. Recent results in the chemistry of transition metal clusters with organic ligands, *Struct. Bond.* **1977**, *32*, 1.
(25) Hieber, W. Metal carbonyls, forty years of research, *Adv. Organomet. Chem.* **1970**, *8*, 1.
(26) Joannis, A. *Compt. Rend* **1891**, *113*, 797.
(27) Joannis, A. *Compt. Rend.* **1892**, *114*, 587.
(28) Joannis, A. *Ann. Chim.* **1906**, *7*, 75.
(29) Kraus, C. A. *J. Am. Chem. Soc.* **1907**, *29*, 1557.

(30) Kraus, C. A. *J. Am. Chem. Soc.* **1922**, *44*, 1216.
(31) Kraus, C. A. *Trans. Am. Electrochem. Soc.* **1924**, *45*, 175.
(32) Peck, E. B. *J. Am. Chem. Soc.* **1918**, *40*, 335.
(33) Smith, H. *J. Am. Chem. Soc.* **1917**, *39*, 1299.
(34) Zintl, E., Goubeau, J.; Dullenkopf, W. *Z. physik. Chem. A* **1931**, *154*, 1.
(35) Zintl, E., Harder, A. *Z. physik. Chem. A* **1931**, *154*, 47.
(36) Zintl, E. *Z. physik. Chem. B* **1932**, *16*, 195.
(37) Zintl, E.; Dullenkopf, W. *Z. physik. Chem. B* **1932**, *16*, 183.
(38) Zintl, E.; Kaiser, H. *Z. Anorg. Allg. Chem.* **1933**, *211*, 113.
(39) Corbett, J. D. *Chem. Rev.* **1985**, *85*, 383.
(40) von Schnering, H. G. *Angew. Chem., Int. Ed. Engl.* **1981**, *20*, 33; *Angew. Chem.* **1981**, *93*, 44.
(41) Grossbruchhaus, V.; Rehder, D. *Inorg. Chim. Acta* **1990**, *172*, 141.
(42) Barr, A. M.; Kerlogue, M. D.; Norman, N. C.; Webster, P. M.; Farrugia, L. J. *Polyhedron* **1989**, *8*, 2495.
(43) Cullen, W. R.; Patmore, D. J.; Sams, J. R.; Newlands, M. J.; Thompson, L. K. *Chem. Commun.* **1971**, 952.
(44) Weber, U.; Huttner, G.; Scheidsteger, O.; Zsolnai, L. *J. Organomet. Chem.* **1985**, *289*, 357.
(45) Panster, P.; Malisch, W. *J. Organomet. Chem.* **1977**, *134*, C32–C36.
(46) Clegg, W.; Compton, N. A.; Errington, R. J.; Fisher, G. A.; Hockless, D. C. R.; Norman, N. C.; Williams, N. A. L.; Stratford, S. E.; Nichols, S. J.; Jarrett, P. S.; Orpen, A. G. *J. Chem. Soc., Dalton Trans.* **1992**, 193.
(47) Clegg, W.; Compton, N. A.; Errington, R. J.; Fisher, G. A.; Norman, N. C.; Marder, B. *J. Chem. Soc., Dalton Trans.* **1991**, 2887.
(48) Clegg, W.; Compton, N. A.; Errington, R. J.; Norman, N. C. *J. Organomet. Chem.* **1987**, *335*, C1.
(49) Errington, R. J.; Farrugia, L. J.; Fisher, G. A.; Niklaus, A.; Norman, N. C. *J. Chem. Soc., Dalton Trans.* **1993**, 1201.
(50) Wallis, J. M.; Müller, G.; Schmidbaur, H. *J. Organomet. Chem.* **1987**, *325*, 159.
(51) Clegg, W.; Compton, N. A.; Errington, R. J.; Norman, N. C. *J. Chem. Soc., Dalton Trans.* **1988**, 1671.
(52) Clegg, W., Compton, N. A.; Errington, R. J.; Norman, N. C. *Polyhedron* **1987**, *6*, 2031.
(53) Whitmire, K. H.; Shieh, M.; Cassidy, J. *Inorg. Chem.* **1989**, *28*, 3164.
(54) Cassidy, J. M.; Whitmire, K. H. *Inorg. Chem.* **1991**, *30*, 2788.
(55) Shieh, M.; Liou, Y.; Peng, S.-M.; Lee, G.-H. *Inorg. Chem.* **1993**, *32*, 2212.
(56) Shieh, M.; Liou, Y.; Jeng, B.-W. *Organometallics* **1993**, *12*, 4926.
(57) Clegg, W.; Compton, N. A.; Errington, R. J.; Hockless, D. C. R.; Norman, N. C.; Ramshaw, M.; Webster, P. M. *J. Chem. Soc., Dalton Trans.* **1990**, 2375.
(58) Wallis, J. M.; Müller, G.; Schmidbaur, H. *Inorg. Chem.* **1987**, *26*, 458.
(59) Cullen, W. R.; Patmore, D. J.; Sams, J. R. *Inorg. Chem.* **1973**, *12*, 867.
(60) Etzrodt, G.; Boese, R.; Schmid, G. *Chem. Ber.* **1979**, *112*, 2574.
(61) Von Seyerl, J.; Huttner, G. *J. Organomet. Chem.* **1980**, *193*, 207.
(62) Haaland, A. *Angew. Chem., Int. Ed. Engl.* **1989**, *28*, 992; *Angew. Chem.* **1989**, *101*, 1017.
(63) Long, G., private communication, 1987.
(64) Leigh, J. S.; Whitmire, K. H. *Angew. Chem., Int. Ed. Engl.* **1988**, *27*, 396; *Anew. Chem.* **1988**, *100*, 399.
(65) Martinengo, S.; Ciani, G. *J. Chem. Soc., Chem. Commun.* **1987**, 1589.
(66) Henderson, P.; Rossignoli, M.; Burns, R. C.; Scudder, M. L.; Craig, D. C. *J. Chem. Soc., Dalton Trans.* **1994**, 1641.
(67) Bachman, R. E.; Miller, S. K.; Whitmire, K. H. *Inorg. Chem.* **1994**, *33*, 2075.

(68) Bachman, R. E.; Miller, S. K.; Whitmire, K. H. *Organometallics* **1995**, *14*, 796.
(69) Lorenz, L.-P., Effinger, G.; Hiller, W. *Chem. Ber.* **1990**, *123*, 251.
(70) Luo, S.; Whitmore, K. H. *Inorg. Chem.* **1989**, *28*, 1424.
(71) Ferrer, M.; Rossell, O.; Seco, M.; Braunstein, P. *J. Organomet. Chem.* **1989**, *364*, C5.
(72) Trinh-Toan; Dahl, L. F. *J. Am. Chem. Soc.* **1971**, *93*, 2654.
(73) Whitmore, K. H.; Lagrone, C. B.; Churchill, M. R.; Fettinger, J. C.; Biondi, L. V. *Inorg. Chem.* **1984**, *23*, 4227.
(74) Churchill, M. R.; Fettinger, J. C.; Whitmire, K. H.; Lagrone, C. B. *J. Organomet. Chem.* **1986**, *303*, 99.
(75) Martinengo, S.; Fumagalli, A.; Ciani, G.; Moret, M. *J. Organomet. Chem.* **1988**, *347*, 413.
(76) Gómez-Sal, P.; Martin, A.; Mena, M.; Yélamos, C. *J. Chem. Soc., Chem. Commun.* **1995**, 2185.
(77) Simon, G. L.; Dahl, L. F. *J. Am. Chem. Soc.* **1973**, *95*, 2175.
(78) Fenske, D.; Basoglu, R.; Hachgenei, J.; Rogel, F. *Angew. Chem., Int. Ed. Engl.* **1984**, *23*, 160; *Angew. Chem.* **1984**, *96*, 160.
(79) Scherer, O. J.; Dave, T.; Braun, J.; Wolmershäuser, G. *J. Organomet. Chem.* **1988**, *350*, C20.
(80) Röttinger, E.; Vahrenkamp, H. *Angew. Chem., Int. Ed. Engl.* **1978**, *17*, 273; *Angew. Chem.* **1978**, *90*, 294.
(81) Röttinger, E.; Vahrenkamp, H. *J. Organomet. Chem.* **1981**, *213*, 1.
(82) Rheingold, A. L.; Foley, M. J.; Sullivan, P. J. *J. Am. Chem. Soc.* **1982**, *104*, 4727.
(83) Foust, A. S.; Dahl, L. F. *J. Am. Chem. Soc.* **1970**, *92*, 7337.
(84) Ciani, G.; Moret, M.; Fumagalli, A.; Martinengo, S. *J. Organomet. Chem.* **1989**, *362*, 291.
(85) King, R. B. *New J. Chem.* **1989**, *13*, 293.
(86) Beurich, H.; Madach, T.; Richter, F.; Vahrenkamp, H. *Angew. Chem., Int. Ed. Engl.* **1979**, *18*, 690; *Angew. Chem.* **1979**, *91*, 751.
(87) Cowley, A. H.; Norman, N. C.; Pakulsi, M.; Bricker, D. L.; Russell, D. H. *J. Am. Chem. Soc.* **1985**, *107*, 8211.
(88) Cowley, A. H.; Norman, N. C.; Pakulski, M. *J. Am. Chem. Soc.* **1984**, *106*, 6844.
(89) Arif, A. M.; Cowley, A. H.; Norman, N. C.; Pakulski, M. *Inorg. Chem.* **1986**, *25*, 4836.
(90) DiMaio, A.-J.; Bitterwolf, T. E.; Rheingold, A. L. *Organometallics* **1990**, *9*, 551.
(91) Knoll, K., Huttner, G.; Zsolnai, L.; Jibril, I.; Wasiucionek, M. *J. Organomet. Chem.* **1985**, *294*, 91.
(92) Deck, W.; Schwarz, M.; Vahrenkamp, H. *Chem. Ber.* **1987**, *120*, 1515.
(93) Field, J. S.; Haines, R. J.; Smit, D. N.; Natarjan, K.; Schiedsteger, O.; Huttner, G. *J. Organomet. Chem.* **1982**, *240*, C23.
(94) Cardin, C. J.; Colbran, S. B.; Johnson, B. F. G.; Lewis, J.; Raithby, P. R. *J. Chem. Soc., Chem. Commun.* **1986**, 1288.
(95) Colbran, S. B.; Lahoz, F. J.; Raithby, P. R.; Lewis, J.; Johnson, B. F. G.; Cardin, C. J. *J. Chem. Soc., Dalton Trans.* **1988**, 173.
(96) Queisser, J.; Oesen, H.; Fenske, D.; Lehari, B. *Z. Anorg. Allg. Chem.* **1994**, *620*, 1821.
(97) Dietrich, W.-R.; Vahrenkamp, H. *J. Chem. Res. (S)* **1985**, 200.
(98) Luo, S.; Whitmire, K. H. *J. Organomet. Chem.* **1989**, *376*, 297.
(99) Feasey, N. D.; Knox, S. A. R.; Orpen, A. G. *J. Chem. Soc., Chem. Commun.* **1982**, 75.
(100) Chisholm, M. H.; Folting, K.; Huffman, J. C.; Leonelli, J.; Marchant, N. S.; Smith, C. A.; Taylor, L. C. E. *J. Am. Chem. Soc.* **1985**, *107*, 3722.
(101) Ramage, D. L.; Geoffroy, G. L.; Rheingold, A. L.; Haggerty, B. S. *Organometallics* **1992**, *11*, 1242.
(102) Puga, J.; Sänchez-Delgado, A. J.; Braga, D. *J. Chem. Soc., Chem. Commun.* **1986**, 1631.
(103) Austin, R. G.; Urry, G. *Inorg. Chem.* **1977**, *16*, 3359.

(104) Buchholz, D.; Huttner, G.; Zsolnai, L.; Imhof, W. *J. Organomet. Chem.* **1989**, *377*, 25.
(105) Lindner, E.; Weiss, G. A. *J. Organomet. Chem.* **1983**, *244*, C9.
(106) Brauer, D. J.; Hietkamp, S.; Sommer, H.; Stelzer, O.; Müller, G.; Krüger, C. *J. Organomet. Chem.* **1985**, *288*, 35.
(107) De, R. L.; Vahrenkamp, H. *Z. Naturforsch.* **1986**, *41b*, 273.
(108) Vahrenkamp, H.; Wucherer, E. J.; Wolters, D. *Chem. Ber.* **1983**, *116*, 1219.
(109) Huttner, G.; Scheider, J.; Müller, H.-D.; Mohr, G.; von Seyerl, J.; Wohlfahrt, L. *Angew. Chem., Int. Ed. Engl.* **1979**, *18*, 76; *Angew. Chem.* **1979**, *92*, 82.
(110) MacLaughlin, S. A.; Carty, A. J.; Taylor, N. J. *Can. J. Chem.* **1982**, *60*, 87.
(111) van Gastel, F.; Corrigan, J. F.; Doherty, S.; Taylor, N. J.; Carty, A. J. *Inorg. Chem.* **1992**, *31*, 4492.
(112) Lavigne, G.; Bonnet, J. J. *Inorg. Chem.* **1981**, *20*, 2713.
(113) Cullen, W. R.; Rettig, S. J.; Zhang, H. *Organometallics* **1991**, *10*, 2965.
(114) Cullen, W. R.; Rettig, S. J.; Zheng, T.-C. *Organometallics* **1992**, *11*, 277.
(115) Cullen, W. R.; Rettig, S. J.; Zheng, T.-C. *Organometallics* **1992**, *11*, 928.
(116) Ahlrichs, R.; Fenske, D.; Oesen, H.; Schneider, U. *Angew. Chem., Int. Ed. Engl.* **1992**, *31*, 323; *Angew. Chem.* **1992**, *104*, 312.
(117) Gorzellik, M.; Nuber, B.; Bohn, T.; Ziegler, M. L. *J. Organomet. Chem.* **1992**, *429*, 173.
(118) Huttner, G.; Mohr, G.; Pritzlaff, B.; von Seyerl, J.; Zsolnai, L. *Chem. Ber.* **1982**, *115*, 2044.
(119) Blechsmitt, K.; Zahn, T.; Ziegler, M. L. *Angew. Chem., Int. Ed. Engl.* **1985**, *24*, 702; *Angew. Chem.* **1985**, *97*, 686.
(120) Arnold, L. J.; Mackay, K. M.; Nicholson, B. K. *J. Organomet. Chem.* **1990**, *387*, 197.
(121) Campana, C. F.; Dahl, L. F. *J. Organomet. Chem.* **1977**, 209.
(122) Rheingold, A. L.; Geib, S. J.; Shieh, M.; Whitmire, K. H. *Inorg. Chem.* **1987**, *26*, 463.
(123) Arif, A. M.; Cowley, A. H.; Pakulski, M. *J. Chem. Soc., Chem. Commun.* **1987**, 622.
(124) Johnson, B. F. G.; Lewis, J.; Raithby, P. R.; Whitton, A. J. *J. Chem. Soc., Chem. Commun.* **1988**, 401.
(125) Kilthau, T.; Nuber, B.; Ziegler, M. L. *Chem. Ber.* **1995**, *128*, 197.
(126) Vizi-Orosz, A. *J. Organomet. Chem.* **1976**, *111*, 61.
(127) Blechsmitt, K.; Pfisterer, H.; Zahn, T.; Ziegler, M. L. *Angew. Chem., Int. Ed. Engl.* **1985**, *24*, 66; *Angew. Chem.* **1985**, *97*, 73.
(128) Ziegler, M. L.; Blechsmitt, K.; Nuber, B.; Zahn, T. *Chem. Ber.* **1988**, *121*, 159.
(129) Caballero, C.; Nuber, B.; Ziegler, M. L. *J. Organomet. Chem.* **1990**, *386*, 209.
(130) Angermaier, K.; Schmnidbaur, H. *J. Chem. Soc., Dalton Trans.* **1995**, 559.
(131) Schmidbaur, H.; Weidenhiller, G.; Steigelmann, O. *Angew. Chem., Int. Ed. Engl.* **1991**, *30*, 433; *Angew. Chem.* **1991**, *103*, 442.
(132) Deck, W.; Vahrenkamp, H. *Z. Anorg. Allg. Chem.* **1991**, *598/599*, 83.
(133) Shieh, M.; Mia, F.-D.; Peng, S.-M.; Lee, G.-H. *Inorg. Chem.* **1993**, *32*, 2785.
(134) Whitmire, K. H.; Lagrone, C. B.; Rheingold, A. L. *Inorg. Chem.* **1986**, *25*, 2472.
(135) Wallis, J. M.; Müller, G.; Riede, J.; Schmidbauer, H. *J. Organomet. Chem.* **1989**, *369*, 165.
(136) Ang, H. G.; Hay, C. M.; Johnson, B. F. G.; Lewis, J.; Raithby, P. R.; Whitton, A. J. *J. Organomet. Chem.* **1987**, *1987*, C5.
(137) Hay, C. M.; Johnson, B. F. G.; Lewis, J.; Raithby, P. R.; Whitton, A. J. *J. Chem. Soc., Dalton Trans.* **1988**, 2091.
(138) Johnson, B. F. G.; Lewis, J.; Whitton, A. J. *J. Chem. Soc., Dalton Trans.* **1990**, 3129.
(139) Whitmore, K. H.; Leigh, J. S.; Gross, M. E. *J. Chem. Soc., Chem. Commun.* **1987**, 926.
(140) Kruppa, W.; Bläser, D.; Boese, R.; Schmid, G. *Z. Naturforsch.* **1982**, *37B*, 209.
(141) Gibson, C. P.; Dahl, L. F. *Organometallics* **1988**, *7*, 543.
(142) Sun, W.; Yang, S.; Wang, H.; Yin, Y.; Yu, K. *Polyhedron* **1992**, *11*, 1143.
(143) Legzdins, P.; Nurse, C. R.; Rettig, S. J. *J. Am. Chem. Soc.* **1983**, *105*, 3727.

(144) Barr, M. E.; Bjarnason, A.; Dahl, L. F. *Organometallics* **1994**, *13*, 1981.
(145) Fjare, D. E.; Gladfelter, W. L. *J. Am. Chem. Soc.* **1981**, *103*, 1572.
(146) Gourdon, A.; Jeannin, Y. *Organometallics* **1986**, *5*, 2406.
(147) Koerner von Gustorf, E.; Wagner, R. *Angew. Chem., Int. Ed. Engl.* **1971**, *10*, 910; *Angew. Chem.* **1971**, *83*, 968.
(148) Barnett, B. L.; Krüger, C. *Angew. Chem., Int. Ed. Engl.* **1971**, *10*, 910; *Angew. Chem.* **1971**, *83*, 969.
(149) Fischler, I.; Wagner, R.; Koerner von Gustorf, E. A. *J. Organomet. Chem.* **1976**, *112*, 155.
(150) Nametkin, N. S.; Tyurin, V. D.; Trusov, V. V.; Nekhaev, A. I.; Batsanov, A. S.; Struchov, Y. T. *J. Organomet. Chem.* **1986**, *302*, 243.
(151) Aime, S.; Milone, L.; Rosetti, R.; Stanghellini, P. L. *Gazz. Chim. Ital.* **1975**, *105*, 617.
(152) NGuini Effa, J.-B.; Djebailli, B.; Lieto, J.; Aune, J.-P. *J. Chem. Soc., Chem. Commun.* **1983**, 408.
(153) Ragaini, F.; Song, J.-S.; Ramage, D. L.; Geoffroy, G. L.; Yap, G. A. P.; Rheingold, L. *Organometallics* **1995**, *14*, 387.
(154) Andrews, M. A.; Kaesz, H. D. *J. Am. Chem. Soc.* **1979**, *101*, 7255.
(155) Blohm, M. L.; Fjare, D. E.; Gladfelter, W. L. *J. Am. Chem. Soc.* **1986**, *108*, 2301.
(156) Han, S.-H.; Song, J.-S.; Macklin, P. D.; Nguyen, S. T.; Geoffroy, G. L.; Rheingold, L. *Organometallics* **1989**, *8*, 2127.
(157) Wang, D.; Shen, H.; Richmond, M. G.; Schwartz, M. *Organometallics* **1995**, *14*, 3636.
(158) Sappa, E.; Milone, L. *J. Organomet. Chem.* **1974**, *61*, 388.
(159) Bhaduri, S.; Gopalkrishnan, K. S.; Sheldrick, G. M.; Clegg, W.; Stalke, D. *J. Chem. Soc., Dalton Trans.* **1983**, 2339.
(160) Han, S.-H.; Geoffroy, G. L. *Organometallics* **1987**, *6*, 2380.
(161) Smieja, J. A.; Gozum, J. E.; Gladfelter, W. L. *Organometallics* **1987**, *6*, 1311.
(162) Stevens, R. E.; Guettler, R. D.; Gladfelter, W. L. *Inorg. Chem.* **1990**, *29*, 451.
(163) Stevens, R. E.; Gladfelter, W. L. *J. Am. Chem. Soc.* **1982**, *104*, 6454.
(164) Han, S.-H.; Geoffroy, G. L.; Rheingold, A. L. *Inorg. Chem.* **1987**, *26*, 3426.
(165) Bhaduri, S.; Khwaja, H.; Sapre, N.; Sharma, K.; Basu, A.; Jones, P. G.; Carpenter, G. *J. Chem. Soc., Dalton Trans.* **1990**, 1313.
(166) Johnson, B. F. G.; Lewis, J.; Mace, J. M. *J. Chem. Soc., Chem. Commun.* **1984**, 186.
(167) Fjare, D. E.; Keyes, D. G.; Gladfelter, W. L. *J. Organomet. Chem.* **1983**, *250*, 383.
(168) Basu, A.; Bhaduri, S.; Khwaja, H. *J. Organomet. Chem.* **1987**, *319*, C28.
(169) Bernhardt, W.; Vahrenkamp, H. *Angew. Chem., Int. Ed. Engl.* **1984**, 381.
(170) Süss-Fink, G.; Rheinwald, G.; Stoeckli-Evans, H.; Bolm, C.; Kaufmann, D. *Inorg. Chem.* **1996**, *35*, 3081.
(171) Basu, A.; Bhaduri, S.; Khwaja, H.; Jones, P. G.; Meyer-Bäse, K.; Sheldrick, G. M. *J. Chem. Soc., Dalton Trans.* **1986**, 2501.
(172) Burgess, K.; Johnson, B. F. G.; Lewis, J.; Raithby, P. R. *J. Chem. Soc., Dalton Trans.* **1982**, 2085.
(173) Yin, C. C.; Deeming, A. J. *J. Chem. Soc., Dalton Trans.* **1974**, 1013.
(174) Otsuka, S.; Nakamura, A.; Yoshida, T. *Inorg. Chem.* **1968**, *7*, 261.
(175) Borm, J.; Huttner, G.; Zsolnai, L. *Angew. Chem., Int. Ed. Engl.* **1985**, *24*, 1069; *Angew. Chem.* **1985**, *97*, 1069.
(176) Delgado, E.; Jeffery, J. G.; Simmons, N. D.; Stone, F. G. A. *J. Chem. Soc., Dalton Trans.* **1986**, 869.
(177) Huttner, G.; Schneider, J.; Mohr, G.; von Seyerl, J. *J. Organomet. Chem.* **1980**, *191*, 161.
(178) Buccholz, D.; Huttner, G.; Imhof, W. *J. Organomet. Chem.* **1990**, *388*, 307.

(179) Knoll, K.; Huttner, G.; Zsolnai, L.; Orama, O.; Wasiucionek, M. *J. Organomet. Chem.* **1986**, *310*, 225.
(180) Kristjánsdóttir, S.; Moody, A. E.; Weberg, R. T.; Norton, J. R. *Organometallics* **1988**, *7*, 1983.
(181) Sunick, D. L.; White, P. S.; Schauer, C. K. *Organometallics* **1993**, *12*, 245.
(182) Bautista, M. T.; White, P. S.; Schauer, C. K. *J. Am. Chem. Soc.* **1991**, *113*, 8963.
(183) Honrath, U.; Shu-Tang, L.; Vahrenkamp, H. *Chem. Ber.* **1985**, *118*, 132.
(184) Winter, A.; Zsolnai, L.; Huttner, G. *J. Organomet. Chem.* **1982**, *234*, 337.
(185) Brauer, D. J.; Hietkamp, S.; Sommer, H.; Stelzer, O. *J. Organomet. Chem.* **1985**, *281*, 187.
(186) Eber, B.; Huttner, G.; Imhof, W.; Duran, J. C.; Jeannin, Y. *J. Organomet. Chem.* **1991**, *426*, 87.
(187) Eber, B.; Huttner, G.; Emmerich, C.; Duran, J. C.; Heim, B.; Jeannin, Y. *J. Organomet. Chem.* **1991**, *419*, 43.
(188) Knoll, K.; Huttner, G.; Zsolnai, L.; Orama, O. *J. Organomet. Chem.* **1987**, *327*, 379.
(189) Knoll, K.; Huttner, G.; Fässler, T.; Zsolnai, L. *J. Organomet. Chem.* **1987**, *327*, 255.
(190) Knoll, K.; Huttner, G.; Zsolnai, L. *J. Organomet. Chem.* **1986**, *312*, C57.
(191) Buccholz, D.; Huttner, G.; Imhof, W.; Orama, O. *J. Organomet. Chem.* **1990**, *388*, 321.
(192) Knoll, K.; Huttner, G. *J. Organomet. Chem.* **1987**, *329*, 369.
(193) Knoll, K.; Huttner, G.; Zsolnai, L.; Orama, O. *Angew. Chem., Int. Ed. Engl.* **1986**, *25*, 1119; *Angew. Chem.* **1986**, *98*, 1099.
(194) Evertz, K.; Huttner, G. *Chem. Ber.* **1987**, *120*, 937.
(195) Knoll, K.; Huttner, G.; Evertz, K. *J. Organomet. Chem.* **1987**, *333*, 97.
(196) Castiglioni, M.; Giordano, R.; Sappa, E. *J. Organomet. Chem.* **1991**, *407*, 377.
(197) Buccholz, D.; Huttner, G.; Zsolnai, L. *J. Organomet. Chem.* **1990**, *381*, 97.
(198) Richter, F.; Müller, M.; Gärtner, N.; Vahrenkamp, H. *Chem. Ber.* **1984**, *117*, 2438.
(199) Huttner, G.; Frank, A.; Mohr, G. *Z. Naturforsch.* **1976**, *31b*, 1161.
(200) Müller, M.; Vahrenkamp, H. *Chem. Ber.* **1983**, *116*, 2311.
(201) Iwasaki, F.; Mays, M. J.; Raithby, P. R.; Taylor, P. L.; Wheatley, P. J. *J. Organomet. Chem.* **1981**, *213*, 185.
(202) Field, J. S.; Haines, R. J.; Smit, D. N. *J. Organomet. Chem.* **1982**, *224*, C49.
(203) Rowley, S. P.; White, P. S.; Schauer, C. K. *Inorg. Chem.* **1992**, *31*, 3158.
(204) Mays, M. J.; Taylor, P. L.; Henrick, K. *Acta Cryst.* **1982**, *B38*, 2261.
(205) Natarajan, K.; Scheidsteger, O.; Huttner, G. *J. Organomet. Chem.* **1981**, *221*, 301.
(206) Kewk, K.; Taylor, N. J.; Carty, A. J. *J. Am. Chem. Soc.* **1984**, *106*, 4636.
(207) Field, J. S.; Haines, R. J.; Smit, D. N. *J. Chem. Soc., Dalton Trans.* **1988**, 1315.
(208) Cullen, W. R.; Rettig, S. J.; Zhang, H. *Organometallics* **1992**, *11*, 1000.
(209) Cherkas A. A.; Corrigan, J. F.; Doherty, S.; MacLaughlin, S. A.; van Gastel, F.; Taylor, N. J.; Carty, A. J. *Inorg. Chem.* **1993**, *342*, 1662.
(210) Corrigan, J. F.; Doherty, S.; Taylor, N. J.; Carty, A. J. *Organometallics* **1993**, *12*, 1365.
(211) Corrigan, J. F.; Doherty, S.; Taylor, N. J.; Carty, A. J. *Organometallics* **1993**, *12*, 993.
(212) Johnson, B. F. G.; Layer, T. M.; Lewis, J.; Raithby, P. R.; Wong, W.-T. *J. Chem. Soc., Dalton Trans.* **1993**, 973.
(213) Charalambous, E.; Heuer, L.; Johnson, B. F. G.; Lewis, J.; Li, W.-S.; McPartlin, M.; Massey, A. D. *J. Organomet. Chem.* **1994**, *468*, C9.
(214) Mays, M. J.; Raithby, P. R.; Taylor, P. R.; Henrick, K. *J. Organomet. Chem.* **1982**, *224*, C45.
(215) Arif, A. M.; Cowley, A. H.; Pakulski, M.; Hursthouse, M. B.; Karauloz, A. *Organometallics* **1985**, *4*, 2227.
(216) Warren, D. S.; Gimarc, B. M. *J. Am. Chem. Soc.* **1992**, *114*, 5378.

(217) Gremaud, G.; Jungbluth, H.; Stoeckli-Evans, H.; Süss-Fink, G. *J. Organomet. Chem.* **1990**, *388*, 351.
(218) Herberhold, M.; Schamel, K.; Herrmann, G.; Gieren, A.; Ruiz-Pérez, C.; Hübner, T. *Z. Anorg. Allg. Chem.* **1988**, *562*, 49.
(219) Betz, H.; Gieren, A.; Hübner, T. *Acta Cryst. A (Fnd. Cryst.)* **1984**, *40*, C265.
(220) Süss-Fink, G.; Guldner, K.; Herberhold, M.; Gieren, A.; Hübner, T. *J. Organomet. Chem.* **1985**, *279*, 447.
(221) Johnson, B. F. G.; Lewis, J.; Massey, A. D.; Braga, D.; Grepioni, F. *J. Organomet. Chem.* **1989**, *369*, C43.
(222) Jackson, P. A.; Johnson, B. F. G.; Lewis, J.; Massey, A. D.; Braga, D.; Gradella, C.; Grepioni, F. *J. Organomet. Chem.* **1990**, *391*, 225.
(223) Fenske, D.; Hachgenei, J. *Angew. Chem., Int. Ed. Engl.* **1986**, *25*, 175; *Angew. Chem.* **1986**, *98*, 165.
(224) Lang, H.; Huttner, G.; Sigwarth, B.; Jibril, I.; Zsolnai, L.; Orama, O. *J. Organomet. Chem.* **1986**, *304*, 137.
(225) Vizi-Orosz, A.; Galamb, V.; Pályi, G.; Markó, L.; Bor, G.; Natile, G. *J. Organomet. Chem.* **1976**, *107*, 235.
(226) Taylor, N. J. *J. Chem. Soc., Chem. Commun.* **1985**, 478.
(227) Gourdon, A.; Jeannin, Y. *J. Organomet. Chem.* **1986**, *304*, C1.
(228) Sunick, D. L.; White, P. S.; Schauer, C. K. *Inorg. Chem.* **1993**, *32*, 5665.
(229) Colbran, S. B.; Johnson, B. F. G.; Lewis, J.; Sorrell, R. M. *J. Chem. Soc., Chem. Commun.* **1986**, 525.
(230) Natarajan, K.; Zsolnai, L.; Huttner, G. *J. Organomet. Chem.* **1981**, *220*, 365.
(231) van Hal, J. W.; Whitmire, K. H.; Zouchoune, B.; Halet, J.-F.; Saillard, J.-Y. *Inorg. Chem.* **1995**, *34*, 5455.
(232) Rheingold, A. L.; Sullivan, P. J. *J. Chem. Soc., Chem. Commun.* **1983**, 39.
(233) Whitmire, K. H.; Leigh, J. S.; Luo, S.; Shieh, M.; Fabiano, M. D.; Rheingold, A. L. *New J. Chem.* **1988**, *12*, 397.
(234) Cowley, A. H.; Pellerin, B.; Atwood, J. L.; Bott, S. G. *J. Am. Chem. Soc.* **1990**, *112*, 6734.
(235) Fermin, M C.; Ho, J.; Stephan, D. W. *Organometallics* **1995**, *14*, 4247.
(236) Ho, J.; Rousseau, R.; Stephan, D. W. *Organometallics* **1994**, *13*, 1918.
(237) Duttera, M. R.; Day, V. W.; Marks, T. J. *J. Am. Chem. Soc.* **1984**, *106*, 2907.
(238) Arif, A. M.; Cowley, A. H.; Pakulski, M.; Norman, N. C.; Orpen, A. G. *Organometallics* **1987**, *6*, 189.
(239) Flynn, K. M.; Murray, B. D.; Olmstead, M. M.; Power, P. P. *J. Am. Chem. Soc.* **1983**, *105*, 7460.
(240) Bartlett, R. A.; Rasika Dias, H. V.; Flynn, K. M.; Olmstead, M. M.; Power, P. P. *J. Am. Chem. Soc.* **1987**, *109*, 5699.
(241) Flynn, K. M.; Barlett, R. A.; Olmstead, M. M.; Power, P. P. *Organometallics* **1986**, *5*, 813.
(242) Lang, H.; Zsolnai, L.; Huttner, G. *Angew. Chem., Int. Ed. Engl.* **1983**, 976.
(243) Halet, J.-F.; Saillard, J.-Y. *J. Organomet. Chem.* **1987**, *327*, 365.
(244) Huttner, G.; Müller, H.-D.; Frank, A.; Lorenz, H. *Angew Chem., Int. Ed. Engl.* **1975**, *14*, 705; *Angew. Chem.* **1975**, *87*, 714.
(245) Strube, A.; Huttner, G.; Zsolnai, L. *Angew. Chem., Int. Ed. Engl.* **1988**, *27*, 1529; *Angew. Chem.* **1988**, *100*, 1586.
(246) Herrmann, W. A.; Koumbouris, B.; Zahn, T.; Ziegler, M. L. *Angew. Chem., Int. Ed. Engl.* **1984**, *23*, 812; *Angew. Chem.* **1984**, *96*, 802.
(247) Herrmann, W. A.; Koumbouris, B.; Schäfer, A.; Zahn, T.; Ziegler, M. L. *Chem. Ber.* **1985**, *118*, 2472.
(248) von Seyerl, J.; Sigwarth, B.; Huttner, G. *Chem. Ber.* **1981**, *114*, 1407.

(249) von Seyerl, J.; Sigwarth, B.; Huttner, G. *Chem. Ber.* **1981**, *114*, 727.
(250) Huttner, G.; Schmid, H.-G. *Angew. Chem., Int. Ed. Engl.* **1975**, *14*, 433; *Angew. Chem.* **1975**, *87*, 454.
(251) Huttner, G.; von Seyerl, J.; Marsili, M.; Schmid, H.-G. *Angew. Chem., Int. Ed. Engl.* **1975**, 14, 434; *Angew. Chem.* **1975**, *87*, 455.
(252) von Seyerl, J.; Moering, U.; Wagner, A.; Frank, A.; Huttner, G. *Angew. Chem., Int. Ed. Engl.* **1978**, *17*, 844; *Angew. Chem.* **1978**, *90*, 912.
(253) Huttner, G.; Sigwarth, B.; von Seyerl, J.; Zsolnai, L. *Chem. Ber.* **1982**, *115*, 2035.
(254) Arnold, F. P., Jr.; Ridge, D. P.; Rheingold, A. L. *J. Am. Chem. Soc.* **1995**, *117*, 4427.
(255) Weber, U.; Zsolnai, L.; Huttner, G. *J. Organomet. Chem.* **1984**, *260*, 281.
(256) von Seyerl, J.; Huttner, G. *Angew. Chem., Int. Ed. Engl.* **1978**, *17*, 843; *Angew. Chem.* **1978**, *90*, 911.
(257) Bringewski, F.; Huttner, G.; Imhof, W.; Zsolnai, L. *J. Organomet. Chem.* **1992**, *439*, 33.
(258) Davies, S. J.; Compton, N. A.; Huttner, G.; Zsolnai, L.; Garner, S. E. *Chem. Ber.* **1991**, *124*, 2731.
(259) Koide, Y.; Schauer, C. K. *Organometallics* **1993**, *12*, 4854.
(260) Koide, Y.; Bautista, M. T.; White, P. S.; Schauer, C. K. *Inorg. Chem.* **1992**, *31*, 3690.
(261) Delbaere, L. T. J.; Kruczynski, L. J.; McBride, D. W. *J. Chem. Soc., Dalton Trans.* **1973**, 307.
(262) Lang, H.; Huttner, G.; Zsolnai, L.; Mohr, G.; Sigwarth, B.; Weber, U.; Orama, O.; Jibril, I. *J. Organomet. Chem.* **1986**, *304*, 157.
(263) Churchill, M. R.; Fettinger, J. C.; Whitmire, K. H. *J. Organomet. Chem.* **1985**, *284*, 13.
(264) Scherer, O. J.; Braun, J.; Walther, P.; Heckmann, G.; Wolmershäuser, G. *Angew. Chem., Int. Ed. Engl.* **1991**, *30*, 852; *Angew. Chem.* **1991**, *103*, 861.
(265) Doedens, R. J. *Inorg. Chem.* **1969**, *8*, 570.
(266) Dekker, M.; Knox, G. R. *J. Chem. Soc., Chem. Commun.* **1967**, 1243.
(267) Tasi, M.; Powell, A. K.; Vahrenkamp, H. *Chem. Ber.* **1991**, *124*, 1549.
(268) Landesberg J. M.; Katz, L.; Olsen, C. *J. Org. Chem.* **1972**, *37*, 930.
(269) Clegg, W.; Sheldrick, G. M.; Stalke, D.; Bhaduri, S.; Khwaja, H. K. *Acta Cryst.* **1984**, *C40*, 2045.
(270) Bockman, T. M.; Kochi, J. K. *J. Am. Chem. Soc.* **1987**, *109*, 7725.
(271) Alper, H.; Paik, H.-N. *Nouv. J. Chim.* **1978**, *2*, 245.
(272) Baikie, P. E.; Mills, O. S. *J. Chem. Soc., Chem. Commun.* **1967**, 516.
(273) Song, J.-S.; Geoffroy, G. L.; Rheingold, A. L. *Inorg. Chem.* **1992**, *31*, 1505.
(274) Meij, R.; Stufkens, D. J.; Vrieze, K.; Brouwers, A. M. F.; Schagen, J. D.; Zwinselman, J. J.; Overbeek, A. R.; Stam, C. H. *J. Organomet. Chem.* **1979**, *170*, 337.
(275) Clegg, W.; Sheldrick, G. M.; Stalke, D.; Bhaduri, S.; Gopalkrishnan, K. *Acta Cryst.* **1984**, *C40*, 927.
(276) Bruce, M. I.; Humphrey, M. G.; Shawkataly, O. B.; Snow, M. R.; Tiekink, E. R. T. *J. Organomet. Chem.* **1986**, *315*, C51.
(277) Smieja, J. A.; Gladfelter, W. L. *Inorg. Chem.* **1986**, *25*, 2667.
(278) McGlinchey, M. J.; Stone, F. G. A. *J. Chem. Soc., Chem. Commun.* **1970**, 1265.
(279) Bautista, M. T.; Jordan, M. R.; White, P. S.; Schauer, C. K. *Inorg. Chem.d* **1993**, *32*, 5429.
(280) Vahrenkamp, H.; Wolters, D. *J. Organomet. Chem. 1982*, *224*, C17.
(281) Fässler, T.; Buccholz, D.; Huttner, G.; Zsolnai, L. *J. Organomet. Chem.* **1989**, *369*, 297.
(282) Bartsch, R.; Hietkamp, S.; Morton, S.; Stelzar, O. *J. Organomet. Chem.* **1981**, *222*, 263.
(283) Cook, S. L.; Evans, J.; Gray, L. R.; Webster, M. *J. Organomet. Chem.* **1982**, *236*, 367.
(284) Kouba, J. K.; Muetterties, E. L.; Thompson, M. R.; Day, V. W. *Organometallics* **1983**, *2*, 1065.
(285) Knoll, K.; Huttner, G.; Zsolnai, L. *J. Organomet. Chem.* **1986**, *307*, 237.

(286) Ohst, H. H.; Kochi, J. K. *Inorg. Chem.* **1986**, *25*, 2066.
(287) Ohst, H. H.; Kochi, J. K. *J. Am. Chem. Soc.* **1986**, *108*, 2897.
(288) Ohst, H. H.; Kochi, J. K. *J. Chem. Soc., Chem. Commun.* **1986**, 121.
(289) Bautista, M. T.; White, P. S.; Schauer, C. K. *J. Am. Chem. Soc.* **1994**, *116*, 2143.
(290) Seyferth, D.; Withers, H. P., Jr. *Organometallics* **1982**, *1*, 1294.
(291) Winter, A.; Jibril, I.; Huttner, G. *J. Organomet. Chem.* **1983**, *247*, 259.
(292) Fackler, J. P., Jr.; Mazany, A. M.; Seyferth, D.; Withers, H. P., Jr.; Wood, T. G.; Campana, C. F. *Inorg. Chim. Acta* **1984**, *82*, 31.
(293) Lindner, E.; Weiss, G. A.; Hiller, W.; Fawzi, R. *J. Organomet. Chem.* **1983**, *255*, 245.
(294) Kruger, G. J.; Lotz, S.; Linford, L.; van Dyk, M.; Raubenheimer, H. G. *J. Organomet. Chem.* **1985**, *280*, 241.
(295) Imhof, W.; Huttner, G.; Eber, B.; Günauer, D. *J. Organomet. Chem.* **1992**, *428*, 379.
(296) Süss-Fink, G.; Thewalt, U.; Klein, H.-P. *J. Organomet. Chem.* **1982**, *224*, 59.
(297) Huttner, G.; Mohr, G.; Frank, A.; Schubert, U. *J. Organomet. Chem.* **1976**, *118*, C73.
(298) Jacob, M.; Weiss, E. *J. Organomet. Chem.* **1977**, *131*, 263.
(299) Eveland, J. R.; Saillard, J.-Y.; Whitmire, K. H. *Inorg. Chem.* **1997**, *36*, 330.
(300) Whitmire, K. H.; Raghuveer, K. S.; Churchill, M. R.; Fettinger, J. C.; See, R. F. *J. Am. Chem. Soc.* **1986**, *108*, 2778.
(301) Whitmire, K. H.; Shieh, M.; Lagrone, C. B.; Robinson, B. H.; Churchill, M. R.; Fettinger, J. C.; See, R. F. *Inorg. Chem.* **1987**, *26*, 2798.
(302) Kahlal, S.; Halet, J.-F.; Saillard, J.-Y.; Whitmire, K. H. *J. Organomet. Chem.* **1994**, *478*, 1.
(303) Albright, T. A.; Yee, K. A.; Saillard, J.-Y.; Kahlal, S.; Halet, J.-F.; Leigh, J. S.; Whitmire, K. H. *Inorg. Chem.* **1991**, *30*, 1179.
(304) Whitmire, K. H.; Albright, T. A.; Kang, A.-K.; Churchill, M. R.; Fettinger, J. C. *Inorg. Chem.* **1986**, *25*, 2799.
(305) Jones, R. A.; Whittlesey, B. R. *J. Am. Chem. Soc.* **1985**, *107*, 1078.
(306) Mani, D.; Vahrenkamp, H. *Angew. Chem., Int. Ed. Engl.* **1985**, *24*, 424; *Angew. Chem.* **1985**, *97*, 428.
(307) Vahrenkamp, H.; Wolters, D. *Organometallics* **1982**, *1*, 874.
(308) Jaeger, T.; Aime, S.; Vahrenkamp, H. *Organometallics* **1986**, *5*, 245.
(309) Ohst, H. H.; Kochi, J. K. *Organometallics* **1986**, *5*, 1359.
(310) Richmond, M. G.; Korp, J. D.; Kochi, J. K. *J. Chem. Soc., Chem. Commun.* **1985**, 1102.
(311) Halet, J.-F.; Hoffmann, R.; Saillard, J.-Y. *Inorg. Chem.* **1985**, *24*, 1695.
(312) Halet, J.-F.; Saillard, J.-Y. *New. J. Chem.* **1987**, *11*, 315.
(313) Brauer, D. J.; Hasselkuß, G.; Hietkamp, S.; Sommer, H.; Stelzer, O. *Z. Naturforsch.* **1985**, *40b*, 961.
(314) van Gastel, F.; Agocs, L.; Cherkas, A. A.; Corrigan, J. F.; Doherty, S.; Ramachandran, R.; Taylor, M. J.; Carty, A. J. *J. Cluster Sci.* **1991**, *2*, 131.
(315) Natarajan, K.; Zsolnai, L.; Huttner, G. *J. Organomet. Chem.* **1981**, *209*, 85.
(316) Fernandez, J. M.; Johnson, B. F. G.; Lewis, J.; Raithby, P. R. *J. Chem. Soc., Chem. Commun.* **1978**, 1015.
(317) Lanfranchi, M.; Tiripicchio, A.; Sappa, E.; Carty, A. J. *J. Chem. Soc., Dalton Trans.* **1986**, 2737.
(318) Hansert, B.; Powell, A. K.; Vahrenkamp, H. *Chem. Ber.* **1991**, *124*, 2697.
(319) Jaeger, J. T.; Vahrenkamp, H. *Organometallics* **1988**, *7*, 1746.
(320) Jaeger, J. T.; Field, J. S.; Collison, D.; Speck, G. P.; Peake, B. M.; Hähnle, J.; Vahrenkamp, H. *Organometallics* **1988**, *7*, 1753.
(321) Feng, T.; Lau, P.; Imhof, W.; Huttner, G. *J. Organomet. Chem.* **1991**, *414*, 89.
(322) Vahrenkamp, H.; Wucherer, E. *Angew. Chem., Int. Ed. Engl.* **1981**, *20*, 680; *Angew. Chem.* **1981**, *93*, 715.

(323) Ryan, R. C.; Dahl, L. F. *J. Am. Chem. Soc.* **1975**, *97*, 6904.
(324) Ryan, R. C.; Pittman, C. U., Jr.; O'Connor, J. P.; Dahl, L. F. *J. Organomet. Chem.* **1980**, *193*, 247.
(325) Richmond, M. G.; Kochi, J. K. *Inorg. Chem.* **1986**, *25*, 656.
(326) Richmond, M. G.; Kochi, J. K. *Inorg. Chem.* **1986**, *25*, 1334.
(327) Queisser, J.; Frenske, D. Z. *Anorg. Allg. Chem.* **1994**, *620*, 58.
(328) Burkhardt, E. W.; Mercer W. C.; Geoffroy, G. L.; Rheingold, A. L.; Fultz, W. C. *J. Chem. Soc., Chem. Commun.* **1983**, 1251.
(329) Lower, L. D.; Dahl, L. F. *J. Am. Chem. Soc.* **1976**, *98*, 5046.
(330) Fenske, D.; Hachgenei, J.; Rogel, F. *Angew. Chem., Int. Ed. Engl.* **1984**, *23*, 982; *Angew. Chem.* **1984**, *96*, 959.
(331) Rieck, D. F.; Rae, A. D.; Dahl, L. F. *J. Chem. Soc., Chem. Commun.* **1993**, 585.
(332) Rieck, D. F.; Gavney, J. A., Jr.; Norman, R. L.; Hayashi, R. K.; Dahl, L. F. *J. Am. Chem. Soc.* **1992**, *114*, 10369.
(333) Eichhöfer, A.; Fenske, D.; Holstein, W. *Angew. Chem., Int. Ed. Engl.* **1993**, *32*, 242; *Angew. Chem.* **1993**, *105*, 245.
(334) Fenske, D.; Merzweiler, K.; Ohmer, J. *Angew. Chem., Int. Ed. Engl.* **1988**, *27*, 1512; *Angew. Chem.* **1988**, *100*, 1572.
(335) Fenske, D.; Fleischer, H.; Persau, C. *Angew. Chem., In. Ed. Engl.* **1989**, *28*, 1665; *Angew. Chem.* **1989**, *101*, 1740.
(336) Fenske, D.; Persau, C. *Z. Anorg. Allg. Chem.* **1991**, *593*, 61.
(337) Whitmire, K. H.; Eveland, J. R. *J. Chem. Soc., Chem. Commun.* **1994**, 1335.
(338) Furet, E.; Le Beuze, A.; Halet, J.-F.; Saillard, J.-Y. *J. Am. Chem. Soc.* **1994**, *116*, 274.
(339) Furet, E.; Le Beuze, A.; Halet, J.-F.; Saillard, J.-Y. *J. Am. Chem. Soc.* **1994**, *117*, 4936.
(340) Rösch, N.; Ackerman, L.; Pacchioni, G. *Inorg. Chem.* **1993**, *32*, 2963.
(341) Longoni, G. *Pure Appl. Chem.* **1990**, *62*, 1183.
(342) King, R. B. *Revue Roumaine de Chimie* **1991**, *38*, 353.
(343) Ahlrichs, R.; Fenske, D.; Fromm, K.; Krautscheid, H.; Krautschied, U.; Treutler, O. *Chem. Eur. J.* **1996**, *2*, 238.
(344) Rieck, D. F.; Montag, R. A.; McKechnie, T. S.; Dahl, L. F. *J. Am. Chem. Soc.* **1986**, *108*, 1330.
(345) DesEnfants, R. E., II; Gavney, J. A., Jr.; Hayashi, R. K.; Rae, A. D.; Dahl, L. F.; Bjarnason, A. *J. Organomet. Chem.* **1990**, *383*, 543.
(346) Albano, V. G.; Demartin, F.; Iapalucci, M. C.; Laschi, F.; Longoni, G.; Sironi, A.; Zanello, P. *J. Chem. Soc., Dalton Trans.* **1991**, 739.
(347) Albano, V. G.; Demartin, F.; Iapalucci, M. C.; Longoni, G.; Sironi, A.; Zanotti, V. *J. Chem. Soc., Chem. Commun.* **1990**, 547.
(348) Albano, V. G.; Demartin, F.; Iapalucci, M. C.; Longoni, G.; Monari, M.; Zanello, P. *J. Chem. Soc., Dalton Trans.* **1992**, 497.
(349) Grohmann, A.; Riede, J.; Schmidbaur, H. *Nature* **1990**, *345*, 140.
(350) Angermaier, K.; Schmidbaur, H. *Inorg. Chem.* **1995**, *34*, 3120.
(351) Martinengo, S.; Ciani, G.; Sironi, A.; Heaton, B. T.; Mason, J. *J. Am. Chem. Soc.* **1979**, 101, 7095.
(352) Bonfichi, R.; Ciani, G.; Sironi, A.; Martinengo, S. *J. Chem. Soc., Dalton Trans.* **1983**, 253.
(353) Anson, C. E.; Attard, J. P.; Johnson, B. F. G.; Lewis, J.; Mace, J. M.; Powell, D. B. *J. Chem. Soc., Chem. Commun.* **1986**, 1715.
(354) Creighton, J. A.; Pergola, R. D.; Heaton, B. T.; Martinengo, S.; Strona, L.; Willis, D. A. *J. Chem. Soc., Chem. Commun.* **1982**, 864.
(355) Brint, R. P.; O'Cuill, K.; Spalding, T. R.; Deeney, F. A. *J. Organomet. Chem.* **1983**, *247*, 61.

(356) Tachikawa, M.; Stein, J.; Muetterties, E. L.; Teller, R. G.; Beno, M. A.; Gebert, E.; Williams, J. M. *J. Am. Chem. Soc.* **1980**, *102*, 6648.
(357) Fjare, D. E.; Gladfelter, W. L. *Inorg. Chem.* **1981**, *20*, 3533
(358) Brint, R. P.; Collins, M. P.; Spalding, T. R.; Deeney, F. A. *J. Organomet. Chem.* **1983**, *258*, C57.
(359) Gourdon, A.; Jeannin, Y. *J. Organomet. Chem.* **1992**, *440*, 353.
(360) Fjare, D. E.; Gladfelter, W. L. *J. Am. Chem. Soc.* **1984**, *106*, 4799.
(361) Braga, D.; Johnson, B. F. G.; Lewis, J.; Mace, J. M.; McPartlin, M.; Puga, J.; Nelson. W. J. H.; Raithby, P. R.; Whitmire, K. H. *J. Chem. Soc., Chem. Commun.* **1982**, 1081.
(362) Collins, M. A.; Johnson, B. F. G.; Lewis, J.; Mace, J.; Morris, J.; McPartlin, M.; Nelson, W. J. H.; Puga, J.; Raithby, P. R. *J. Chem. Soc., Chem. Commun.* **1983**, 689.
(363) Blohm, M. L.; Gladfelter, W. L. *Organometallics* **1985**, *4*, 45.
(364) Attard, J. P.; Johnson, B. F. G.; Lewis, J.; Mace, J. M.; Raithby, P. R. *J. Chem. Soc., Chem. Commun.* **1985**, 1526.
(365) Hourihane, R.; Spalding, T. R.; Ferguson, G.; Deeney, T.; Zanello, P. *J. Chem. Soc., Dalton Trans.* **1993**, 43.
(366) Gourdon, A.; Jeannin, Y. *J. Organomet. Chem.* **1985**, *290*, 199.
(367) Blohm, M. L.; Fjare, D. E.; Gladfelter, W. L. *Inorg. Chem.* **1983**, *1983*, 1004.
(368) Martinengo, S.; Ciani, G.; Sironi, A. *J. Chem. Soc., Chem. Commun.* **1984**, 1577.
(369) Martinengo, S.; Ciani, G.; Sironi, A. *J. Am. Chem. Soc.* **1982**, *104*, 328.
(370) Martinengo, S.; Ciani, G.; Sironi, A. *J. Chem. Soc., Chem. Commun.* **1986**, 1742.
(371) Sunick, D. L.; White, P. S.; Schauer, C. K. *Angew. Chem., Int. Ed. Engl.* **1994**, *33*, 75; *Angew. Chem.* **1994**, *106*, 108.
(372) MacLaughlin, S. A.; Taylor, N. J.; Carty, A. J. *Inorg. Chem.* **1983**, *22*, 1409.
(373) Colbran, S. B.; Hay, C. M.; Johnson, B. F. G.; Lahoz, F. J.; Lewis, J.; Raithby, P. R. *J. Chem. Soc., Chem. Commun.* **1986**, 1766.
(374) Chini, P.; Ciani, G.; Martinengo, S.; Sironi, A.; Longhetti, L.; Heaton, B. T. *J. Chem. Soc., Chem. Commun.* **1979**, 188.
(375) Vidal, J. L.; Walker, W. E.; Pruett, R. L.; Schoening, R. C. *Inorg. Chem.* **1979**, *18*, 129.
(376) Heaton, B. T.; Strona, L.; Della Pergola, R.; Vidal, J. L.; Schoening, R. C. *J. Chem. Soc., Dalton Trans.* **1983**, 1941.
(377) Gansow, O. A.; Gill, D. S.; Bennis, F. J.; Hutchinson, J. R.; Vidal, J. L.; Schoening, R. C. *J. Am. Chem. Soc.* **1980**, *102*, 2449.
(378) Vidal, J. L.; Walker, W. E.; Schoening, R. C. *Inorg. Chem.* **1981**, *20*, 238.
(379) Fenske, D.; Holstein, W. *Angew. Chem., Int. Ed. Engl.* **1994**, *33*, 1290; *Angew. Chem.* **1994**, *106*, 1311.
(380) Zeller, E.; Beruda, H.; Schmidbaur, H. *Chem. Ber.* **1993**, *126*, 2033.
(381) Bachman, R. E.; Schmidbaur, H. *Inorg. Chem.* **1996**, *35*, 1399.
(382) Beruda, H.; Zeller, E.; Schmidbaur, H. *Chem. Ber.* **1993**, *126*, 2037.
(383) Zeller, E.; Schmidbaur, H. *J. Chem. Soc., Chem. Commun.* **1993**, 69.
(384) Vidal, J. L. *Inorg. Chem.* **1981**, *20*, 243.
(385) Vidal, J.; Troup, J. M. *J. Organomet. Chem.* **1981**, *213*, 351.
(386) Teo, B. K.; Hall, M B.; Fenske, R. F.; Dahl, L. F. *Inorg. Chem.* **1975**, *14*, 3103.
(387) Huttner, G.; Sigwarth, B.; Scheidsteger, O.; Zsolnai, L.; Orama, O. *Organometallics* **1985**, *4*, 326.
(388) Doedens, R. J.; Ibers, J. A. *Inorg. Chem.* **1969**, *8*, 2709.
(389) Goh, L. Y.; Wong, R. C. S.; Sinn, E. *Organometallics* **1993**, *12*, 888.
(390) Goh, L. Y.; Chu, C. K.; Wong, R. C. S.; Hambley, T. W. *J. Chem. Soc., Dalton Trans.* **1989**, 1951.
(391) Goh, L. Y.; Chen, W.; Wong, R. C. S.; Karaghiosoff, K. *Organometallics* **1995**, *14*, 3886.

(392) Scherer, O. J.; Sitzmann, H.; Wolmershäuser, G. *Angew. Chem., Int. Ed. Engl.* **1985**, *24*, 351; *Angew. Chem.* **1985**, *97*, 358.
(393) De, R. L.; Wolters, D.; Vahrenkamp, H. *Z. Naturforsch.* **1986**, *41b*, 283.
(394) Vahrenkamp, H.; Wolters, D. *Angew. Chem., Int. Ed. Engl.* **1983**, *22*, 154; *Angew. Chem.* **1983**, *95*, 152.
(395) Campana, C.; Vizi-Orosz, A.; Palyi, G.; Markó, L.; Dahl, L. F. *Inorg. Chem.* **1979**, *18*, 3054.
(396) Vizi-Orosz, A.; Pályi, G.; Markó, L.; Boese, R.; Schmid, G. *J. Organomet. Chem.* **1985**, *288*, 179.
(397) Scherer, O. J.; Sitzmann, H.; Wolmershäuser, G. *J. Organomet. Chem.* **1986**, *309*, 77.
(398) Sullivan, P. J.; Rheingold, A. L. *Organometallics* **1982**, *1*, 1547.
(399) Foust, A. S.; Foster, M. S.; Dahl, L. F. *J. Am. Chem. Soc.* **1969**, *91*, 5633.
(400) Foust, A. S.; Campana, C. F.; Sinclair, J. D.; Dahl, L. F. *Inorg. Chem.* **1979**, *18*, 3047.
(401) Müller, M.; Vahrenkamp, H. *J. Organomet. Chem.* **1983**, *252*, 95.
(402) DiMaio, A.-J.; Rheingold, A. L. *J. Chem. Soc., Chem. Commun.* **1987**, 404.
(403) Harper, J. R.; Rheingold, A. L. *J. Organomet. Chem.* **1990**, *390*, C36.
(404) Clegg, W.; Compton, N. A.; Errington, R. J.; Norman, N. C. *Polyhedron* **1988**, *7*, 2239.
(405) Das, B. K.; Kanatzidis, M. G. *Inorg. Chem.* **1995**, *34*, 6505.
(406) Bantel, H.; Hanser, B.; Powell, A. K.; Tasi, M.; Vahrenkamp, H. *Angew. Chem., Int. Ed. Engl.* **1989**, *28*, 1059; *Angew. Chem.* **1989**, *101*, 1084.
(407) Scheer, M.; Becker, U. *Phosphorus, Sulfur and Silicon* **1994**, *93/94*, 257.
(408) Song, L.-C.; Hu, Q.-M. *J. Organomet. Chem.* **1991**, *414*, 219.
(409) Hou, Z.; Breen, T. L.; Stephan, D. W. *Organometallics* **1993**, *12*, 3158.
(410) Green, J. C.; Green, M. L. H.; Morris, G. E. *J. Chem. Soc., Chem. Commun.* **1974**, 212.
(411) Borm, J.; Zsolnai, L.; Huttner, G. *Angew. Chem., Int. Ed. Engl.* **1983**, *22*, 977; *Angew. Chem.* **1983**, *95*, 1018.
(412) Flynn, K. M.; Hope, H.; Murray, B. D.; Olmstead, M. M.; Power, P. P. *J. Am. Chem. Soc.* **1983**, 105, 7750.
(413) Mathieu, R.; Caminade, A.-M.; Majoral, J.-P.; Attali, S.; Sanchez, M. *Organometallics* **1986**, *5*, 1914.
(414) Caminade, A.-M.; Majoral, J.-P.; Sanchez, M.; Mathieu, R.; Attali, S.; Grand, A. *Organometallics* **1987**, *6*, 1459.
(415) Flynn, K. M.; Olmstead, M. M.; Power, P. P. *J. Am. Chem. Soc.* **1983**, *105*, 2085.
(416) Fenske, D.; Merzweiler, K. *Angew. Chem., Int. Ed. Engl.* **1984**, *23*, 635; *Angew. Chem.* **1984**, *96*, 600.
(417) Schäfer, H.; Binder, D.; Fenske, D. *Angew. Chem., Int. Ed. Engl.* **1985**, *24*, 522; *Angew. Chem.* **1985**, *97*, 523.
(418) Chatt, J.; Hitchcock, P. B.; Pidcock, A.; Warrens, C. P.; Dixon, K. R. *J. Chem. Soc., Dalton Trans.* **1984**, 2237.
(419) Chatt, J.; Hitchcock, P. B.; Pidcock, A.; Warrens, C. P.; Dixon, K. R. *J. Chem. Soc., Chem. Commun.* **1982**, 932.
(420) Huttner, G.; Schmid, H.-G.; Frank, A.; Orama, O. *Angew. Chem., Int. Ed. Engl.* **1976**, 234.
(421) Elmes, P. S.; Leverett, P.; West, B. O. *J. Chem. Soc., Chem. Commun.* **1971**, 747.
(422) Huttner, G.; Weber, U.; Sigwarth, B.; Scheidstger, O. *Angew. Chem., Int. Ed. Engl.* **1982**, *21*, 215; *Angew Chem.* **1982**, *94*, 210.
(423) Fenske, D.; Merzweiler, K. *Angew. Chem., Int. Ed. Engl.* **1986**, *25*, 338; *Angew. Chem.* **1986**, *98*, 357.
(424) Scherer, O. J.; Pfeiffer, K.; Heckmann, G.; Wolmershäuser, G. *J. Organomet. Chem.* **1992**, *425*, 141.

(425) Evans, W. J.; Gonzales, S. L.; Ziller, J. W. *J. Am. Chem. Soc.* **1991**, *113*, 9880.
(426) Weber, L.; Reizig, K.; Bungardt, D.; Boese, R. *Chem. Ber.* **1987**, *120*, 1421.
(427) Weber, L.; Bungardt, D.; Müller, A.; Bögge, H. *Organometallics* **1989**, *8*, 2800.
(428) Leigh, J. S.; Whitmire, K. H. *J. Am. Chem. Soc.* **1989**, *111*, 2726.
(429) Sigwarth, B.; Zsolnai, L.; Berke, H.; Huttner, G. *J. Organomet. Chem.* **1982**, *226*, C5.
(430) Huttner, G.; Weber, U.; Zsolnai, L. *Z. Naturforsch.* **1982**, *37B*, 707.
(431) Plößl, K.; Huttner, G.; Zsolnai, L. *Angew. Chem., Int. Ed. Engl.* **1989**, *28*, 446; *Angew. Chem.* **1989**, *101*, 482.
(432) Barr, M. E.; Dahl, L. F. *Organometallics* **1991**, *10*, 3991.
(433) Scherer, O. J.; Swarowsky, M.; Wolmershäuser, G. *Angew. Chem., Int. Ed. Engl.* **1988**, *27*, 405; *Angew. Chem.* **1988**, *100*, 423.
(434) Scherer, O. J.; Vondung, J.; Wolmershäuser, G. *J. Organomet. Chem.* **1989**, *376*, C35.
(435) Fitzpatrick, N. J.; Groarke, P. J.; Nguyen, M. T. *Polyhedron* **1992**, *11*, 2517.
(436) Di Vaira, M.; Midollini, S.; Sacconi, L. *J. Am. Chem. Soc.* **1979**, *101*, 1757.
(437) Scherer, O. J.; Blath, C.; Heckmann, G.; Wolmershäuser, G. *J. Organomet. Chem.* **1991**, *409*, C15.
(438) Whitmire, K. H.; Churchill, M. R.; Fettinger, J. C. *J. Am. Chem. Soc.* **1985**, *107*, 1056.
(439) Scheer, M.; Friedrich, G.; Schuster, K. *Angew. Chem., Int. Ed. Engl.* **1993**, *32*, 593; *Angew. Chem.* **1993**, *105*, 641.
(440) Cecconi, F.; Dapporto, P.; Midollini, S.; Sacconi, L. *Inorg. Chem.* **1978**, *17*, 3292.
(441) Cecconi, F.; Ghilardi, C. A.; Midollini, S.; Orlandini, A. *J. Chem. Soc., Chem. Commun.* **1982**, 229.
(442) Ghilardi, C. A.; Midollini, S.; Orlandini, A.; Sacconi, L. *Inorg. Chem.* **1980**, *19*, 301.
(443) Di Vaira, M.; Ghilardi, C. A.; Midollini, S.; Sacconi, L. *J. Am. Chem. Soc.* **1978**, *100*, 2550.
(444) Di Vaira, M.; Sacconi, L.; Stoppioni, P. *J. Organomet. Chem.* **1983**, *250*, 183.
(445) Bianchini, C.; Mealli, C.; Meli, A.; Sacconi, L. *Inorg. Chim. Acta* **1979**, *37*, L543.
(446) Midollini, S.; Orlandini, A.; Sacconi, L. *Angew. Chem., Int. Ed. Engl.* **1979**, *18*, 81; *Angew. Chem.* **1979**, *91*, 93.
(447) Capozzi, G.; Chiti, L.; Di Vaira, M.; Peruzzini, M.; Stoppioni, P. *J. Chem. Soc., Chem. Commun.* **1986**, 1799.
(448) Dapporto, P., Sacconi, L.; Stoppioni, P.; Zanobini, F. *Inorg. Chem.* **1981**, *20*, 3834.
(449) Foust, A. S.; Foster, M. S.; Dahl, L. F. *J. Am. Chem. Soc.* **1969**, *91*, 5631.
(450) Bernal, I.; Brunner, H.; Meier, W.; Pfister, H.; Wachter, J.; Ziegler, M. L. *Angew. Chem., Int. Ed. Engl.* **1984**, *23*, 438; *Angew. Chem.* **1984**, *96*, 428.
(451) Scherer, O. J.; Wiedeman, W.; Wolmershäuser, G. *Chem. Ber.* **1990**, *123*, 3.
(452) Scherer, O. J.; Braun, J.; Walther, P.; Wolmershäuser, G. *Chem. Ber.* **1992**, *125*, 2661.
(453) Scherer, O. J.; Werner, B.; Heckmann, G.; Wolmershäuser, G. *Angew. Chem., Int. Ed. Engl.* **1991**, *30*, 553; *Angew. Chem.* **1991**, *103*, 562.
(454) Di Vaira, M.; Mani, F.; Moneti, S.; Peruzzini, M.; Sacconi, L.; Stoppioni, P. *Inorg. Chem.* **1985**, *24*, 2230.
(455) Di Vaira, M.; Midollini, S.; Sacconi, L.; Zanobini, F. *Angew. Chem., Int. Ed. Engl.* **1978**, *17*, 676; *Angew. Chem.* **1978**, *90*, 720.
(456) King, R. B.; Wu, F.-J.; Holt, E. M. *J. Organomet. Chem.* **1986**, *314*, C27.
(457) Barth, A.; Huttner, G.; Fritz, M.; Zsolnai, L. *Angew. Chem., Int. Ed. Engl.* **1990**, *29*, 929; *Angew. Chem.* **1990**, *102*, 956.
(458) DiMaio, A.-J.; Rheingold, A. L. *Inorg. Chem.* **1990**, *29*, 798.
(459) Scherer, O. J.; Swarowsky, M.; Swarowsky, H.; Wolmershäuser, G. *Angew. Chem., Int. Ed. Engl.* **1988**, *27*, 694; *Angew. Chem.* **1988**, *100*, 738.
(460) Scherer, O. J.; Vondung, J.; Wolmershäuser, G. *Angew. Chem., Int. Ed. Engl.* **1989**, *28*, 1355; *Angew. Chem.* **1989**, *101*, 1359.

(461) Barr, M. E.; Smith, S. K.; Spencer, B.; Dahl, L. F. *Organometallics* **1991**, *10*, 3983.
(462) Scheer, M.; Herrmann, E.; Sieler, J.; Oehme, M. *Angew. Chem., Int. Ed. Engl.* **1991**, *30*, 969; *Angew. Chem.* **1991**, *103*, 1023.
(463) Schmid, G.; Kempny, H.-P. *Z. Anorg. Allg. Chem.* **1977**, *432*, 160.
(464) Rheingold, A. L.; Fountain, M. E. *Organometallics* **1984**, *3*, 1417.
(465) Scheer, M.; Becker, U.; Huffman, J. C.; Chisholm, M. H. *J. Organomet. Chem.* **1993**, *461*, C1.
(466) Scherer, O. J.; Swarowsky, M.; Wolmershäuser, G. *Organometallics* **1989**, *8*, 841.
(467) Scheer, M.; Troitzsch, C.; Jones, P. G. *Angew. Chem., Int. Ed. Engl.* **1992**, *31*, 1377; *Angew. Chem.* **1992**, *104*, 1395.
(468) Scheer, M.; Troitzsch, C.; Hilfert, L.; Dargatz, M.; Kleinpeter, E.; Jones, P. G.; Sieler, J. *Chem. Ber.* **1995**, *128*, 251.
(469) Ginsberg, A. P.; Lindsell, W. E. *J. Am. Chem. Soc.* **1971**, *93*, 2082.
(470) Ginsberg, A. P.; Lindsell, W. E.; McCullough, K. J.; Sprinkle, C. R.; Welch, A. J. *J. Am. Chem. Soc.* **1986**, *108*, 403.
(471) Scheer, M.; Dargatz, M.; Rufińska, A. *J. Organomet. Chem.* **1992**, *440*, 327.
(472) Scherer, O. J., Pfeiffer, K.; Wolmershäuser, G. *Chem. Ber.* **1992**, *125*, 2367.
(473) Gatehouse, B. M. *Chem. Commun.* **1969**, 948.
(474) Elmes, P. S.; West, B. O. *J. Organomet. Chem.* **1971**, *32*, 365.
(475) Scherer, O. J.; Brück, T. *Angew. Chem., Int. Ed. Engl.* **1987**, *26*, 59; *Angew. Chem.* **1987**, *99*, 59.
(476) Jemmis, E. D.; Reddy, A. C. *Organometallics* **1988**, *7*, 1561.
(477) Tremel, W.; Hoffmann, R.; Kertesz, M. *J. Am. Chem. Soc.* **1989**, *111*, 2030.
(478) Baudler, M.; Etzbach, T. *Angew. Chem., Int. Ed. Engl.* **1991**, *30*, 580; *Angew. Chem.* **1991**, *103*, 590.
(479) Rink, B.; Scherer, O. J.; Heckman, G.; Wolmershäuser, G. *Chem. Ber.* **1992**, *125*, 1011.
(480) Baudler, M.; Akpapoglou, Z.; Ouzounis, D.; Wasgestian, F.; Meinigke, B.; Budzikiewicz, H.; Münster, H. *Angew. Chem., Int. Ed. Engl.* **1988**, *27*, 280; *Angew. Chem.* **1988**, *100*, 288.
(481) Detzel, M.; Mohr, T.; Scherer, O. J.; Wolmershäuser, G. *Angew. Chem., Int. Ed. Engl.* **1994**, *33*, 1110; *Angew. Chem.* **1994**, *106*, 1142.
(482) Detzel, M.; Friedrich, G.; Scherer, O. J.; Wolmershäuser, G. *Angew. Chem., Int. Ed. Engl.* **1995**, *34*, 1321; *Angew. Chem.* **1995**, *107*, 1454.
(483) Rink, B.; Scherer, O. J.; Wolmershäuser, G. *Chem. Ber.* **1995**, *128*, 71.
(484) Rheingold, A. L.; Churchill, M. R. *J. Organomet. Chem.* **1983**, *243*, 165.
(485) Scherer, O. J.; Blath, C.; Wolmershäuser, G. *J. Organomet. Chem.* **1990**, *387*, C21.
(486) Scherer, O. J.; Swarowsky, H.; Wolmershäuser, G.; Kaim, W.; Kohlmann, S. *Angew. Chem., Int. Ed. Engl.* **1987**, *26*, 1153; *Angew. Chem.* **1987**, *99*, 1178.
(487) Scherer, O. J.; Schwalb, J.; Swarowsky, H.; Wolmershäuser, G.; Kaim, W.; Gross, R. *Chem. Ber.* **1988**, *121*, 443.
(488) Reddy, A. C.; Jemmis, E. D.; Scherer, O. J.; Winter, R.; Heckmann, G.; Wolmershäuser, G. *Organometallics* **1992**, *11*, 3894.
(489) Hey, E.; Lappert, M. F.; Atwood, J. L.; Bott, S. G. *J. Chem. Soc., Chem. Commun.* **1987**, 597.
(490) Tang, H.; Hoffman, D. M.; Albright. T. A.; Deng, H.; Hoffman, R. *Angew. Chem., Int. Ed. Engl.* **1993**, *32*, 1616; *Angew. Chem.* **1993**, *105*, 1682.
(491) Dehnicke, K. *Angew. Chem., Int. Ed. Engl.* **1992**, *31*, 424; *Angew. Chem.* **1992**, *104*, 437.
(492) Cecconi, F.; Ghilardi, C. A.; Midollini, S.; Orlandini, A. *Inorg. Chem.* **1986**, *25*, 1776.
(493) Rheingold, A. L.; Fountain, M. E. *Organometallics* **1986**, *5*, 2410.
(494) Detzel, M.; Pfeiffer, K.; Scherer, O. J.; Wolmershäuser, G. *Angew. Chem., Int. Ed. Engl.* **1993**, *32*, 914; *Angew. Chem.* **1993**, *105*, 936.

(495) Hönle, W.; von Schnering, H. G.; Fritz, G.; Schneider, H.-W. *Z. Anorg. Allg. Chem.* **1990**, *584*, 51.
(496) Fritz, G.; Layher, E.; Hönle, W.; von Schnering, H. G. *Z. Anorg. Allg. Chem.* **1991**, *595*, 67.
(497) Fritz, G.; Layher, E.; Schneider, H.-W. *Z. Anorg. Allg. Chem.* **1991**, *598/599*, 111.
(498) Fritz, G.; Schneider, H.-W.; Hönle, W.; von Schnering, H. G. *Z. Anorg. Allg. Chem.* **1990**, *585, 51.*
(499) Fritz, G.; Mayer, B. *Z. Anorg. Allg. Chem.* **1992**, *610*, 51.
(500) Charles, S.; Fettinger, J. C.; Eichhorn, B. W. *Inorg. Chem.* **1996**, *35*, 1540.
(501) Charles, S.; Eichhorn, B. W.; Rheingold, A. L.; Bott, S. G. *J. Am. Chem. Soc.* **1994**, *116*, 8077.
(502) Bolle, U.; Tremel, W. *J. Chem. Soc., Chem., Commun.* **1994**, 217.
(503) Eichhorn, B. W.; Haushalter, R. C.; Huffman, J. C. *Angew. Chem., Int. Ed. Engl.* **1989**, *28*, 1032; *Angew .Chem.* **1989**, *101*, 1081.
(504) Charles, S.; Eichhorn, B. W.; Bott, S. G. *J. Am. Chem. Soc.* **1993**, *115*, 5837.
(505) Goh, L. Y.; Wong, R. C. S.; Sinn, E. *J. Chem. Soc., Chem. Commun.* **1990**, 1484.
(506) Scherer, O. J.; Höbel, B.; Wolmershäuser, G. *Angew. Chem., Int. Ed. Engl.* **1992**, *31*, 1027; *Angew. Chem.* **1992**, *104*, 1042.
(507) Rheingold, A. L.; Fountain, M. E.; DiMaio, A.-J. *J. Am. Chem. Soc.* **1987**, *109*, 141.
(508) Scherer, O. J.; Winter, R.; Heckmann, G.; Wolmershäuser, G. *Angew. Chem., Int. Ed. Engl.* **1991**, *30*, 850; *Angew. Chem.* **1991**, *103*, 860.
(509) DiMaio, A.-J.; Rheingold, A. L. *Organometallics* **1987**, *6*, 1138.
(510) Kubat-Martin, K. A.; Spencer, B.; Dahl, L. F. *Organometallics* **1986**, *6*, 2580.
(511) Whitmire, K. H. Unpublished results.
(512) Eveland J. R.; Saillard, J.-Y.; Whitmire, K. H. *Inorg. Chem.* **1997**, *36*, 4387.
(513) Ang, H. G.; West, B. O. *Aust. J. Chem.* **1967**, *20*, 1133.
(514) Bush, M. A.; Cook, V. R.; Woodward, P. *Chem. Commun.* **1967**, 630.
(515) Forster, A.; Cundy, C. S.; Green, M.; Stone, F. G. A. *Inorg. Nucl. Chem. Lett.* **1966**, *2*, 233.
(516) Ang, H. G.; West, B. O. *Aust. J. Chem.* **1967**, *20*, 1133.
(517) Lavigne, G.; Bonnet, J. J. *Inorg. Chem.* **1981**, *20*, 2713.
(518) Demartin, F.; Manassero, M.; Sansoni, M.; Garlaschelli, L.; Sartorelli, U. *J. Organomet. Chem.* **1981**, *204*, C10.
(519) Bagga, M. M.; Pauson, P. L.; Preston, F. J.; Reed, R. I. *Chem. Commun.* **1965**, 543.
(520) Bagga, M. M.; Baikie, P. E.; Mills, O. S.; Pauson, P. L. *Chem. Commun.* **1967**, 1106.

ADVANCES IN ORGANOMETALLIC CHEMISTRY, VOL. 42

Transition-Metal Complexes of Arynes, Strained Cyclic Alkynes, and Strained Cyclic Cumulenes

WILLIAM M. JONES

Department of Chemistry
University of Florida
Gainesville, Florida 32611

JERZY KLOSIN

Catalysis Laboratory
The Dow Chemical Company
Midland, Michigan 48674

I

INTRODUCTION

If incorporated into rings with less than nine members, both alkynes[1–3] and cumulenes[4] are strained due to distortion from linearity that is imposed on carbon by *sp* hybridization and, in the case of even-numbered cumulenes (e.g., allene), distortion from orthogonality of adjacent double bonds. As

0065-3055/98 $25.00

a result, the smallest unsubstituted cycloalkyne and cyclic cumulene with lifetimes long enough for isolation are cyclooctyne and 1,2-cyclononadiene.

Stabilizing highly reactive organic molecules by complexing to transition metals is well documented.[5,6] This is particularly effective for relieving strain in cycloalkynes due to the normal distortion from linearity that results from complexing metals to triple bonds,[7] and numerous transition-metal complexes of strained cycloalkynes and arynes have been prepared. Two excellent reviews of this subject appeared in the late 1980s.[2,8] This article brings these two reviews up to date (through 1996). In this coverage, earlier references to complexes of arynes and cyclic alkynes are limited to those that were not included in previous reviews or are important to provide continuity and appropriate historical perspective.

Strain in cyclic cumulenes is similarly relieved by complexing one of the double bonds to a transition metal. Most of the research in this area has appeared since the late 1980s, and to date no review has appeared. The second purpose of this article is to fill this void.

Finally, a word about notation. The relative contributions of classical π-complex and metallacyclopropane resonance forms in coordinated alkenes and alkynes varies widely with changes in both the metal and its ligands.[9] For the sake of simplicity, we use the classical π-complex notation throughout, with the caveat that this notation implies nothing about the importance or lack thereof of the metallacyclopropane form.

II

COMPLEXES OF ARYNES

A. *General Synthesis and Mechanism*

Most early transition-metal complexes of arynes have been prepared by thermally induced β-hydrogen elimination from appropriately substituted aryl sigma complexes [Eq. (1)][2,8]

$$\text{M(R)}-\text{C}_6\text{H}_5 \xrightarrow[\text{- RH}]{\text{Heat}} \text{M}-\text{C}_6\text{H}_4 \quad (1)$$

The eliminations generally show first-order kinetics (see Table I for rates and activation parameters) and, in two crossover experiments [$Cp_2^*Ti(C_6H_5)(CH_3)/Cp_2^*Ti(C_6D_5)(CD_3)$ and $Cp_2Ti(C_6H_5)_2/Cp_2Ti(C_6D_5)_2$],[11] the products showed very little scrambling of deuterium. These data are indicative of unimolecular reactions.

TABLE I

ACTIVATION PARAMETERS AND RATE CONSTANTS (AT 70°C) FOR L_nMR_2 COMPLEXES[a]

Complex	E_a (kcal/mol)	$\Delta H\ddagger$ (kcal/mol)	$\Delta S\ddagger$ (eu)	$\Delta G\ddagger_{25°C}$ (kcal/mol)	$10^4 k_{70°C}$ (s^{-1})	Ref.
$Cp_2Ti(Ph)_2$	22	20.9[b]	1.0[b]	20.6[b]	0.62[b]	10
$Cp^*_2Ti(Ph)(Me)$	23.8(2)	23.0(2)	−9.8(2)	25.9(2)	1.1	11
$Cp^*_2Zr(Ph)_2$	23.1(3)	22.5(3)	−11.8(3)	26.0(4)	0.92	12a
$Cp^*_2Zr(Ph\text{-}d_{10})_2$	26.5(4)	25.8(4)	−5.8(15)	27.5(8)		12a
$Cp_2Zr(Ph)_2$					0.3	13
$Cp_2Zr(p\text{-tolyl})_2$					0.28	13
$Cp^*(C_4H_6)Ta(Ph)(Me)$	25.8(0.5)[b]	25.2(0.2)[b]	−6.2(0.5)[b]	27.1(0.5)	2.72	14
$Cp_2Zr(Me)(SCH_2Ph)$	19.2	18.6(0.1)	−20.6(0.4)	24.7	3.45	15
$Cp^*_2U(Ph)_2$					≥17	16

[a] A major portion of this table has been reproduced from Ref. 12b.

[b] Calculated using data given in the reference and the Eyring equation.

Two mechanisms have been considered for these reactions (Scheme 1). In one the hydrogen is transferred as a hydride[17] in a β-hydrogen elimination step to give an intermediate metal hydride (**B**), which then reductively eliminates RH. In this mechanism, the rate-determining step could, in principle, be either the β-hydrogen elimination step (k_1) or the reductive

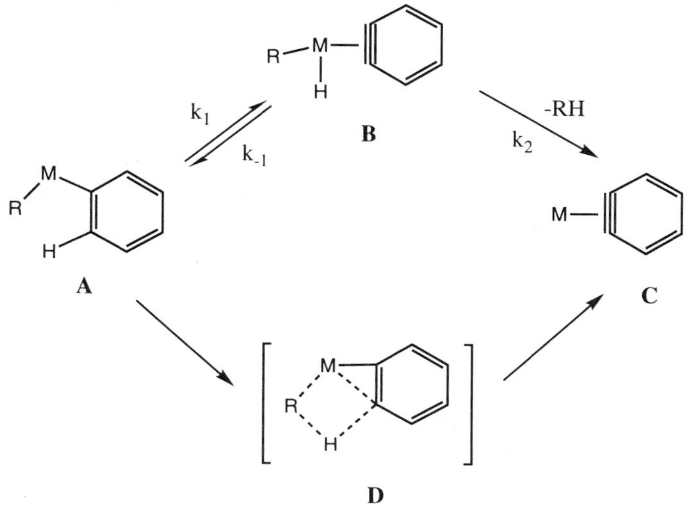

SCHEME 1

elimination (k_2). In the second mechanism, the metal carbon and metal hydrogen bonds break simultaneously via the four-centered transition state **D** to give the aryne complex directly. The principal experimental evidence for the concerted mechanism early transition metals is a large primary kinetic isotope effect [$k_H/k_D \sim 7$ for $Cp_2Ti(Me)(Ph)$][11] and a related thioaldehyde[15] when extrapolated to RT assuming normal Arrhenius behavior.[18] Large primary isotope effects are inconsistent with either form of the two-step mechanism[15] for the following reasons. If reductive elimination is the rate-determining step that is preceded by a rapid preequilibrium, the observed isotope effects are too large because the equilibrium isotope effect in the preequilibrium should be inverse (M—H is weaker than C—H, driving the equilibrium further to the right for the deuterated complex)[19] and primary isotope effects in reductive eliminations typically fall in a range of $k_H/k_D \sim 2.0$–3.5.[20–23] The observed primary isotope effect is also too large for a mechanism with a slow β-hydrogen elimination followed by rapid reductive elimination, but the argument in this case is less obvious because there are no experimental values for primary kinetic isotope effects for β-hydrogen eliminations in well-defined systems. However, in a very elegant study, Doherty and Bercaw[17] determined a kinetic isotope effect of $k_H/k_D = 1.1$ at 105°C (assuming normal Arrhenius behavior, 1.13 at RT)[24] for the olefin insertion reaction in Eq. (2). This is the reverse of β-hydrogen elimination and, since the primary isotope effect in the forward reaction should be greater than in the reverse because M—H and M—D bonds are weaker than C—H and C—D bonds, the primary isotope effect of the β-hydrogen elimination should be less than 1.1. Thus, the observed primary isotope effect for aryne formation is too large for any reasonable two-step mechanism. However, the observed isotope effects are in the range expected for a concerted mechanism with an approximately symmetrical C $\cdots$ H $\cdots$ C transition state.[25] Finally, it should be pointed out that an attractive feature of the concerted mechanism over either two-step reaction sequence for d^0 metals is that it avoids π-complexes with metals that cannot back-bond into the π-system.

$$Cp^*_2Nb(H)(CH_2{=}CH_2) \rightleftharpoons Cp^*_2Nb{-}CH_2CH_3 \quad (2)$$

A tetrakis(trimethylphosphine)ruthenium complex of benzyne has been prepared[6,26] by a reaction similar to that used for the Group 4 and 5 metals: thermally induced β-hydride elimination of methane or benzene from **1** or **2**, respectively [Eq. (3)]. A careful study of the kinetics of the elimination of methane from **1** revealed that dissociation of trimethylphosphine precedes the elimination reaction. This led Bergman to suggest a two-step β-hydrogen elimination/reductive elimination sequence for the mechanism, because a concerted mechanism analogous to that of the early d^0 metals should not require phosphine loss. This was a reasonable suggestion because, unlike the d^0 metals, ruthenium is d^4 and is therefore able to stabilize the aryne in the intermediate metal hydride by back-bonding. Furthermore, this sequence is typical for ortho-metallations (γ-eliminations) of late transition-metal complexes.[27,28] However, the rate of loss of benzene from the niobium complex **6** [Eq. (4)] is retarded by excess trimethylphosphine.[29] Since it is likely that this reaction occurs by a concerted mechanism because there are no d electrons to stabilize the benzyne complex in the intermediate metal hydride, it would appear that, at least in this case, a 16-electron intermediate is required for the concerted elimination. To date, primary kinetic H/D isotope effects have not been reported for either the ruthenium or niobium reactions.

$(Me_3P)_4Ru(R)(C_6H_4\text{-}H)$ —Heat, $- PMe_3$→ $(Me_3P)_3Ru(R)(C_6H_4\text{-}H)$ **3** →

1: $R = CH_3$
2: $R = C_6H_5$

→ $(Me_3P)_3Ru(R)(H)(C_6H_4)$ **4** —$- RH$, $+ PMe_3$→ $(Me_3P)_4Ru(C_6H_4)$ **5** (3)

$Cp^*(RN{\equiv})Nb(Ph)_2(PMe_3)$ **6** —Heat→ $Cp^*(RN{\equiv})Nb(C_6H_4)(PMe_3)$ **7** (4)

Finally, if the two-step mechanism is confirmed for benzyne formation in the case of **5**, this may be the sole example of aryne formation by such a mechanism, although it is also a possibility for a rhenium–benzyne complex prepared by Arnold *et al.*[30]

Metal alkyls of late transition metals readily undergo decomposition by β-hydrogen elimination.[31] In principle, it should therefore be possible to prepare aryne complexes of d^{10} metals by thermolysis of appropriately substituted late transition-metal aryl complexes. However, in practice this

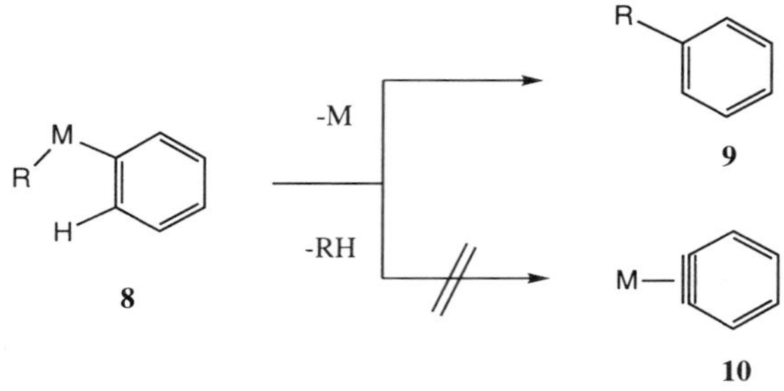

SCHEME 2

has failed,[2] presumably because reductive elimination from **8** is faster than *ortho*-metallation (Scheme 2).

With the exception of a nickel–benzyne complex (**13**) [Eq. (5)],[32,33] which is probably formed by trapping the free intermediate with the coordinatively unsaturated metal,[34] all complexes of nickel and platinum have been prepared by reduction of preformed (2-haloaryl)halometal complexes, which in turn have been synthesized by oxidative addition of M(0) to *o*-dihaloarenes.[2,35–37] Typical of this procedure is preparation of the bis(triethylphosphine)nickel–benzyne complexes **19** shown in Eq. (6).

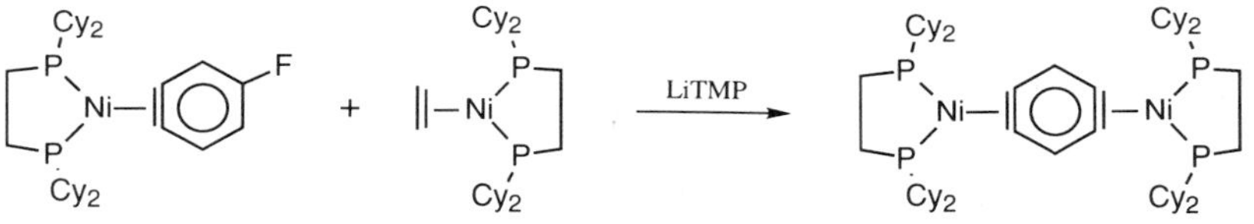

(5)

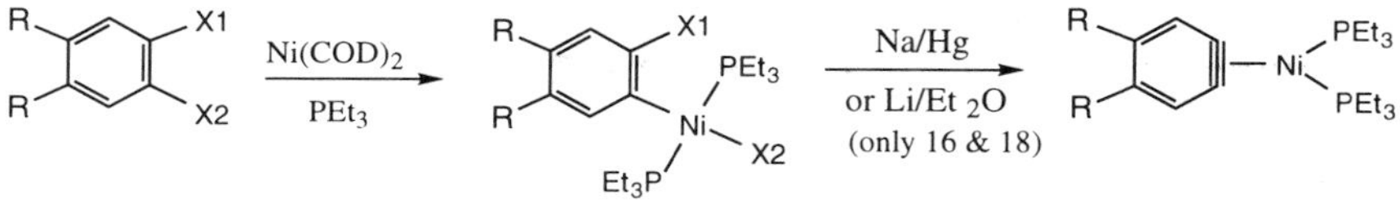

14: X1, X2 = Br
15: X1, X2 = Cl

16: X1, X2 = Br
17: X1, X2 = Cl
18: X1 = Br, X2 = Cl

(6)

The mechanism of the aryne-forming step[38] has not been studied in detail, although the interesting observation has been made that **16** and **18** can be reduced with either lithium in ether or Na/Hg, whereas **17** is inert to the former. This led Bennett to conclude that the first step in the reduction is halide abstraction from the aryne group rather than from the metal.

B. *Early Transition-Metal Complexes of Benzynes*

Titanium, zirconium, tantalum, niobium, and rhenium complexes of benzyne have been discussed in earlier reviews.[2,8]

Reactions of Cp_2Ti–benzyne complexes (including early examples not mentioned[39] and more recent reactions[40]) are summarized in Scheme 3. In all cases the titanium–benzyne complex **21** was generated *in situ* in the presence of the trapping reagent. Coupling with CO_2 and the phosphaacetylene is regiospecific as indicated. Unsymmetrically substituted benzyne complexes **21** (R = Me[41], F[40]) react with alkynes regioselectively to give predominantly metallacycles **25** and **26**, which result from acetylene approaching **21** from the less sterically hindered side. The metallaindanes resulting from coupling between **21** (R = H, Me) and olefins (propene, 1-octene) have been used as precursors for Ziegler-type polymerization of propene.[42]

Attempts to prepare titanaindanes and titanaindenes from thermolysis of **28a** or **28b** in the presence of alkenes and alkynes failed; only the cyclometalated complexes **30a** and **30b** from intramolecular C–H activation in **29** were observed [Eq. (7)].[42] A similar reaction has been observed for substituted Cp_2Zr–benzyne complexes.[2]

Thermolysis of **31** in the presence of $M(CO)_6$ (M = Mo, W) gives the interesting carbene complexes **33a** and **33b** [Eq. (8)].[43] The regiochemistry

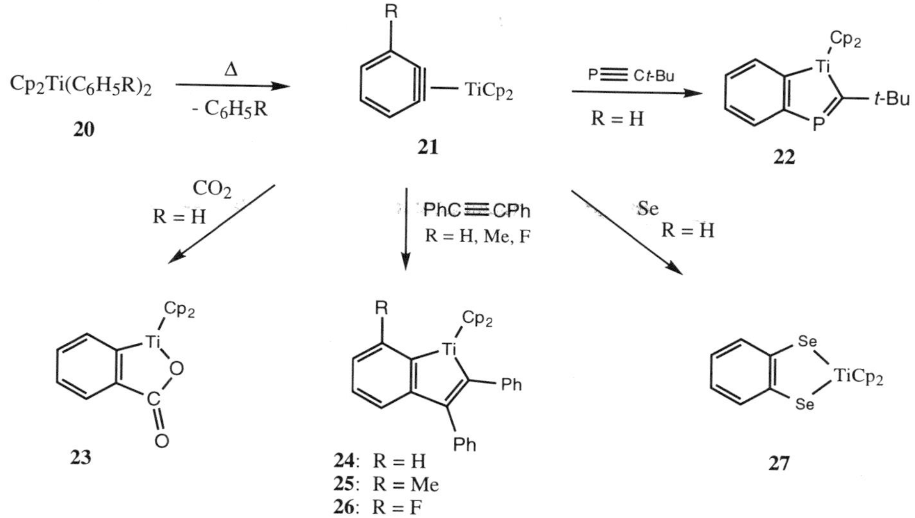

$Cp_2Ti(C_6H_5R)_2$ (**20**) —Δ, $-C_6H_5R$→ **21** (C_6H_3R–$TiCp_2$)

21 —P≡C*t*-Bu, R = H→ **22**

21 —CO_2, R = H→ **23**

21 —PhC≡CPh, R = H, Me, F→ **24**: R = H; **25**: R = Me; **26**: R = F

21 —Se, R = H→ **27**

SCHEME 3

of this reaction with oxygen bonded to the highly oxophilic titanium center is as expected. In a similar reaction, **35** reacts with CO_2 to give **37**. In the absence of a trapping reagent, **35** undergoes intramolecular C–H activation to give **36** quantitatively (Scheme 4).[11]

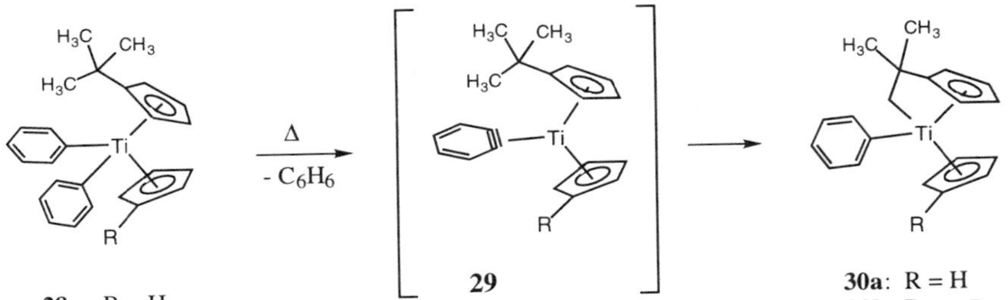

Δ, $- C_6H_6$

28a: R = H
28b: R = *t*-Bu

29

30a: R = H
30b: R = *t*-Bu

(7)

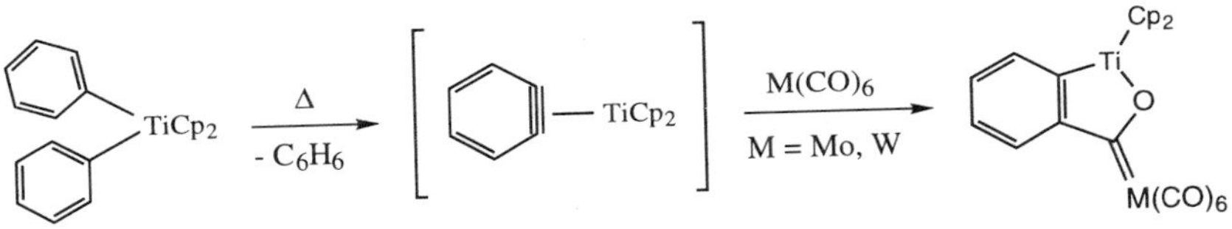

Δ, $- C_6H_6$; $M(CO)_6$, M = Mo, W

31 **32**

33a: R = Mo
33b: R = W

(8)

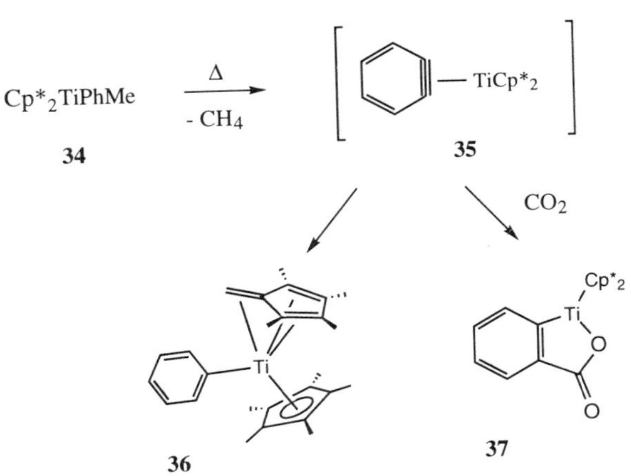

$Cp^*_2TiPhMe$ **34** → (Δ, $- CH_4$) [$TiCp^*_2$] **35** → **36**; **35** → (CO_2) **37**

SCHEME 4

Until 1993, formation of titanium–benzyne complexes had been only inferred from reaction products. However, in 1993, Buchwald *et al.* succeeded in isolating and characterizing **39a** and **39b** by trapping the intermediate aryne complex with PMe_3 [Eq. (9)].[44] Using variable-temperature nuclear magnetic resonance (NMR), the authors were able to show that, in solution, **39a** and **39b** interconvert by PMe_3 dissociation. When heated to 60°C in the presence of 1 equiv 3-hexyne, **39** gives the expected titanaindene. That a coordinatively unsaturated intermediate is required for this reaction was shown by complete inhibition of titanaindene formation by a 10-fold excess of PMe_3.

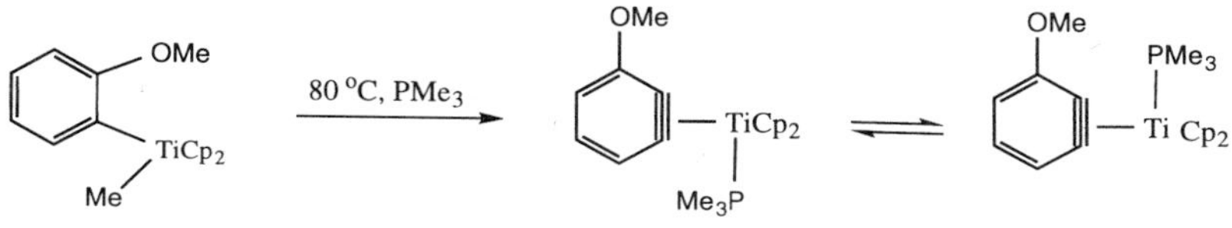

OMe; TiCp2; Me — 80 °C, PMe3 → OMe; TiCp2; Me3P ⇌ OMe; PMe3; Ti Cp2 (9)

38 **39a** **39b**

Reaction of **39** with $P(OMe)_3$ gives **40** in 84% yield [Eq. (10)]. It is interesting that cleavage of the P—O bond occurs, rather than phosphine–phosphite exchange. High oxophilicity of titanium was suggested as a possible driving force for this cleavage. 1-Hexyne reacts with **39** to give metallacycles **41a** and **41b** (3:1) [Eq. (10)]. Heating **39** in the presence of methyl vinyl ketone and ethylene leads to regioselective formation of metalacycles **42** and **43**, respectively [Eq. (11)]. The same regioselectivity is observed in the products obtained from thermolysis of **39** with 3-hexyne and *p*-bromobenzonitrile.[44]

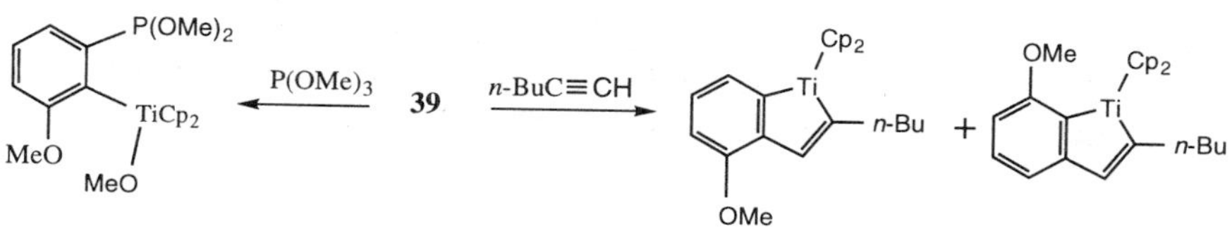

P(OMe)2; TiCp2; MeO; MeO ← P(OMe)3 — 39 — n-BuC≡CH → Cp2; Ti; n-Bu; OMe + OMe; Cp2; Ti; n-Bu

40 **41a** **41b** (10)

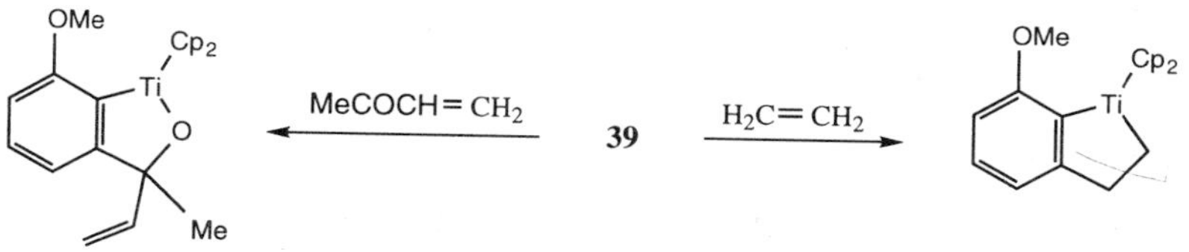

OMe; Cp2; Ti; O; Me ← MeCOCH=CH2 — 39 — H2C=CH2 → OMe; Cp2; Ti (11)

42 **43**

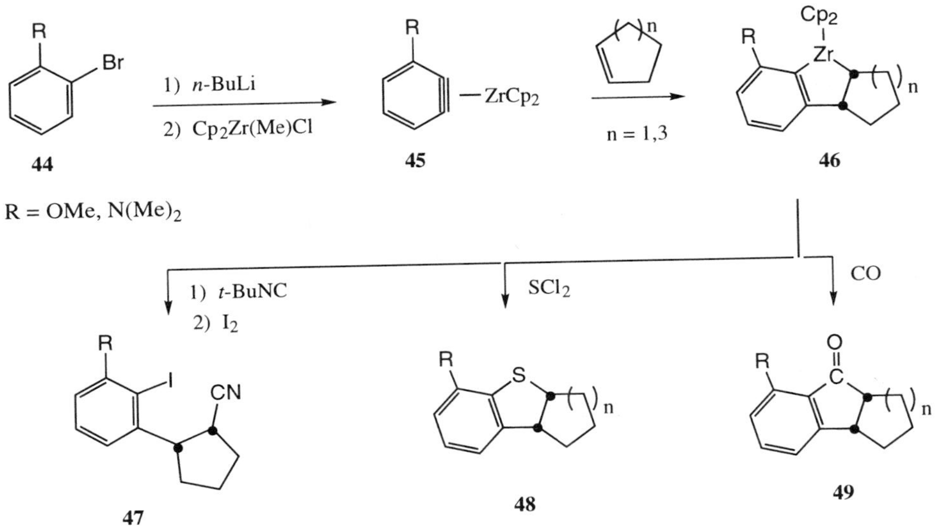

R = OMe, N(Me)$_2$

SCHEME 5

Zirconium–benzyne complexes have been used rather extensively in organic synthesis.[8,45] For this purpose, one particularly important characteristic of zirconium–aryne complexes is that olefin insertion into the Zr—C bond occurs stereospecifically. Thus, when generated *in situ*, the zirconium–benzyne complex (**45**) reacts with cyclic alkenes to give exclusively the *cis*-zirconaindanes (**46**), which upon treatment with electrophiles provide access to a variety of *cis*-difunctionalized cycloalkanes (**47**–**49**) (Scheme 5).[46] For example, carbonylation of intermediate **46** affords tricyclic ketone **49**, reaction with sulfur dichloride gives thiophene **48**, and reaction of **46** with *tert*-butylisocyanide followed by I_2 gives **47** via **50** and, presumably, intermediate **51** [Eq. (12)].

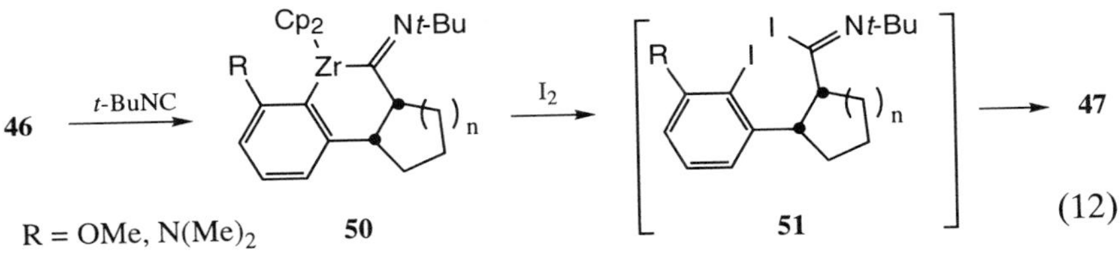

R = OMe, N(Me)$_2$

(12)

Interestingly, cyclic allylic ethers do not give analogous coupling products with zirconium–benzyne complexes.[47] Reaction of **52** with 2,5-dihydrofuran

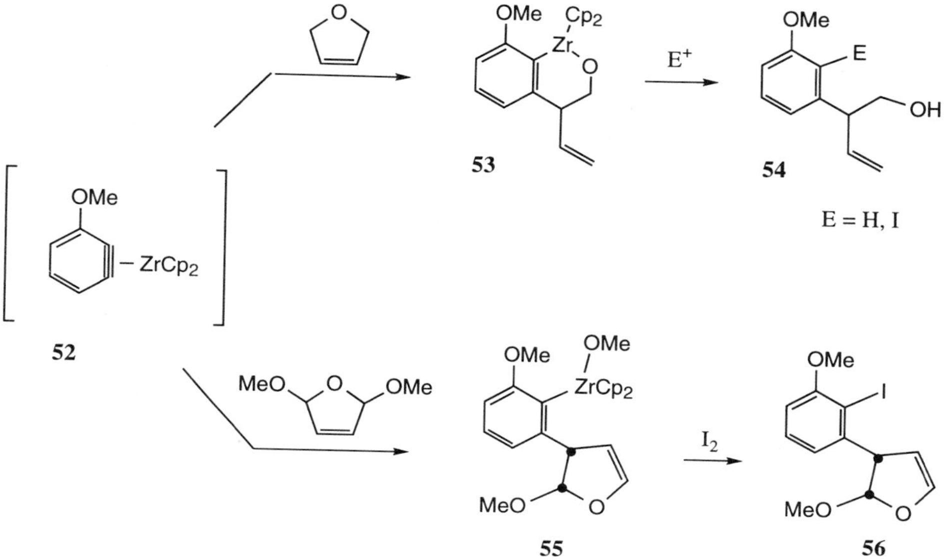

SCHEME 6

gives metallacycle **53**, which when treated with electrophiles (MeOH or I_2) gives the homoallylic alcohol **54** (Scheme 6). Other substituted benzyne and naphthalyne complexes (vide infra) react with 2,5-dihydrofuran in the same fashion. 2,5-Dimethoxy-2,5-dihydrofuran reacts with **52**, forming σ-complex **55** as both a single regioisomer and a single diastereomer. Surprisingly, 2,3-dihydrofuran does not show any reaction with **52** even at elevated temperatures, suggesting that only ethers in which the double bond and the oxygen are separated by a methylene unit are capable of undergoing productive reaction. The mechanism of these transformations has not been investigated.

Zirconacycles resulting from coupling between zirconium–aryne complexes and unsaturated organic substrates have been used by Fagan and Nugent to prepare heterocyclic compounds via transmetallation reactions.[48] Benzisothiazoles,[2,49] benzothiophenes,[2,50] and, more recently, new heterocyclic compounds containing antimony[51] and germanium[52] have been prepared by this method. For example, **57** and **60** react readily with 1 equiv antimony trichloride or phenylantimony dichloride to give antimony metallacycles **58**, **59**, **61**, and **62** in good yields [Eq. (13)].[51]

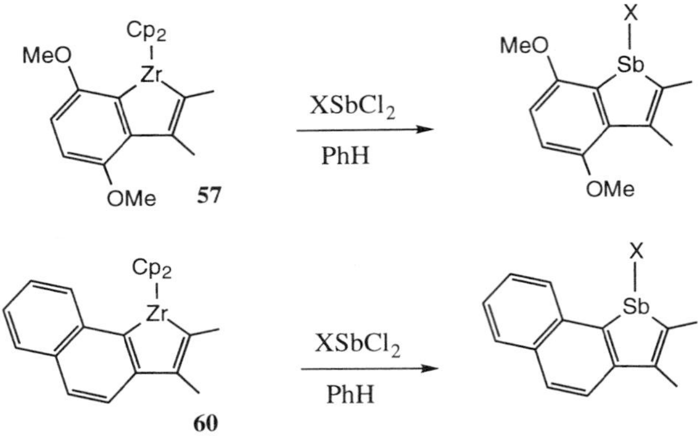

57 $\xrightarrow[\text{PhH}]{\text{XSbCl}_2}$ **58**: X=Cl, 74%; **59**: X=Ph, 86%

60 $\xrightarrow[\text{PhH}]{\text{XSbCl}_2}$ **61**: X=Cl, 74%; **62**: X=Ph, 83% (13)

Similarly, germanium tetrachloride reacts with a variety of substituted zirconaindenes (**63**) to give good yields of 1-germaindenes (**64**) [Eq. (14)].

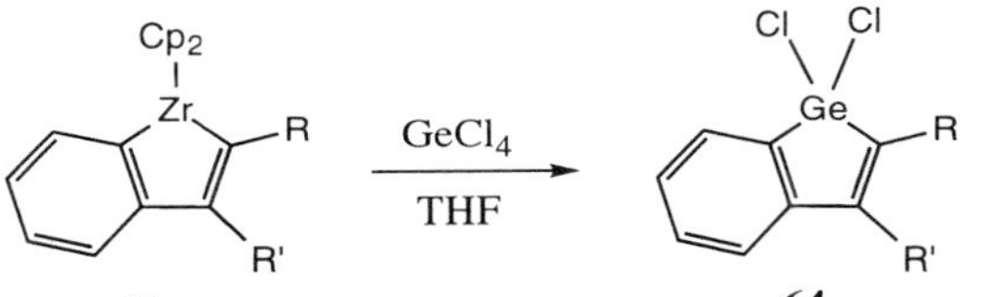

63 $\xrightarrow[\text{THF}]{\text{GeCl}_4}$ **64**

R = R' = Me, Et, Pr, TMS
R = TMS, R' = Me (14)
R = TMS, R' = NEt_2

Sulfur and selenium heterocycles have also been prepared by reaction of zirconium–benzyne complexes with the elemental chalcogens. In these reactions, chalcogens insert into both of the zirconium–carbon bonds (Scheme 7).[53–55] Interestingly, neither *para*-bromo nor *para*-dimethylamino substituents in **65** interfere with their conversion to **66** (60–80% yields).[54] Complexes **66** react further with mono- and bifunctionalized electrophiles (Scheme 7) to yield *ortho*-dichalcogenated benzenic compounds **67–69**.

Thermolysis of *para*-substituted diphenylzirconocene **70a** in the presence of gray selenium powder gives, in addition to the expected complex **71a**, the rearranged product **72a** in a ratio of 60 : 40;[55] whereas **70b** gives only the rearranged product **72b** [Eq. (15)]. Formation of the rearranged product was rationalized by proposing an equilibrium between two isomeric benzyne complexes [Eq. (16)] prior to reaction with selenium.

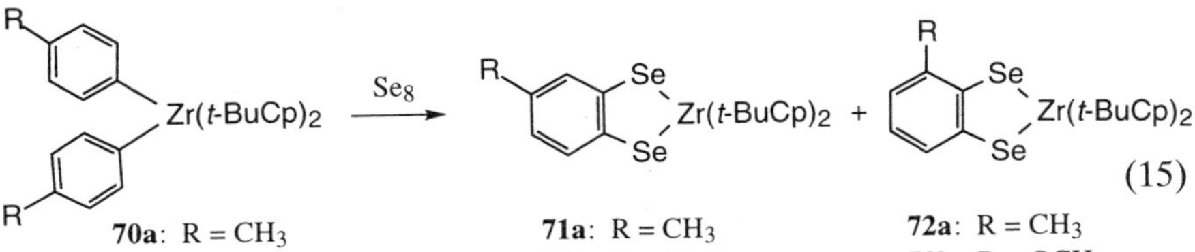

$(RC_6H_4)_2Zr(t\text{-BuCp})_2 \xrightarrow{\text{Se}_8}$ **71** + **72** (15)

70a: R = CH_3
70b: R = OCH_3

71a: R = CH_3
71b: R = OCH_3

72a: R = CH_3
72b: R = OCH_3

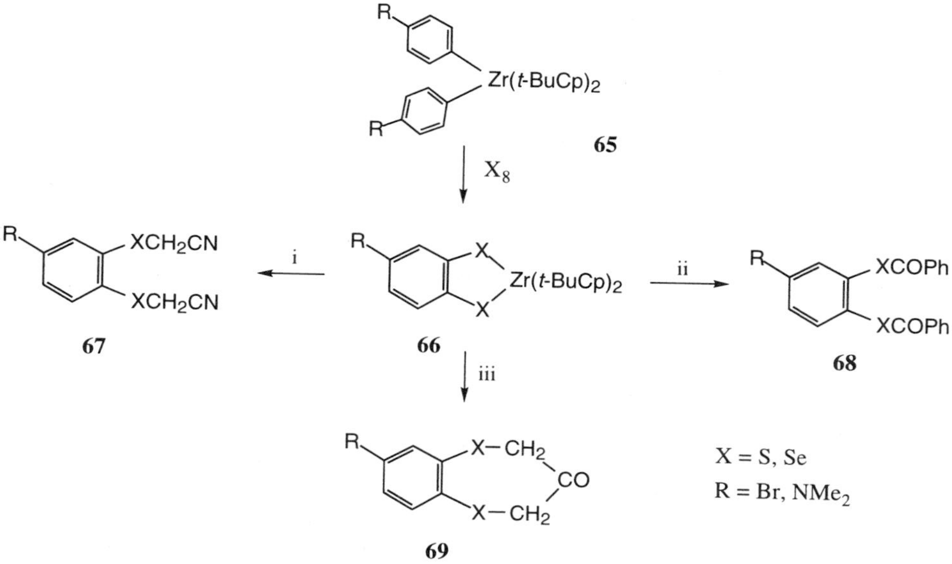

65

X_8

67

i

66

ii

68

iii

69

X = S, Se
R = Br, NMe_2

(i) 2 $CNCH_2Cl$; (ii) 2 PhCOCl; (iii) $ClCH_2COCH_2Cl$.

SCHEME 7

Support for this mechanism is found in the reactions of **74** (R = H) (which can be isolated) with 3-hexyne, acetone, acetonitrile, and selenium to give products expected of an aryne complex.[55] Cyclometallation of other zirconium–benzyne complexes with alkyl-substituted Cps has been observed in a few other cases.[2]

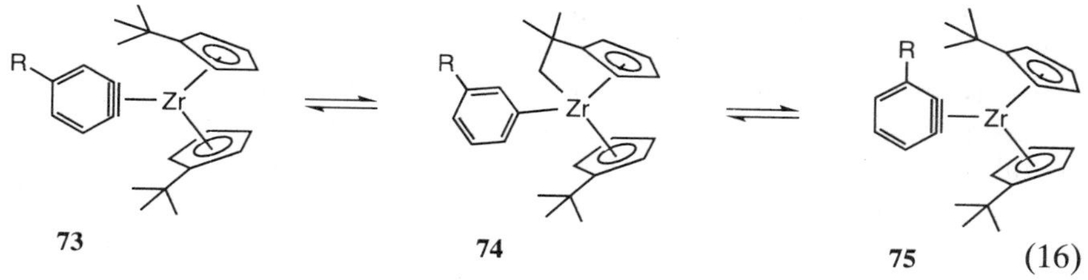

73 74 75 (16)

The selenium-containing complex **72b** has been used to prepare a variety of new heterocycles containing both selenium and germanium. For example, reaction with dichlorodiphenylgermanium and germanium tetrachloride gives **76** and **77**, respectively (Scheme 8).[56]

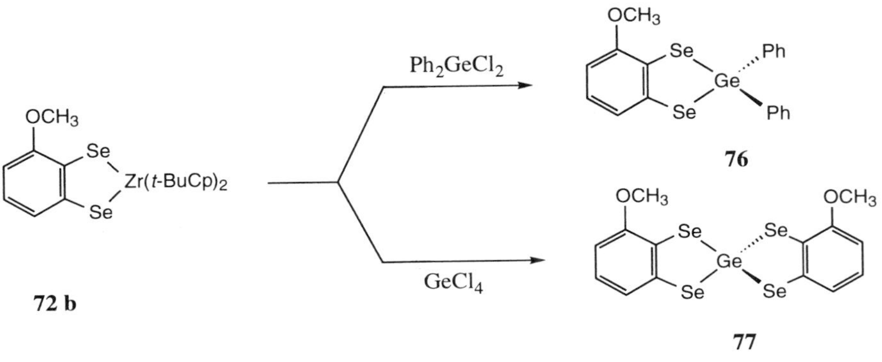

OCH_3 Se Zr(t-BuCp)$_2$ Se — **72 b** — Ph_2GeCl_2 → **76**; $GeCl_4$ → **77**

SCHEME 8

The zirconium–benzyne complex **78** reacts with 2 equiv trimethylaluminum[57] or trimethylgallium[58] to give **79** and **80**, respectively [Eq. (17)]. As shown by X-ray analysis, these complexes contain a planar-tetracoordinate carbon center (C2) at the bridgehead position. Triethylaluminum and diisobutylaluminum hydride react similarly.[57]

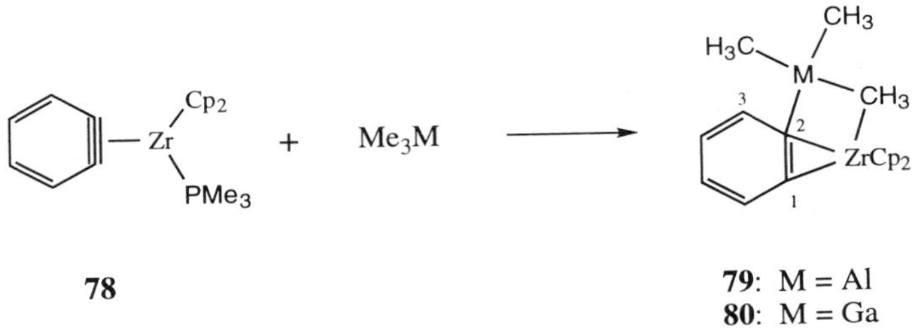

$$\textbf{78} + Me_3M \longrightarrow \textbf{79}: M = Al;\ \textbf{80}: M = Ga \qquad (17)$$

Intramolecular coupling of zirconium–benzyne complexes with olefins followed by an ene reaction has been used as a new route to 3,4-disubstituted indoles.[59] For example, reaction of *N*-allyl-*N*-benzyl-2-bromoaniline (**81**), with *t*-BuLi in the presence of $Cp_2Zr(Me)Cl$, gives **84**, presumably via benzyne complex **83** (Scheme 9). Reaction of **84** with I_2 gives **85** (70% overall from **81**), which, upon dihydrohalogenation with DBU, gives **86**. Reaction of **86** with active enophiles gives 3,4-disubstituted indoles **87**. This approach has been used to construct 6-hydroxyindolines (**89**), which are precursors to pharmacophores (**88**) of potential antitumor antibiotics such as duocarmycin (**90**).[60] A major challenge in these syntheses has been regioselective introduction of a hydroxy group into the proper position on the aromatic ring. This was accomplished as shown in Scheme 10.

Intramolecular coupling of an olefin with a zirconium–benzyme complex

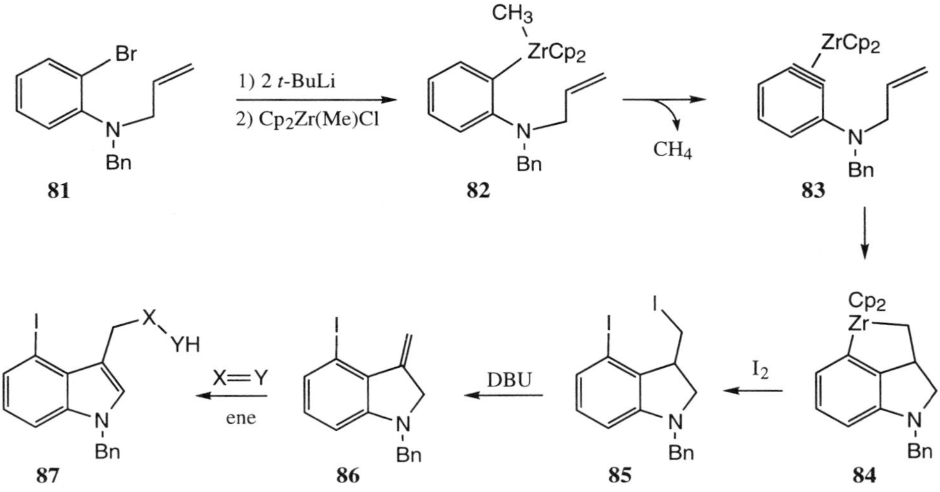

1) 2 *t*-BuLi
2) $Cp_2Zr(Me)Cl$

CH_4

81 **82** **83** **84** **85** **86** **87**

I_2

DBU

X=Y
ene

SCHEME 9

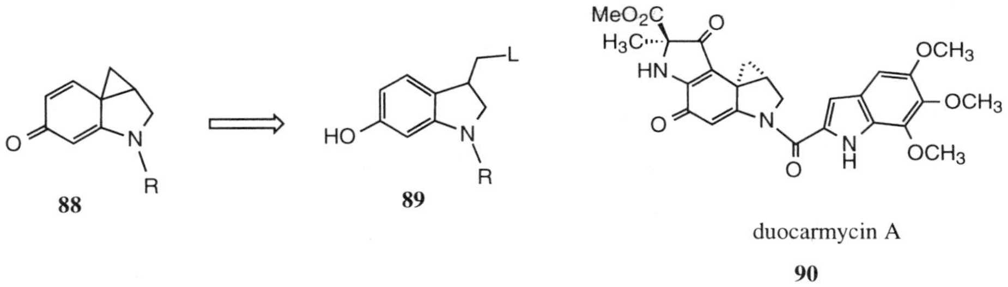

88 **89**

duocarmycin A

90

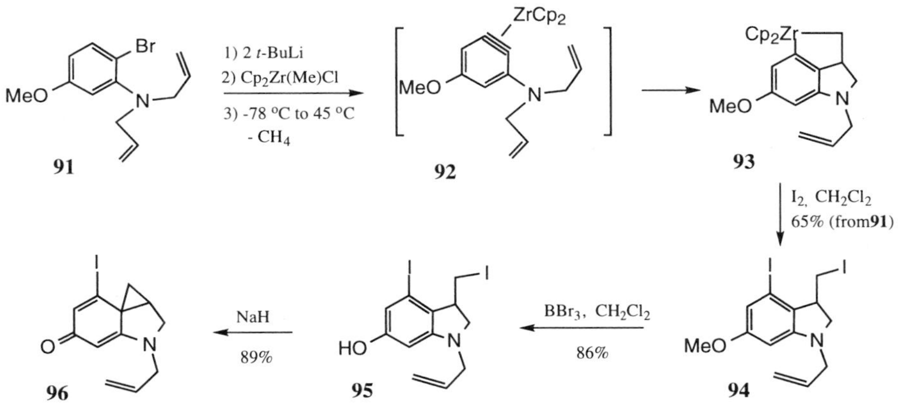

1) 2 *t*-BuLi
2) $Cp_2Zr(Me)Cl$
3) -78 °C to 45 °C
- CH_4

91 **92** **93**

I_2, CH_2Cl_2
65% (from **91**)

94

BBr_3, CH_2Cl_2
86%

95

NaH
89%

96

SCHEME 10

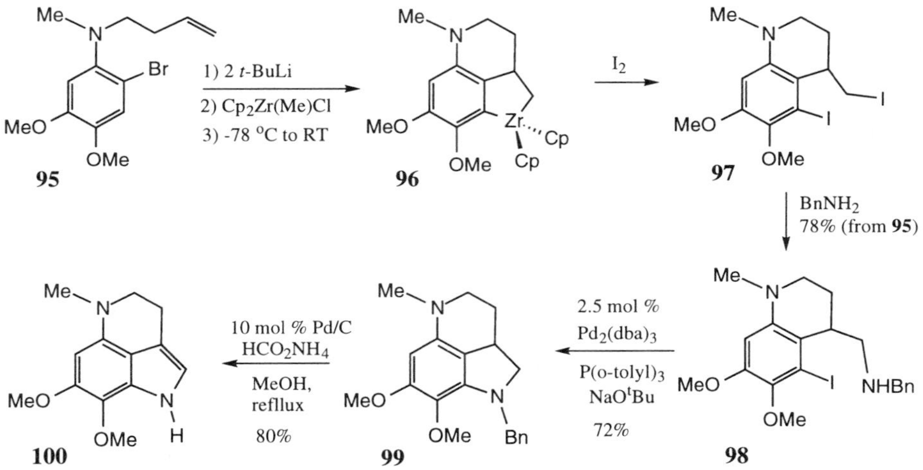

SCHEME 11

followed by a palladium-catalyzed ring closure has been used to prepare **100** toward the total synthesis of damirone A and B and makaluvanime C (Scheme 11).[61] This is different from earlier examples of indole synthesis, in that this sequence illustrates formation of a six-membered ring by the aryne method. This methodology has also been used in a new synthesis of dehydrobufotenine and tetrahydropyrroloquinolines.[61]

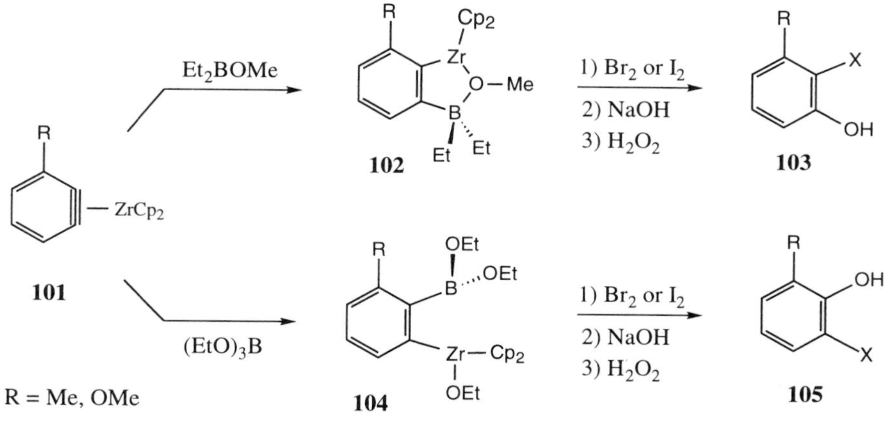

SCHEME 12

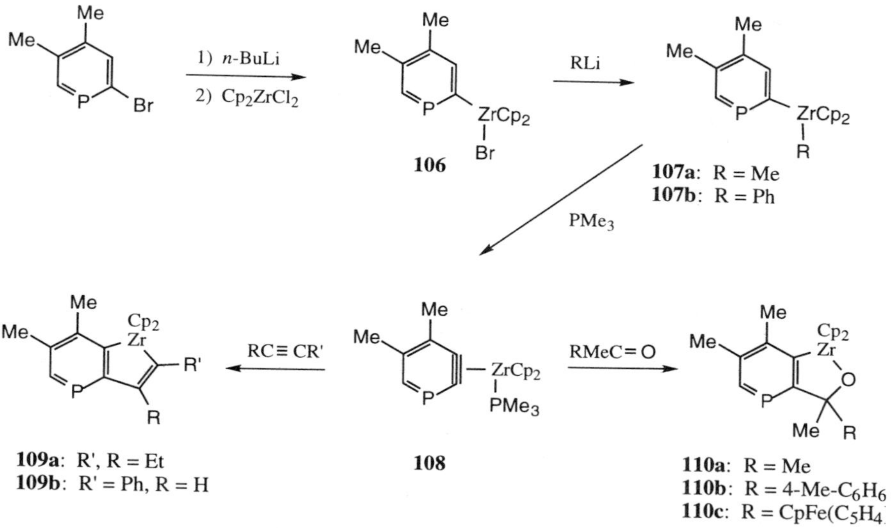

SCHEME 13

Zirconium–aryne complexes have also found application in the regioselective synthesis of halophenols.[62] Reaction of **101** with methyl diethylborinate or triethylborate leads to insertion of a boron–oxygen bond into a Zr–benzyne bond with formation of **102** and **104**, respectively (Scheme 12). Further reactions of **102** and **104** with bromine or iodine followed by excess NaOH and 30% hydrogen peroxide gives halomethyl(methoxy)phenols **103** and **105** in good yields. In most cases only one regioisomer of the halophenol was observed.

A zirconium–phosphabenzyne complex **108**, an orange air-sensitive oil, has been prepared by thermolysis of **107** in the presence of trimethylphosphine (Scheme 13).[63] Reaction of **108** with alkynes and ketones gives metallacycles **109** and **110**, respectively.

A paramagnetic vanadium–benzyne complex (**112**) has been synthesized by thermolysis of $CpVPh_2(PMe_3)_2$ (**111**) at 50°C ($t_{1/2}$ = 3–4 h) [Eq. (18)].[64] An X-ray crystal structure was reported for **112**. Bond lengths for the benzyne ligand are given in Table III. The C1–C2 distance [1.368(5) Å] is similar to those in the other structurally characterized benzyne complexes. A band in the IR spectrum of **112** at 1547 cm^{-1} was assigned to the coordinated triple bond.

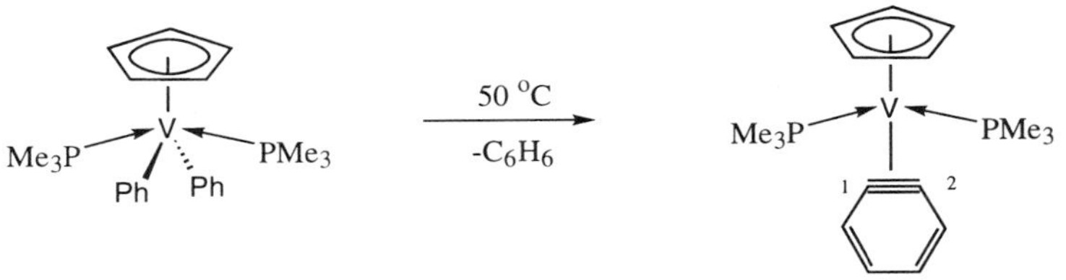

$$\mathbf{111} \xrightarrow[-C_6H_6]{50\ ^{\circ}C} \mathbf{112} \quad (18)$$

In contrast to ruthenium (Section II,C) and zirconocene benzynes, **112** does not activate aromatic C—H bonds, as shown by lack of exchange with C_6D_6 even after heating at 100°C for several days. A number of reactions of **112** are recorded in Scheme 14. Reaction with diphenylacetylene gives the vanadaindene **113**. This reaction is retarded by trimethylphosphine, indicating that the rate-determining step is loss of phosphine. The vanadium benzyne also readily undergoes regioselective reaction with terminal alkenes to give β-substituted vanadaindanes (**114**). The thermal stability of these complexes depends on the substituents in the β-position. If both are alkyl groups (**114c**), the product is stable. However, if one β-substituent is

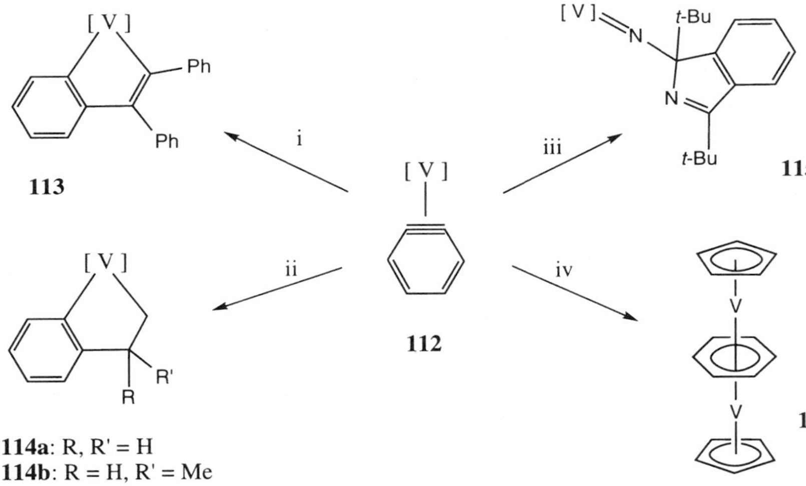

114a: R, R' = H
114b: R = H, R' = Me
114c: R, R' = Me

[V] = $CpV(PMe_3)_2$: (i) PhC≡CPh; (ii) CH_2=CRR'; (iii) *t*-BuCN; (iv) H_2

SCHEME 14

hydrogen, β-hydrogen elimination can occur which ultimately leads to decomposition.[65] For example, **114a** rapidly decomposes to CpV($\eta^2C_2H_4$)(PMe_3)$_2$ and styrene, presumably by reductive elimination of the latter following β-H elimination.

The thermal decomposition products of vanadaindanes are analogous to those of ruthinaindane (Section II,C). Interestingly, and in contrast, the analogous zirconoindane (**117**) is thermally stable at room temperature and decomposes only if heated above 100°C. Reaction of **112** with *t*-BuCN gives the new imido complex **115**. This behavior differs from that of benzyne complexes of Zr, Ru, and Ta in that the latter react with nitriles to give azametallacycles by monoinsertion of the nitrile into the carbon–metal bond. It is believed that an analogous azavanadacycle (**118**) is initially formed from **112**, but undergoes further reaction with a second mole of the nitrile to give the diazavanadacycle **119**, which rearranges to **115** [Eq. (19)]. An X-ray crystal structure of **115** was reported.[64] Finally, reaction of **112** with dihydrogen gives the known triple-decker complex **116** through partial hydrogenation of the benzyne ligand.

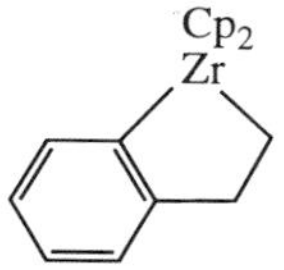

Cp_2Zr

117

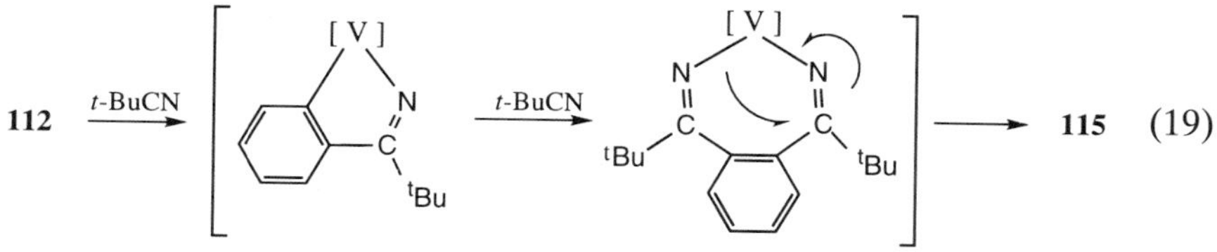

112 —*t*-BuCN→ [[V] N C ᵗBu **118** —*t*-BuCN→ [V] N N C C ᵗBu ᵗBu **119**] ——→ **115** (19)

A mononuclear tantalum–benzyne complex (**121**) has been prepared by thermolysis of **120** [Eq. (20)].[14] An X-ray crystal structure was reported for **121**. Bond lengths for the benzyne unit are given in Table III. Complex **121** exhibits a rich insertion chemistry similar to that of Ti, Zr, and Ru benzyne complexes. Insertion reactions of **121** with ethylene, 2-butyne, acetonitrile, and carbon dioxide give **122**, **123**, **124**, and **125**, respectively (Scheme 15). Diphenylacetylene does not couple with **121**, presumably because of steric constraints. Reagents with acidic protons such as methanol or terminal alkynes cleave the Ta—C bond to give σ-complexes **126** and **127**. Complex **121** is inert to *t*-butyl isocyanide and carbon monoxide, but forms a thermally unstable adduct (characterized spectroscopically) with PMe_3.

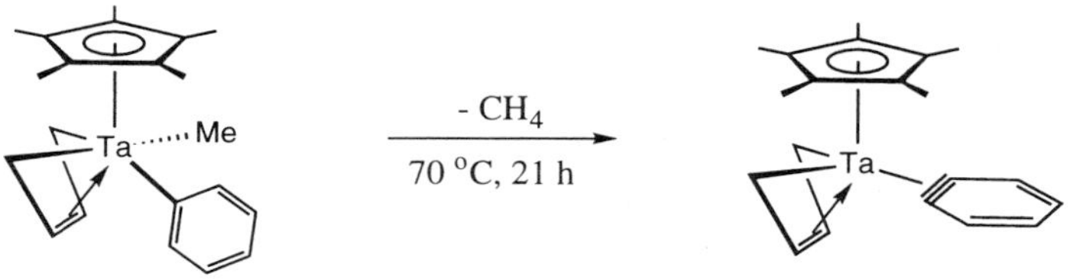

(20)

Complexes **120**, **121**, **123**, **124**, **126**, and **127** have been characterized by X-ray single-crystal analysis. All show a *supine* geometry of the butadiene ligand. It is interesting that although the metal fragment Ta(Cp*)(η^4-C_4H_6) in **121** is isolobal with that of Cp_2M (M = Ti, Zr), benzyne complexes of the latter are very labile and can only be isolated as phosphine adducts.

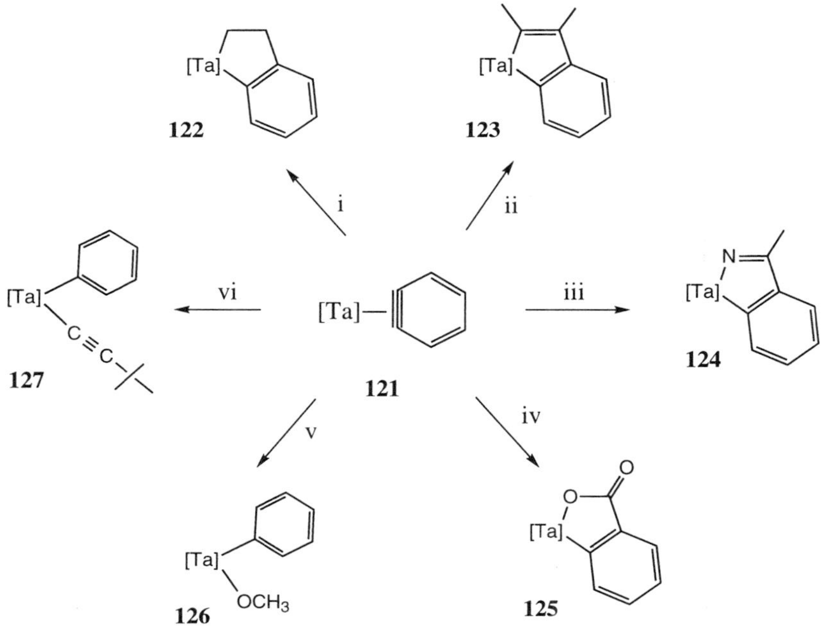

[Ta] = Ta(Cp*)(η^2-C_4H_6): (i) CH_2=CH_2; (ii) 2-butyne; (iii) acetonitrile; (iv) CO_2; (v) methanol; (vi) 3,3-dimethyl-1-butyne

SCHEME 15

A tantalum–benzyne complex possessing a *nido*-carborane as a ligand (**130**) has been prepared[66] by reaction of **128** with phenylmagnesium bromide followed by refluxing in toluene in the presence of PMe_3 (Scheme 16). An X-ray crystal structure was reported for **130**. Bond lengths for the benzyne unit are recorded in Table III.

The tantalum–benzyne complex (**130**) is much less reactive than other early transition-metal aryne complexes. It shows no reaction with acetone, benzophenone, benzaldehyde, acetonitrile, 3-hexyne, or methanol. The lack of reactivity of **130** was attributed to nonlability of the PMe_3 ligand. Indeed, no phosphine exchange was observed when **130** was mixed with an excess of PMe_3-d_9. Refluxing **129** in a mixture of methanol and toluene (3 : 10 v/v) leads to clean formation of **131**. This presumably results from reaction of a 16-electron benzyne complex with the alcohol.

A half-sandwich imido complex of niobium (**132**), which is both isoelectronic and isolobal with the bent metallocene Cp_2ZrCl_2, reacts with two equiv PhMgBr in the presence of PMe_3 to give the 18-e^- complex **6** [Eq. (21)].[29] In contrast to the zirconocene analogue, trimethylphosphine is required to stabilize this σ-complex.

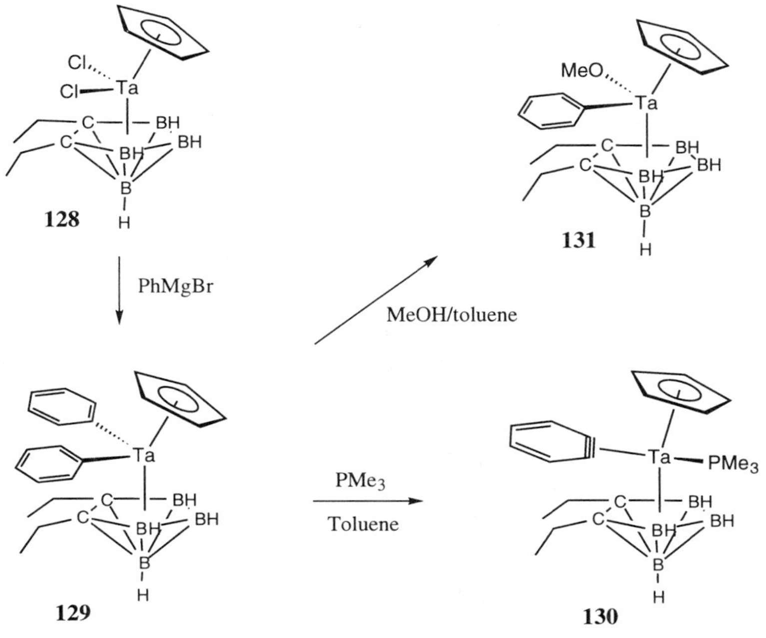

SCHEME 16

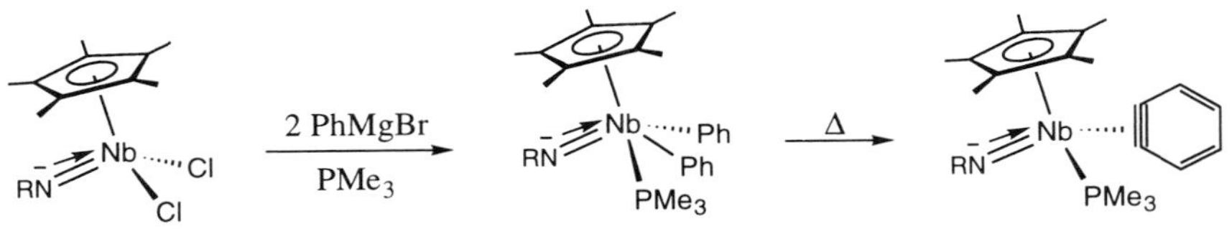

132 —(2 PhMgBr, PMe_3)→ 6 —(Δ)→ 7

R = 2,6-$Pr^i{}_2C_6H_3$

(21)

Thermolysis of **6** at 60°C gives the benzyne complex **7**. This reaction is significantly retarded by excess trimethylphosphine, which indicates that benzyne formation occurs from a 16-e^- intermediate. An X-ray crystal structure of **7** was reported. Bond lengths for the benzyne unit are given in Table III. No reactions of **7** have been reported.

C. *Late Transition-Metal Complexes of Benzyne*

Nickel and platinum complexes of benzyne have been discussed in earlier reviews.[2,8] More recently, a ruthenium complex of benzyne was described. The ruthenium–benzyne complex (**5**) was synthesized by thermolysis of either $(PMe_3)_4Ru(Ph)_2$ (**2**) or $(PMe_3)_4Ru(Me)(Ph)$ (**1**) as shown in Eq. (22).[6,26] The structure of **5** was confirmed by X-ray crystallography. Bond lengths for the benzyne unit are given in Table III. Reversibility of the formation of **5** from **2** was demonstrated by thermolysis of **5** in C_6D_6, which gave the deuterated isotopomer $(PMe_3)_4Ru(\eta^2\text{-}C_6D_4)$ (**133**) (Scheme 17). *p*-Cresol reacts rapidly with **5** to give **134**, which, upon mild heating, eliminates benzene to give the *ortho*-metallation product **135**. Aniline reacts similarly. Reaction with 2-propanol gives the metal hydride **137** and acetone, presumably via a phenyl isopropoxide intermediate followed by β-hydrogen elimination. Addition of water across the Ru—C bond gives **138**. Di-*p*-tolylacetylene, benzonitrile, and benzaldehyde insert into the Ru—C bond of **5** to give stable cycloaddition products **139**, **141**, and **142**, respectively (Scheme 18). Ethylene probably also inserts to give an unstable ruthenaindane, which rapidly undergoes β-hydrogen elimination followed by reductive elimination to give styrene and a 16-e^- complex that is subsequently trapped by additional ethylene to give **140**.

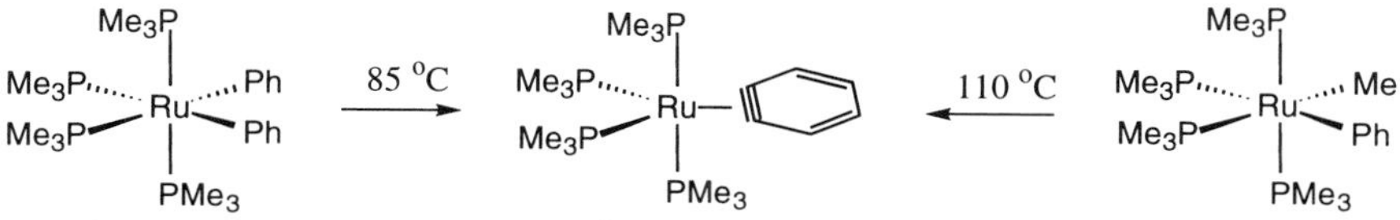

2 —(85 °C)→ 5 ←(110 °C)— 1

(22)

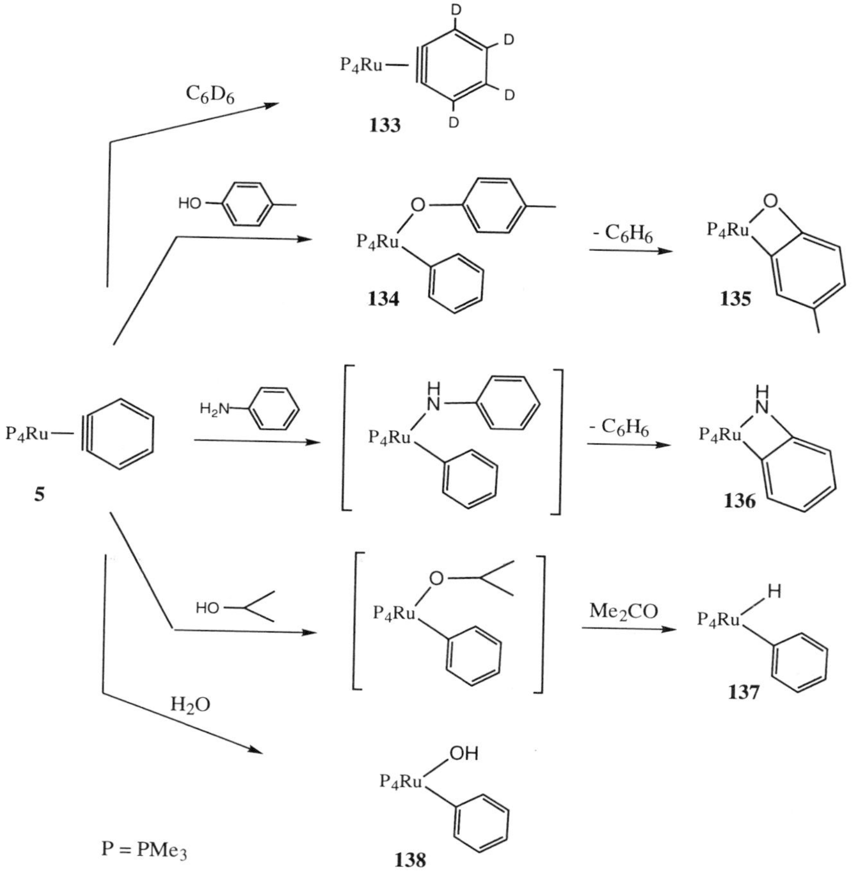

SCHEME 17

The rich addition and insertion chemistry exhibited by **5** is unusual for late transition-metal complexes and, as mentioned, more closely resembles the reactivity of benzyne complexes of electrophilic d^0 metals.

New nickel–benzyne complexes (**143–147**) have been prepared by reaction of *o*-dihaloarenes with $Ni(COD)_2$ in the presence of a trialkylphosphine followed by reduction of the oxidative addition product with either Li or 1% Na/Hg in ether [e.g., Eq. (23)]. The oxidative addition reaction depends on the nature of substituents on the arene and fails to occur when strong electron-donating groups are present. Based on NMR and mass spectrometry (MS) data, the new complexes were formulated as monomeric. It had been

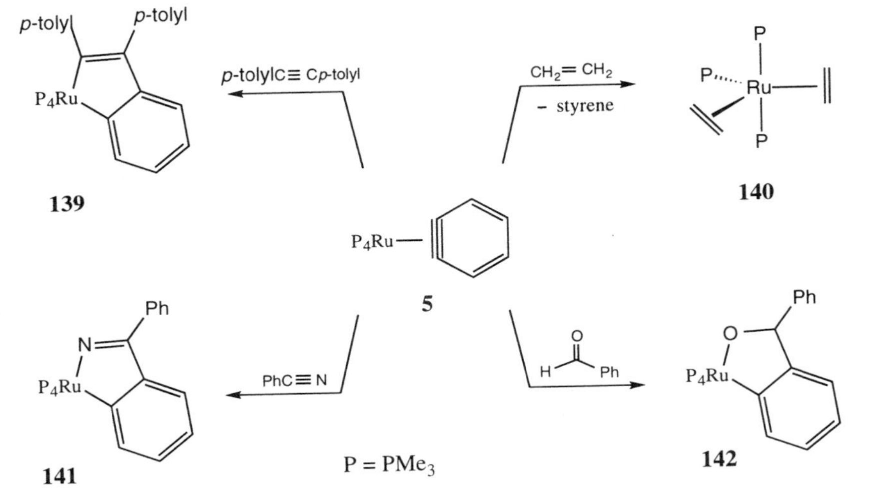

SCHEME 18

previously suggested[67] that lithium reduction of $NiCl(2\text{-}BrC_6H_4)(PEt_3)_2$ gives the dimeric product **148**. However, in view of these more recent results, it was suggested[37] that the monomeric benzyne structure is more likely.

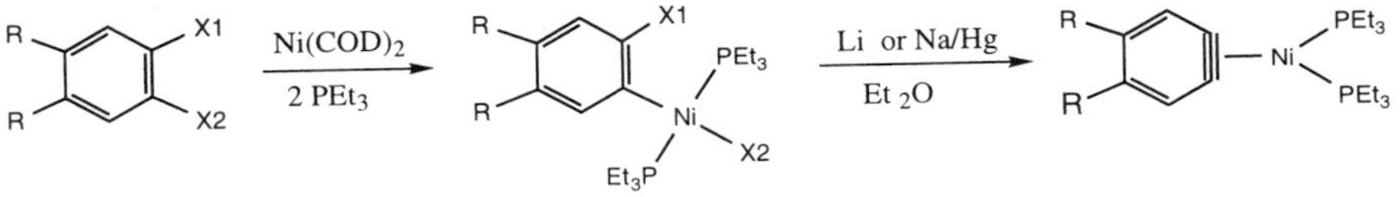

16: X1, X2 = Br
18: X1=Br, X2=Cl

143: R = F
144: R = H

(23)

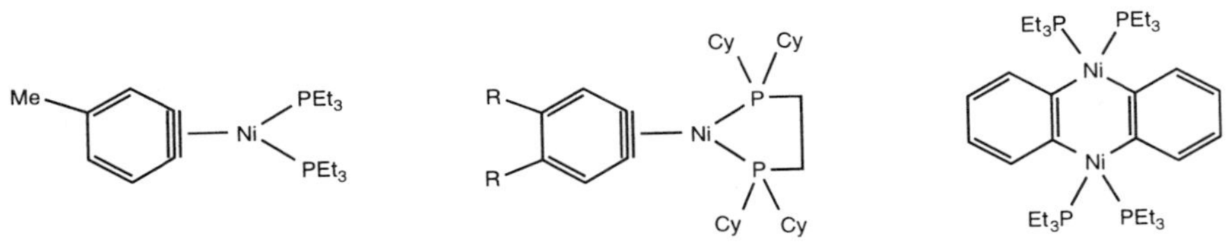

145

146: R = F
147: R = H

148

The nickel complexes **143**, **144**, **146**, and **147**[68] react readily with a variety of alkynes (summarized in Table II) but, in contrast to analogous Ti, Zr, Ta,

TABLE II

INSERTION CHEMISTRY OF Ni = BENZYNE COMPLEXES

Complex	Reagent	Products
143, 146	$EtC{\equiv}CEt$	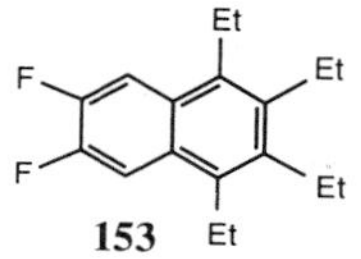 **153** (F, F, Et, Et, Et, Et): 89% (**143**), ca. 100% (**146**)
144	$t\text{-}BuC{\equiv}CH$	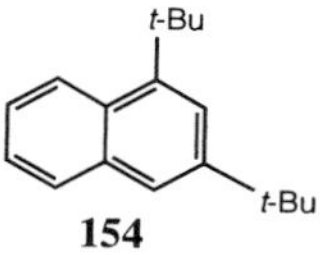 **154** (t-Bu, t-Bu): 56%
143	$MeC{\equiv}CCO_2Me$	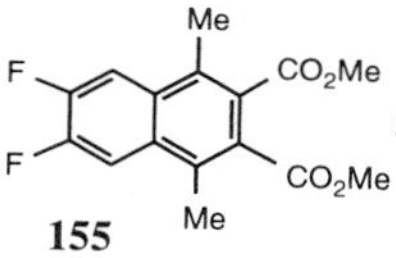 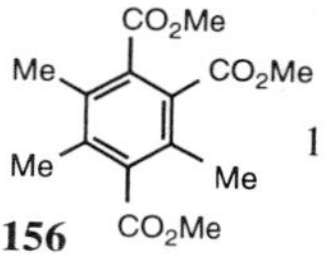 **155** (F, F, Me, Me, CO_2Me, CO_2Me): 56% + **156** (Me, Me, Me, CO_2Me, CO_2Me, CO_2Me): 11%
143, 146	$MeO_2CC{\equiv}CCO_2Me$	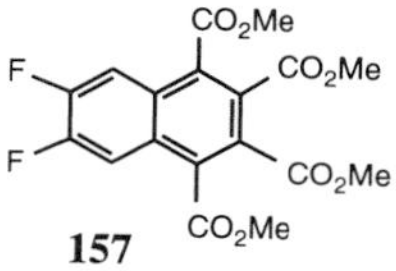 **157** (F, F, CO_2Me ×4): 5% (for **143**), 43% (for **146**) 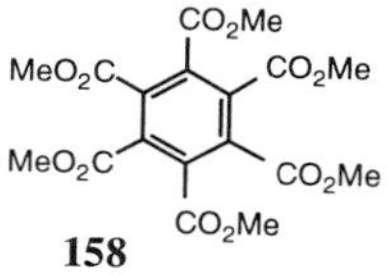 + **158** (MeO_2C, MeO_2C, CO_2Me ×4): 48% (for **143**), 0% (for **146**)
147	$MeO_2CC{\equiv}CCO_2Me$	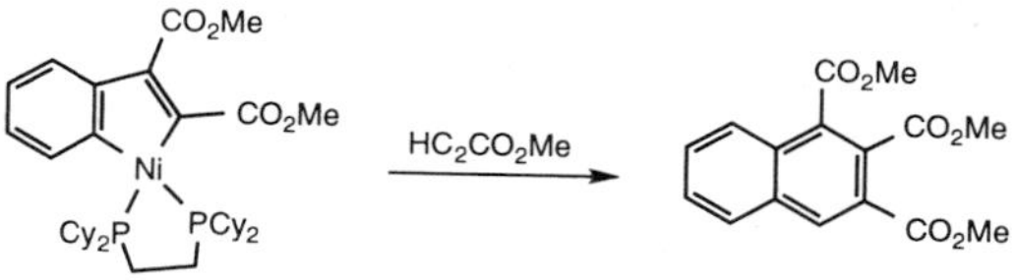 **159** (CO_2Me, CO_2Me, Ni, Cy_2P, PCy_2) $\xrightarrow{HC_2CO_2Me}$ (CO_2Me, CO_2Me, CO_2Me)
143	$CF_3C{\equiv}CCF_3$	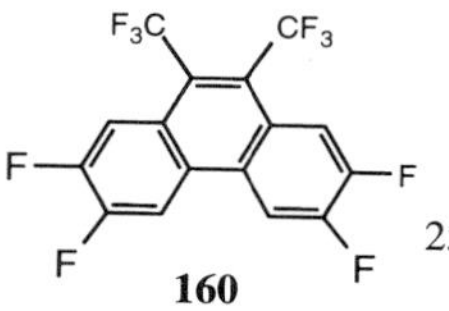 **160** (F_3C, CF_3, F, F, F, F): 25% + **161** (F_3C, F_3C, CF_3 ×4): 16% + **162** (F, F, CF_3 ×4): 10%

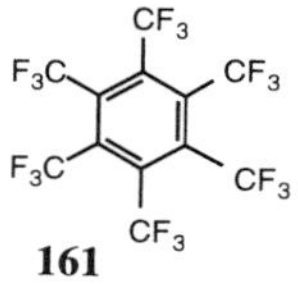

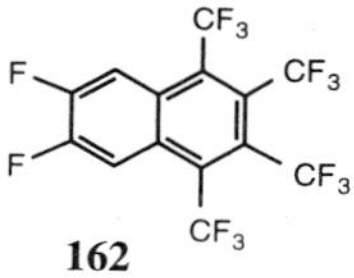

and Ru (vide supra) benzyne complexes,[2,36] in most cases mono-insertion products cannot be isolated. Instead, the presumed mono-insertion product undergoes further reaction with a second mole of alkyne to give substituted naphthalenes. Reductive elimination from an intermediate metallacycloheptatriene as exemplified in Eq. (24) is a reasonable mechanism for naphthalene formation. This sequence finds strong support in Bennett and Wenger's[68] isolation of metallaindene **159** from careful addition of less than 1 equiv dimethylacetylene dicarboxylate to **147** and their observation that it readily reacts with methyl propiolate to give trimethyl 1,2,3-naphthalene tricarboxylate (Table II). Reaction of **143** with electron-deficient alkynes also gives substituted benzenes, presumably from $Ni(PEt_3)_2$-catalyzed alkyne trimerization. This can become the predominant reaction (e.g., reaction of **143** with dimethylacetylene dicarboxylate), although it has been found that substituted benzene formation is suppressed in the reaction of **143** with $CH_3CCCO_2CH_3$ by the addition of PEt_3.

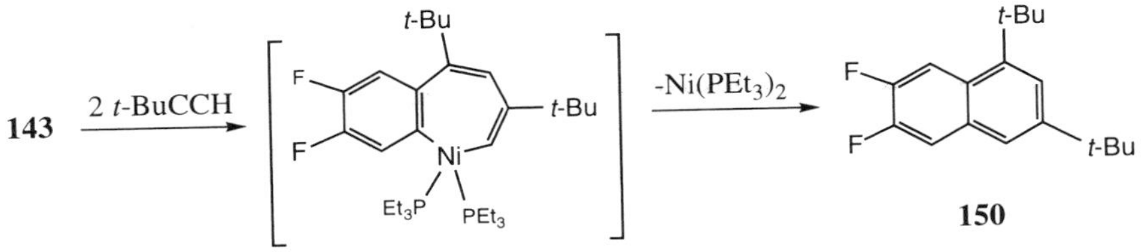

143 —2 *t*-BuCCH→ [**149**] —$-Ni(PEt_3)_2$→ **150** (24)

In one case a substituted phenanthrene (**160**) is also formed. Formally, this must arise from coupling of one acetylene with two benzynes. Insertion of a benzyne moiety into a nickelaindene is a reasonable mechanism.

In Bennett and Wenger's study[68] of the factors that control the two insertions in the naphthalene forming sequence, they concluded that (1) the regiochemistry of the first insertion step is sterically controlled with the substituted carbon bonding to Ni (e.g., **152** favored over **151**); (2) the second insertion occurs between Ni and vinyl rather than Ni and aryl and is sterically controlled with electron-rich alkynes; and (3) with electron-deficient alkynes, both steric and electronic factors influence the regiochemistry of the second insertion step.

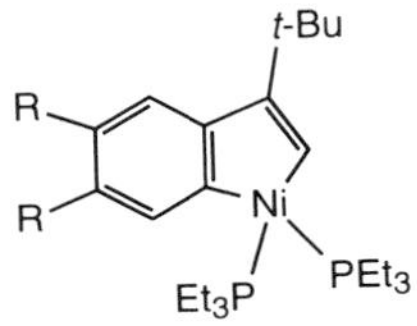

151 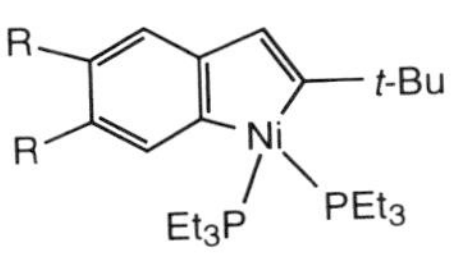 **152**

146 and **147** also react with carbon monoxide,[69] with the interesting finding that the nature of the products varies with the CO concentration (Scheme 19). When CO is allowed to diffuse slowly into a solution of either **146** or **147**, fluoreneones **164** and **165** are formed together with $Ni(CO)_2$(dcpe). This reaction presumably occurs via the acyl intermediate **163**, which, at low CO concentration, reacts with a second mole of the putative benzyne to generate a metallacycle that, upon reductive elimination, gives the fluoreneone. This is analogous to the reaction of **143** with hexafluoro-2-butyne to give the substituted phenanthrene (**160**). At high CO concentrations, it was suggested that insertion of a second CO generates the air-sensitive, nonisolable bis(acyl) complexes **167** and **168**, which, in the presence of oxygen, form stable phthalato complexes **169** and **170**, small amounts of **171** and **172**, and traces of the corresponding phthalic anhydride. Bis(acyl) complexes analogous to **167**/**168** have been isolated from the reaction of Ni(η^2-2-alkyne)(bipy) (alkyne = PhC_2H, PhC_2Ph, MeC_2Me) with CO.[70]

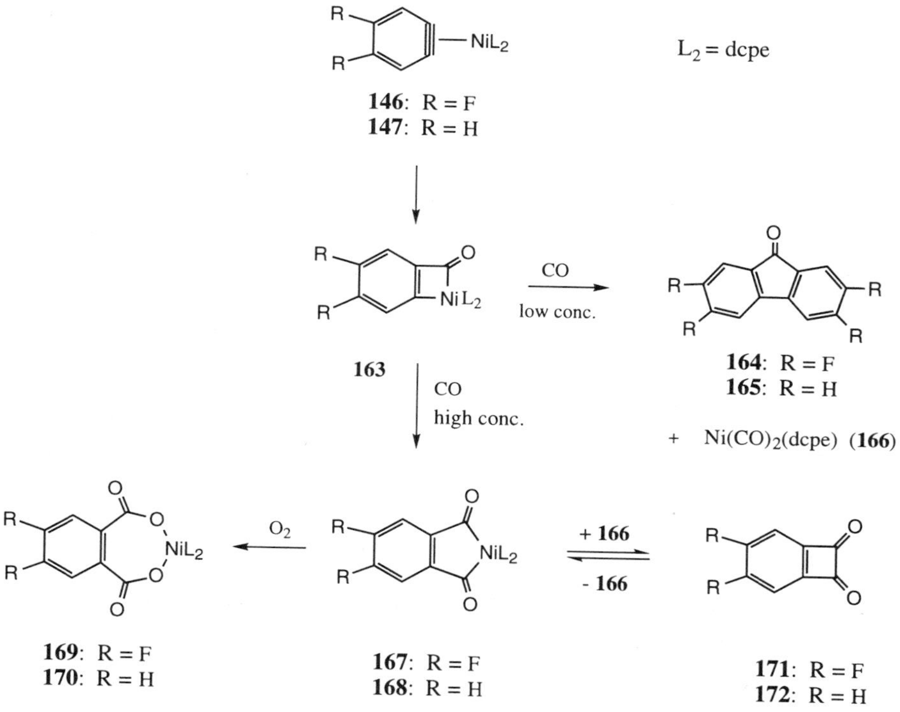

L_2 = dcpe

146: R = F
147: R = H

163

CO, low conc.

164: R = F
165: R = H

+ $Ni(CO)_2$(dcpe) (**166**)

CO, high conc.

O_2

+ **166** / - **166**

169: R = F
170: R = H

167: R = F
168: R = H

171: R = F
172: R = H

SCHEME 19

TABLE III

COMPARISON OF C—C BOND LENGTHS IN METAL–BENZYNE COMPLEXES WITH THOSE IN FREE BENZYNE[a,b]

	173[71]	**174**[72]	**78**[73]	**7**[29]	**121**[14]	**5**[6]	**130**[66]	**112**[64]
C1—C2	1.240	1.346(5)	1.364(8)	1.342(11)	1.370(6)	1.355(3)	1.353(8)	1.368(5)
C2—C3	1.393	1.410(5)	1.389(8)	1.394(12)	1.387(6)	1.382(3)	1.408(8)	1.384(5)
C3—C4	1.402	1.362(2)	1.383(9)	1.393(13)	1.383(8)	1.372(3)	1.401(9)	1.385(5)
C4—C5	1.421	1.403(2)	1.380(9)	1.357(15)	1.37(1)	1.411(4)	1.39(1)	1.404(5)
C5—C6	1.402	1.375(6)	1.377(9)	1.384(14)	1.383(8)	1.363(4)	1.39(1)	1.382(5)
C6—C1	1.393	1.408(6)	1.406(8)		1.387(6)	1.398(4)	1.395(8)	1.398(5)

[a] To facilitate comparison, selected examples are taken from Bennett's review[2] and the same format is used. The reader is referred to this review for a discussion of structural features of transition metal complexes of arynes.

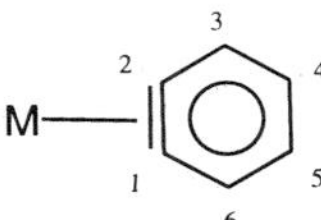

[b] C_6H_4 (**173**), TaCp*Me$_2$(η^2-C_6H_4) (**174**), ZrCp$_2$(η^2-C_6H_4)(PMe$_3$) (**78**), NbCp(NR)(η^2-C_6H_4)(PMe$_3$) (**7**), TaCp*(η^4-C_4H_6)(η^2-C_6H_4) (**121**), Ru(η^2-C_6H_4)(PMe$_3$)$_4$ (**5**), TaCp(Et$_2$C$_2$B$_4$H$_4$)(η^2-C_6H_4)(PMe$_3$) (**130**), VCp(PMe$_3$)$_2$(η^2-C_6H_4)(**112**).

TABLE IV

SELECTED ^{13}C NMR DATA FOR METAL–BENZYNE COMPLEXES[a,b]

Complex	Electron count[c]	*d*(ipso-C)(ppm)	Ref.
TaCp*Me$_2$(η^2-C_6H_4) (**174**)	4e (16 e$^-$)	230.5	74
TaCp*(η^4-C_4H_6)(η^2-C_6H_4) (**121**)	4e (18 e$^-$)	203.0	14
ZrCp$_2$(η^2-C_6H_4)(PMe$_3$) (**78**)	2e (18 e$^-$)	174.3	73
NbCp(NR)(η^2-C_6H_4)(PMe$_3$) (**7**)	2e (18 e$^-$)	162.7	29
		154.1	
Ni(η^2-C_6H_4)(dcpe) (**147**)	2e (16 e$^-$)	145.2	74
Ru(η^2-C_6H_4)(PMe$_3$)$_4$ (**5**)	2e (18 e$^-$)	142.1	6
TaCp(Et$_2$C$_2$B$_4$H$_4$)(η^2-C_6H_4)(PMe$_3$) (**130**)	2e (18 e$^-$)	168.2	66
		141.8	
TiCp$_2$(η^2-C_6H_3-OMe)(PMe$_3$) (**39**)	2e (18 e$^-$)	156.2	44
		149.9	
[Re(PMe$_3$)$_2$(η^2-2-Me-C_6H_3)(2-Me-C_6H_4)$_2$] [PF$_6$] (**175**)	2e (14 e$^-$)	164.4	30
		167.4	

[a] The reader is referred to an earlier review[2] for a discussion of the IR and NMR spectroscopy of metal complexes of arynes.

[b] A major portion of this table has been reproduced from Ref. 14.

[c] Electrons donated by the benzyne ligand, followed by the electron count for the complexes in parentheses.

D. *Benzdiynes*

To date, neither 1,2,4,5-tetradehydrobenzene (**176**), 1,2,3,4-tetradehydrobenzene (**177**), nor any derivative has been observed as a free molecule. However, a number of metal complexes have been reported, including zirconium complexes of **176** and **177**,[75] and nickel[33] and platinum[36] complexes of **176**. The early chemistry of these complexes has been reviewed[2,8,36] and is therefore only briefly summarized here.

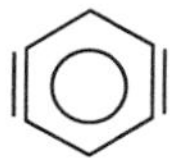

176

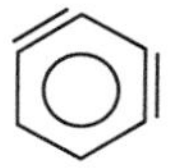

177

In a reaction that is neither regiospecific nor stereospecific, heating **178** in the presence of PMe_3 gives a mixture (**179a–179d**) of four isomeric zirconocene complexes of **176** and **177** (Scheme 20). Blocking two of the *ortho* positions reduces the number of isomers to two; thermolysis of **180** in the presence of PMe_3 gives only **181a** and **181b** [Eq. (25)]. The *anti*-isomer **181a** was characterized by an X-ray crystal analysis.

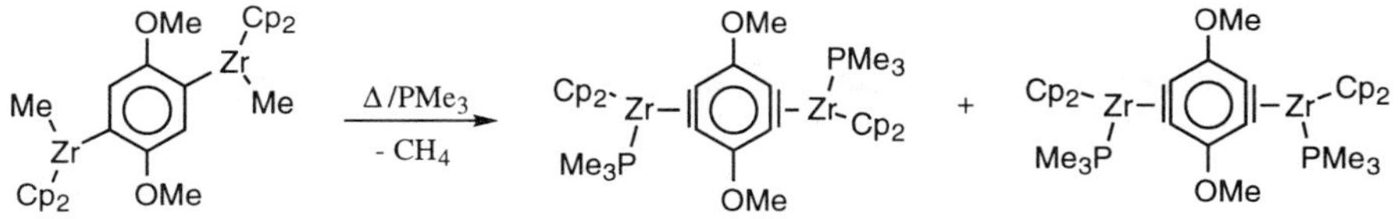
OMe
Cp2
Zr
Me
Me
Zr
Cp2
OMe
Δ /PMe3
- CH4
OMe
PMe3
Cp2
Zr
Zr
Cp2
Me3P
OMe
+
OMe
Cp2
Zr
Zr
Cp2
Me3P
PMe3
OMe

180 **181a** **181b**

(25)

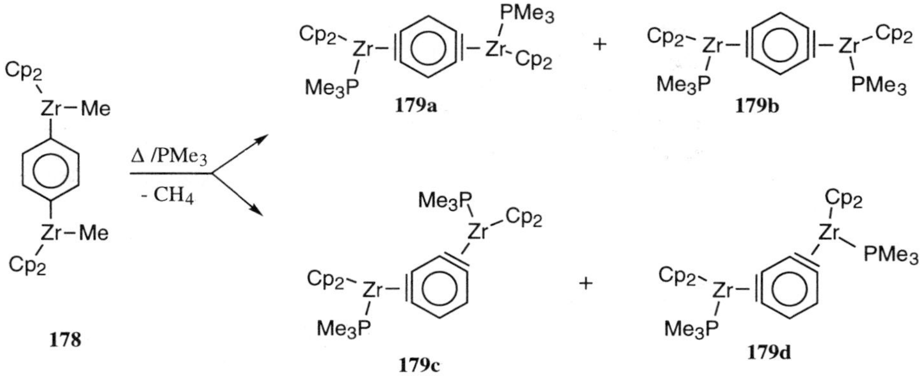
Cp2
Zr−Me
Zr−Me
Cp2
Δ /PMe3
- CH4
PMe3
Cp2
Zr
Zr
Cp2
Me3P
+
Cp2
Zr
Zr
Cp2
Me3P
PMe3
Me3P
Zr
Cp2
Cp2
Zr
Me3P
+
Cp2
Zr
PMe3
Cp2
Zr
Me3P

179a **179b** **178** **179c** **179d**

SCHEME 20

Complexes **181a** and **181b** react with ethylene, acetone, and 2-butyne to give mixtures of regioisomeric zirconacycles. For example, reaction of **181** with 2-butyne gives both **182a** and **182b** [Eq. (26)].[75]

181a, 181b $\xrightarrow{MeC_2Me}$ **182a** + **182b** (26)

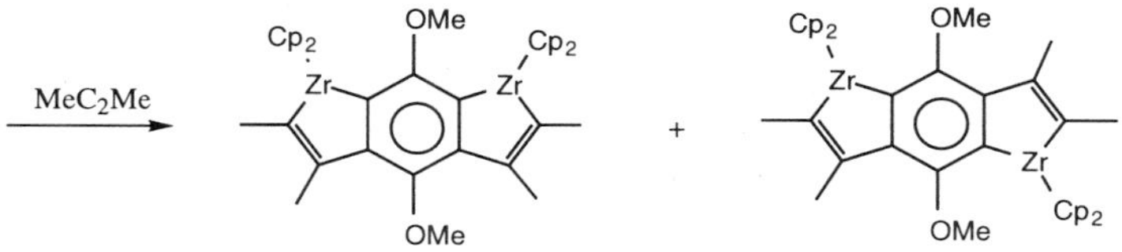

Attempts to prepare a monosubstituted benzdiyne from **184** failed; warming with PMe_3 led to elimination of only 1 mol benzene, giving **185** as the sole product [Eq. (27)].[76]

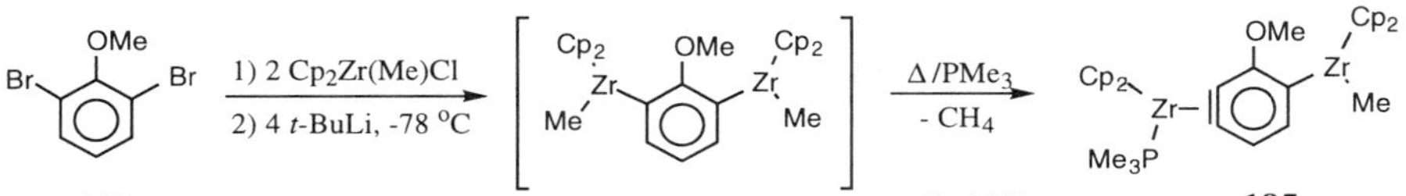

183 $\xrightarrow[\text{2) 4 } t\text{-BuLi, -78 °C}]{\text{1) 2 } Cp_2Zr(Me)Cl}$ [**184**] $\xrightarrow[-CH_4]{\Delta/PMe_3}$ **185** (42-54%) (27)

However, **185** mimics a benzdiyne in that it undergoes regioselective bis-insertion with two equivalents of 3-hexyne, acetonitrile, and acetone to form bis-zirconacycles **186**, **187**, and **188**, respectively (Scheme 21). This presumably occurs by sequential, and regiospecific, coupling and elimination of methane to give a new benzyne followed by a second coupling. The resulting bis-zirconacycles **186–188** react with acid to give 3,5-disubstituted anisoles (**189–191**) in high yield. Attempts to use **185** to prepare unsymmetrically substituted bis-zirconacycles was not satisfactory because methane elimination from the first-formed coupling product is competitive with the initial coupling. As a result, reaction of **185** with one equiv of the coupling reagent gives a mixture of the bis-insertion product, the mono-insertion product, and unreacted **185**.

Synthesis of an unsymmetrically substituted bis-zirconacycle was achieved by the sequence shown in Eq. (28).[76]

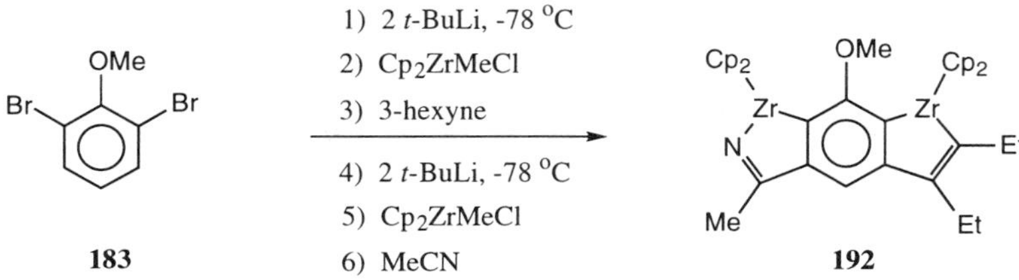

183 → **192** (28)

1) 2 *t*-BuLi, -78 °C
2) $Cp_2ZrMeCl$
3) 3-hexyne
4) 2 *t*-BuLi, -78 °C
5) $Cp_2ZrMeCl$
6) MeCN

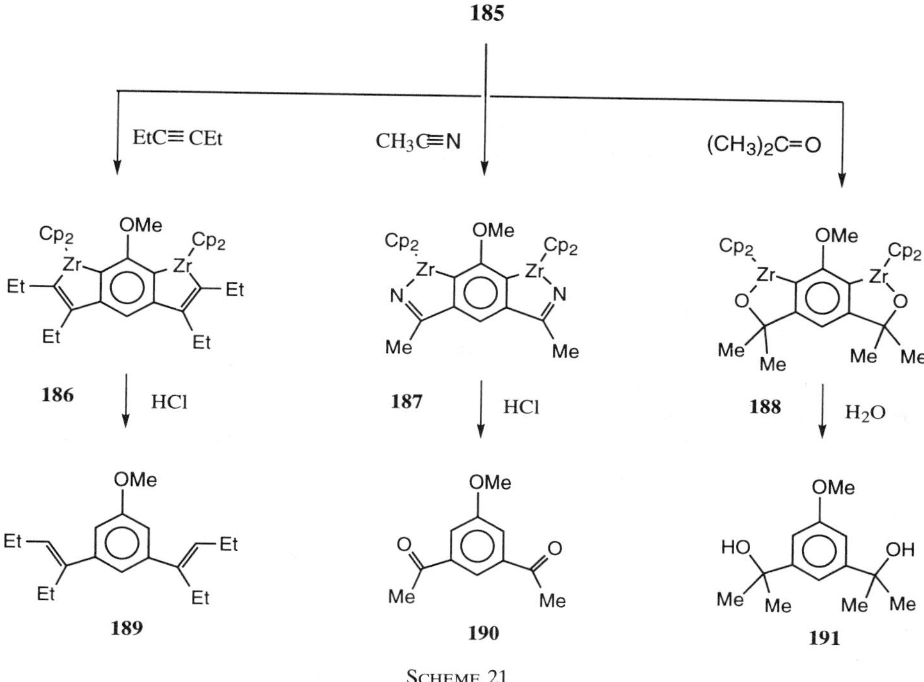

SCHEME 21

A nickel–benzdiyne complex **13** has been prepared in good yield (60–70%) by treating **11** with lithium tetramethylpiperidide (LiTMP) in the presence of **12** [Eq. (29)].[33] The benzdiyne complex was characterized by X-ray diffraction analysis.

$$\mathbf{11} + \mathbf{12} \xrightarrow{\text{LiTMP}} \mathbf{13} \qquad (29)$$

Platinum–benzdiyne complexes (**195a**, **195b**) have been prepared in low yields (10–30%) by Na/Hg reduction of **194** [Eq. (30)].[36] No reactions have been reported for **13**, **195a**, or **195b**.

Br, COCl, ClOC, Br — 1) $Pt(PPh_3)_4$; 2) - CO; 3) L_2 → Br, $PtClL_2$, L_2ClPt, Br — 43% Na/Hg, THF → L_2Pt—[benzdiyne]—PtL_2

193

194a: $L_2 = 2\ P\textit{i}\text{-}Pr_3$
194b: $L_2 = Cy_2PCH_2CH_2PCy_2$

195a: $L_2 = 2\ P\textit{i}\text{-}Pr_3$
195b: $L_2 = Cy_2PCH_2CH_2PCy_2$

(30)

E. *1,3,5-Benztriyne*

Successful synthesis of benzdiyne complexes **171**, **181**, and **13** raises the interesting question of whether metal–aryne complexes with even more highly unsaturated organic ligands such as benztriyne **196** could be prepared by the same methodology. However, in the only such attempt reported to date, it was found that **198**, prepared from **197**, when heated in the presence of trimethylphosphine gave an unexpected product (70%, by NMR) that shows spectroscopic properties consistent with **199** rather than the benztriyne complex **200** (Scheme 22).[77] An X-ray diffraction analysis of **202**, formed from thermolysis of **201** (Scheme 23), provides strong support for this unusual structure. In **202**, the zirconium atoms and bridging methyl group are only slightly displaced from the plane containing the six-membered ring. These two compounds (**199** and **202**) are similar to Erker's "anti-van't Hoff Lebel" complexes that result from reaction of Me_3Al or Me_3Ga with the zirconocene complexes of benzyne (Section II,B) and cyclohexyne (Section III,C). The bridging methyl group in **202** is readily replaced by chlorine or fluorine with $AlCl_3$ or $BF_3{\cdot}OEt_2$, respectively. Reaction of **202** with acetone or iodine gives **203** and **204**, respectively (Scheme 23). It fails to react cleanly with alkynes, nitriles, or isonitriles.

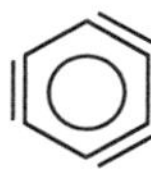

196

F. *Naphthalynes*

The first two transition-metal complexes of 2-naphthalyne[78] have been prepared[79] to be used as intermediates in the synthesis of polysubstituted naphthalenes. The naphthalyne intermediate **206**, prepared from **205** [Eq. (31)], couples with a wide array of nitriles to give zirconocycles (**207**) that have been converted into substituted naphthalenes (**208**) by reaction with HCl and iodonaphthalenes (**209**) by reaction with I_2 followed by HCl. The

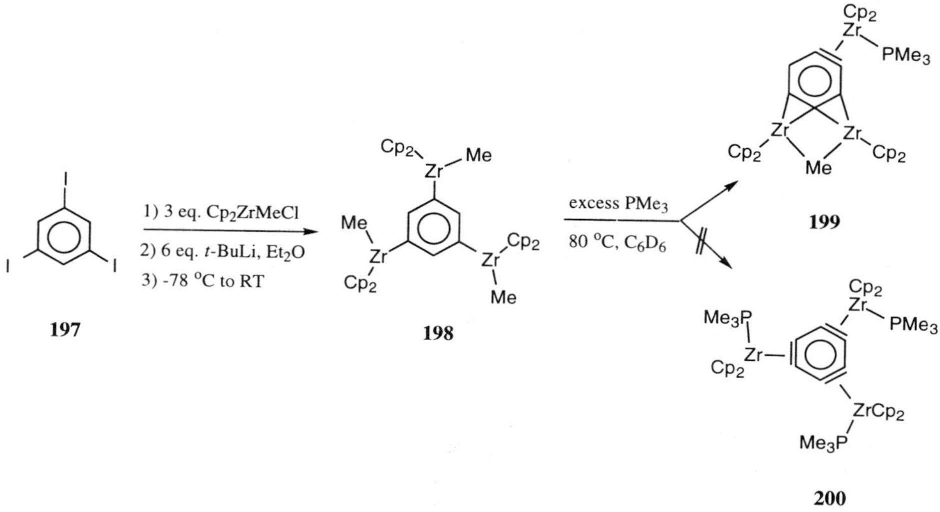

SCHEME 22

latter (R = MEM) can, in turn, be further oxidized with Jones' reagent to give naphthoquinone **210** in good yield (Scheme 24).[79] The reactions with nitriles, which are analogous to similar reactions of zirconium–benzyne complexes,[49,80] are highly regiospecific and show significant tolerance to-

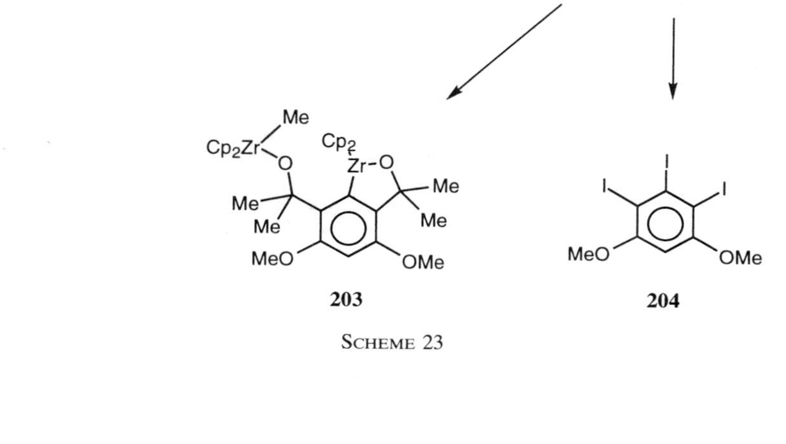

SCHEME 23

ward nitriles carrying a variety of functional groups, including double bonds, heteroaromatics, and protected ketones. Naphthalyne complexes also react with cyclic allylic ethers[62] and triethylborate[47] in reactions analogous to those of the benzyne analogues (Section II,B).

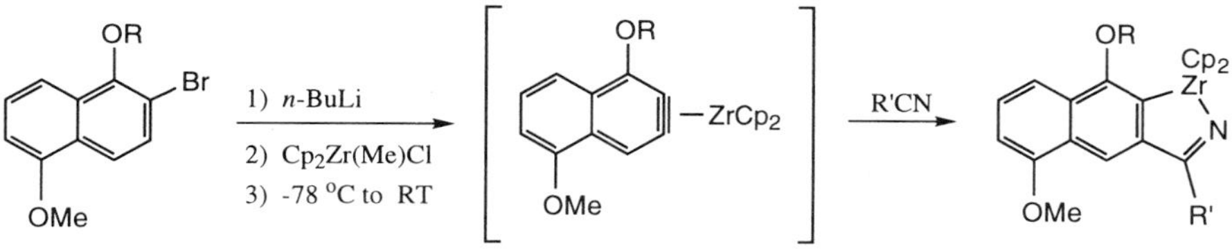

205 → 1) n-BuLi 2) $Cp_2Zr(Me)Cl$ 3) -78 °C to RT; R = Bn, MEM → [206] → R'CN → 207 (31)

Two nickel complexes of 2-naphthalyne (**213** and **214**) (Scheme 25) have been prepared by reduction of **211** and **212** with 1% Na/Hg[37] by the same method used to synthesize nickel complexes of benzyne[35,74] (Section II,C). The reactions of **213** and **214** are generally similar to those of their benzyne counterparts.[2] Complex **214** couples with carbon dioxide and tetrafluoroethylene to give **215** and **218**, respectively, whereas **213** reacts with alkynes to give anthracenes **216** and **217**, albeit in low yield. The reaction conditions necessary to induce coupling between **214** and tetrafluoroethylene (−60°C to RT) are significantly milder than those of the corresponding reaction between ethylene and $Ni(\eta^2\text{-}C_6H_4)(dcpe)$ (**147**) (3 bar, 80°C).[74] Complex

207 → 1N HCl; R = Bn, MEM → 208

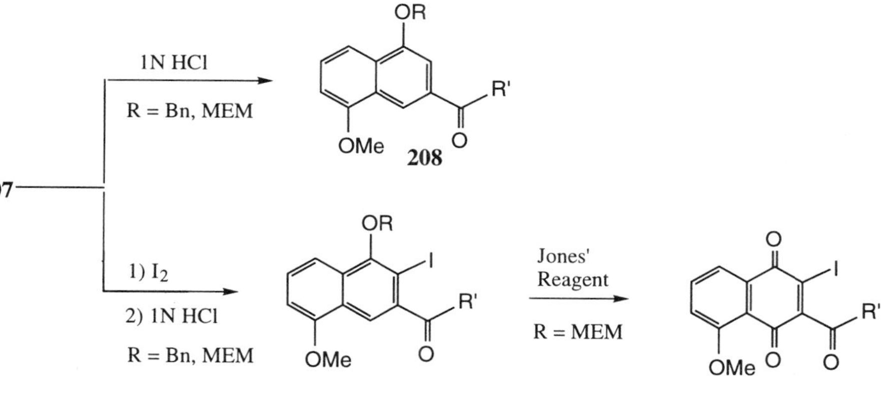

207 → 1) I_2 2) 1N HCl; R = Bn, MEM → 209 → Jones' Reagent; R = MEM → 210

SCHEME 24

218 reacts further with dimethyl acetylene dicarboxylate in refluxing THF to give the seven-membered nickelacycle **219** (Scheme 25). X-ray crystal structures were reported for **215**, **218**, and **219**.

G. *Synthesis and Reactivity of Tropyne Complexes*

To date, benzyne's next higher homologue, the cycloheptadienynylium ion (**220**)—which has been given the common name "tropyne"—is still unknown as a free intermediate. However, a number of metal complexes have been prepared. The bis(triphenylphosphine)platinum(0) complex of the parent (**223**) was synthesized by hydride abstraction from a mixture of **222a** and **222b** (Scheme 26), which was prepared by dehydrobromination of **221a–c** with lithium diisopropylamide (LDA) in the presence of $Pt(PPh_3)_3$ (Scheme 26; see Section IV,C).[81,82] The tropyne complex **223**, a red, air-stable crystalline material,[81] is much less reactive than its benzyne counter-

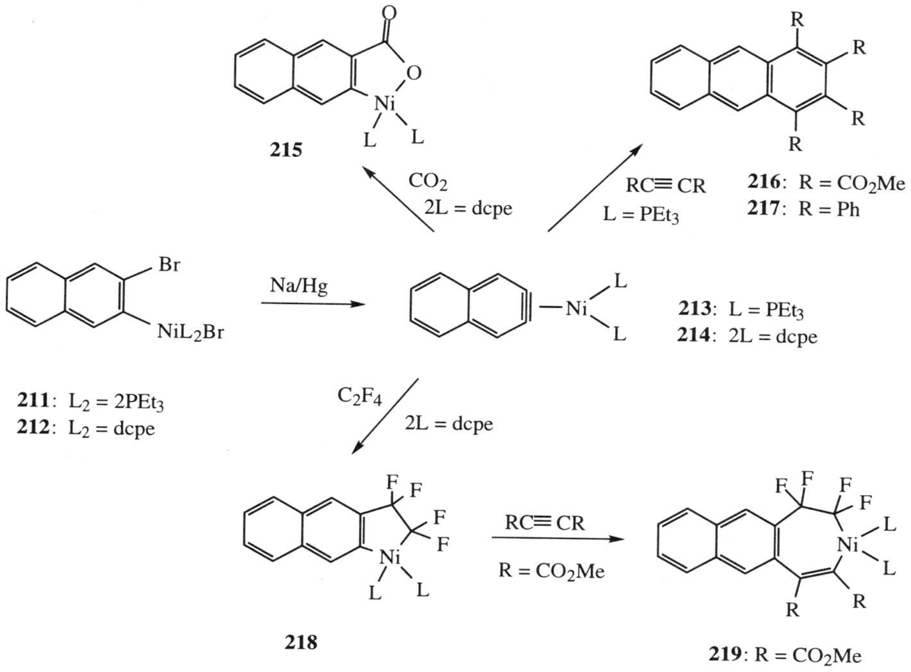

211: $L_2 = 2PEt_3$
212: L_2 = dcpe

213: L = PEt_3
214: 2L = dcpe

215

216: R = CO_2Me
217: R = Ph

218

219: R = CO_2Me

SCHEME 25

part, It shows no reaction with acetone, methanol (70°C for 12 h), or acetonitrile. However, it reacts instantaneously with HCl or HBr in THF to form cycloheptatrienylidene complexes **224a** and **224b** and is readily reduced with $KBEt_3H$ to regenerate **222a** and **222b**.

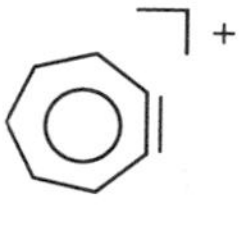

220

Hydride abstraction from **225** gives the dibenzannuleted tropyne complex **226** (Scheme 27).[85] As in the case of **223**, **226** reacts rapidly with HBr to give **227**, which, in solution, isomerizes slowly to **228**. $KBEt_3H$ and bis(dicyclohexylphosphine)ethane attack the 4-position of the tropyne to give alkyne complexes **225** and **229**, respectively.

A palladium complex of tropyne (**230**) has been prepared by the same method used for **223**.[86] Interestingly, in contrast to all of the platinum tropyne complexes (**223**, **226**, **235**), which are stable at room temperature, the palladium complex is thermally labile, decomposing at −35°C with a half-life of about 3 h. Reactions of **230** with $KBEt_3H$, $LiAl(O\text{-}t\text{-}Bu)_3H$, HBr, or cyclopentadiene gave only intractable mixtures.

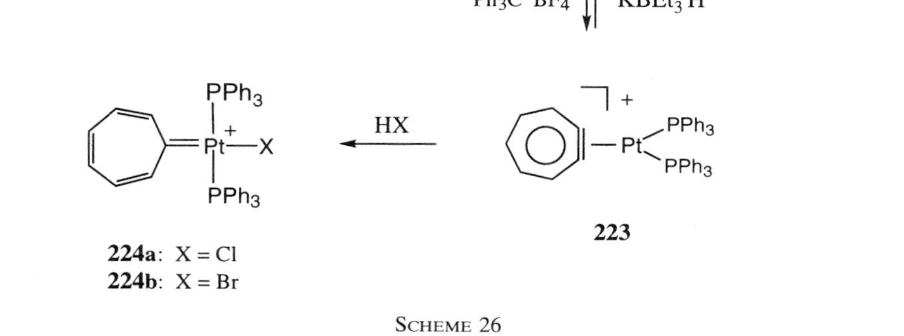

Br; LDA; $Pt(PPh_3)_3$; PPh_3; Pt; PPh_3; +; Pt; PPh_3; PPh_3

221 a-c **222a** **222b**

$Ph_3C^+BF_4^-$ $KBEt_3H$

PPh_3; Pt; X; PPh_3; HX; Pt; PPh_3; PPh_3

224a: X = Cl
224b: X = Br

223

SCHEME 26

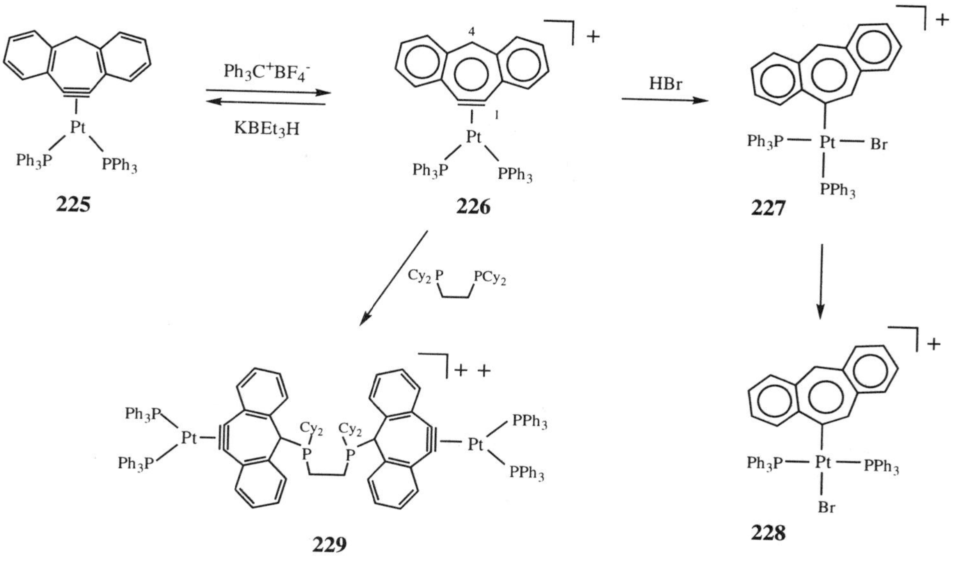
Ph3C+BF4-
KBEt3H
HBr
Pt
Ph3P
PPh3
Br
Cy2P PCy2
225
226
227
228
229

SCHEME 27

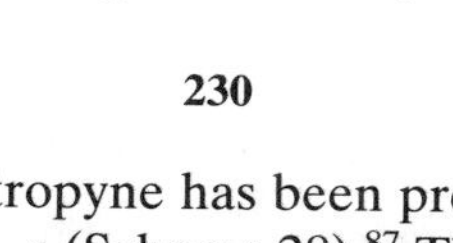
Pd
PPh3
PPh3

230

A zirconium complex of tropyne has been prepared by hydride abstraction from a mixture of **231a–c** (Scheme 28).[87] This compound is extremely thermally unstable, decomposing to a complex mixture of products above −50°C. Identification of both **230** and **232** rests exclusively on low-temperature, multinuclear NMR spectra. Attempts to couple **232** with reagents such as alkenes, alkynes, nitriles, or ketones gave only complex mixtures of products.

The bimetallic tropyne complex **235** has been synthesized in two ways: by reacting **223** with (η^6-*p*-xylene)$Mo(CO)_3$ in THF and by hydride abstraction from a mixture of **236** and **237** (Scheme 29).[88] Transfer of $Mo(CO)_3$ from *p*-xylene to the tropyne complex is complete in a few minutes at room temperature. Reaction of **235** with $LiAl(O\text{-}t\text{-}Bu)_3H$ or $KBEt_3H$ results in hydride transfer to the ring. The reaction is regioselective, giving predominately the bimetallic complex **238**, which has a cycloheptatetraene ring with

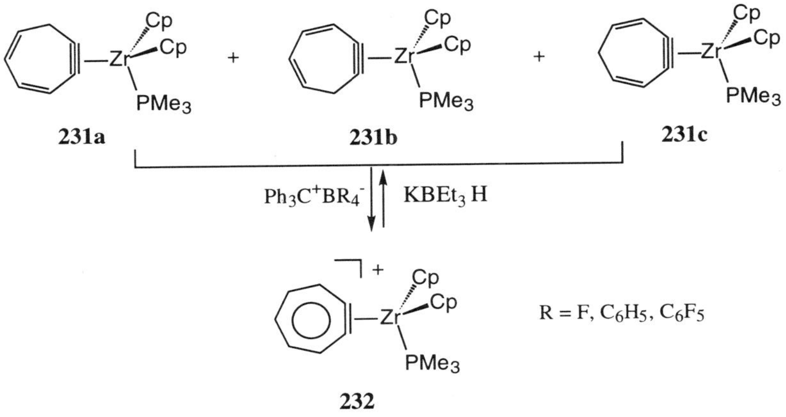

Cp Cp PMe3 **231a** + Cp Cp PMe3 **231b** + Cp Cp PMe3 **231c**

$Ph_3C^+BR_4^-$ ⇅ $KBEt_3H$

Cp Cp PMe3 **232**

R = F, C_6H_5, C_6F_5

SCHEME 28

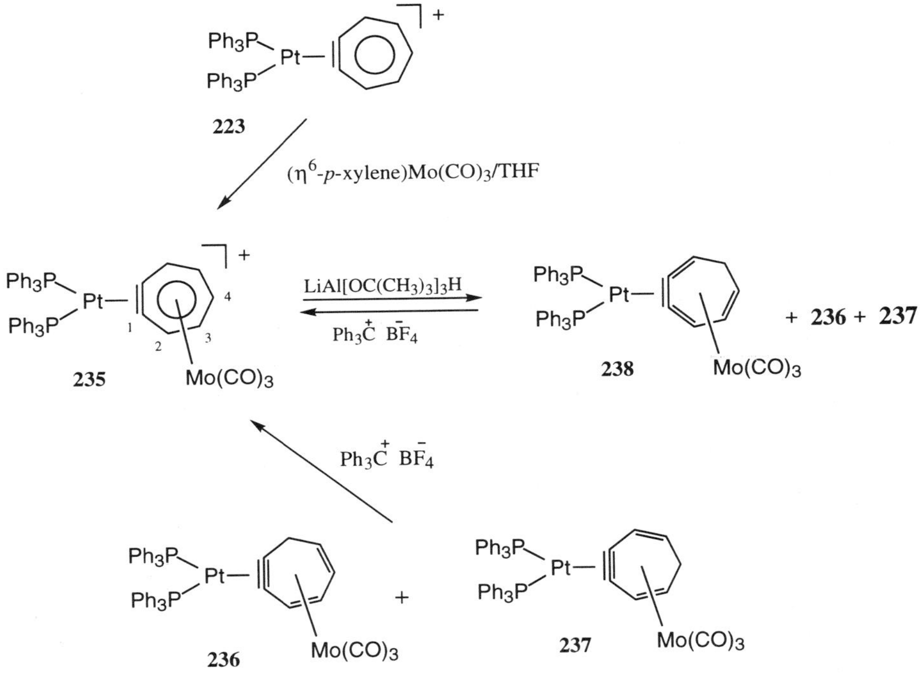

Ph_3P, Ph_3P Pt **223**

(η^6-*p*-xylene)$Mo(CO)_3$/THF

235 $Mo(CO)_3$

$LiAl[OC(CH_3)_3]_3H$ / $Ph_3\overset{+}{C}\ B\overset{-}{F}_4$

238 $Mo(CO)_3$ + **236** + **237**

$Ph_3\overset{+}{C}\ B\overset{-}{F}_4$

236 $Mo(CO)_3$ + **237** $Mo(CO)_3$

SCHEME 29

TABLE V

^{1}H NMR Data for Tropyne Complexes

Complex no.	H2	H3	H4	H5	H6
223	7.66 (m)	8.34 (m)	8.64 (t)		
235	5.55 (t)	6.10 (m)	5.65 (t)		
226			10.43 (s)		
230	7.83 (d)	8.32 (m)	8.52 (t)		
232	8.85 (d)	7.85 (m)	8.00 (t)	7.85 (m)	8.95 (d)

Letters in parentheses refer to the multiplicity of the NMR peaks—m (multiplet), t (triplet), d (doublet), s (singlet).

a unique 1,2,3-butatriene moiety. This is discussed in more detail in Section IV,C. Small amounts of **236** and **237** are also formed.

H. *Structure and Spectroscopy of Tropyne Complexes*

The bis(triphenylphosphine)platinum complex **223** is the only tropyne complex to date that has been characterized by X-ray diffraction analysis. The seven-membered ring of the tropyne ligand is planar. All carbon–carbon bond distances in the ring are equivalent within experimental error with an average bond length of 1.38(2) Å. The carbon–carbon triple bond length [1.37(2) Å] is equal within experimental error to that in transition-metal–benzyne complexes (Table III) but is longer than the triple bonds in bis(triphenylphosphine)platinum(0) cyclohexyne [1.297 Å],[89] bis(triphenylphosphine)platinum(0) cycloheptyne [1.283 Å],[89] or the dibenzannelated didehydrotropone complex **259** [1.283(15) Å].[85] The chemical shift of all hydrogens bonded directly to the tropyne ligand (Table V) and all carbons except C1 (Table VI) in all of the tropyne complexes are upfield from those of the tropylium ion (^{1}H: δ 9.55 ppm; $^{13}C\{^1H\}$: δ 160.6 ppm). This upfield

TABLE VI

$^{13}C\{^1H\}$ NMR Data for Tropyne Complexes

Complex no.	C1	C2	C3	C4	C5	C6	C7
223	174.6 (dd)	135.6 (m)	150.0 (d)	141.8 (s)			
235	111.28 (dd)	87.08 (d)	100.75 (d)	97.96 (s)			
226	160.15 (dd)			155.9 (s)			
230	173.34 (dd)	139.47 (t)	149.9 (s, br)	143.05 (s)			
232	235.5 (d)	156.0 (s)	140.3 (s)	143.2 (s)		152.4 (s)	214.1 (d)

Letters in parentheses refer to the multiplicity of the NMR peaks—m (multiplet), t (triplet), d (doublet), dd (doublet of doublets), s (singlet), and br (broad).

shift has been attributed to electron donation from the metal into the π-framework of the tropyne.[81] Such an interaction, which would strengthen the metal–carbon bonds, should be possible since the tropylium ion has a vacant low-energy orbital of proper symmetry to interact with the HOMO of the transition metal. A large ^{195}Pt–^{13}C coupling constant (458.5 Hz) between platinum and C1 in **223** and carbonyl stretching frequencies in **235** that are 40 cm^{-1} lower than those of $(\eta^7\text{-}C_7H_7)Mo(CO)_3$[88] are also consistent with "communication" between the platinum atom and the π-system of the tropyne ring. Interestingly, the $Mo(CO)_3$ in **235** also causes upfield shifts of both the proton and carbon resonances of the tropyne ligand relative to other tropyne complexes (Tables V and VI). The $^{13}C\{^1H\}$ NMR spectrum of **235** has only one carbonyl resonance at −100°C, indicating a low barrier to rotation of the $Mo(CO)_3$ fragment around the molybdenum–tropyne axis, a barrier that is significantly lower than in the corresponding cycloheptadienyne complex (Section IV,C). The ^{195}Pt $\{^1H\}$ and $^{31}P\{^1H\}$ NMR data for tropyne complexes are recorded in Table VII.

TABLE VII
^{195}Pt$\{^1H\}$ NMR AND $^{31}P\{^1H\}$ NMR DATA FOR TROPYNE COMPLEXES

Complex no.	^{195}Pt$\{^1H\}$ (ppm)	$^{31}P\{^1H\}$ (ppm)
223	−3788 (t, $^1J_{Pt\text{-}P}$ = 3121.8 Hz)	21.5 (s)
235	−4087.6 (t, $^1J_{Pt\text{-}P}$ = 3282.5 Hz)	19.96 (s)
226	−4060.2 (t, $^1J_{Pt\text{-}P}$ = 3189.2 Hz)	20.25 (s)
230		24.2 (s)
232		−4.6 (s)

Letter 's' (in parentheses) refers to the multiplicity of the NMR peaks and is defined as 'singlet'.

III

COMPLEXES OF STRAINED CYCLOALKYNES[1]

A. *Three- and Four-Membered Rings*

To date, no mononuclear transition metal complexes of three- or four-membered cycloalkynes have been reported. A number of polynuclear complexes of cyclobutyne have been prepared.[90] However, since this article is limited to complexes in which the aryne, alkyne, or cumulene is coordinated to a single metal atom, a review of this topic is not included.

B. *Five-Membered Rings*[91]

The bis(triphenylphosphine)platinum(0) complex of cyclopentyne (**240**) [Eq. (32)] is the only complex of a five-membered cycloalkyne that is discussed in an earlier review.[2] The mechanism of its preparation is significant because **239** was isolated and shown to react with Na/Hg to give **240**, thus suggesting that π-complexes may play a role in the formation of d^{10} complexes of larger cycloalkynes[2,36,82] and, possibly, cyclic allenes (Section IV)[82] where trapping of the free intermediate had often been presumed.

1,2-dibromocyclopentene $\xrightarrow{Pt(PPh_3)_3}$ **239** [$(\eta^2$-1,2-dibromocyclopentene$)Pt(PPh_3)_2$] $\xrightarrow{Na/Hg}$ **240** [$(\eta^2$-cyclopentyne$)Pt(PPh_3)_2$] (32)

More recently, a zirconocene complex of 3,3-dimethylcyclopentyne (**242**) was prepared by warming **241** to 90°C ($t_{1/2}$ = 240 min) [Eq. (33)].[92] Without the *gem*-dimethyl group, the reaction failed; the parent showed no reaction when heated at 120°C for 36 h. Buchwald *et al.*[92] suggested that the *gem*-dimethyl accelerates the reaction by sterically forcing the zirconium closer to the vinyl hydrogen. Selected parameters from an X-ray crystal structure of **242** are given in Table VIII. The most distinctive parameter is a bend-back angle (deviation of the C≡C—C angle from 180°) of 67.2°, which is the largest that has been recorded to date, nearly 13° larger than that of the cyclohexyne counterpart.

241 [$Cp_2Zr(CH_3)$(5,5-dimethylcyclopentenyl)] $\xrightarrow[PMe_3]{90\ ^{\circ}C}$ **242** [$(\eta^2$-3,3-dimethylcyclopentyne$)ZrCp_2(PMe_3)$] (33)

TABLE VIII

SELECTED PARAMETERS FOR CYCLOAKYNE–METAL COMPLEXES[a,b]

Compound	d(C≡C)(Å)	Bend-back angle [°]	C–M–C angle [°]	d(M–C) [Å]	Ref.
284	1.279(13)	34.0	36.6	2.036(9)	2; Ref. 68,69
285	1.283(5)	41.3	36.5	2.050(4)	2; Ref. 48,67
286	1.297(8)	52.7	37.1	2.039(5)	2; Ref. 67
287	1.295(25)	54.4	32.1	2.204(13)	2; Ref. 60
275	1.307(8)	49(8)[c]	31.2(2)	2.2215(7)[c]	100
222a	1.31(2)	47.8[c]	37.5(6)	2.03(2)[c]	82
259	1.283(15)	44.2(14)	36.6(4)	2.043(13)[c]	85
236	1.292(14)	45.9[c]	37.2(4)	2.029(10)	88
273	1.321	50.6	37.6	2.052	99
250	1.272(5)	53.1(4)[c]	39.7(2)	1.879(4)[c]	93
279	1.295(6)	67.2(5)[c]	34.5(2)	2.815(4)[c]	92
282	1.35(2)	58.5(10)[c]	38.8(5)	2.03(1)[c]	108
260	1.258(6)	45.5(4)[c]	36.0(2)	2.035(4)	86

[a] To facilitate comparison, selected examples are taken from Bennett and Schwemlein's review[2] and the same format is used. The reader is referred to this review for an excellent discussion of the IR and ^{13}C NMR spectra of metal complexes of strained cycloalkynes.

[b] [$Pt(PPh_3)_2(C_8H_{12})$] **284**. [$Pt(PPh_3)_2(C_7H_{10})$] **285**. [$Pt(PPh_3)_2(C_6H_8)$] **286**. $Cp_2Zr(C_6H_8)PMe_3$ **287**, Cp_2Zr(COT)PMe_2Ph **275**, [$Pt(PPh_3)_2$(cycloheptadienyne)] **222a**, [$Pt(PPh_3)_2$(dibenzotropyneone) **259**, [$Pt(PPh_3)_2Mo(CO)_3$(cycloheptadienyne) **236**, [$Pt(PPh_3)_2$(homoadamantyne)] **273**, [Ni(dcpe)(C_6H_8)] **250**, Cp_2Zr(dimethylcyclopentyne)PMe_3 **279**, {$Pt(PPh_3)_2$[2.2]paracyclophane-l-yne} **282**, [$Pd(PPh_3)_2$(cyclohepta-3,5-dien-l-yne) **260**.

[c] Average value.

Dissociation of PMe_3 from **242** is apparently more rapid than from the cyclohexyne analogue, which was suggested to account for its greater reactivity; **242** reacts with olefins at RT [Eq. (34)], whereas the cyclohexyne complex requires warming. The regiochemistry of the olefin addition products from the five- and six-membered rings also differs. The cyclohexyne complex reacts with 1-hexene to give a ca. 1:1 mixture of regioisomers, whereas the cyclopentyne complex gives only the isomer with the zirconium bonded to the CH_2 (**244**). Furthermore, in all coupling reactions with the cyclopentyne complex, the zirconium in the product is also bonded to the carbon α to the *gem*-dimethyl group. It was proposed that this results from approach of the olefin to the less hindered side of **243** oriented in such a way as to place the substituent in the sterically less congested β-position. With chiral allyl ethers the cyclopentyne complex reacts diastereoselectively as well as regioselectively. The degree of diastereoselectivity varies with the ether substitutents, but in a particularly impressive example **242** reacts with **245** (80% *ee*) followed by CO to give cyclopentenone **246** in 50–54% isolated yield as a single diastereoisomer (80% *ee*) [Eq. (35)]. Buchwald *et al.*[92] point out that this "represents the first transition-metal-induced intermolecular carbon–carbon bond formation via a formal reductive coupling of an olefin and an alkyne with ~100% asymmetric induction caused by an existing chiral center on one of the substrates."

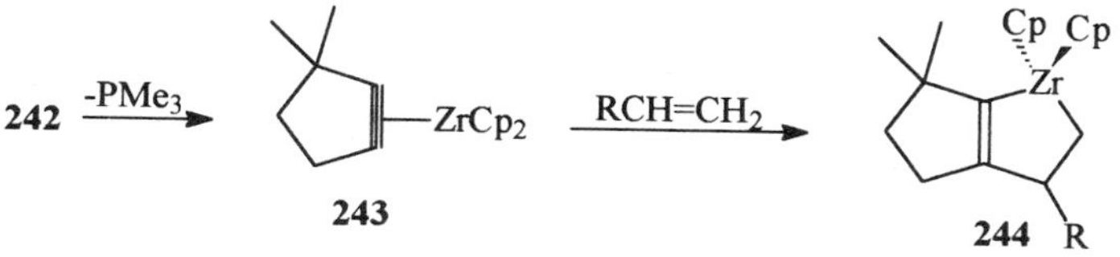

(34)

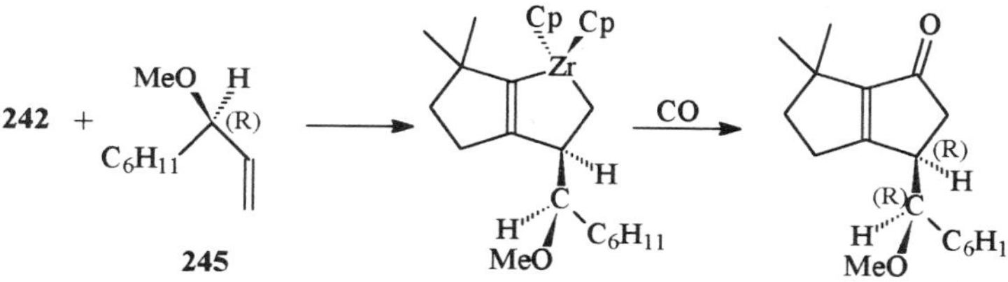

(35)

C. *Six-Membered Rings*

Synthesis, reactions, and physical properties of stable mononuclear platinum and zirconium complexes of cyclohexyne reported prior to the late 1980s have been comprehensively covered in earlier reviews.[2,8] More recently, reaction of the zirconocene complex of cyclohexyne with trimethylaluminum and trimethylgallium has been reported to give **247** and **248**, respectively [Eq. (36)].[57,58] These products are novel because four atoms (carbons C1 and C3, the transition metal, and the main group metal) are covalently bonded to an sp^2 hybridized carbon (C2) in a planar tetracoordinate fashion. Synthesis of this type of complex, which Erker describes as "anti-van't Hoff/LeBel," does not require the strained cycloalkyne ring; zirconocene complexes of acyclic alkynes react similarly.[57,58]

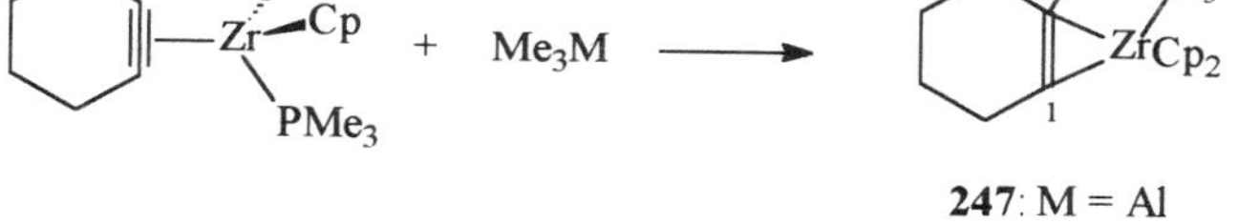

(36)

247: M = Al
248: M = Ga

Three nickel complexes of cyclohexyne (**249–251**) have been prepared by reduction of 1,2-dibromocyclohexene in the presence of $Ni(\eta^2\text{-}C_2H_4)L_2$ [Eq. (37)].[93] Two reasonable routes were suggested for this reaction: NiL_2 could simply trap transient cyclohexyne, or Na/Hg could reduce an initially formed insertion product (**252**) (Scheme 30). At this time, both remain viable since reduction of **252** (L,L = dcpe) with Na/Hg gives **250**. This has precedent in nickel/benzyne chemistry[74] and is in contrast to the platinum/cyclohexyne analogue, where the insertion product is stable to the reduction conditions. Reduction of an initially formed, but undetectable, π-complex of 1,2-dibromocyclohexene, analogous to reduction of the corresponding π-complex of 1,2-dibromocyclopentene,[2] is a third possibility that has not been excluded.

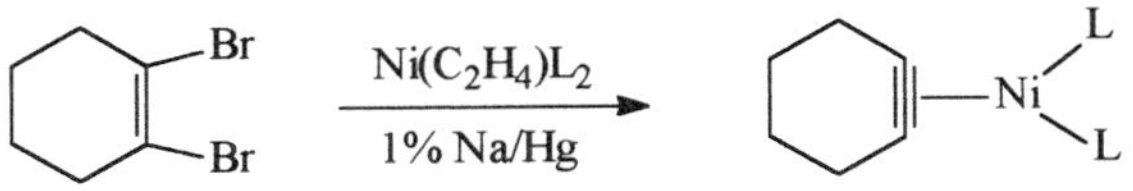

Br, Br — $Ni(C_2H_4)L_2$ / 1% Na/Hg → Ni(L)(L) (37)

249: L = PPh_3
250: L,L = dcpe
251: L = PEt_3

An x-ray crystal structure was reported for **250**. Selected parameters are recorded in Table VIII. The nickel complex is similar to its platinum counterpart in that it has a nearly planar coordination geometry, and most of the parameters are very similar for the two. The most striking difference is in the M–C and M–P bond distances, differences that are comparable to those found in acyclic complexes.[94] In the IR spectrum, the triple bond

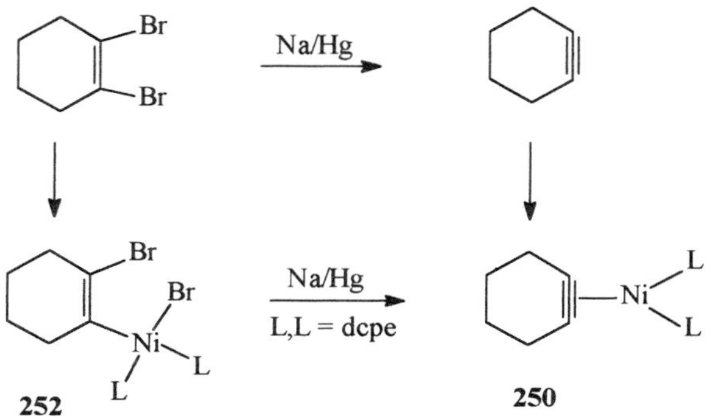

Br, Br — Na/Hg →

Br, Br, Ni, L, L — Na/Hg / L,L = dcpe → Ni, L, L

252 **250**

SCHEME 30

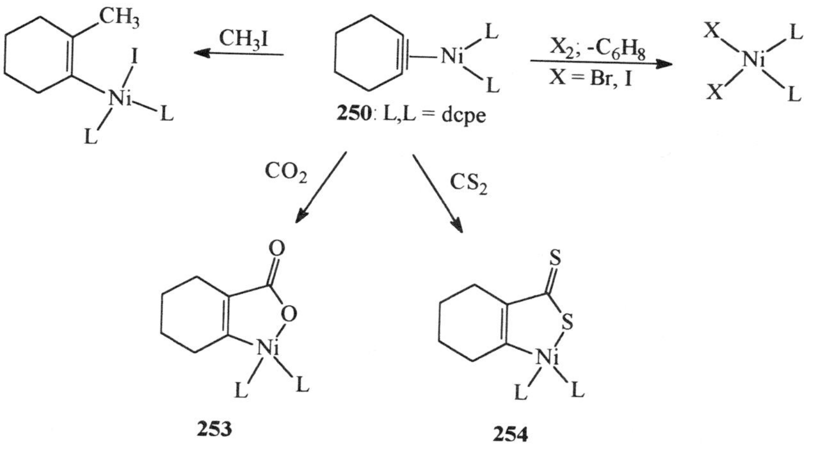

SCHEME 31

of **249** appears at 1735 cm^{-1}, 14 cm^{-1} higher than the platinum counterpart, which is consistent with complexes of comparable acyclic alkynes.[95,96]

Reactions of **250** with four electrophiles are recorded in Scheme 31. In general, the products are more stable than those from complexes with monodentate ligands. Reaction of **249** with CS_2, **249** or **250** with alkynes, and **249–251** with ethylene gives products in which the C_6H_8 has been lost but its fate has not been determined; attempts to trap free cyclohexyne failed.[93] Loss of the organic ligand appears to occur more readily in nickel complexes than in those with platinum.

A zircononium complex of cyclohex-3-eneyne (**257**) (two isomers) has been prepared by β-hydrogen elimination from **256** [Eq. (38)].[97] This reaction was studied in an attempt to prepare a zirconium complex of 1,2,3-cyclohextriene (**255**) but, in the absence of a suitable blocking group (Section IV,B), H2 was eliminated to the exclusion of H6 and only **257** was formed.

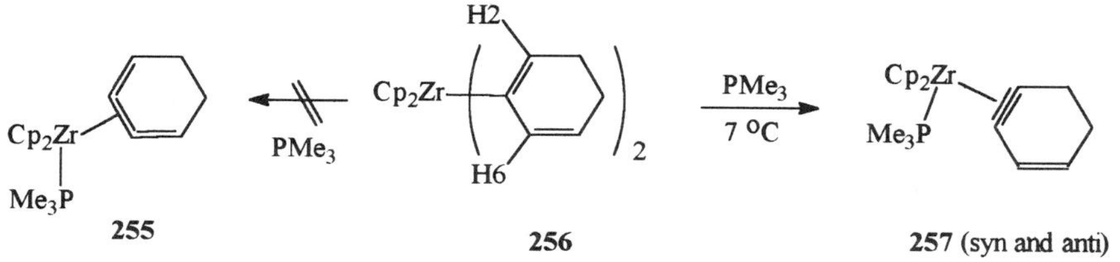

(38)

D. *Seven-Membered Rings*

Stable platinum, zirconium, and cobalt complexes of seven-membered cycloalkynes have been discussed in earlier reviews.[2,8]

More recently, a number of mononuclear Pt,[81,82,85] Pd,[86] and Zr[87] complexes of cycloheptadienyne have been prepared, in most cases as precursors to complexes of tropyne (e.g., **223**) [Eq. (39)]. The chemistry of tropynes is discussed in Sections II,G and II,H.

$$\mathbf{222b} \xrightarrow{Ph_3C^+} \mathbf{223} \quad (39)$$

All of the new Pt and Pd complexes were prepared by LDA-induced elimination of HBr from an appropriate vinyl halide, as illustrated for the preparation of a mixture of **222a** and **222b** [Eq. (40)]. The mechanism of formation of these complexes is not clear. For some time it had been presumed that free cycloheptadienyne was trapped by the metal. However, as discussed earlier for formation of the bis(triphenylphosphine)platinum(0) complex of cyclopentyne (**240**), elimination from an initially formed π-complex[2,82] (**258**) [Eq. (41)] is also possible. Elimination from either of the two possible insertion products was excluded by their isolation and the finding that neither reacts with LDA to give **222**.[98]

$$\mathbf{221a\text{-}c} \xrightarrow[Pt(PPh_3)_3]{LDA} \mathbf{222a} + \mathbf{222b} \quad (40)$$

$$\mathbf{258} \xrightarrow[?]{LDA} \mathbf{222a} \quad (41)$$

X-ray crystal structures have been reported for mononuclear platinum–cycloheptadienyne complexes **222a**[82] and **259**[85] and the palladium analogue of **222a**.[86] Selected structural parameters are given in Table VIII.

Reactions of a series of dibenzannelated platinum/cycloheptadienyne complexes **259** (X = CO), **225** (X = CH_2), and **261** (X = O) are summarized in Scheme 32.[85] Attempts to protonate **259** to give hydroxy tropyne **262** [Eq. (42)] failed; reaction occurred instead at the metal center to give **263** and **264** (Scheme 32). Two of the dibenzannelated complexes (**259** and **261**) undergo exchange with $(Ph_3P)_2CH_2CH_2(PPh_3)_2$ without displacing the alkyne. Reaction of **261** with TCNE gives the coupling product **267**, whereas reaction of **259** with the same reagent leads to trimer **266**. The last has an interesting structure because steric crowding not only distorts the central benzene ring, but also forces one of the dibenzotropones to position itself

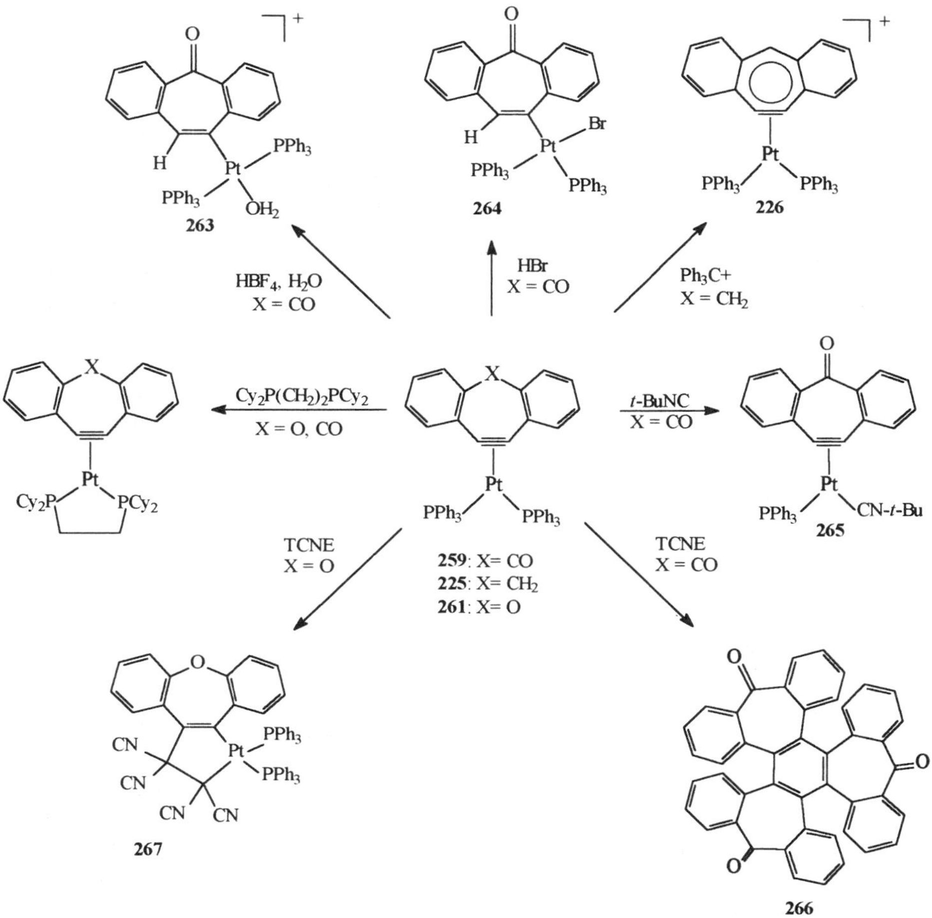

SCHEME 32

uniquely with one phenyl ring above and the other below the middle benzene ring. None of the dibenzannelated complexes react with alkynes or weak acids such as ethanol or acetonitrile, even when warmed to 80°C for 2 days.

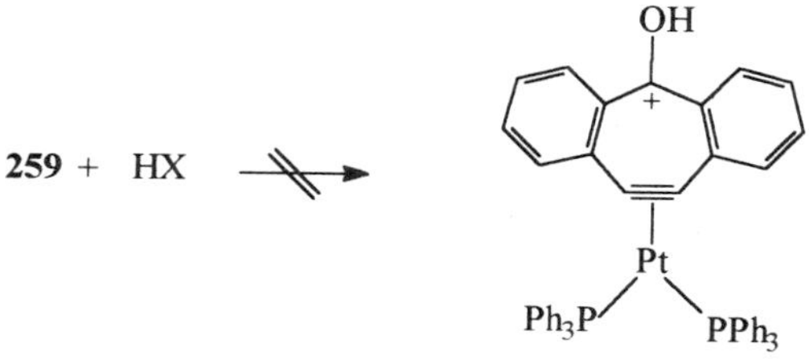

(42)

Three zirconium/cycloheptadienyne complexes (**231a–c**) have been prepared by β-hydrogen elimination from a mixture of cycloheptatrienyl complexes **269–271** (Scheme 33) and have been used as intermediates for the preparation of a zirconaazulene.[87] The alkyne complexes are formed to the exclusion of the allene isomer **268**. This is believed to be due to the proximity of the β-vinyl hydrogen that is a result of both the shorter double bond and its forced coplanarity with the metal. Allene formation from **269** might be induced by blocking the vinyl position (see Sections IV,B and IV,C), but this has not been tested.

A complex of homoadamantyne (**273**) has been prepared by dehalogenation of **272** in the presence of tris(triphenylphosphine)platinum(0) [Eq. (43)].[99] Trapping free homoadamantyne was presumed to be the mechanism of this reaction. However, reduction of neither an initially formed π-complex[2,82] nor a C–Br insertion product[93] was excluded. Interestingly, the free alkyne, generated by dehydrohalogenation of the vinyl bromide, was successfully trapped with diphenyl-*iso*-benzofuran.

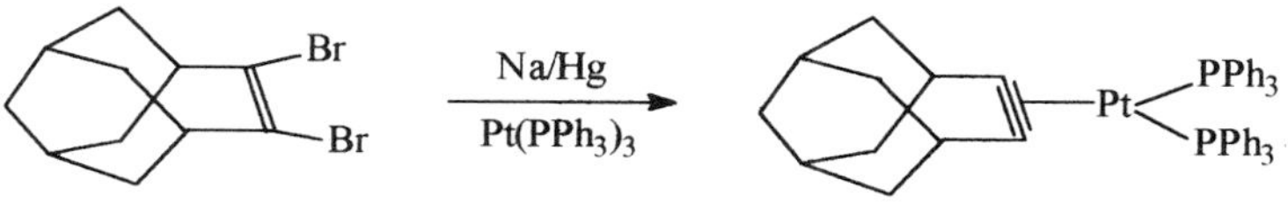

(43)

The properties of the complex of homoadamantyne, which is formally a cycloheptyne, more closely resemble platinum complexes of cyclohexyne than those of the seven-membered ring alkyne. The bend-back angle (50.6°) (Table VIII) is significantly larger than that of the corresponding cycloheptyne complex (41.3°) and nearly as large as that of the cyclohexyne analogue

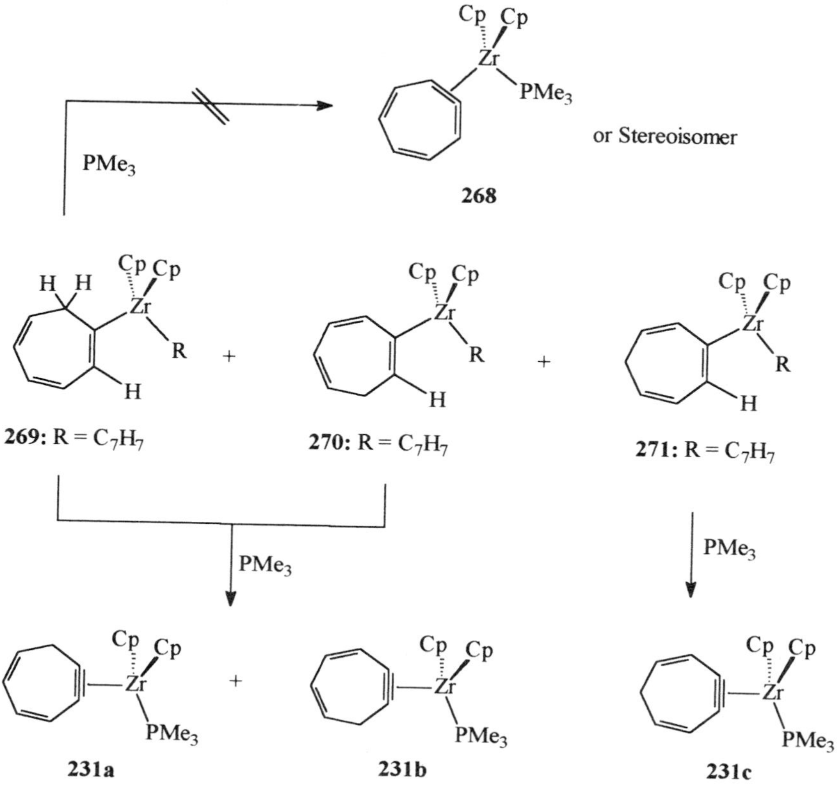

SCHEME 33

(52.7°). The triple-bond stretching frequency (1735 cm^{-1}) is also closer to that of the cyclohexyne complex (1721 cm^{-1}) than to that of the corresponding cycloheptyne (1770 cm^{-1}).[2]

Pt/Mo complexes of two cycloheptadienynes (**236** and **237**) have been prepared by $Mo(CO)_3$ transfer from (η^6-*p*-xylene)$Mo(CO)_3$ to **222a** and **222b** and by hydride reduction of the corresponding Pt/Mo complex of tropyne (**235**) (Scheme 34)[88] (see Sections II,G and II,H). The platinum cycloheptadienyne complexes can be regenerated from **236** and **237** with PPh_3 or phenanthroline.

From its X-ray, the bend-back angle of the Pt/Mo complex **236** (47.8°) (Table VIII) is significantly larger than that of the corresponding cycloheptyne complex (41.3°). This is also true for **222a**, but the difference is somewhat less. These larger bend-back angles presumably reflect somewhat

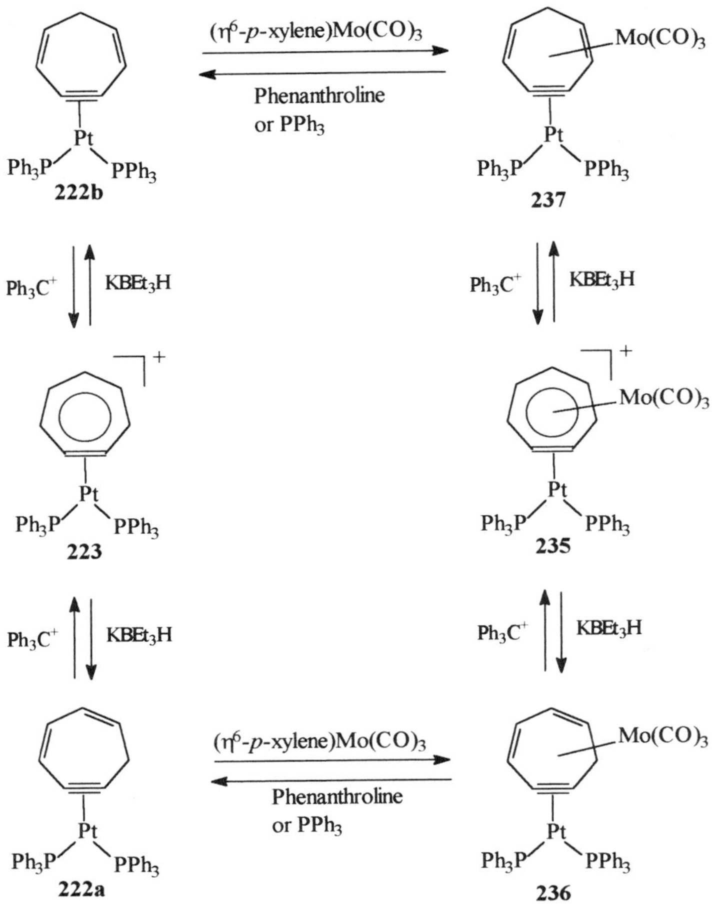

SCHEME 34

more strain originating from the presence of the additional double bonds. In the IR spectrum, the triple bond of **236** appears at 1610 cm^{-1}, 100 cm^{-1} lower than in **222a**, indicating that coordination of molybdenum results in reduction in bond order of the triple bond. The IR spectrum of **236** also shows three very strong terminal carbonyl bands that appear at 15 cm^{-1} lower frequency than those of the tricarbonylmolybdenum complex of cycloheptatriene. This suggests that the seven-membered ring in **222a** is more electron-rich than cycloheptatriene.[100] Irreversible exchange of $Mo(CO)_3$ from cycloheptatriene to **222a** and **222b** [Eq. (44)] is consistent with this suggestion. The ^{13}C NMR of **236** shows three distinct absorptions at −20°C that disappear at RT, clear evidence for rapid rotation about the Mo–cycloheptadienyne axis at the higher temperature.

222a + **222b** + [cycloheptadienyne]–$Mo(CO)_3$ ⇄ **236** + **237** + [cycloheptatriene] (44)

E. *Eight-Membered Rings*

Stable zirconium, platinum, molybdenum, and tungsten complexes of cyclooctyne, a zirconium complex of cycloocta-5-enyne, and a bimetallic molybdenum complex of cyclocta-3,7-dienyne have been discussed in earlier reviews.[2,8] More recently, two stable zirconocene complexes of cycloocta-trienyne (**275** and **276**) have been prepared[101] by β-hydride elimination from **274** in the presence of PMe_2R [Eq. (45)].

Cp_2Zr(cyclooctatetraenyl)$_2$ (**274**) —RT, PMe_2R→ cyclooctatrienyne–$Zr(Cp)_2(PMe_2R)$ (45)

274

275: R = Ph
276: R = Me

It has been predicted that the triple bond in uncomplexed cyclooctatrienyne should force the ring to be flat.[102] However, this cannot be the case in solution for either **275** or **276** because the ^1H NMR spectra of both complexes show temperature-dependent ring inversion. Surprisingly, the activation barrier for this process in each is about the same as that of cyclooctatetraene, suggesting that the complexed triple bond behaves very much like a double bond. An X-ray crystal structure of **275** confirmed the boat conformation with no significant flattening when compared with COT. The bend-back angle of **275** (49°) is significantly larger than that of the bis(triphenylphosphine)–platinum(0) complex of cyclooctyne (34°).[103,104] Selected structural parameters for **275** are given in Table VIII.

F. *[2.2]Paracyclophane-1-yne and [2.2]Paracyclophane-1,9-diyne*

Normally, a discussion of strained cyclic alkynes (which is the focus of this section) would not include rings with more than seven or eight members.[1–3] However, a notable exception is [2.2]paracyclophane-1-yne (**277**), formally a 12-membered ring but one that is so strained that, if it is not coordinated to a metal, it has only been implied as a transient intermediate.[105,106]

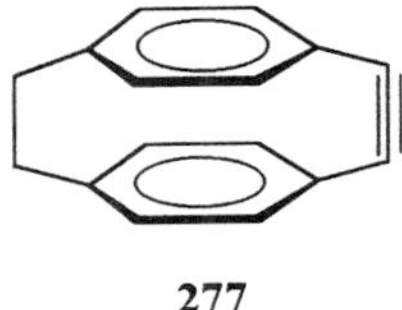

277

A zirconium (**277**)[107] and a platinum (**282**)[108] complex of this cycloalkyne and the even more highly strained zirconium diyne complex **279**[107] have been reported. Both of the stable zirconium complexes were prepared by Buchwald's method[73] as illustrated for **278** in Scheme 35. In each case, substitution of 2-butyne for PMe_3 gave insertion products (e.g., **280**) (a mixture of regioisomers from the diyne complex). Crystals suitable for X-ray could not be obtained for any of the zirconium cyclophyne complexes. The platinum complex of **227** (**282**)[108] was prepared by reduction of **281** in the presence of $Pt(PPh_3)_3$ (Scheme 36). The organic ligand in this complex appears to be rather tightly bound; both PMe_3 and $Ph_2PCH_2CH_2PMe_2$ displace PPh_3, leaving the cycloalkyne unchanged. Br_2 and HCl react with **282** to give **281** and **283**, respectively.

n-BuLi

Li

1) Cp_2ZrClR
2) 2-Butyne
-78 °C to RT

1) Cp_2ZrClR
2) PMe_3
-78 °C to RT

Cp Cp Zr Me Me

280

278

SCHEME 35

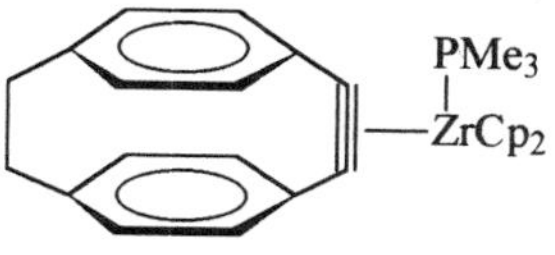

PMe_3 — $ZrCp_2$

278

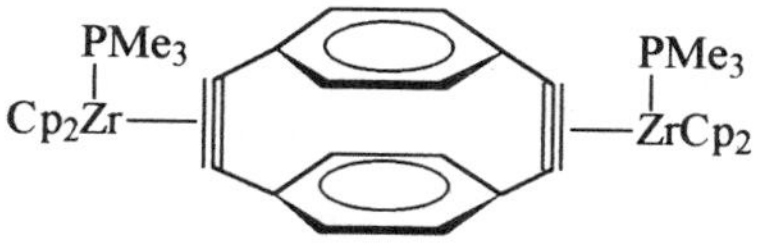

PMe_3, Cp_2Zr — ... — $ZrCp_2$, PMe_3

279 (two diastereoisomers)

An X-ray crystal structure of **282** was obtained.[108] Significant parameters are included in Table VIII. Particularly notable is the large bend-back angle of 58–59°, which is consistent with the presence of the highly strained [2.2]paracyclophane-1-yne moiety.[109]

IV

COMPLEXES OF STRAINED CYCLIC CUMULENES

A. *Three-, Four-, and Five-Membered Ring Allenes and Higher Cumulenes*

To date there are no examples of the very highly strained cyclopropadiene (**288**), 1,2-cyclobutadiene (**289**), or 1,2-cyclopentadiene (**290**), either free or complexed to transition metals. Complexes of a valence isomeric form of cyclopropadiene (**291**) have been prepared,[110] but as with the free hydrocarbon,[111,112] there is no evidence for any converting to the allene form [Eq. (46)].

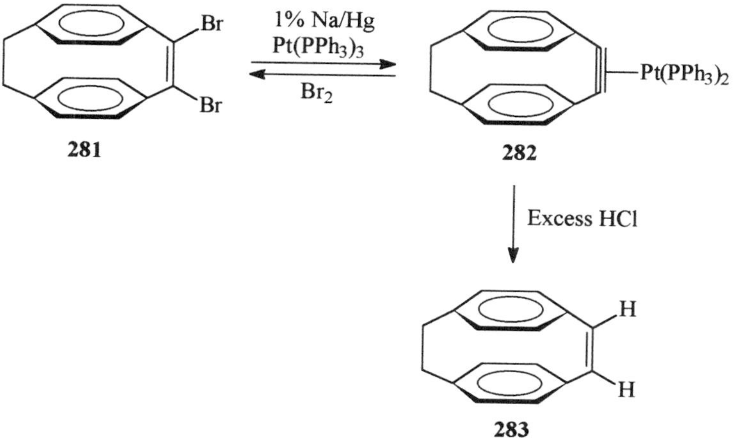

Br, Br

281

1% Na/Hg, $Pt(PPh_3)_3$ ⇄ Br_2

$Pt(PPh_3)_2$

282

Excess HCl

H, H

283

SCHEME 36

288 289 290

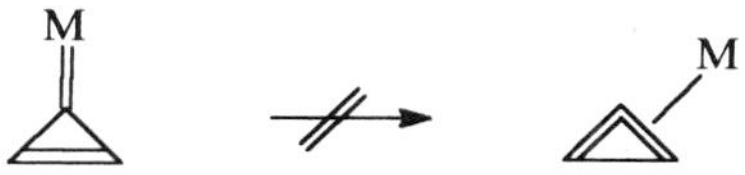

291 292 (46)

Rosenthal has discovered that 2,3,4-metallacyclopentatrienes **294a**[113] and **294b**[114a] can be prepared by reaction of di-*t*-butyl-1,3-butadiyne with **293a** or **293b** [Eq. (47)] and that the bis(triphenylphosphine)nickel(0) complexes of these cumulenes are formed on reaction of **295** with Cp_2M (M=ZrPy; Ti) [Eq. (48)].[114b] In the mononuclear complexes the metal serves to both relieve strain because of its size and to stabilize the cumulene by coordination to the central double bond. In the mononuclear complexes, evidence for the cumulene structures rather than the bis-alkyne structures **297a** and **297b** includes the following: (1) Neither **294a** nor **294b** shows infrared stretching vibrations for coordinated (1700–1800 cm^{-1}) or free (2050–2200 cm^{-1}) triple bonds. (2) The bond lengths of C2–C3, C3–C4, and C4–C5 are equivalent in the zirconium complex **294a** indicative of comparable bond orders. (3) The same carbon–carbon bond lengths for the titanium complex, although not equivalent, are consistent with "double bond order."

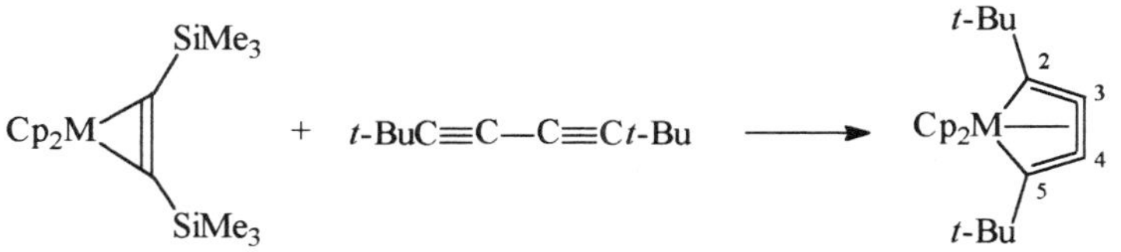

Cp_2M ($SiMe_3$)($SiMe_3$) + *t*-BuC≡C—C≡C*t*-Bu ⟶ Cp_2M (*t*-Bu)(*t*-Bu) (47)

293a: M = ZrPy
293b: M = Ti

294a: M = ZrPy
294b: M = Ti

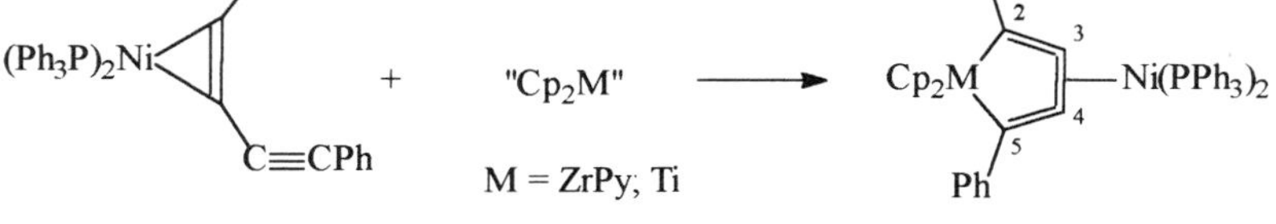

$(Ph_3P)_2Ni$ (Ph)(C≡CPh) + "Cp_2M" ⟶ Cp_2M (Ph)(Ph) —$Ni(PPh_3)_2$ (48)

M = ZrPy; Ti

295

296a: M = ZrPy
296b: M = Ti

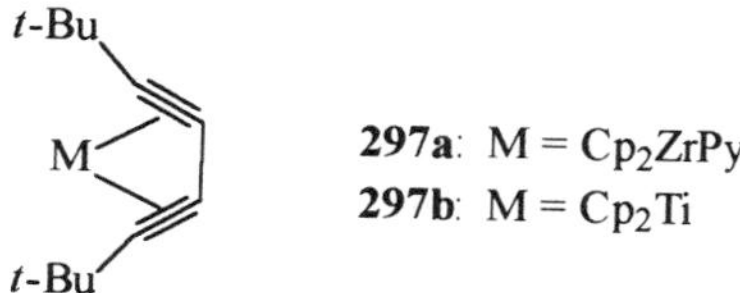

X-ray crystal structures were reported for all four of these new complexes (**294a,b**; **296a,b**). In all cases, all five atoms of the cyclopentatriene ring lie in the same plane, which, in the two mononuclear complexes, places the central carbon–carbon double bond in a position to coordinate to the metal center. The C–C–C internal bond angles of both the titanium and zirconium mononuclear complexes are about 150° despite the larger size of the zirconium atom. The angle is significantly larger than the cumulene angle in metal complexes of seven-membered ring allenes (ca. 130–140°),[115,116] a bimetallic seven-membered carbocyclic butatriene (**238**) (139°)[88] (Section IV,C), **296a** (139°), or the angle calculated for uncomplexed 1,2,3-cyclopentatriene (about 117°).[117] Coordination of the central double bond of the cumulene to the metal was offered as one reason for this relatively small deviation from linearity. Finally, the central bond length in **296b** [1.419(6) Å] is 0.08 Å longer than in its mononuclear counterpart (**294b**) and is in the range expected of double bonds coordinated to nickel.

Complex **294b** has also been detected as a transient from photolysis of the diacetylide **298** [Eq. (49)].[114b]

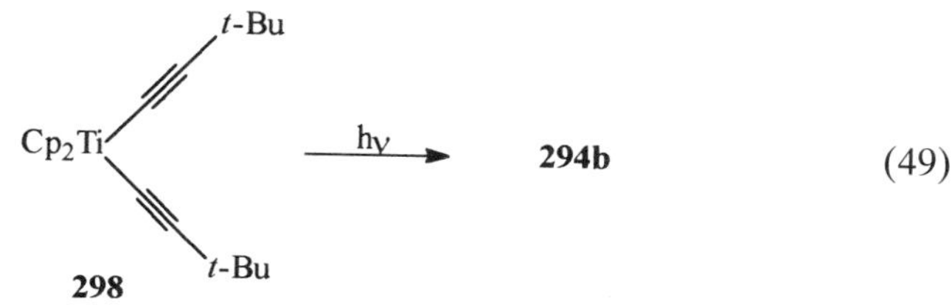

B. *Six-Membered Ring Allenes and Higher Cumulenes*

Unsuccessful attempts to prepare stable complexes of six-membered ring allenes that succeeded for their seven-membered counterparts (Section IV,C) include reaction of **299** with KO*t*-Bu in the presence of tris(triphenylphosphine)platinum(0) [Eq. (50)][118] and abstraction of methoxy from **301** (Scheme 37)[119,120] (L = CO) with trimethylsilyl triflate. The latter gave products suggestive of an intermediate complex of 1,2-cyclohexadiene, but it could not be detected or trapped. Based on fluxionality studies in the seven-membered counterpart (Section IV,C), equilibration of **302** with a highly reactive allyl cation form **303** was suggested as the reason for this instability. When tributylphosphine was substituted for one of the carbonyls in an attempt to stabilize the allene form relative to the allyl cation form, a solid was isolated that reacted with ethanol and $NaHCO_3$ at 0°C to give **304**.[121] This solid was too unstable to be purified and characterized (rapid decomposition in all NMR solvents), although it is probably the allene complex **302**. The substituted zirconocene complex **306**, prepared by elimination of 1-methylcyclohexene from **305** in the presence of PMe_3 (Scheme 38), is the only transition-metal complex of a 1,2-cyclohexadiene that has been isolated and characterized to date.[116] Without the methyl substituent a complex of cyclohexyne is formed (Section III,C). Preparation of **306** required a somewhat higher temperature than for the seven-membered ring analogue (53°C vs 30°C). This may be due to either or both greater ring strain in the product and a less favorable dihedral angle between the metal and the β-hydrogen. An X-ray crystal structure of **306** reflects its ring strain in that the internal bond angle [125.1(3)°] is more than 35° more acute than the most highly bent acyclic allene complex.[122,123] The Zr–C2 bond in **306** is shorter than the Zr–C1 bond because of the bulk of the PMe_3 group[8,46,92,124] and the different hybridization of the two carbons. As expected, the coordinated double bond (C1–C2) is longer [1.443(5) Å] than the uncoordinated double bond [C2–C3; 1.333(5) Å].

299 (1-bromocyclohexene) + KO*t*-Bu + $Pt(PPh_3)_3$ ⇸ **300** [$(Ph_3P)_2Pt$–cyclohexadiene complex] (50)

The chemistry of **306** differs from that of its seven-membered analogue in that its reaction with *trans*-stilbene gives reductive coupling product **308** in addition to the stilbene adduct of zirconocene, the sole product of the latter (vide infra). This was attributed to differences in ring strain.[116]

To date, no examples of metal complexes of either carbocyclic or metallacyclic higher cumulenes of six-membered rings have been reported, although free 1,2,3-cyclohexatriene has been trapped.[125]

C. *Seven-Membered Ring Allenes and Higher Cumulenes*

The first metal complex of a strained cyclic allene was reported in 1972 by Visser and Ramakers,[118] who found that treating 1-bromocycloheptene

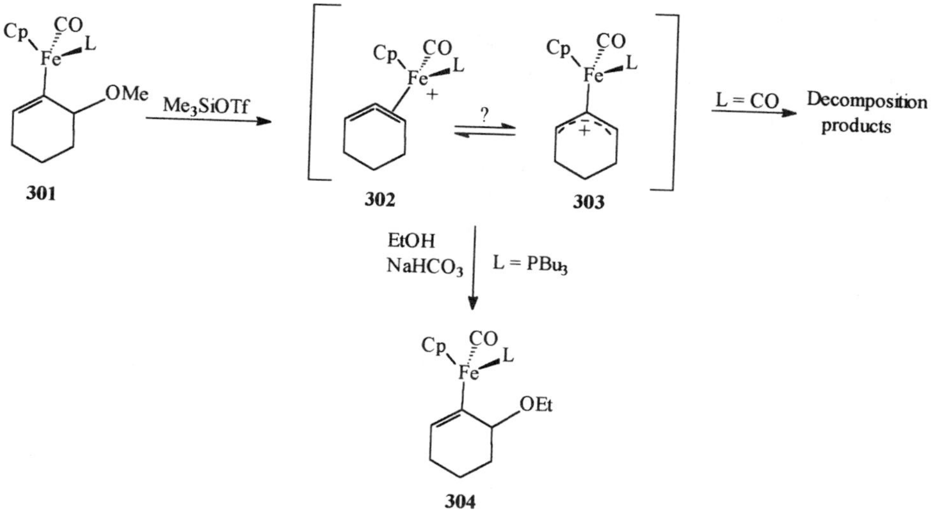

Cp
CO
L
Fe
OMe
Me_3SiOTf
+
?
L = CO
Decomposition products
301
302
303
EtOH
$NaHCO_3$
$L = PBu_3$
OEt
304

SCHEME 37

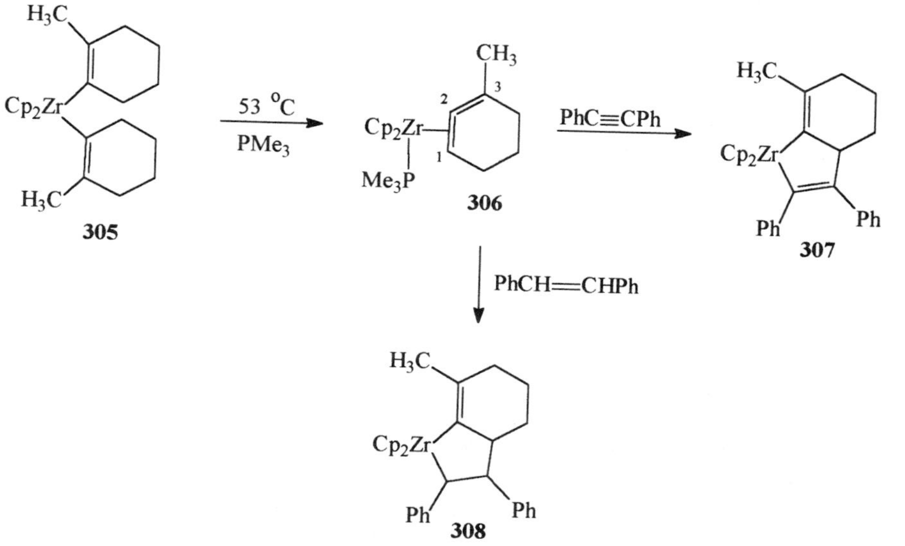

H_3C
Cp_2Zr
53 °C
PMe_3
CH_3
Me_3P
PhC≡CPh
Ph
305
306
307
PhCH═CHPh
308

SCHEME 38

with KOt-Bu in the presence of tris(triphenylphosphine)platinum(0) gives the 1,2-cycloheptadiene complex **310** [Eq. (51)]. The mechanism of this reaction—for which simple trapping of 1,2-cycloheptadiene by the metal had been presumed—was not examined until much later when evidence for elimination of HBr from an initially formed π-complex was found.[82] More recently, iron[119] (**312**; M = Fe) [Eq. (52)], ruthenium[120] (**312**; M = Ru) [Eq. (52)], and zirconium[116] (**315**) [Eq. (53)] complexes of 1,2-cycloheptadiene have been reported. For preparation of the zirconium complex, the methyl substituent was required in **314** to prevent cycloheptyne formation.

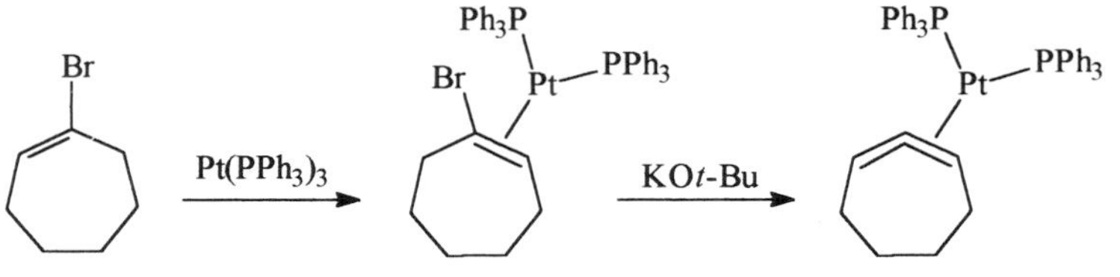

Br; Pt(PPh$_3$)$_3$ →; Ph$_3$P, Br, Pt—PPh$_3$; KOt-Bu →; Ph$_3$P, Pt—PPh$_3$ (51)

309 **310**

Cp, CO, M—CO, OMe; Me$_3$SiOTf (M=Fe,Ru), HBF$_4$ (M=Ru) →; Cp, CO, +M—CO; ← BF$_3$, M=Fe; Cp, CO, F, M—CO

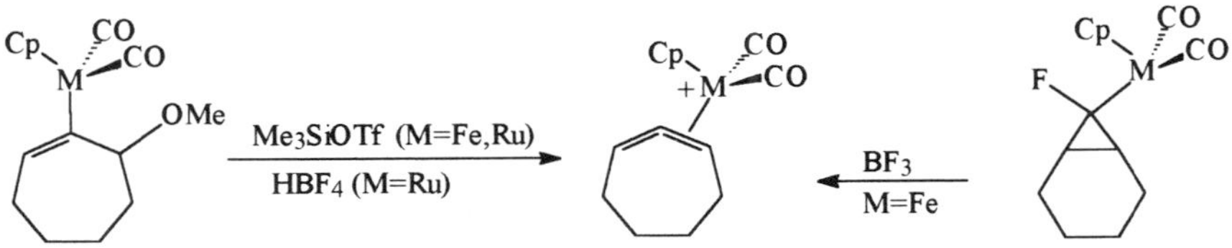

311 **312** **313**

(52)

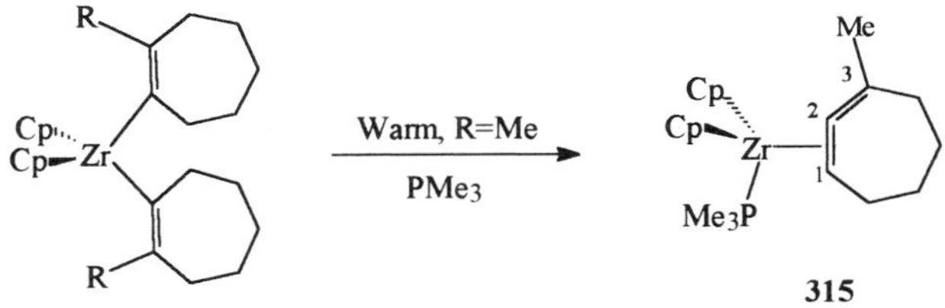

R, Cp, Cp—Zr, R; Warm, R=Me, PMe$_3$ →; Me, 3, 2, 1, Cp, Cp—Zr, Me$_3$P (53)

314 **315**

The iron complex of 1,2-cycloheptadiene was synthesized to study the relationship between **312** and a valence isomeric form (**316**) that has been referred to as the "allyl cation" form (Scheme 39).[119] Because of ring strain, the latter had been predicted by EHMO calculations to lie rather close in energy to an allene ground state.[126] The NMR of **312** confirmed the allene form as the ground state and fluxionality experiments on the terminal allene hydrogens showed rapid equilibration. However, the activation barrier for this equilibration (13.9 kcal/mol),[119] which is lower than the barrier in acyclic complexes,[127] was found to also be lower than the barrier to exchange of the diastereotopic carbonyls on the iron. This requires that chirality be retained throughout the interconversion and led to the suggestion[128] that fluxionality occurs via the classical, albeit distorted, Vrieze–Rosenblum transition state[127] **317** rather than **316**. Spin saturation transfer was used to determine a fluxionality barrier of approximately 17–18 kcal/mol for the more stable phosphine-substituted iron complex **318** (Scheme 40).[128] By partial separation of the two diastereoisomers of **318** (**318a** and **318b**) and monitoring of their interconversion, a higher energy fluxionality process with an activation barrier of about 23 kcal/mol was detected. Since this requires loss of chirality of the allene moiety, allyl cation **320** was proposed as an intermediate or transition state. This allyl cation is about 5 kcal/mol higher than the chiral Vrieze–Rosenblum[127] transition state **319**.

Fluxionality in the Rp^+ [Rp = dicarbonyl(η^5-cyclopentadienyl)ruthenium] complex of 1,2-cycloheptadiene closely parallels that of its iron counterpart.[129] For iron, $\Delta H‡ = 14.6$ kcal/mol, $\Delta S‡ = 0.65$ eu; for ruthenium $\Delta H‡ = 14.2$ kcal/mol, $\Delta S‡ = -1.15$ eu.

X-ray crystal structures have been reported for [(η2-cyclohepta-1,2-diene)(η^5-cyclopentadienyl)carbonyltriphenylphosphineiron(II) hexafluo-

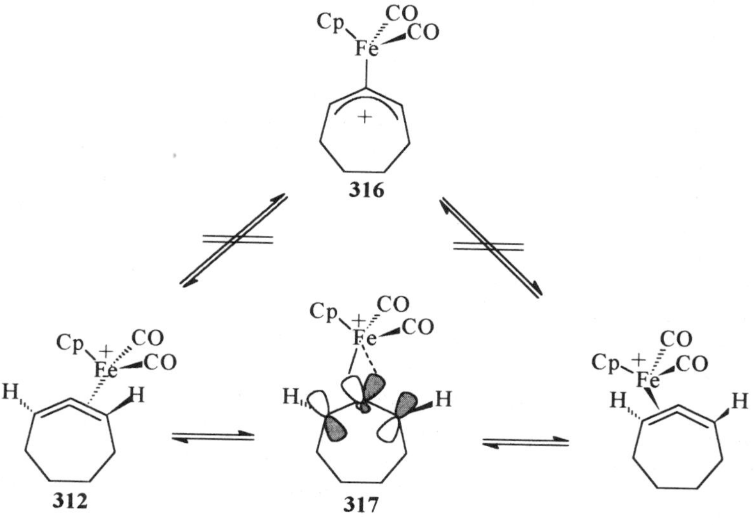

SCHEME 39

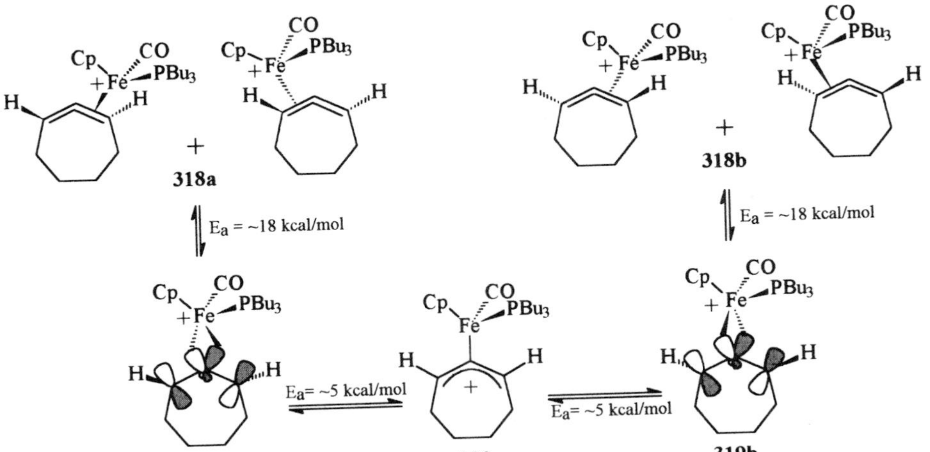

SCHEME 40

rophosphate][115] and the zirconocene complex **315**.[116] A notable feature of the former is a 41.9(3)° deviation of the central allene bond angle from linearity. This is somewhat larger than for most acyclic allene complexes,[122,123] but is not large enough to suggest very much strain. As expected, the double bond complexed to iron [1.385(6) Å] is longer than the uncomplexed double bond [1.303(7) Å]. The Zr–C2 bond in **315** is shorter than Zr–Cl because of different hybridizations of the carbons and the bulk of the PMe_3.[8,39,46,92,124] The allene deviates from linearity by 48°, which is somewhat more severe than for the Fp^+ complex [41.9(3)°] and a platinum complex of 1,2,4,6-cycloheptatetraene (45°) (Section IV,C).[130] This is consistent with more metallacyclopropane character in the zirconium complex.

Reactions of complexes of 1,2-cycloheptadienes have received only cursory attention. 1,2-cycloheptadiene is readily displaced from bis(triphenylphosphine)platinum(0)[118] [Eq. (54)], but no reagent has been found that will displace the allene from iron.[119] Reaction of the iron complex with alcohol in the presence of base (e.g., **312** → **322**) is typical of Fp+ complexes of acyclic allenes.[131,132] The thermal chemistry of **312** is unusual in its decomposition to **324** (Scheme 41). This is probably attributable to the presence of the triflate counterion, since the corresponding fluoroborate salt is stable when warmed to 40°C for 16 h.[119] A mechanism to **324** via carbene complex **321** appears likely.

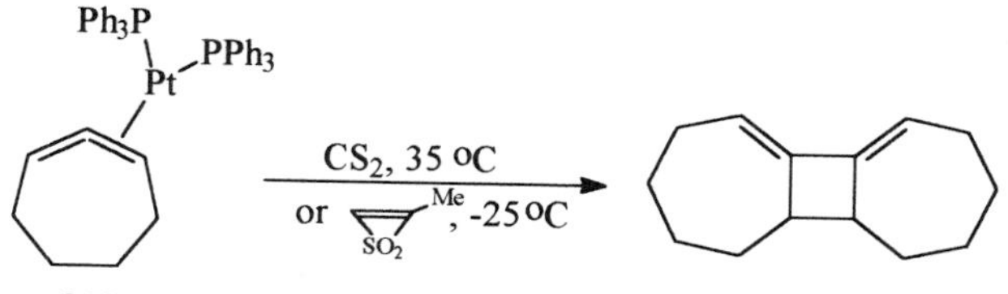

(54)

The zirconium complex **315** reacts with diphenylacetylene to give the coupling product **326** [Eq. (55)].[116] The regiochemistry of the reaction parallels that of an acyclic allene complex.[133] Complex **315** also reacts with stilbene, but in this case, the seven-membered ring is displaced from the metal. The fate of the seven-membered ring is unknown.

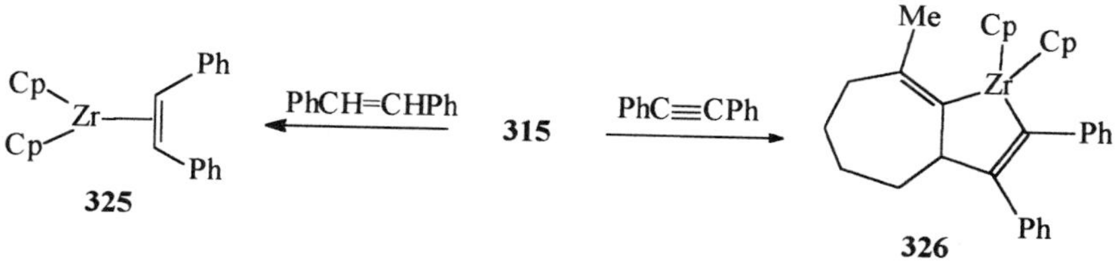

(55)

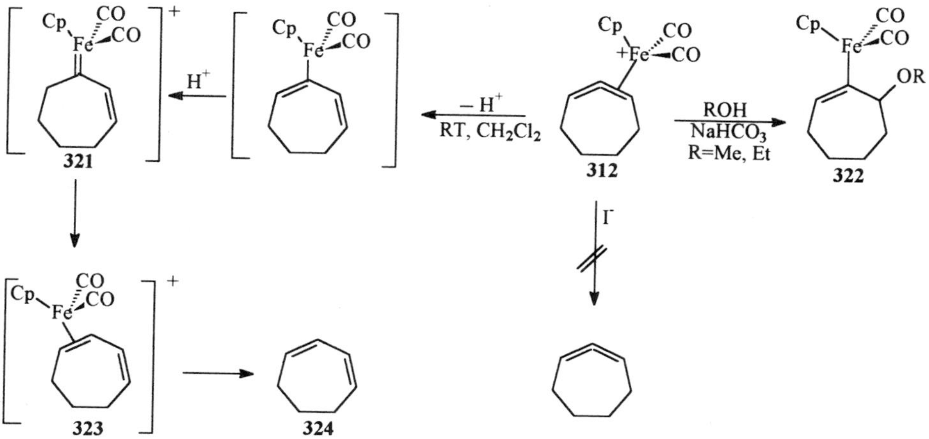

SCHEME 41

Transition-metal complexes of the more highly unsaturated seven-membered allene 1,2,4,6-cycloheptatetraene have been isolated in two forms depending on the metal and its ligands: an allene form (**329**), which is a complex of the parent cycloheptatetraene **327**,[134] and a carbene form (**330/331**), which is a complex of the controversial cycloheptatrienylidene **328**.[4,135,136] The carbene form corresponds to the "allyl cation" (**320**) that was suggested as an intermediate/transition state for fluxionality of **318**. Scheme 42 lists all such complexes that have been prepared to date. The allene form is the ground state for all Pt(0) complexes (**343**, **344**, **345**, **347**)[82,83,130,137,138] and one dibenzannelated W(II) complex[139] (**346**), whereas the carbene form is the ground state for all $Fe(II)^+$ complexes (**332–336**),[140–143] all $Ru(II)^+$ complexes (**337–340**),[144] one $Pt(II)^+$ complex (**342**),[137] and one W(0) complex (**341**).[139]

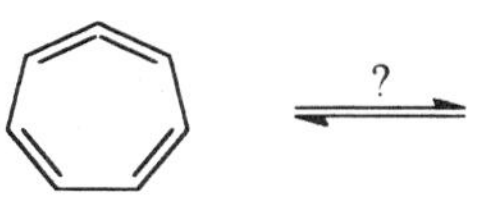

?

327 **328** (56)

M M M⁻

+

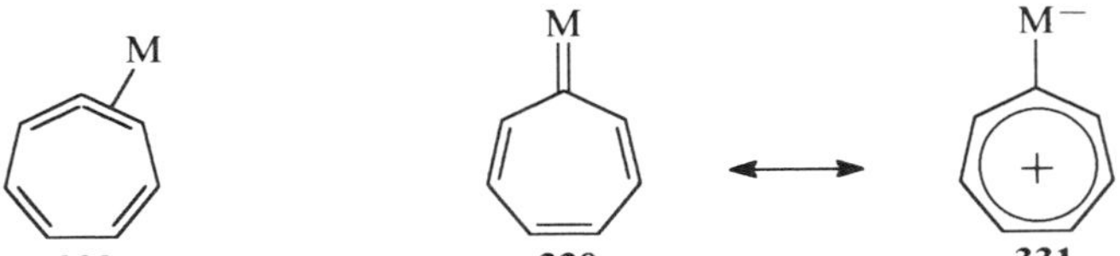

329 **330** **331**

Structural preferences in these complexes have been studied using extended Huckel calculations.[137,145] For the Pt(0) complex (PH_3 substituted for PPh_3), the allene form was found to be 12 kcal/mol lower energy than its carbene form, whereas the opposite is the case for the Pt(II) complex; **348** is preferred over **349** by 38 kcal/mol. These ground-state structures correspond to the experimental structures for **343** and **342** and are consistent with the large experimental difference (at least 27 kcal/mol) between **343** and its carbene form.[137] The origin of the structural preferences for the allene versus the carbene form was explored using allene as a model for cycloheptatetraene and the allyl cation (central carbon as the "carbene" carbon) as a model for cycloheptatrienylidene.[126,137] The allene form was found to be preferred in the Pt(0) complex because a high-energy Pt(0) LUMO fragment orbital is unfavorable for bonding to the sp^2 orbital of the "carbene" carbon, whereas back-bonding from the Pt(0) b_2 orbital into the allene LUMO is quite favorable. However, the Pt(II) complex (as well

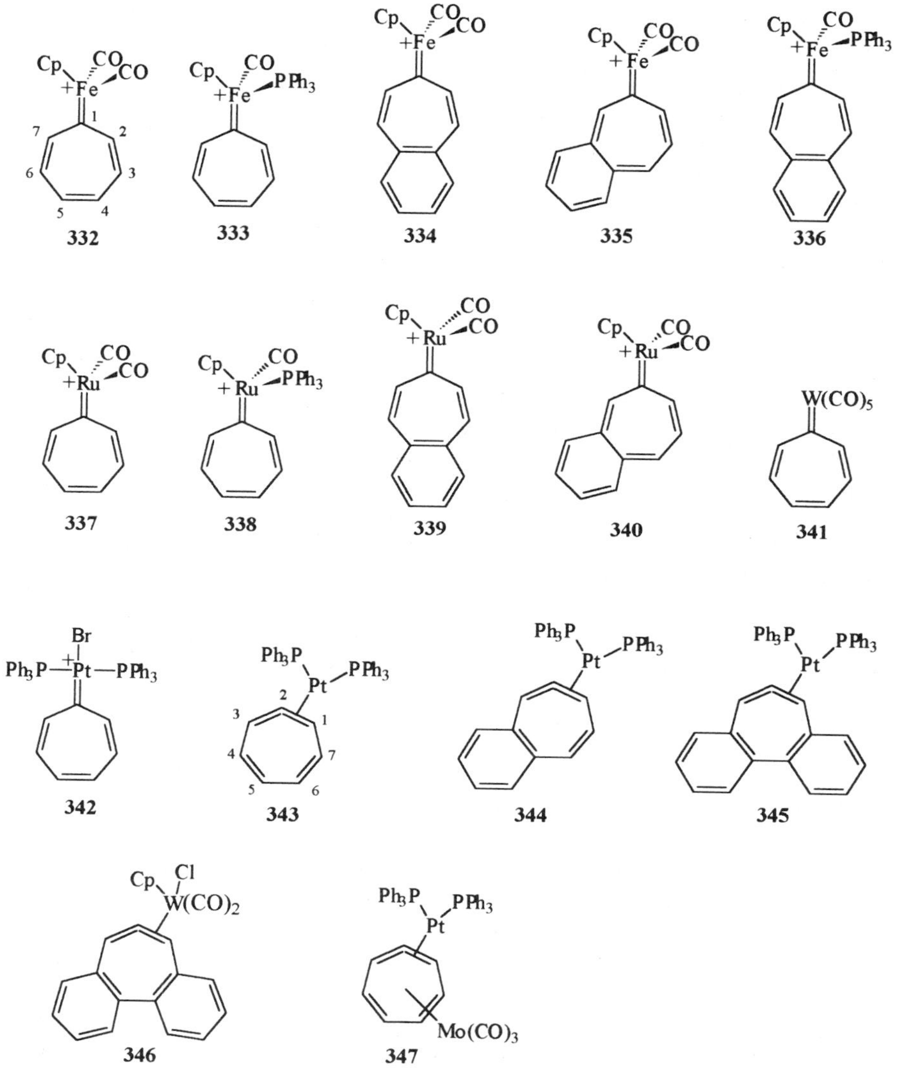

332 333 334 335 336

337 338 339 340 341

342 343 344 345

346 347

SCHEME 42

as the Fp$^+$ complex and, presumably, **333–341**) prefers the carbene form because of a strong interaction between the sp^2 orbital of the "carbene" carbon (the HOMO) and the metal LUMO, orbitals that show both good overlap and close energies.

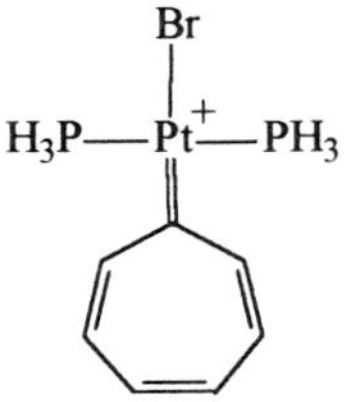

All "carbene" complexes were prepared by hydride abstraction from an appropriately substituted 1,3,5-cycloheptatriene sigma complex [e.g., **350** in Eq. (57)]. All but one of the "carbene" complexes are cations. Although it is likely that the tropylium ion resonance form represented by **331** contributes significantly to the hybrid, three experimental observations provide evidence for some contribution from the carbene form as well: (1) Significant bond alternation between the carbons remote from the metal in **332**; C3–C4 = 1.411(8) Å, C4–C5 = 1.358(9) Å[141,143]; (2) significant downfield chemical shifts for the "carbene" carbon in the ^{13}C NMR spectra (in the range of 200–280 ppm),[143,144] although less than in analogous more typical carbene complexes, which often show downfield shifts of more than 300 ppm[143]; and (3) a higher barrier to rotation around the carbon metal bond in **333** than in **336** in which benzannelation should reduce back-bonding from the metal.[142]

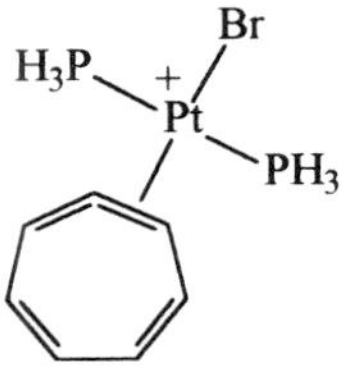

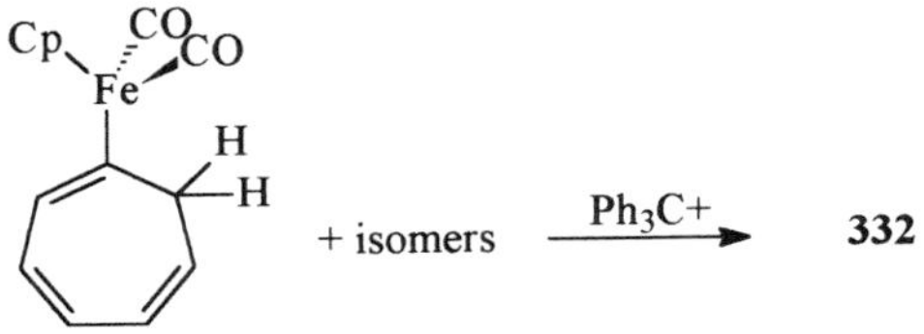

The chemical shifts of the carbene carbons of identically substituted iron and ruthenium complexes have been compared. In general, the ruthenium complexes appear about 20 ppm upfield from their iron analogues.[144]

In none of the complexes with carbene ground states has any detectable trace of the allene complex been observed. Even benzannelated complexes **335**[141] and **340**[144] exist exclusively in the carbene form [Eq. (58)]. Attempts to prepare dibenzannelated complexes of iron analogous to **346** failed.[146]

335: M = Fe
336: M = Ru

(58)

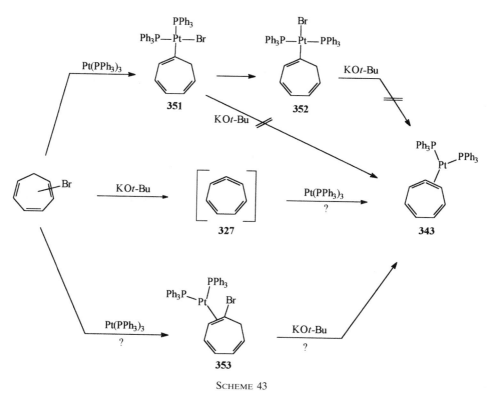

Pt(0) complexes **343**, **344**, and **345**, all of which have allene ground states, were prepared as illustrated and in Scheme 43 for **343** by treating appropriate bromocycloheptatrienes with KO*t*-Bu in the presence of tris (triphenylphosphine)platinum(0). The mechanism of this reaction is in ques-

$Pt(PPh_3)_3$ **351** **352** KO*t*-Bu **327** **343** **353** ?

SCHEME 43

tion. It is known that cycloheptatetraenes (e.g., **327**) are formed under the reaction conditions,[82] and for some time it was assumed that the allene is trapped by the metal. However, as discussed in Sections II,A, III,B, and III,C for the mechanism of the preparation of platinum and nickel complexes of benzene and cycloalkynes, both insertion products (**351** and **352**) and a π-complex (**353**) must also be considered as possible intermediates. In the case of the parent, both of the insertion intermediates can be ruled out because neither gives **343** when treated with KO*t*-Bu. The π-complex could be neither isolated nor detected and therefore remains a possible intermediate.[82,83]

The tungsten–allene complex **346** was prepared by ring opening of **355** [Eq. (59)].[139] It is not known if the preferred structure in **346** is due to the nature of the metal and its other ligands, dibenzannelation on the cycloheptatetraene ring, or both.

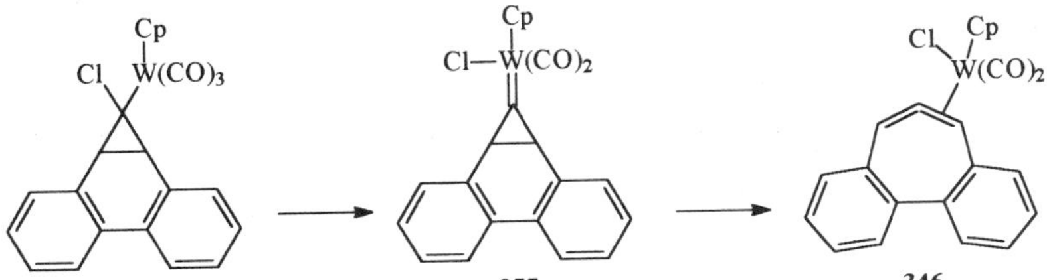

Cp Cl W(CO)$_3$ — **354** ⟶ Cp Cl—W(CO)$_2$ — **355** ⟶ Cl Cp W(CO)$_2$ — **346** (59)

Fluxionality studies have been carried out on **343** (Scheme 44).[137] No coalescence of the two allene hydrogens (H1 and H6) was observed in temperature ranges where the complex is stable (ca. 100°C). However, spin saturation transfer was observed between 60 and 100°C, which yielded activation parameters ($\Delta H\ddagger$ = 26.8 ± 1.3 kcal/mol; $\Delta S\ddagger$ = 15.1 ± 1.3 eu) for migration of the metal between the two double bonds. At 80°C, **343**

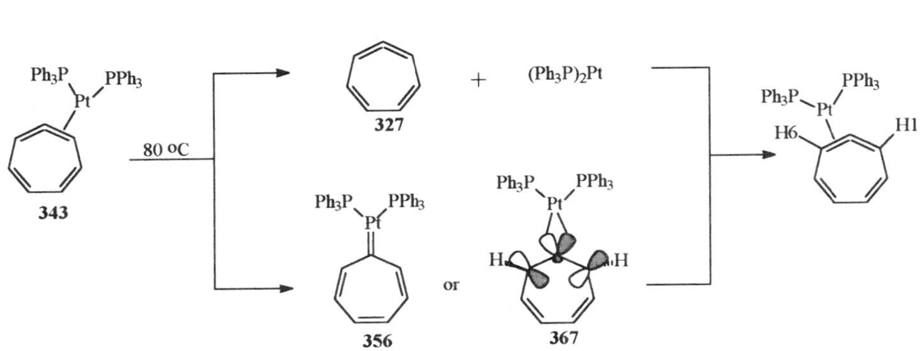

Ph$_3$P PPh$_3$ Pt — **343** —80 °C⟶ **327** + (Ph$_3$P)$_2$Pt; or Ph$_3$P PPh$_3$ Pt — **356** or Ph$_3$P PPh$_3$ Pt H H — **367** ⟶ Ph$_3$P PPh$_3$ Pt H6 H1

SCHEME 44

undergoes this process by both an intermolecular and an intramolecular mechanism.[137,138] These two mechanisms were distinguished by selectively irradiating H1 in the ^1H NMR spectrum of the isotopomer of **343** that is complexed to magnetically inactive platinum (i.e., not irradiating the platinum satellites that result from coupling to ^{195}Pt) and observing spin saturation transfer to both the central peak and the two satellites of H6. This requires scrambling of the platinum atoms between allenes, an intermolecular process. This mechanism disappeared when the temperature was lowered to 60°C; specific irradiation of the central peak of H1 showed transfer to the central peak of H6, but no detectable transfer to the satellite peaks.[138] This requires either an intramolecular process or its equivalent. Complexes **356** and **357** were proposed as possible transition states, with circumstantial evidence favoring the former.

Magnetization transfer was used to detect fluxionality in the bimetallic complex **347** (Scheme 45).[138] Both metals move in this case, and the reaction occurs at a considerably lower temperature than is required for **343**. An intramolecular mechanism via an achiral intermediate (or transition state) as pictured in **358** was proposed based on selective isotopomer irradiation as described for **343**. Stronger bonding of the $Mo(CO)_3$ moiety to the tropylium ion in the transition state than to the distorted cycloheptatriene

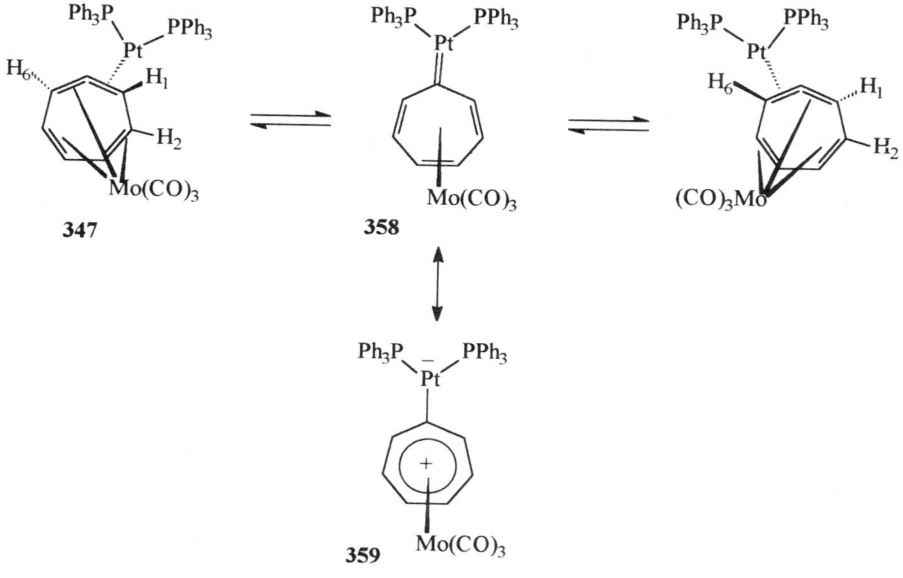

SCHEME 45

in the ground state was suggested as the reason for the rate acceleration in **347** as compared with **343**.

X-ray crystal structures have been reported for five cycloheptatrienylidene/cycloheptatetraene complexes, **332**,[141,143] **335**,[141,143] **342**,[147] **343**,[130] and **346**.[139] In both of the iron complexes, the seven-membered rings are essentially planar and adopt a conformation about the Fe–C bond that is orthogonal to the theoretical ground state of the analogous methylene complex (CH_2 bisects the two COs).[148,149] This conformation places the acceptor p-π orbital of the CHT in an orientation that is suitable to back-bond with the donor $1a'$ d-hybrid orbital of the iron, but not with the more favorable a'' orbital. This unusual preference is presumably due to a combination of the steric bulk of the seven-membered ring and tropylium stabilization that reduces the importance of the metal–carbon double bond. The mean value of the internal bond angles of the seven-membered rings in the two iron complexes are $129 \pm 3°$ and $129 \pm 4°$, respectively. These are consistent with the idealized angle of 128.6° for a regular heptagon.

The internal allene bond angles of the platinum[130] (**343**) and tungsten[139] (**346**) complexes of cycloheptatetraene deviate from linearity by 45.1(9)° and 44.9(19)°, respectively. This is similar to the zirconium complex of 1,2-cycloheptadiene.[116] As expected, the Pt–C2 bond [2.000(7) Å] is shorter than Pt–C1 [2.111(8) Å] because of the different hybridization of the two allene carbons.

Chemical reactions of metal complexes of cycloheptatrienylidene and cycloheptatetraene remain essentially unexplored. Attempts to prepare metallaazulene precursors by reductive coupling of complexes of either cycloheptatrienylidene or cycloheptatetraene with alkynes failed [Eq. (60)]. For example, although photolysis of **332** in the presence of phenylacetylene may have proceeded via the dihydroferraazulene **363** (Scheme 46), it could not be observed because of rapid rearrangement to a product that, based on ^{1}H NMR, has been tentatively assigned structure **365**.[150] Attempts to couple alkynes to **343** failed,[97] although a second mole of cycloheptatetraene couples to **343** to give the dimeric product **366** [Eq. (61)].[84] The regiochemistry of this reaction is opposite to that normally found in allene dimerizations induced by Group VIII metals.[151] It is also contrary to HOMO-LUMO considerations,[152] which led to the suggestion of a transition state with significant diradical character.

M

360 + RC≡CR $\xrightarrow{?}$ **361** (M, R, R) $\xleftarrow{?}$ RC≡CR + **362** (M) (60)

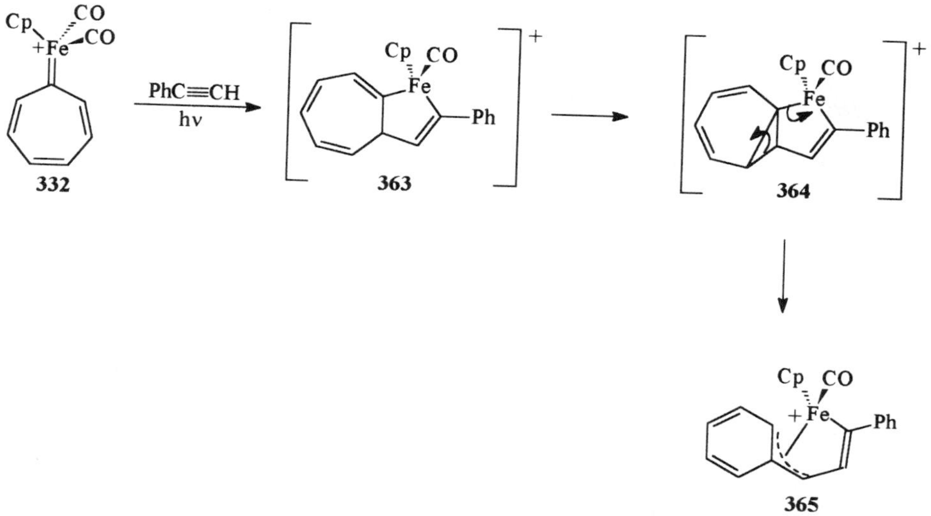

SCHEME 46

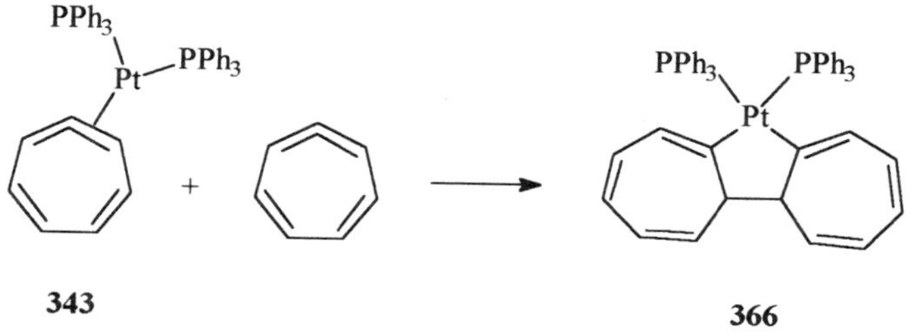

(61)

Three metal stabilized seven-membered rings containing 1,2,3-butatriene units have been prepared [Eq. (62)]. In the metallacycle **367**, independently synthesized by Hsu *et al.*[153] and Burlakov *et al.*[114] by essentially the same method, zirconium serves as both a member of the seven-membered ring and as a cumulene-stabilizing metal. An X-ray crystal structure excluded the alternative structure **368**, which was proposed as a possible intermediate in the reaction. The X-ray of **367** is consistent with the cumulene formulation in that bond lengths for C2–C3 [1.337(6) Å], C3–C4 [1.298(6) Å], and C4–C5 [1.337(6) Å] indicate bonds of roughly similar bond order. The bond angles for C2–C3–C4 [148.8(5)°] and C3–C4–C5 [160.1(5)°] are also consistent with a severely distorted (from linearity) cumulene.

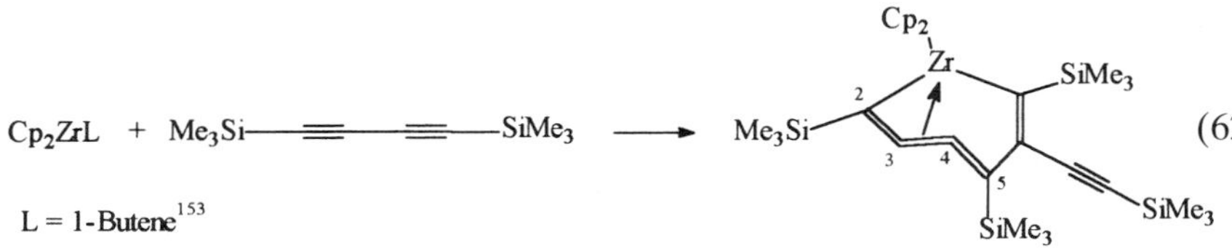

$$Cp_2ZrL + Me_3Si{-}C{\equiv}C{-}C{\equiv}C{-}SiMe_3 \longrightarrow \mathbf{367} \quad (62)$$

L = 1-Butene[153]
L = Py; $Me_3SiC{\equiv}CSiMe_3$[114]

367

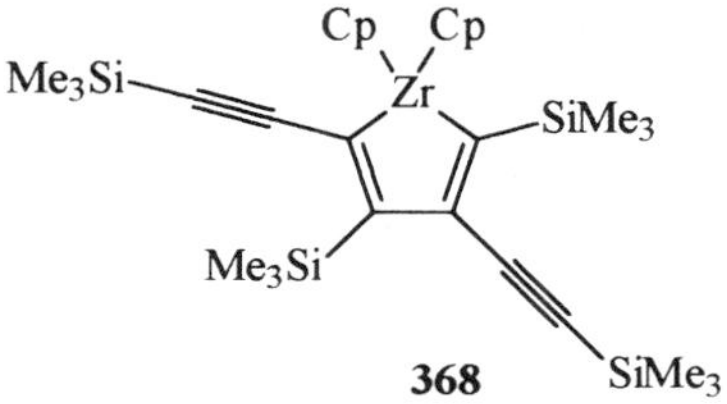

368

A low yield of the cycloheptatetraenone **370** has been prepared by dehydrobomination of 4-bromotropone in the presence of tris(triphenylphosphine)platinum(0) [Eq. (63)].[154] The structure assignment is based on ^1H NMR (including decoupling experiments), ^{31}P NMR, and high-resolution MS (FAB). The low yield precluded further study.

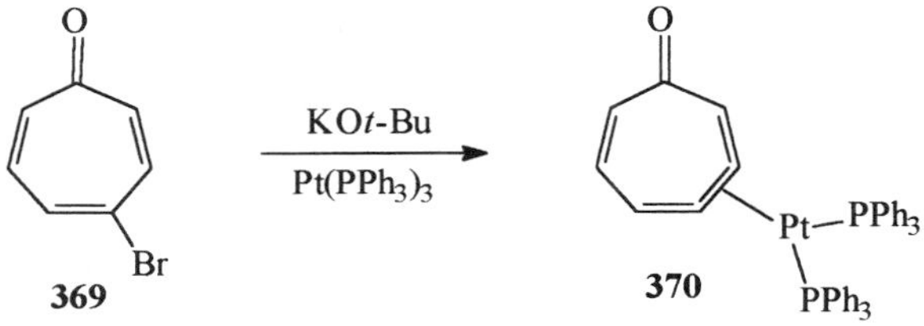

$$\mathbf{369} \xrightarrow[Pt(PPh_3)_3]{KOt\text{-}Bu} \mathbf{370} \quad (63)$$

369 **370**

Finally, a bimetallic seven-membered cumulene (**238**) has been prepared by reduction of the bimetallic tropyne complex (**235**), which, in turn, was regenerated by hydride abstraction with Ph_3C+ [Eq. (64)].[99] The $^{13}C\ \{^1H\}$ NMR spectrum of **238** displays three carbonyls at $-35°C$ that coalesce upon warming, indicating rapid spinning of the $Mo(CO)_3$ fragment around the metal-seven membered ring axis. An X-ray crystal structure of **238** revealed six carbon atoms belonging to the seven-membered ring in the same plane with the saturated carbon atom 0.68(3) Å above the plane. The cumulene angles in **372** [139(2)° and 130(2)° for C7–C1–C2 and C1–C2–C3, respectively] are similar to the central allene bond angle of the bis(triphenylphosphine)platinum(0) complex of 1,2,4,6-cycloheptatetraene (**238**) [134.9(9)°][130] and are significantly more acute than in **367**. This is consistent with a somewhat less strained ring in the latter due to the presence of the large zirconium heteroatom.

Ph_3P PPh_3 Pt $(CO)_3Mo$ + $\xrightarrow[Ph_3C^+]{KBEt_3H}$ Ph_3P PPh_3 Pt $(CO)_3Mo$ (64)

235 **238**

REFERENCES

(1) For reviews including theoretical discussions of uncomplexed strained cycloalkynes, see: (a) Carlson, H. A.; Quelch, G. E.; Schaefer, H. F., *J. Am. Chem. Soc.* **1992**, *114*, 5344; (b) Wentrup, C.; Blanch, R.; Briehl, H.; Gross, G. *J. Am. Chem. Soc.* **1988**, *110*, 1874; (c) Nakagawa, M. In *The Chemistry of the Carbon–Carbon Triple Bond;* Patai, S., Ed.; Wiley: Chichester, U.K., 1978, pp. 635–712; (d) Greenbert, A.; Liebman, J. F. *Strained Organic Molecules;* Academic Press: New York, 1978, pp. 133–138; (e) Bloomquist, A. T., ; Liu, L. H. *J. Am. Chem. Soc.* **1973**, *95*, 790; (f) Krebs, A.; Kimling, *Angew. Chem., Int. Ed. Engl.* **1971**, *10*, 509; (g) Krebs, A. In *Chemistry of Acetylenes;* Viehe, H. G., Ed.; Marcel Dekker: New York, 1969, pp. 987–1062.

(2) Bennett, M. A.; Schwemlein, H. P. *Angew. Chem., Int. Ed. Engl.* **1989**, *28*, 1296.

(3) March, J. *Advanced Organic Chemistry*, 2nd ed., John Wiley & Sons: New York, 1992, p. 159.

(4) Johnson, R. P. *Chem. Rev.* **1989**, *89*, 1111.

(5) Coleman, J. P.; Hegedus, L. S.; Norton, J. R.; Finke, R. G. *Principles and Applications of Organotransition Metal Chemistry*, University Science Books: Mill Valley, California, 1987, Chapter 3.

(6) Hartwig, J. F.; Bergman, R. G.; Andersen, R. A. *J. Am. Chem. Soc.* **1991**, *113*, 3404. Ref. 5; pp. 156–158.

(7) Ref. 5; pp. 156–158.

(8) Buchwald, S. L.; Nielsen, R. B. *Chem. Rev.* **1988**, *88*, 1047.

(9) Ref. 5, pp. 150–154.

(10) Boekel, C. P.; Teuben, J. H.; de Liefde Meijer, H. J. *J. Organometal. Chem.* **1975**, *102*, 161.

(11) Luinstra, G. A.; Teuben, J. H. *Organometallics* **1992**, *11*, 1793.

(12) (a) Schock, L. E.; Brock, C. P.; Marks, T. J. *Organometallics* **1987**, *6*, 232; (b) Schock, L. E.; Marks, T. J. *J. Am. Chem. Soc.* **1988**, *110*, 7701.

(13) Erker, G. *J. Organometal. Chem.* **1977**, *134*, 189.

(14) Mashima, K.; Tanaka, Y.; Nakamura, A. *Organometallics* **1995**, *14*, 5642.

(15) Buchwald, S. L.; Nielsen, R. B. *J. Am. Chem. Soc.* **1988**, *110*, 3171.

(16) Fagan, P. J.; Manriquez, J. M.; Maatta, E. A.; Seyam, A. M.; Marks, T. J. *J. Am. Chem. Soc.* **1981**, *103*, 6650.

(17) Doherty, N. M.; Bercaw, J. E. *J. Am. Chem. Soc.* **1985**, *107*, 2670.

(18) A primary isotope effect at room temperature of 13.5 has been reported[12] for thermolysis of $Cp_2Zr(Ph)_2$. This abnormally large effect may be due to tunneling.

(19) Ref. 5, p. 331.

(20) Michilin, R. A.; Faglia, S.; Uguagliati, P. *Inorg. Chem.* **1983**, *22*, 1831.

(21) Abis, L.: Sen, A.; Halpern, J. *J. Am. Chem. Soc.* **1978**, *100*, 2915.

(22) McCarthy, T. J.; Nuzzo, R. G.; Whitesides, G. M. *J. Am. Chem. Soc.* **1981**, *103*, 3396.

(23) Moore, S. S.; DiCosimo, R.; Sowinski, A. F.; Whitesides, G. M. *J. Am. Chem. Soc.* **1981**, *103*, 948.
(24) Lowry, T. H.; Richardson, K. S. *Mechanism and Theory in Organic Chemistry*, 2nd ed.; Harper & Row: New York, 1981, p. 206.
(25) Cf. Carey, F. A.; Sundberg, R. J. *Advanced Organic Chemistry*, 2nd ed., Part A. Plenum Press, New York, 1984, pp. 190–194.
(26) Hartwig, J. F.; Bergman, R. G.; Andersen, R. A. *J. Am. Chem. Soc.* **1989**, *111*, 2717.
(27) (a) Bennett, M. A.; Milner, D. L. *J. Am. Chem. Soc.* **1969**, *91*, 6983; (b) Tulip, T. H.; Thorn, D. L. *J. Am. Chem. Soc.* **1981**, *103*, 2448; (c) Webster, D. E. *Adv. Organomet. Chem.* **1977**, *15*, 147; (d) Ryabov, A. D. *Chem. Rev.* **1990**, *90*, 403.
(28) Foley, F.; DiCosimo, R.; Whitesides, G. M. *J. Am. Chem. Soc.* **1980**, *102*, 6713.
(29) Cockcroft, K. J.; Gibson, C. V.; Howard, J. A. K.; Poole, A. D.; Siemeling, U.; Wilson, C. *J. Chem. Soc., Chem. Commun.* **1992**, 1668.
(30) Arnold, J.; Wilkinson, G.; Hussain, B.; Hursthouse, M. B. *Organometallics* **1989**, *8*, 415.
(31) Ref. 5, p. 386.
(32) A mononuclear nickel complex, which was prepared in very low yields by trapping free benzyne, was also observed spectroscopically.[2]
(33) Bennett, M. A.; Drage, J. S.; Griffiths, K. D.; Roberts, N. K.; Robertson, G. B.; Wiekramasinghe, W. A. *Angew. Chem., Int. Ed. Engl.* **1988**, *27*, 941.
(34) Shepard, K. L. *Tetrahedron Lett.* **1975**, 3371.
(35) Bennett, M. A., Wenger, E. *Organometallics* **1995**, *14*, 1267.
(36) Bennett, M. A. *Pure & Appl. Chem.* **1989**, *10*, 1695.
(37) Bennett, M. A.; Hockless, D. C. R.; Wenger, E. *Organometallics* **1995**, *14*, 2091.
(38) For a discussion of the mechanism of the oxidative addition reaction, see Ref. 35.
(39) Binger, P.; Biedenbach, B.; Mynott, R.; Regitz, M. *Chem. Ber.* **1988**, *121*, 1455.
(40) Butler, I. R., Cullen, W. R.; Einstein, F. W. B.; Jones, R. H. *J. Organomet. Chem.* **1993**, *463*, C6.
(41) Rausch, M. D.; Mintz, E. A. *J. Organomet. Chem.* **1980**, *190*, 65.
(42) Erker, G.; Korek, U.; Petrenz, R.; Rheingold, A. L. *J. Organomet. Chem.* **1991**, *421*, 215.
(43) Erker, G.; Dorf, U.; Lecht, R.; Ashby, M. T.; Aulbach, M.; Schlund, R.; Krüger, C.; Mynott, R. *Organometallics* **1989**, *8*, 2037.
(44) Cámpora, J.; Buchwald, S. L. *Organometallics* **1993**, *12*, 4182.
(45) Broene, R. D.; Buchwald, S. L. *Science* **1993**, *261*, 1696.
(46) Cuny, G. D.; Gutierrez, A.; Buchwald, S. L. *Organometallics* **1991**, *10*, 537.
(47) Cuny, G. D., Buchwald, S. L. *Organometallics* **1991**, *10*, 363.
(48) (a) Fagan, P. J.; Nugent, W. A. *J. Am. Chem. Soc.* **1988**, *110*, 2310; (b) Fagan, P. J.; Nugent, W. A.; Calabrese, J. C. *J. Am. Chem. Soc.* **1994**, *116*, 1880.
(49) Buchwald, S. L.; Watson, B. T.; Lum, R. T.; Nugent, W. A. *J. Am. Chem. Soc.* **1987**, *109*, 7137.
(50) Buchwald, S. L.; Fang, Q. *J. Org. Chem.* **1989**, *54*, 2793.
(51) Buchwald, S. L.; Fisher, R. A.; Foxman, B. M., *Angew. Chem., Int. Ed. Engl.* **1990**, *29*, 771.
(52) Kanj, A.; Meunier, P.; Gautheron, B.; Dubac, J.; Daran, J. C. *J. Organomet. Chem.* **1993**, *454*, 51.
(53) Meunier, P.; Gautheron, B.; Mazouz, A. *J. Organomet. Chem.* **1987**, *320*, C39.
(54) Bodiguel, J.; Meunier, P.; Gautheron, B. *Appl. Organomet. Chem.* **1991**, *5*, 479.
(55) Legrand, G.; Meunier, P.; Petersen, J. L.; Tavares, P.; Bodiguel, J.; Gautheron, B.; Dousse, G. *Organometallics* **1995**, *14*, 162.
(56) Tavarès, P.; Meunier, P.; Kubicki, M. M.; Gautheron, B.; Dousse, G.; Lavayssière, H.; Stagé, J. *J. Heteroatom. Chem.* **1993**, *4*, 393.
(57) Erker, G.; Albrecht, M. *Organometallics* **1992**, *11*, 3517.

(58) Erker, G.; Albrecht, M.; Kruger, C.; Werner, S. *J. Am. Chem. Soc.* **1992**, *114*, 8531.
(59) Tidwell, J. H.; Senn, D. R.; Buchwald, S. L. *J. Am. Chem. Soc.* **1991**, *113*, 4685.
(60) Tidwell, J. H.; Buchwald, S. L. *J. Org. Chem.* **1992**, *57*, 6380.
(61) Peat, A. J.; Buchwald, S. L. *J. Am. Chem. Soc.* **1996**, *118*, 1028.
(62) de Rege F. M. G.; Buchwald, S. L. *Tetrahedron* **1995**, *51*, 4291.
(63) Le Floch, P.; Kolb, A.; Mathey, F. *J. Chem. Soc., Chem. Commun.* **1994**, 2065.
(64) Karel, J.; Buijink, F.; Kloetstra, K. R.; Meetsma, A.; Teuben, J. H. *Organometallics* **1996**, *15*, 2523.
(65) (a) Hessen, B.; Teuben, J. H.; Lemmen, T. H.; Huffman, J. C.; Caulton, K. G. *Organometallics* **1985**, *4*, 946; (b) Hessen, B.; Lemmen, T. H.; Luttikhedde, H. J. G.; Teuben, J. H.; Petersen, J. L.; Jagner, S.; Huffman, J. C.; Caulton, K. G. *Organometallics* **1987**, *6*, 2354; (c) Hessen, B., Buijink, J. K. F.; Meetsma, A.; Teuben, J. H.; Helgesson, G.; Hakansson, M.; Jagner, S.; Spek, A. L. *Organometallics* **1993**, *12*, 2268.
(66) Houseknecht, K. L.; Stockman, K. E.; Sabat, M.; Finn, M. G.; Grimes, R. N. *J. Am. Chem. Soc.* **1995**, *117*, 1163.
(67) Dobson, J. E.; Miller, R. G.; Wiggen, J. P. *J. Am. Chem. Soc.* **1971**, *93*, 554.
(68) Bennett, M. A.; Wenger, E. *Organometallics* **1996**, *15*, 5536.
(69) Bennett, M. A.; Hockless, D. C. R.; Humphrey, M. G.; Schultz, M.; Wenger, E. *Organometallics* **1996**, *15*, 928.
(70) Hoberg, H.; Herrera, A. *Angew. Chem., Int. Ed. Engl.* **1980**, *19*, 927.
(71) Bock, C. W.; Trachtman, G. M. *J. Phys. Chem.* **1984**, *88*, 1467.
(72) McLain, S. J.; Schrock, R. R.; Sharp, P. R.; Churchill, M. R.; Youngs, W. J. *J. Am. Chem. Soc.* **1979**, *103*, 263.
(73) Buchwald, S. L.; Watson, B. T.; Huffman, J. C. *J. Am. Chem. Soc.* **1986**, *108*, 7411.
(74) Bennett, M. A.; Hambley, T. W.; Roberts, N. K.; Robertson, G. B. *Organometallics* **1985**, *4*, 1992.
(75) Buchwald, S. L.; Lucas, E. A.; Bewan, J. C. *J. Am. Chem. Soc.* **1987**, *109*, 4396.
(76) Hsu, D. P.; Lucas, E. A.; Buchwald, S. L. *Tetrahedron Lett.* **1990**, *39*, 5563.
(77) Buchwald, S. L.; Lucas, E. A.; Davis, W. M. *J. Am. Chem. Soc.* **1989**, *111*, 397.
(78) LeHoullier, C. S.; Gribble, G. W. *J. Org. Chem.* **1983**, *48*, 2364, and references therein.
(79) Buchwald, S. L.; King, S. M. *J. Am. Chem. Soc.* **1991**, *113*, 258.
(80) Buchwald, S. L.; Sayers, A.; Watson, B. T.; Dewan, J. C. *Tetrahedron Lett.* **1987**, *28*, 3245.
(81) Lu, Z.; Abboud, K. A.; Jones, W. M. *J. Am. Chem. Soc.* **1992**, *114*, 10991.
(82) Lu, Z.; Abboud, K. A.; Jones, W. M. *Organometallics* **1993**, *12*, 1471.
(83) Winchester, W. R.; Jones, W. M. *Organometallics* **1985**, *4*, 2228.
(84) Winchester, W. R.; Gawron, M.; Palenik, G. J.; Jones, W. M. *Organometallics* **1985**, *4*, 1894.
(85) Klosin, J.; Abboud, K. A.; Jones, W. M. *Organometallics* **1995**, *14*, 2892.
(86) Klosin, J.; Abboud, K. A.; Jones, W. M. *Organometallics* **1996**, *15*, 2465.
(87) Lu, Z.; Jones, W. M. *Organometallics* **1994**, *13*, 1539.
(88) Klosin, J.; Abboud, K. A.; Jones, W. M. *Organometallics* **1996**, *15*, 596.
(89) Robertson, G. B.; Whimp, P. O. *J. Am. Chem. Soc.* **1975**, *97*, 1051.
(90) (a) Adams, R. D.; Qu, X.; Wu, W. *Organometallics* **1995**, *14*, 1377; (b) Adams, R. D.; Chen, L.; Qu, X. *Organometallics* **1994**, *13*, 1992; (c) Adams, R. D.; Qu, X.; Wu, W. *Organometallics* **1994**, *13*, 1272; (d) Adams, R. D.; Qu, X.; Wu, W. *Organometallics* **1993**, *12*, 4117; (e) Adams, R. D.; Chen, G., Qu, X.; Wu, W.; Yamamoto, J. H. *Organometallics* **1993**, *12*, 3426; (f) Adams, R. D.; Chen, G.; Qu, X.; Wu, W.; Yamamoto, J. H. *Organometallics* **1993**, *12*, 3029; (g) Adams, R. D.; Chen, G.; Qu, X.; Wu, W.; Yamamoto, J. H. *J. Am. Chem. Soc.* **1992**, *114*, 10977.
(91) (a) For an example of a cobalt cluster complex of fur-3-yne, see: Barnes, C. E.; King,

W. D.; Orvis, J. A. *J. Am. Chem. Soc.* **1995**, *117*, 1855; (b) for an example of an osmium cluster complex of ferrocyne, see: Cullen, W. R.; Rettig, S. J.; Zheng, T. *Organometallics* **1992**, *11*, 928; (c) for examples of ditungsten complexes of cyclopentyne through cyclooctyne, see: Chisholm, M. H.; Floting, K.; Huffman, J. C.; Lucas, E. A. *Organometallics* **1991**, *10*, 535.

(92) Buchwald, S. L.; Lum, R. T.; Fisher, R. A.; Davis, W. M. *J. Am. Chem. Soc.* **1989**, *111*, 9113.

(93) Bennett, M. A.; Johnson, J. A.; Willis, A. C. *Organometallics* **1996**, *15*, 68.

(94) (a) Rosenthal, U.; Oehme, G.; Gorls, H.; Burlakov, V. V.; Polyakov, A. V.; Yanovskii, A. I.; Struchkov, Y. T. *J. Organomet. Chem.* **1990**, *389*, 251; (b) Rosenthal, U.; Schulz, W., Gorls, H. Z. *Anorg. Allgem. Chem.* **1987**, *550*, 169; (c) Porschke, K. R.; Tsay, Y. H.; Kruger, C. *Angew. Chem., Int. Ed. Engl.* **1985**, *24*, 323; (d) Porschke, K. R.; Mynott, R.; Angermund, K.; Kruger, C. Z. *Z. Naturforsch. B*, **1985**, *40*, 199.

(95) Rosenthal, U. Z. *Anorg. Allgem. Chem.* **1981**, *482*, 179.

(96) Greaves, E. O.; Lock, C. J. L.; Maitlis, P. M. *Can. J. Chem.* **1968**, *46*, 3879.

(97) Yin, J.; Abboud, K. A.; Jones, W. M. *J. Am. Chem. Soc.* **1993**, *115*, 8859.

(98) Lu, Z. Ph.D. Dissertation, University of Florida, Gainesville, 1992.

(99) Komatsu, K.; Kamo, H.; Tsuji, R.; Masuda, H.; Takeuchi, K. *J. Chem. Soc., Chem. Commun.* **1991**, 71.

(100) Cotton, F. A.; Wilkinson, G. *Advanced Inorganic Chemistry*. John Wiley & Sons: New York, 1988, p. 60.

(101) Yin, J.; Klosin, J.; Abboud, K. A.; Jones, W. M. *Tetrahedron Lett.* **1995**, 3107.

(102) Huang, N. Z.; Sondheimer, F., *Acc. Chem. Res.* **1982**, *15*, 96.

(103) Robertson, G. B.; Whimp, P. O. *Aust. J. Chem.* **1980**, *33*, 1373.

(104) Manojlovic-Muir, L.; Muir, K. W.; Walker, R. *Acta Crystallogr. Sect.* **1979**, *B 35*, 2416.

(105) (a) Psiorz, M.; Hopf, H. *Angew. Chem.* **1982**, *92*, 639; (b) Chan, C. W., Wong, H. N. C. *J. Am. Chem. Soc.* **1985**, *107*, 3790; **1988**, *110*, 462.

(106) (a) de Meijere, A.; Heinze, J.; Meerholz, K.; Reiser, O.; Konig, B. *Angew. Chem., Int. Ed. Engl.* **1990**, *29*, 1418; (b) Konig, B.; Heinze, J.; Meerholz, K.; de Meijere, A. *Angew. Chem., Int. Ed. Engl.* **1991**, *30*, 1361.

(107) Konig, B.; Bennett, M. A.; de Meijere, A. *Synlett* **1994**, 653.

(108) Albrecht, K.; Hockless, D. C. R.; Konig, B.; Neumann, H.; Bennett, M. A.; de Meijere, A. *J. Chem. Soc., Chem. Commun.* **1996**, 543.

(109) (a) Chan, C. W.; Wong, H. N. C. *J. Am. Chem. Soc.* **1988**, *110*, 462; (b) Stobbe, M.; Reiser, O.; Nader, R.; de Meijere, A. *Chem. Ber.* **1987**, *120*, 1667; (c) Chan, C. W.; Wong, H. N. C. *J. Am. Chem. Soc.* **1985**, *107*, 4790.

(110) Kawada, Y.; Jones, W. M. *J. Organomet. Chem.* **1980**, *192*, 87, and references cited.

(111) Cf. Stowe, M. E.; Wells, E. E.; Lester, E. W.; Jones, W. M. *J. Am. Chem. Soc.* **1968**, *90*, 1849.

(112) Cooper, D. L.; Murphy, S. C. *Astrophys. J.* **1988**, *333*, 482.

(113) (a) Rosenthal, U.; Ohff, A.; Baumann, W.; Kempe, R.; Tillack, A.; Burlakov, V. V. *Angew. Chem., Int. Ed. Engl.* **1994**, *33*, 1605; (b) Rosenthal, U.; Ohff, A.; Baumann, W.; Kempe, R.; Tillack, A.; Burlakov, V. V. *Angew. Chem., Int. Ed. Engl.* **1994**, *33*, 1608.

(114) (a) Burlakov, V. V.; Ohff, A.; Lefeber, C.; Tillack, A.; Baumann, W.; Kempe, R.; Rosenthal, U. *Chem. Ber.* **1995**, *128*, 967; (b) Pulst, S.; Arndt, P.; Heller, B.; Baumann, W.; Kempe, R.; Rosenthal, U. *Angew. Chem., Int. Ed. Engl.* **1996**, *35*, 1112.

(115) Oon, S. M.; Koziol, A. E.; Jones, W. M.; Palenik, G. J. *J. Chem. Soc., Chem. Commun.* **1987**, 491.

(116) Yin, J.; Abboud, K. A.; Jones, W. M. *J. Am. Chem. Soc.* **1993**, *115*, 3810.

(117) Angus, R. O. Jr.; Johnson, R. P. *J. Org. Chem.* **1984**, *49*, 2880.

(118) Visser, J. P.; Ramakers, J. E. *J. Chem. Soc., Chem. Commun.* **1972**, 178.
(119) Manganiello, F. J.; Oon, S. M.; Radcliffe, M. D.; Jones, W. M. *Organometallics* **1985**, *4*, 1069.
(120) Omrcen, T.; Conti, N. J.; Jones, W. M. *Organometallics* **1991**, *10*, 913.
(121) Oon, S. M. Ph.D. Dissertation, University of Florida, Gainesville, 1988.
(122) Cf. (a) Jacobs, T. L. In *The Chemistry of the Allenes;* Landor, S. R., Ed.; Academic Press: New York, 1982, Vol. II, Chapter 4; (b) Gallop, M. R.; Roper, W. R. *Adv. Organomet. Chem.* **1986**, *25*, 121.
(123) (a) Otsuka, S.; Nakamura, A. *Adv. Organomet. Chem.* **1976**, *14*, 245; (b) Bowden, F. L.; Giles, R. *Coord. Chem. Rev.* **1976**, *20*, 81.
(124) Fisher, R. A.; Buchwald, S. L. *Organometallics* **1991**, *10*, 537.
(125) Shakespeare, W. C.; Johnson, R. P. *J. Am. Chem. Soc.* **1990**, *112*, 8578.
(126) Winchester, W. R. Ph.D. Dissertation, University of Florida, Gainesville, 1985.
(127) (a) Foxman, B.; Marten, D.; Rosan, A.; Raghu, S.; Rosenblum, M. J. *J. Am. Chem. Soc.* **1977**, *99*, 2160; (b) Vrieze, K.; Volger, H. C.; Gronert, M.; Pratt, A. P. *J. Organomet. Chem.* **1969**, *10*, 19; (c) Vrieze, K.; Volger, H. C.; Pratt, A. P. *J. Organomet. Chem.* **1970**, *21*, 467.
(128) Oon, S. M.; Jones, W. M. *Organometallics* **1988**, *7*, 2172.
(129) Omrcen, T. Ph.D. Dissertation, University of Florida, Gainesville, 1992.
(130) Abboud, K. A.; Lu, Z.; Jones, W. M. *Acta Crystallogr.* **1992**, *C48*, 909.
(131) Wojcicki, A. *J. Organomet. Chem.* **1975**, *94*, 311.
(132) Lichtenberg, D.; Wojcicki, A. *J. Am. Chem. Soc.* **1972**, *94*, 8271.
(133) Yin, J.; Jones, W. M. *Tetrahedron* **1995**, 4395.
(134) Cf. Harris, J. W.; Jones, W. M. *J. Am. Chem. Soc.* **1983**, *104*, 7329.
(135) Matzinger, S., Bally, T.; Patterson, E. V.; McMahon, R. J. *J. Am. Chem. Soc.* **1996**, *118*, 1535, and references cited.
(136) Kirmse, W.; Loosen, K.; Sluma, H. D. *J. Am. Chem. Soc.* **1981**, *103*, 5935.
(137) Lu, Z.; Jones, W. M.; Winchester, W. R. *Organometallics* **1993**, *12*, 1344.
(138) Klosin, J.; Zheng, X.; Jones, W. M. *Organometallics* **1996**, *15*, 3788.
(139) Feher, J.; Gergens, D. D.; Ziller, J. W. *Organometallics* **1993**, *12*, 2810.
(140) Allison, N. T.; Kawada, Y.; Jones, W. M. *J. Am. Chem. Soc.* **1978**, *100*, 5224.
(141) Riley, P. E.; Davis, R. E.; Allison, N. T.; Jones, W. M. *J. Am. Chem. Soc.* **1980**, *102*, 2458.
(142) Manganiello, F. J.; Radcliffe, M. D.; Jones, W. M. *J. Organomet. Chem.* **1982**, *228*, 273.
(143) Riley, P. E.; Davis, R. E.; Allison, N. T.; Jones, W. M. *Inorg. Chem.* **1982**, *21*, 1321.
(144) Lisko, J. R.; Jones, W. M. *Organometallics* **1986**, *5*, 1890.
(145) For Huckel calculations, cf. Hoffmann, R. *J. Chem. Phys.* **1963**, *39*, 1397.
(146) Allison, N. J. Ph.D. Dissertation, University of Florida, Gainesville, 1978.
(147) Klosin, J.; Abboud, K. A.; Jones, W. M. *Acta Crystallogr.* **1996**, *C52*, 1101.
(148) Schilling, B. E. R.; Hoffmann, R.; Lichtenberger, D. L. *J. Am. Chem. Soc.* **1979**, *101*, 585.
(149) Brookhart, M.; Tucker, J. R.; Flood, T. C.; Jenson, J. *J. Am. Chem. Soc.* **1980**, *101*, 1203.
(150) Ficarro, S. M. M. S. Thesis, University of Florida, Gainesville, 1982.
(151) (a) Ingrosso, G.; Immirizi, A.; Parri, L. *J. Organomet. Chem.* **1973**, *60*, C35; (b) Jolly, P. W.; Kruger, C.; Salz, R.; Skeutowski, J. C. *J. Organomet. Chem.* **1979**, *165*, C39.
(152) McKinney, R. J.; Thorn, D. L.; Hoffmann, R.; Stockis, A. *J. Am. Chem. Soc.* **1981**, *103*, 2595.
(153) Hsu, D. P.; Davis, W. M.; Buchwald, S. L. *J. Am. Chem. Soc.* **1993**, *115*, 10394.
(154) Klosin, J. Ph.D. Dissertation, University of Florida, Gainesville, 1995.

ADVANCES IN ORGANOMETALLIC CHEMISTRY, VOL. 42

Bridged Silylene and Germylene Complexes

HIROSHI OGINO and HIROMI TOBITA

Department of Chemistry
Graduate School of Science
Tohoku University
Sendai 980-77, Japan

I
INTRODUCTION

The coordination chemistry of reactive silicon(II) and germanium(II) species, that is, silylene and germylene, has been a fascinating topic since before the 1960s, and considerable efforts have been devoted to the synthe-

0065-3055/98 $25.00

sis of the transition-metal complexes having these species as ligands. Among them, bridged silylene and germylene complexes form the largest group. Since the first report on the synthesis of $\{Mn(CO)_5\}_2(\mu\text{-}GeH_2)$ by Stone *et al.* in 1963,[1] more than 180 bridged silylene and germylene complexes have been synthesized by a variety of methods. For the central metals, 12 elements of Groups 4 and 6–10, that is, Ti, W, Mn, Re, Fe, Ru, Os, Co, Rh, Ir, Pd, and Pt, are known. Among them, the complexes contain overwhelmingly Groups 8, 9, and 10 metals.

Most of the silylene- or germylene-bridged complexes are formed by the replacement of two σ-bonded groups in silanes and germanes by transition-metal fragments. However, this does not mean that the metal–silicon or–germanium bonds in these complexes are simply composed of σ-bonds between the sp^3-hybridized silicon or germanium atom and two transition-metal atoms. In fact, a majority of complexes, which have a metal–metal bond and are classified into the **B**-type complex in Section II, possess very acute M–E–M bond angles (E = Si, Ge) ranging from 58° to 77° (see Tables VI, VII). Therefore, it seems more realistic to consider the bonding of the core of this complex as the σ- and π-type orbital interactions between the sp^2-hybridized silylene or germylene ligand and the dinuclear metal unit (see Section IV,A). This bonding mode is supported by the structural and spectroscopic properties, especially the large downfield shift of the ^{29}Si NMR signals of silylene ligands, and also some reactivity studies.

Reactivity is an intriguing aspect of these complexes not only because of the previously mentioned unique bonding mode, but also because these complexes can serve as good models for surface-bound divalent silicon and germanium species. Unfortunately, only a limited number of the reactions with small molecules and catalytic reactions have been reported so far. In contrast, the dynamic behaviors and isomerization reactions of these complexes, which give plentiful information about their chemical properties, are well documented.

There have been several review articles from wider viewpoints, dealing in part with bridged silylene and germylene complexes.[2–16] But this article may be the first one that handles only bridged silylene and germylene complexes. Therefore, we attempt to cover most of the known μ-silylene and μ-germylene complexes, which are summarized in Tables I–V, up to the end of 1996. Sections II and III are intended as an overview of the formation reactions of these complexes, so that there is overlap with previous review articles. As for physical properties and reactions of these complexes described in Sections IV and V, we emphasize the recent developments. Finally, the dinuclear metal complexes with bridging M–H–Si interactions have been already reviewed in excellent articles of Graham[10] and Schubert[11] in 1986 and 1990, respectively, so that only the formation of these complexes are presented in this article.

TABLE I

FORMATION OF BRIDGED SILYLENE COMPLEXES WITHOUT METAL–METAL BONDS (**A**-TYPE COMPLEXES)

Reaction type	Reaction	Product complex	Ref.
F	$Cp_2TiCl_2 + SiH_3K$	$\{Cp_2Ti\}_2(\mu\text{-}SiH_2)_2$	53
D	$Cp^*Mn(CO)_2(THF) + SiH_4$	$\{Cp^*Mn(CO)_2(H)\}_2(\mu\text{-}SiH_2)$	43
H, K	$Fe(CO)_4(SiMe_2 \cdot HMPA)$, 115°C, 10^{-4} torr, 20 min or $Et_4N[Fe(CO)_4SiMe_2Cl] + AlCl_3$	$Fe_2(CO)_8(\mu\text{-}SiMe_2)_2$	57,59
B	$Fe(CO)_5 + H_2SiPh_2$, *hν*, 20 h	$Fe_2(CO)_8(\mu\text{-}SiPh_2)_2$	17,18
F	$CpFe(CO)_2SiMeHCl + Na[Fe(CO)_2Cp]$	$CpFe(CO)_2(\mu\text{-}SiMeH)Fe(CO)_2Cp$	54
F	$CpFe(CO)_2Si$*p*-TolHCl + $Na[Fe(CO)_2Cp]$	$CpFe(CO)_2(\mu\text{-}Si$*p*-TolH$)Fe(CO)_2Cp$	23
F	$CpFe(CO)_2SiCl_2H + Na[Fe(CO)_2Cp]$	$CpFe(CO)_2(\mu\text{-}SiClH)Fe(CO)_2Cp$	62
J	$CpFe(CO)_2(\mu\text{-}SiMeH)Fe(CO)_2Cp + CCl_4$, 25°C, 60 min	$CpFe(CO)_2(\mu\text{-}SiMeCl)Fe(CO)_2Cp$	54,62
J	$CpFe(CO)_2(\mu\text{-}SiClH)Fe(CO)_2Cp + CCl_4$, 25°C, 35 min	$CpFe(CO)_2(\mu\text{-}SiCl_2)Fe(CO)_2Cp$	62
J	$CpFe(CO)_2(\mu\text{-}SiMeCl)Fe(CO)_2Cp + AgBF_4$, 25°C, 24 h	$CpFe(CO)_2(\mu\text{-}SiMeF)Fe(CO)_2Cp$	62
J	$CpFe(CO)_2(\mu\text{-}SiCl_2)Fe(CO)_2Cp + AgBF_4$, 70°C, 4h	$CpFe(CO)_2(\mu\text{-}SiF_2)Fe(CO)_2Cp$	62
J	$CpFe(CO)_2(\mu\text{-}SiMeH)Fe(CO)_2Cp + Me_2CO_2$	$CpFe(CO)_2(\mu\text{-}SiMeOH)Fe(CO)_2Cp$	23
J	$CpFe(CO)_2(\mu\text{-}Si$*p*-TolH$)Fe(CO)_2Cp + Me_2CO_2$	$CpFe(CO)_2(\mu\text{-}Si$*p*-TolOH$)Fe(CO)_2Cp$	23
B	$CpFe(CO)_2SiCl_2H + Co_2(CO)_8$	$CpFe(CO)_2(\mu\text{-}SiCl_2)Co(CO)_4$	36
H	$Cp_2Fe_2(CO)_4$ 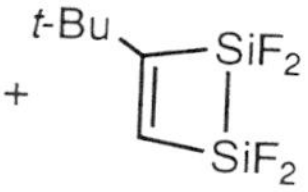 + (*t*-Bu, SiF_2, SiF_2 ring), 85 °C, 24 h	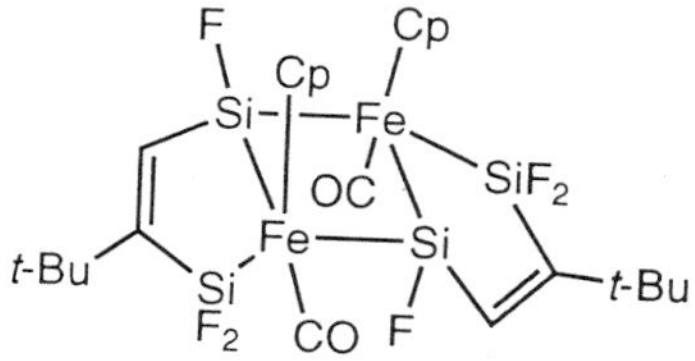 (F, Cp, Cp, Si, Fe, OC, SiF_2, *t*-Bu, Fe, Si, Si F_2, CO, F, *t*-Bu)	38
D, F	$Pt(PEt_3)_3 + H_3SiPh$ or $(Et_3P)_2PtCl_2 + Ph_2SiHLi$ or $(Et_3P)_2PtCl_2 + H_3SiPh + Na$	$(Et_3P)_2Pt(\eta^2,\eta^2\text{-}Ph(H)SiSi(H)Ph)Pt(PtEt_3)_2$ + etc.	46–48
D	$Pt(PEt_3)_3 + H_3Si$*p*-Tol	$(Et_3P)_2Pt(\eta^2,\eta^2$-*p*-Tol(H)SiSi(H)*p*-Tol$)Pt(PtEt_3)_2$	46
D	$Pt(PEt_3)_3 + H_3SiMes$	$(Et_3P)_2Pt(\eta^2,\eta^2\text{-}Mes(H)SiSi(H)Mes)Pt(PtEt_3)_2$	46
D	$(Et_3P)_2PtCl_2 + H_3SiCy + Na$	$(Et_3P)_2Pt(\eta^2,\eta^2\text{-}Cy(H)SiSi(H)Cy)Pt(PtEt_3)_2$, mixture of *cis* and *trans* isomers + etc.	48

TABLE II

FORMATION OF BRIDGED SILYLENE COMPLEXES WITH METAL-METAL BONDS (**B**-TYPE COMPLEXES)[a]

Reaction type	Reaction	Product complex	Ref.
D	$Mn(CO)_5SiMe_2H + Pt(C_2H_4)(PPh_3)_2$ or $Pt(PPh_3)_4$, RT	$(CO)_4Mn(\mu\text{-}SiMe_2)(\mu\text{-}H)Pt(PPh_3)_2$	45
D	$Mn(CO)_5SiPh_2H + Pt(C_2H_4)(PPh_3)_2$ or $Pt(PPh_3)_4$, RT	$(CO)_4Mn(\mu\text{-}SiPh_2)(\mu\text{-}H)Pt(PPh_3)_2$	45
D	$Mn(CO)_5SiCl_2H + Pt(C_2H_4)(PPh_3)_2$ or $Pt(PPh_3)_4$, RT	$(CO)_4Mn(\mu\text{-}SiCl_2)(\mu\text{-}H)Pt(PPh_3)_2$	45
J	$(CO)_4Mn(\mu\text{-}SiPh_2)(\mu\text{-}H)Pt(PPh_3)_2$ + CO, RT, 30 min	$(CO)_4Mn(\mu\text{-}SiPh_2)(\mu\text{-}H)Pt(PPh_3)(CO)$	45
J	$(CO)_4Mn(\mu\text{-}SiPh_2)(\mu\text{-}H)Pt(PPh_3)_2$ + $2PEt_3$, RT, 5 h	$(CO)_4Mn(\mu\text{-}SiPh_2)(\mu\text{-}H)Pt(PEt_3)_2$	45
K	$Fe_2(CO)_9 + HClSiMe_2 + R_3N$	$Fe_2(CO)_8(\mu\text{-}SiMe_2)$	65
B	$Fe_2(CO)_9 + H_2SiMes_2$, hexane reflux, 36 h	$Fe_2(CO)_8(\mu\text{-}SiMes_2)$	20
K	$Fe_3(CO)_{12} + H_2Si\{O(2,6\text{-}i\text{-}Pr_2C_6H_3)\}_2$, toluene reflux, 16 h	$Fe_2(CO)_8[\mu\text{-}Si\{O(2,6\text{-}i\text{-}Pr_2C_6H_3)\}_2]$	20
K	$Fe(CO)_4Si(NMe_2)_2 \cdot NHMe_2$ + $Pt(C_2H_4)(PPh_3)_2$, RT, 4 h	$Fe(CO)_4\{\mu\text{-}Si(NMe_2)_2\}Pt(PPh_3)_2$	67
K	$Fe(CO)_4Si(NMe_2)_2 \cdot NHMe_2$ + $Pt(C_2H_4)(Pp\text{-}Tol_3)_2$, RT, 4 h	$Fe(CO)_4\{\mu\text{-}Si(NMe_2)_2\}Pt(Pp\text{-}Tol_3)_2$	67
K	$Fe(CO)_4Si(NMe_2)_2 \cdot NHMe_2$ + $Pt(C_2H_4)(Ph_2PCH_2PPh_2)$, RT 4 h	$Fe(CO)_4\{\mu\text{-}Si(NMe_2)_2\}Pt(Ph_2PCH_2PPh_2)$	67
E, I	$Fe_2(CO)_9 + HMe_2SiSiMe_2H$, RT, 20 h or $Fe_2(CO)_8(\mu\text{-}SiMe_2)_2$, $h\nu$	$\{Fe(CO)_3\}_2(\mu\text{-}CO)(\mu\text{-}SiMe_2)_2$	51
B	$Fe_2(CO)_9 + H_3SiPh$, RT, 20 h	$\{Fe(CO)_3\}_2(\mu\text{-}CO)(\mu\text{-}SiPhH)_2$	19
B	$Fe(CO)_5 + H_2SiPhMe$, $h\nu$, 15 h	$\{Fe(CO)_3\}_2(\mu\text{-}CO)(\mu\text{-}SiPhMe)_2$	17,18
J	$Fe_2(CO)_8(\mu\text{-}SiMe_2)_2 + Fe(CO)_4(H)(SiMe_2Cl)$, $h\nu$, 3 h	$Fe_2(CO)_6(\mu\text{-}SiMe_2)_2(\mu\text{-}SiMeCl)$	64
I	$CpFe(CO)_2(\mu\text{-}SiMeH)Fe(CO)_2Cp$, $h\nu$, 6 h	$Cp_2Fe_2(CO)_2(\mu\text{-}CO)(\mu\text{-}SiMeH)$ (mixture of three isomers)	54
I, J	$CpFe(CO)_2(\mu\text{-}SiMeCl)Fe(CO)_2Cp$, $h\nu$ or $CpFe(CO)_2(\mu\text{-}SiMeH)Fe(CO)_2Cp + CCl_4$, RT	$Cp_2Fe_2(CO)_2(\mu\text{-}CO)(\mu\text{-}SiMeCl)$ (mixture of three isomers)	54
B	$CpFe(CO)_2SiMe_3 + H_3Sit\text{-}Bu$, $h\nu$, 0°C, 4 h	$Cp_2Fe_2(CO)_2(\mu\text{-}CO)(\mu\text{-}Sit\text{-}BuH)$ (*cis* isomer)	21, 22
B	$CpFe(CO)_2SiMe_3 + H_3Si(CMe_2)_2H$, $h\nu$, 0°C, 2 h	$Cp_2Fe_2(CO)_2(\mu\text{-}CO)[\mu\text{-}Si\{(CMe_2)_2H\}H]$ (*cis* isomer)	21,22
J	$Cp_2Fe_2(CO)_2(\mu\text{-}CO)(\mu\text{-}Sit\text{-}BuH)$ + CCl_4, RT, 17.5 h	$Cp_2Fe_2(CO)_2(\mu\text{-}CO)(\mu\text{-}Sit\text{-}BuCl)$ (mixture of two isomers)	60,61
J	$Cp_2Fe_2(CO)_2(\mu\text{-}CO)(\mu\text{-}Sit\text{-}BuH)$ + $CHBr_3$, RT, 80 min	$Cp_2Fe_2(CO)_2(\mu\text{-}CO)(\mu\text{-}Sit\text{-}BuBr)$ (mixture of two isomers)	60,61
J	$Cp_2Fe_2(CO)_2(\mu\text{-}CO)(\mu\text{-}Sit\text{-}BuH)$ + CH_2I_2, RT, 3 days	$Cp_2Fe_2(CO)_2(\mu\text{-}CO)(\mu\text{-}Sit\text{-}BuI)$ (mixture of two isomers)	60,61
J	$Cp_2Fe_2(CO)_2(\mu\text{-}CO)(\mu\text{-}Sit\text{-}BuCl)$ + MeLi, RT, 8.5 h	$Cp_2Fe_2(CO)_2(\mu\text{-}CO)(\mu\text{-}Sit\text{-}BuMe)$ (mixture of two isomers)	60,61
J	$Cp_2Fe_2(CO)_2(\mu\text{-}CO)(\mu\text{-}Sit\text{-}BuI)$ + *N*-methylimidazole (NMI), 40°C, 30 min	$[Cp_2Fe_2(CO)_2(\mu\text{-}CO)(\mu\text{-}Sit\text{-}Bu \cdot NMI)]I$[b] (*cis* isomer)	61,63

TABLE II (*continued*)

Reaction type	Reaction	Product complex	Ref.
J	$Cp_2Fe_2(CO)_2(\mu\text{-}CO)(\mu\text{-}Si\textit{t}\text{-}BuI)$ + 4-(dimeth-ylamino)pyridine (DMAP), 40°C, 30 min	$[Cp_2Fe_2(CO)_2(\mu\text{-}CO)(\mu\text{-}Si\textit{t}\text{-}Bu \cdot DMAP)]I^b$ (*cis* isomer)	61
B	$CpFe(CO)_2Me + H_2Si(2,4,6\text{-}\textit{i}\text{-}Pr_3C_6H_2)_2$, $h\nu$, 2.3 h	$Cp_2Fe_2(\mu\text{-}CO)_2\{\mu\text{-}Si(2,4,6\text{-}\textit{i}\text{-}Pr_3C_6H_2)_2\}$	24
B	$CpFe(CO)_2Me + H_2Si(2,6\text{-}Et_2C_6H_3)_2$, $h\nu$, 2.3 h	$Cp_2Fe_2(\mu\text{-}CO)_2\{\mu\text{-}Si(2,6\text{-}Et_2C_6H_3)_2\}$	24
B	$CpFe(CO)_2Me + H_2SiMes_2$, $h\nu$, 3 h	$Cp_2Fe_2(\mu\text{-}CO)_2(\mu\text{-}SiMes_2)$	24
I	$Cp_2Fe_2(\mu\text{-}CO)_2\{\mu\text{-}Si(2,6\text{-}Et_2C_6H_3)_2\}$ + CO, RT	$Cp_2Fe_2(CO)_2(\mu\text{-}CO)\{\mu\text{-}Si(2,6\text{-}Et_2C_6H_3)_2\}$ (*trans* isomer)	24
I	$Cp_2Fe_2(\mu\text{-}CO)_2(\mu\text{-}SiMes_2)$ + CO, RT	$Cp_2Fe_2(CO)_2(\mu\text{-}CO)(\mu\text{-}SiMes_2)$ (*trans* isomer)	24
B	$CpFe(CO)_2Me + H_3SiMes$, $h\nu$, 6 h	$Cp_2Fe_2(CO)_2(\mu\text{-}CO)(\mu\text{-}SiMesH)$ (mix-ture of three isomers)	23
J	$Cp_2Fe_2(CO)_2(\mu\text{-}CO)(\mu\text{-}SiMesH) + Me_2CO_2$	$Cp_2Fe_2(CO)_2(\mu\text{-}CO)(\mu\text{-}SiMesOH)$ (mix-ture of three isomers)	23
G	$Cp(CO)_2FeSiMe_2SiMe_2Fe(CO)_2Cp$, $h\nu$, 40 min	$Cp_2Fe_2(CO)_2(\mu\text{-}CO)\{\mu\text{-}SiMe(SiMe_3)\}$ (mixture of three isomers)	55,56
G	$Cp(CO)_2FeSiMe_2SiMe_2Fe(CO)_2Cp$, $h\nu$, 10 h	$Cp_2Fe_2(CO)_2(\mu\text{-}SiMe_2)_2$ (mixture of two isomers)	55
B	$Cp^*Fe(CO)_2SiMe_3 + H_3Si\textit{p}\text{-}Tol$, $h\nu$	$Cp^*_2Fe_2(CO)_2(\mu\text{-}CO)(\mu\text{-}Si\textit{p}\text{-}TolH)$ (mixture of two isomers)	25
B	$Cp^*Fe_2(CO)_4(\mu\text{-}CO)(\mu\text{-}PPh_2) + H_2SiPh_2$ or $HPh_2SiSiPh_2H$, RT–80°C, 6–9 days	$Cp^*Fe_2(CO)_4(\mu\text{-}SiPhH)(\mu\text{-}PPh_2)$	26
B	$Cp^*Fe_2(CO)_4(\mu\text{-}CO)(\mu\text{-}PPh_2) + H_2Si\textit{p}\text{-}Tol_2$, RT–80°C, 9 days	$Cp^*Fe_2(CO)_4(\mu\text{-}Si\textit{p}\text{-}TolH)(\mu\text{-}PPh_2)$	26
E	$Cp^*Fe_2(CO)_4(\mu\text{-}CO)(\mu\text{-}PPh_2) + HMe_2SiSiMe_2H$, RT–80°C, 6 days	$Cp^*Fe_2(CO)_4(\mu\text{-}SiMeH)(\mu\text{-}PPh_2)$	26
E	$Cp^*Fe_2(CO)_4(\mu\text{-}CO)(\mu\text{-}PPh_2) + HPh_2SiSiPh_2H$, RT–80°C, 6 days	$Cp^*Fe_2(CO)_4(\mu\text{-}SiPhH)(\mu\text{-}PPh_2)$	26
I	$\{Cp^*Ru(\mu\text{-}\eta^2\text{-}HSiPh_2)\}_2(\mu\text{-}H)(H)$, 110°C, 4 h	$Cp^*_2Ru_2(\mu\text{-}H)_2(\mu\text{-}SiPh_2)_2$	38
I	$\{Cp^*Ru(\mu\text{-}\eta^2\text{-}HSiPh_2)\}_2(\mu\text{-}H)(H)$ + CO (10 atm), 150°C or $Cp^*_2Ru_2(\mu\text{-}H)_2(\mu\text{-}SiPh_2)_2$ + CO (10 atm), 150°C	$Cp^*_2Ru_2(\mu\text{-}CO)_2(\mu\text{-}SiPh_2)_2$ (*trans* isomer)	38
D	$[Cp^*Ru(\mu\text{-}OMe)]_2 + H_2SiPh_2$, RT, 30 min	$Cp^*_2Ru_2(\mu\text{-}SiPhOMe)(\mu\text{-}OMe)(\mu\text{-}H)$	50
E	$Ru_3(CO)_{12}$, $[Me_3ERu(CO)_4]_2$ (E = Si, Ge), or $(Me_3Si)_2Ru(CO)_4 + HSi_2Me_5$, $h\nu$	$\{Ru(CO)_3(SiMe_3)\}_2(\mu\text{-}SiMe_2)_2$	52
E	$Os_3(CO)_{12} + HSi_2Me_5$, 160°C, 44 h or $[Me_3SiOs(CO)_4]_2$ or $(Me_3Si)_2Os(CO)_4 + HSi_2Me_5$, $h\nu$	$\{Os(CO)_3(SiMe_3)\}_2(\mu\text{-}SiMe_2)_2$	52
E	$[Me_3ERu(CO)_4]_2$ (E = Si, Ge) + $HMe_2SiSiMe_2H$, RT, 12–19 days	$Ru_2(CO)_6(\mu\text{-}SiMe_2)_3$	52
E	$Os(CO)_4H_2 + HSi_2Me_5$, 80°C, 18 h	$Os_3(CO)_9(\mu\text{-}SiMe_2)_3$	52

(*continues*)

TABLE II (*continued*)

Reaction type	Reaction	Product complex	Ref.
B	$Co_2(CO)_8$ + CpFe(CO)(MeNC)(SiH_3), 25°C, 4 days	$Co_2(CO)_6(\mu\text{-CO})[\mu\text{-SiH}\{Fe(CO)(MeNC)Cp\}]$	30
B	$Co_2(CO)_8$ + CpFe(CO)(*t*-BuNC)(SiH_3), 25°C, 3 days	$Co_2(CO)_6(\mu\text{-CO})[\mu\text{-SiH}\{Fe(CO)(t\text{-BuNC})Cp\}]$	30
E	$Co_2(CO)_8$ + $HMe_2SiSiMe_2H$, RT	$Co_2(CO)_6(\mu\text{-SiMe}_2)_2$	51
C	$Rh_2(Ph_2PCH_2PPh_2)_2(H)_2(CO)_2$ + H_3SiPh, −30°C, 10 min	$Rh_2(Ph_2PCH_2PPh_2)_2(H)_2(CO)_2(\mu\text{-SiHPh})$	31,39
C	$Rh_2(Ph_2PCH_2PPh_2)_2(H)_2(CO)_2$ + $2H_3SiPh$, 60°C, 24 h	$Rh_2(Ph_2PCH_2PPh_2)_2(CO)_2(\mu\text{-SiHPh})_2$	31,39
C	$Rh_2(Ph_2PCH_2PPh_2)_2(H)_2(CO)_2$ + $2H_3SiEt$, 60°C, 24 h	$Rh_2(Ph_2PCH_2PPh_2)_2(CO)_2(\mu\text{-SiHEt})_2$	31,39
C	$Rh_2(Ph_2PCH_2PPh_2)_2(H)_2(CO)_2$ + $2H_3Si\textit{n}\text{-}C_6H_{13}$, 60°C, 24 h	$Rh_2(Ph_2PCH_2PPh_2)_2(CO)_2(\mu\text{-SiH}\textit{n}\text{-}C_6H_{13})_2$	31
C	$Rh_2(Ph_2AsCH_2AsPh_2)_2(H)_2(CO)_2$ + $2H_3SiPh$, 60°C, 24 h	$Rh_2(Ph_2AsCH_2AsPh_2)_2(CO)_2(\mu\text{-SiHPh})_2$	31
C	$Rh_2(Ph_2AsCH_2AsPh_2)_2(H)_2(CO)_2$ + $2H_3SiEt$, 60°C, 24 h	$Rh_2(Ph_2AsCH_2AsPh_2)_2(CO)_2(\mu\text{-SiHEt})_2$	31
C	$Rh_2(Ph_2AsCH_2AsPh_2)_2(H)_2(CO)_2$ + $2H_3Si\textit{n}\text{-}C_6H_{13}$, 60°C, 24 h	$Rh_2(Ph_2AsCH_2AsPh_2)_2(CO)_2(\mu\text{-SiH}\textit{n}\text{-}C_6H_{13})_2$	31
B	$Rh_2(Ph_2PCH_2PPh_2)_2(CO)_3$ + $2H_3SiPh$, RT, 1 h	$Rh_2(Ph_2PCH_2PPh_2)_2(CO)_2(\mu\text{-CO})(\mu\text{-SiHPh})$	31
B	$Rh_2(Ph_2PCH_2PPh_2)_2(CO)_3$ + $2H_3SiEt$, RT, 1 h	$Rh_2(Ph_2PCH_2PPh_2)_2(CO)_2(\mu\text{-CO})(\mu\text{-SiHEt})$	31
B	$Rh_2(Ph_2PCH_2PPh_2)_2(CO)_3$ + $2H_3Si\textit{n}\text{-}C_6H_{13}$, RT, 1 h	$Rh_2(Ph_2PCH_2PPh_2)_2(CO)_2(\mu\text{-CO})(\mu\text{-SiH}\textit{n}\text{-}C_6H_{13})$	31
C	$\{(i\text{-Pr}_2PCH_2CH_2P\textit{i}\text{-Pr}_2)Rh\}_2(\mu\text{-H})_2$ + H_2SiPh_2 (≥2 equiv), RT, or $\{(i\text{-Pr}_2PCH_2CH_2P\textit{i}\text{-Pr}_2)Rh\}_2(\mu\text{-H})(\mu\text{-}\eta^2\text{-HSiPh}_2)$ + H_2SiPh_2, RT	$\{(i\text{-Pr}_2PCH_2CH_2P\textit{i}\text{-Pr}_2)Rh\}_2(\mu\text{-SiPh}_2)_2$	40,41
C	$\{(i\text{-Pr}_2PCH_2CH_2P\textit{i}\text{-Pr}_2)Rh\}_2(\mu\text{-H})_2$ + $H_3Si\textit{n}\text{-Bu}$ (2 equiv), RT	$\{(i\text{-Pr}_2PCH_2CH_2P\textit{i}\text{-Pr}_2)Rh\}_2(\mu\text{-SiH}\textit{n}\text{-Bu})_2$ (mixture of two isomers)	42
C	$\{(i\text{-Pr}_2PCH_2CH_2P\textit{i}\text{-Pr}_2)Rh\}_2(\mu\text{-H})_2$ + $H_3Si\textit{p}\text{-Tol}$ (2 equiv), RT	$\{(i\text{-Pr}_2PCH_2CH_2P\textit{i}\text{-Pr}_2)Rh\}_2(\mu\text{-SiH}\textit{p}\text{-Tol})_2$ (mixture of two isomers)	42
J	$\{(i\text{-Pr}_2PCH_2CH_2P\textit{i}\text{-Pr}_2)Rh\}_2(\mu\text{-H})(\mu\text{-}\eta^2\text{-HSiPh}_2)$ + CO (1 equiv)	$\{(i\text{-Pr}_2PCH_2CH_2P\textit{i}\text{-Pr}_2)Rh\}_2(\mu\text{-CO})(\mu\text{-SiPh}_2)$	41
J	$\{(i\text{-Pr}_2PCH_2CH_2P\textit{i}\text{-Pr}_2)Rh\}_2(\mu\text{-H})(\mu\text{-}\eta^2\text{-HSiMe}_2)$ + CO (1 equiv)	$\{(i\text{-Pr}_2PCH_2CH_2P\textit{i}\text{-Pr}_2)Rh\}_2(\mu\text{-CO})(\mu\text{-SiMe}_2)$	41
J	$\{(i\text{-Pr}_2PCH_2CH_2P\textit{i}\text{-Pr}_2)Rh\}_2(\mu\text{-H})(\mu\text{-}\eta^2\text{-HSiMePh})$ + CO (1 equiv)	$\{(i\text{-Pr}_2PCH_2CH_2P\textit{i}\text{-Pr}_2)Rh\}_2(\mu\text{-CO})(\mu\text{-SiMePh})$	41
B	$Ir_2(Ph_2PCH_2PPh_2)_2(CO)_3$ + $2H_2SiMe_2$, RT, 10 min	$Ir_2(Ph_2PCH_2PPh_2)_2(H)_2(CO)_2(\mu\text{-SiMe}_2)$	32
B	$Ir_2(Ph_2PCH_2PPh_2)_2(CO)_3$ + $2H_2SiEt_2$, RT, 30 min	$Ir_2(Ph_2PCH_2PPh_2)_2(H)_2(CO)_2(\mu\text{-SiEt}_2)$	32
B	$Ir_2(Ph_2PCH_2PPh_2)_2(CO)_3$ + $2H_2SiPh_2$, RT, 2 h	$Ir_2(Ph_2PCH_2PPh_2)_2(H)_2(CO)_2(\mu\text{-SiPh}_2)$	32
B	$Ir_2(Ph_2PCH_2PPh_2)_2(CO)_3$ + $2H_2SiPh_2$, RT, 1 h	$Ir_2(Ph_2PCH_2PPh_2)_2(H)_2(CO)_2(\mu\text{-SiHPh})$	32

[a] Including the complexes that formally have metal–metal bonds on the basis of valence electron count.
[b] A silylyne complex.

TABLE III

FORMATION OF BRIDGED SILYLENE COMPLEXES WITH THE HYDROGEN ATOM(S) BRIDGING THE SILICON–METAL BOND(S) (**C**-TYPE COMPLEXES)

Reaction type	Reaction	Product complex	Ref.
D	$Cp_2TiMe_2 + H_3SiPh$ (excess), RT 24 h	$\{Cp_2Ti(\mu\text{-}\eta^2\text{-}HSiHPh)\}_2$	49
D	$Cp_2TiMe_2 + H_3SiPh$ (1 equiv), RT, 2 h	$Cp_2Ti(\mu\text{-}H)(\mu\text{-}\eta^2\text{-}HSiHPh)TiCp_2$	49
C	$Mn_2(\mu\text{-}H)_2(CO)_6(\mu\text{-}Ph_2PCH_2PPh_2) + H_2SiPh_2$, 65°C	$Mn_2(\mu\text{-}\eta^3\text{-}H_2SiPh_2)(CO)_6(\mu\text{-}Ph_2PCH_2PPh_2)$	37
B	$Re_2(CO)_{10} + H_2SiPh_2$, $h\nu$	$\{Re(CO)_4(H)\}_2(\mu\text{-}SiPh_2)$	33
B	$Fe_2(CO)_9 + H_2SiPh_2$, RT, 20 h	$Fe_2(CO)_7(\mu\text{-}\eta^2\text{-}HSiPh_2)(SiPh_2H)$	19
B	$Fe_2(CO)_7(\mu\text{-}\eta^2\text{-}HSiPh_2)(SiPh_2H)$, $h\nu$	$Fe_2(CO)_6(\mu\text{-}\eta^2\text{-}HSiPh_2)_2$	19
C	$Cp^*_2Ru_2(\mu\text{-}H)_4 + Et_2SiH_2$, −78°C–RT	$\{Cp^*Ru(\mu\text{-}\eta^2\text{-}HSiEt_2)\}_2(\mu\text{-}H)(H)$	38
C	$Cp^*_2Ru_2(\mu\text{-}H)_4 + Ph_2SiH_2$, −78°C–RT	$\{Cp^*Ru(\mu\text{-}\eta^2\text{-}HSiPh_2)\}_2(\mu\text{-}H)(H)$	38
K	$Cp^*_2Ru_2(\mu\text{-}H)_2(\mu\text{-}SiPh_2CH{=}CHSiPh_2) + H_2$ (5 atm), 80°C	$Cp^*_2Ru_2(\mu\text{-}H)(H)(\mu\text{-}SiPh_2CH{=}CH_2)(\mu\text{-}\eta^2\text{-}HSiPh_2)$	66
D	$[Cp^*Ru(\mu\text{-}OMe)]_2 + H_2SiCyMe$, RT, 30 min	$\{Cp^*Ru(\mu\text{-}\eta^2\text{-}HSiCyMe)\}_2(\mu\text{-}H)(H)$	50
C	$\{(i\text{-}Pr_2PCH_2CH_2Pi\text{-}Pr_2)Rh\}_2(\mu\text{-}H)_2 + H_2SiPh_2$, RT	$\{(i\text{-}Pr_2PCH_2CH_2Pi\text{-}Pr_2)Rh\}_2(\mu\text{-}H)(\mu\text{-}\eta^2\text{-}HSiPh_2)$	40,41
C	$\{(i\text{-}Pr_2PCH_2CH_2Pi\text{-}Pr_2)Rh\}_2(\mu\text{-}H)_2 + H_2SiMe_2$, RT	$\{(i\text{-}Pr_2PCH_2CH_2Pi\text{-}Pr_2)Rh\}_2(\mu\text{-}H)(\mu\text{-}\eta^2\text{-}HSiMe_2)$	41
C	$\{(i\text{-}Pr_2PCH_2CH_2Pi\text{-}Pr_2)Rh\}_2(\mu\text{-}H)_2 + H_2SiMePh$, RT	$\{(i\text{-}Pr_2PCH_2CH_2Pi\text{-}Pr_2)Rh\}_2(\mu\text{-}H)(\mu\text{-}\eta^2\text{-}HSiMePh)$	41
I	$\{(i\text{-}Pr_2PCH_2CH_2Pi\text{-}Pr_2)Rh\}_2(\mu\text{-}SiHn\text{-}Bu)_2 + 2H_2$, RT	$\{(i\text{-}Pr_2PCH_2CH_2Pi\text{-}Pr_2)Rh(H)\}_2(\mu\text{-}\eta^2\text{-}HSiHn\text{-}Bu)_2$ (mixture of two isomers)	42
I	$\{(i\text{-}Pr_2PCH_2CH_2Pi\text{-}Pr_2)Rh\}_2(\mu\text{-}Sip\text{-}Tol)_2 + 2H_2$, RT	$\{(i\text{-}Pr_2PCH_2CH_2Pi\text{-}Pr_2)Rh(H)\}_2(\mu\text{-}\eta^2\text{-}HSiHp\text{-}Tol)_2$ (mixture of two isomers)	42
C	$\{(i\text{-}Pr_2PCH_2CH_2Pi\text{-}Pr_2)Rh\}_2(\mu\text{-}H)_2 +$ 4–5 equiv $H_3Sin\text{-}Bu$, RT	$\{(i\text{-}Pr_2PCH_2CH_2Pi\text{-}Pr_2)Rh\}_2(\mu\text{-}\eta^2\text{-}HSiHn\text{-}Bu)_2(\mu\text{-}SiHn\text{-}Bu)$	42
C	$\{(i\text{-}Pr_2PCH_2CH_2Pi\text{-}Pr_2)Rh\}_2(\mu\text{-}H)_2 +$ 4–5 equiv $H_3Sip\text{-}Tol$, RT	$\{(i\text{-}Pr_2PCH_2CH_2Pi\text{-}Pr_2)Rh\}_2(\mu\text{-}\eta^2\text{-}HSiHp\text{-}Tol)_2(\mu\text{-}SiHp\text{-}Tol)$	42
D, K	$Pt(C_2H_4)_2(PCy_3) + H_2SiMe_2$, RT or $\{Pt(PCy_3)(\mu\text{-}H)(SiMe_2Ph)\}_2$, hexane reflux, 0.5 h	$\{Pt(PCy_3)(\mu\text{-}\eta^2\text{-}HSiMe_2)\}_2$	44
D	$Pt(C_2H_4)_2(PMet\text{-}Bu_2) + H_2SiMe_2$, RT	$\{Pt(PMet\text{-}Bu_2)(\mu\text{-}\eta^2\text{-}HSiMe_2)\}_2$	44
D	$Pt(C_2H_4)_2(Pi\text{-}Pr_2Ph) + H_2SiMe_2$, RT	$\{Pt(Pi\text{-}Pr_2Ph)(\mu\text{-}\eta^2\text{-}HSiMe_2)\}_2$	44
D	$Pt(C_2H_4)_2(PPh_3) + H_2SiMe_2$, RT	$\{Pt(PPh_3)(\mu\text{-}\eta^2\text{-}HSiMe_2)\}_2$	44
D	$Pt(C_2H_4)_2(PCy_3) + H_2SiPh_2$, RT	$\{Pt(PCy_3)(\mu\text{-}\eta^2\text{-}HSiPh_2)\}_2$	44
D	$Pt(C_2H_4)_2(PMet\text{-}Bu_2) + H_2SiPh_2$, RT	$\{Pt(PMet\text{-}Bu_2)(\mu\text{-}\eta^2\text{-}HSiPh_2)\}_2$	44
D	$Pt(C_2H_4)_2(Pi\text{-}Pr_2Ph) + H_2SiPh_2$, RT	$\{Pt(Pi\text{-}Pr_2Ph)(\mu\text{-}\eta^2\text{-}HSiPh_2)\}_2$	44
D	$Pt(C_2H_4)_2(PPh_3) + H_2SiPh_2$, RT	$\{Pt(PPh_3)(\mu\text{-}\eta^2\text{-}HSiPh_2)\}_2$	44

TABLE IV

FORMATION OF BRIDGED GERMYLENE AND GERMYLYNE COMPLEXES WITHOUT METAL–METAL BONDS (**A**-TYPE COMPLEXES)

Reaction type	Reaction	Product complex	Ref.
J	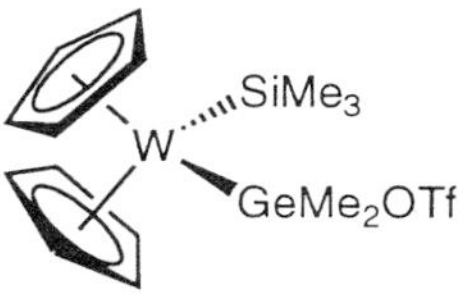 $Cp_2W(SiMe_3)(GeMe_2OTf)$, 80 °C	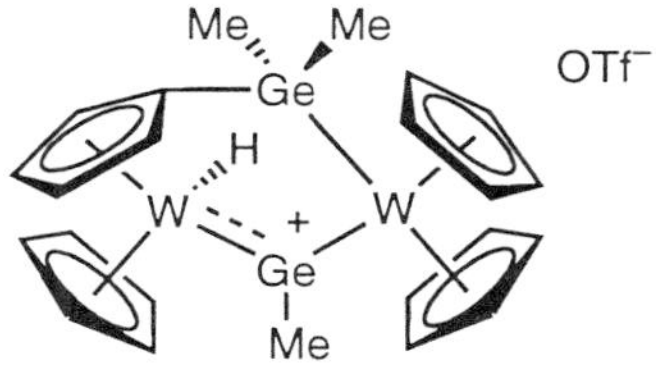 Me, Me, Ge, H, W, +, W, Ge, Me, OTf^-	99
J	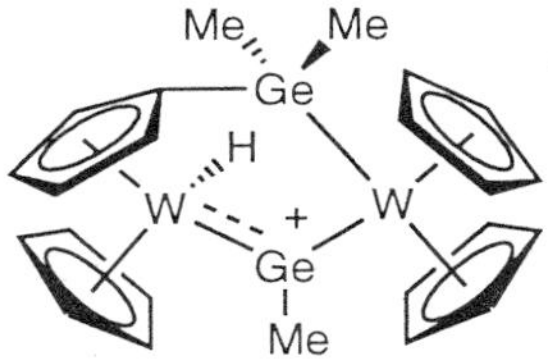 Me, Me, Ge, H, W, +, W, Ge, Me, OTf^- + MeMgBr	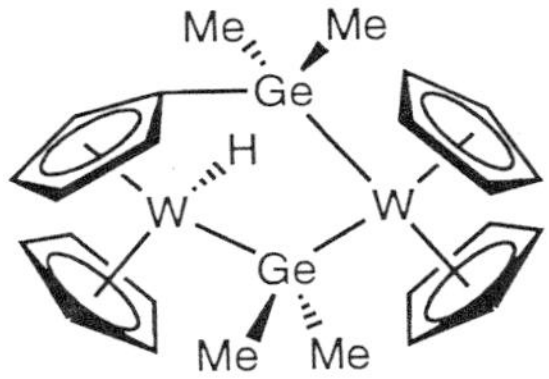 Me, Me, Ge, H, W, W, Ge, Me, Me	99
B	$Mn(CO)_5H + GeH_4$, RT, 8 days or $Mn_2(CO)_{10} + GeH_4$, 140°C	$Mn(CO)_5(\mu\text{-}GeH_2)Mn(CO)_5$	1
D	$Na[Mn(CO)_5] + Me_2GeCl_2$	$Mn(CO)_5(\mu\text{-}GeMe_2)Mn(CO)_5$	82
E	$Fe(CO)_4(GeH_3)(GeR_3)$, RT, R_3 = MeH_2, Me_2H, Me3	$Fe_2(CO)_8(\mu\text{-}GeH_2)_2$	87,88
E	$Fe(CO)_4(GeMeH_2)(GeR_3)$, RT, R_3 = MeH_2, Me_2H, Me_3	$Fe_2(CO)_8(\mu\text{-}GeMeH)_2$	85,88
E	$Fe(CO)_4(GeMe_2H)(GeR_3)$, RT, R_3 = Me_2H, Me_3	$Fe_2(CO)_8(\mu\text{-}GeMe_2)_2$	86,88
D	$Na[CpFe(CO)_2] + Me_2GeCl_2$	$CpFe(CO)_2(\mu\text{-}GeMe_2)Fe(CO)_2Cp$	82
D	R^1, R^2, R^3, R^4, Ge, Br, Br + $Na[ML_n]$	R^1, R^2, R^3, R^4, Ge, ML_n, ML_n	83
D	R^1, R^2, R^3, R^4, Ge, Br, Br + $Na[ML_n]$ + $Na[M'L_n]$	R^1, R^2, R^3, R^4, Ge, $M'L_n$, ML_n	83

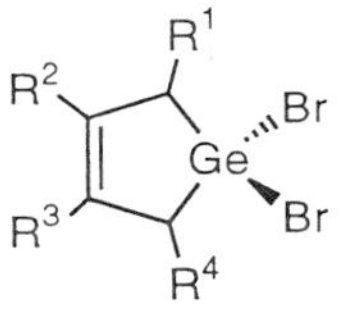

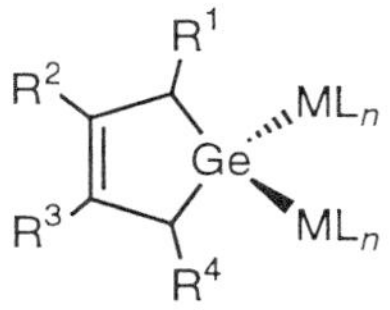

R^1	R^2	R^3	R^4	ML_n
H	Me	Me	H	$Mn(CO)_5$
H	Me	H	H	$Mn(CO)_5$
H	H	H	H	$Mn(CO)_5$
H	Me	Me	H	$Fe(CO)_2Cp$
H	Me	Me	H	$Co(CO)_4$

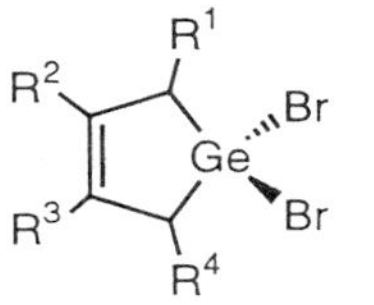

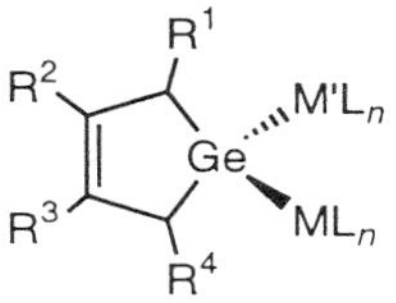

TABLE IV (*continued*)

Reaction type	Reaction					Product complex	Ref.
	R^1	R^2	R^3	R^4	ML_n	$M'L_n$	
	H	Me	Me	H	$Co(CO)_4$	$Mn(CO)_5$	
	H	Me	Me	H	$Co(CO)_4$	$Fe(CO)_2Cp$	
	H	H	H	H	$Co(CO)_4$	$Mn(CO)_5$	
D	$Na_2[Os(CO)_4] + Me_2GeCl_2$, −40°C–RT					$Os_2(CO)_8(\mu\text{-}GeMe_2)_2$	84
D	$Na[Co(CO)_4] + Me_2GeCl_2$, RT, 10 min					$Co(CO)_4(\mu\text{-}GeMe_2)Co(CO)_4$	80
D	$Na[Co(CO)_4] + MeGeI_3$, RT, 10 min					$Co(CO)_4(\mu\text{-}GeMeI)Co(CO)_4$	80
C	$3Co_2(CO)_8 + 2GeCl_4$, RT, 10 min					$Co(CO)_4(\mu\text{-}GeCl_2)Co(CO)_4$	80

II
FORMATION OF BRIDGED SILYLENE COMPLEXES

A. *General Considerations*

Most of bridged silylene complexes consist of two metal fragments and one bridging silylene unit. They can be classified into the complexes without a metal–metal bond **A** and those with a metal–metal bond **B**. The physical properties and perhaps the chemical properties of them are fairly different owing to the difference of the bonding mode between the μ-silylene unit and metals in these two types of complexes. The bonding between the metals and silicon in **A**-type complexes, especially when there is no other bridging ligand, are closely related to that in the usual mononuclear silyl complexes, so that they appear to behave as metal-substituted silyl complexes. In contrast, in **B**-type complexes, the silylene bridge(s), two metals, and other bridging ligand(s) (if any) construct a multicenter bond, so that the properties of these complexes are unique. In most cases, **A**-type complexes can be used as precursors of **B**-type complexes.

Another important class of bridged silylene complexes that should be handled separately is complexes that have hydrogen atoms bridging the metal–silylene bonds **C**. This bonding can be regarded as a metal–hydrogen–silicon 3*c*–2*e* bond, or the η^2-coordination of Si–H σ-bonds to metals that has been well established for mononuclear complexes.[10,11] These **C**-type complexes are now considered to play important roles in the formation of **A**- or **B**-type complexes.

The reaction that is most extensively employed for the preparation of bridged silylene complexes is undoubtedly the oxidative addition of di-,

TABLE V

FORMATION OF BRIDGED GERMYLENE COMPLEXES WITH METAL–METAL BONDS (**B**-TYPE COMPLEXES)

Reaction type	Reaction	Product complex	Ref.
J	$W(CO)_5{=}Ge(SMes)_2 + BCl_3$	$W_2(CO)_{10}(\mu\text{-}GeCl_2)$	96
J	$W(CO)_5{=}Ge(SMes)_2 + BBr_3$	$W_2(CO)_{10}(\mu\text{-}GeBr_2)$	96
H	$Mn(CO)_5(\mu\text{-}GeMe_2)Mn(CO)_5$, *hν*, 1 h	$Mn_2(CO)_8(\mu\text{-}CO)(\mu\text{-}GeMe_2)$	82
C	$Mn(CO)_5GeMe_2Cl$, *hν*, 4.5 h	$Mn_2(CO)_8(\mu\text{-}GeMe_2)_2$	81,82
B	$Fe_3(CO)_{12} + H_2GeMe_2$, 65°C, 5h	$Fe_2(CO)_6(\mu\text{-}GeMe_2)_3$	70,71
H	$Fe_2(CO)_8(\mu\text{-}GeMe_2)_2$, *hν*, 3 days	$Fe_2(CO)_6(\mu\text{-}CO)(\mu\text{-}GeMe_2)_2$	86
C, H	$CpFe(CO)_2GeMe_2Cl$, *hν*, 4 h or $CpFe(CO)_2(\mu\text{-}GeMe_2)Fe(CO)_2Cp$, *hν*, 20 min	$Cp_2Fe_2(CO)_2(\mu\text{-}CO)(\mu\text{-}GeMe_2)$ (mixture of two isomers)	81,82
H	$CpFe(CO)_2(\mu\text{-}GePh_2)Fe(CO)_2Cp$, *hν*, 45 min	$Cp_2Fe_2(CO)_2(\mu\text{-}CO)(\mu\text{-}GePh_2)$ (mixture of two isomers)	92
G	$CpFe(CO)_2SiMe_2GeMe_2Fe(CO)_2Cp$, *hν*, 2 h	$Cp_2Fe_2(CO)_2(\mu\text{-}CO)(\mu\text{-}GeMeSiMe_3)$ (mixture of three isomers)	91
G	$CpFe(CO)_2SiMe_2GeMe_2Fe(CO)_2Cp$, *hν*, 40 h	$Cp_2Fe_2(CO)_2(\mu\text{-}SiMe_2)(\mu\text{-}GeMe_2)$ (mixture of two isomers)	91
B	$CpFe(CO)_2SiMe_3 + H_3Ge\textit{t}\text{-}Bu$, *hν*, 4 h	$Cp_2Fe_2(CO)_2(\mu\text{-}CO)(\mu\text{-}Ge\textit{t}\text{-}BuH)$ $Cp_2Fe_2(CO)_2(\mu\text{-}Ge\textit{t}\text{-}BuH)_2$	78
B	$CpFe(CO)_2SiMe_3 + H_2Ge\textit{p}\text{-}Tol_2$, *hν*, 100 min	$Cp_2Fe_2(CO)_2(\mu\text{-}CO)(\mu\text{-}Ge\textit{p}\text{-}Tol_2)$ $Cp_2Fe_2(CO)_2(\mu\text{-}Ge\textit{p}\text{-}Tol_2)_2$	79
B	$Cp^*Fe(CO)_2SiMe_3 + H_3Ge\textit{p}\text{-}Tol$, *hν*, 100 min	$Cp^*_2Fe_2(CO)_2(\mu\text{-}CO)(\mu\text{-}Ge\textit{p}\text{-}TolH)$ $Cp^*_2Fe_2(CO)_2(\mu\text{-}Ge\textit{p}\text{-}TolH)_2$ $Cp^*_2Fe_2(CO)_2(\mu\text{-}\eta^2\text{-}HGe\textit{p}\text{-}TolH)_2$	79
B	$Cp^*Fe(CO)_2SiMe_3 + H_3Ge\textit{t}\text{-}Bu$, *hν*, 3 h	$Cp^*_2Fe_2(CO)_2(\mu\text{-}Ge\textit{t}\text{-}BuH)_2$ (major product) $Cp^*_2Fe_2(CO)_2(\mu\text{-}CO)(\mu\text{-}Ge\textit{t}\text{-}BuH)$	68
B	$Cp^*Fe(CO)_2GeH_2\textit{t}\text{-}Bu$, *hν*, 2°C, 15 min	$Cp^*_2Fe_2(CO)_2(\mu\text{-}\eta^2\text{-}HGe\textit{t}\text{-}BuH)_2$ $Cp^*_2Fe_2(CO)_2(\mu\text{-}Ge\textit{t}\text{-}BuH)_2$	68
I	$Cp_2Fe_2(CO)_2(\mu\text{-}CO)(\mu\text{-}Ge\textit{t}\text{-}BuH) + CH_2Cl_2$	$Cp_2Fe_2(CO)_2(\mu\text{-}CO)(\mu\text{-}Ge\textit{t}\text{-}BuCl)$ (mixture of two isomers)	78
I	$Cp_2Fe_2(CO)_2(\mu\text{-}CO)(\mu\text{-}Ge\textit{t}\text{-}BuH) + CH_3I$, RT	$Cp_2Fe_2(CO)_2(\mu\text{-}CO)(\mu\text{-}Ge\textit{t}\text{-}BuI)$ (mixture of two isomers)	95
I	$Cp_2Fe_2(CO)_2(\mu\text{-}CO)(\mu\text{-}Ge\textit{t}\text{-}BuI)$ + DMAP + AgOTf, RT	$[Cp_2Fe_2(CO)_2(\mu\text{-}CO)(\mu\text{-}Ge\textit{t}\text{-}Bu\cdot DMAP)]OTf$[a]	95
H	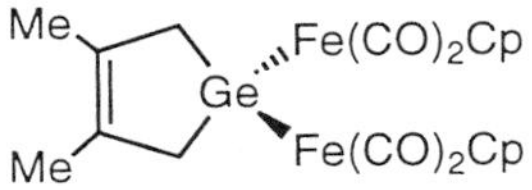 (3,4-dimethylgermacyclopent-3-ene)Ge[Fe(CO)$_2$Cp]$_2$ (Me, Me, Ge, Fe(CO)$_2$Cp, Fe(CO)$_2$Cp), hν, 45 min	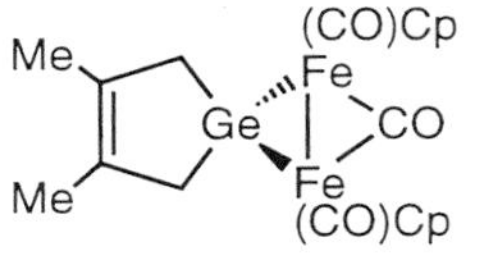 (3,4-dimethylgermacyclopent-3-ene)Ge bridging Fe(CO)Cp–Fe(CO)Cp with μ-CO (Me, Me, Ge, (CO)Cp, Fe, CO, Fe, (CO)Cp) (mixture of two isomers)	83
F	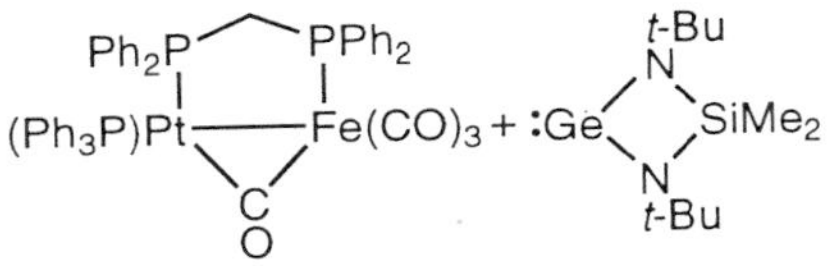 $Ph_2P\text{–}CH_2\text{–}PPh_2$ bridged $(Ph_3P)Pt(\mu\text{-}CO)Fe(CO)_3$ + :Ge(N*t*-Bu)$_2$SiMe$_2$, r.t.	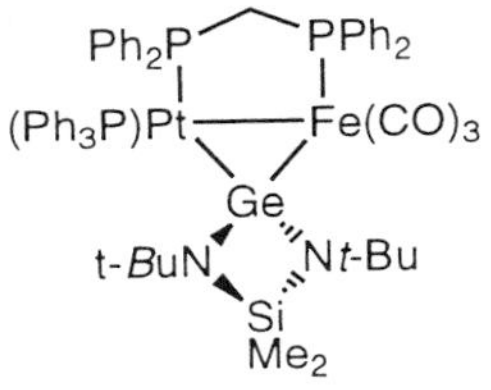 $Ph_2P\text{–}CH_2\text{–}PPh_2$ bridged $(Ph_3P)Pt(\mu\text{-}Ge)Fe(CO)_3$, Ge(N*t*-Bu)$_2$SiMe$_2$ (t-BuN, N*t*-Bu, Si, Me$_2$)	90

TABLE V (*continued*)

FORMATION OF BRIDGED GERMYLENE COMPLEXES WITH METAL–METAL BONDS (**B**-TYPE COMPLEXES)

Reaction type	Reaction	Product complex	Ref.
F	$(Ph_3P)Pt(\mu\text{-}CO)(\mu\text{-}Ph_2PN(H)PPh_2)Fe(CO)_3$ + $:Ge(N\text{-}t\text{-}Bu)_2SiMe_2$, r.t.	$(Ph_3P)Pt(\mu\text{-}Ph_2PN(H)PPh_2)(\mu\text{-}Ge\{N(t\text{-}Bu)\}_2SiMe_2)Fe(CO)_3$	90
B	$Ru_3(CO)_{12}$ + $HGeMe_3$, 100°C, 96 h	$\{Ru(CO)_3(GeMe_3)\}_2(\mu\text{-}GeMe_2)_2$ [major product: $Ru(CO)_4(GeMe_3)_2$]	72
B	$Os_3(CO)_{12}$ + $HGeMe_3$, 150°C, 260 h	$\{Os(CO)_3(GeMe_3)\}_2(\mu\text{-}GeMe_2)_2$ [major product: $Os(CO)_4(GeMe_3)_2$]	72
H	$Ru(CO)_4(GeMe_3)_2$, 140°C, 16 days	$Ru_3(CO)_9(\mu\text{-}GeMe_2)_3$ $Ru_2(CO)_6(\mu\text{-}GeMe_2)_3$ (trace amount)	72
H	$Os(CO)_4(GeMe_3)_2$, 160°C, 20 days	$Os_3(CO)_9(\mu\text{-}GeMe_2)_3$ $Os_2(CO)_6(\mu\text{-}GeMe_2)_3$ (trace amount)	72
H	$Os_2(CO)_8(\mu\text{-}GeMe_2)_2$, 100°C, 4 days or $h\nu$, 16.5 h	$Os_3(CO)_9(\mu\text{-}GeMe_2)_3$ (major product) $Os_2(CO)_6(\mu\text{-}GeMe_2)_3$ $Os_3(CO)_{11}(\mu\text{-}GeMe_2)_2$ $Os_4(CO)_{12}(\mu\text{-}GeMe_2)_4$	84
B	$Co_2(CO)_8$ + H_3GeMe, 10–20°C	$Co_2(CO)_6(\mu\text{-}CO)\{\mu\text{-}GeMeCo(CO)_4\}$	76
B	$Co_2(CO)_8$ + $2H_3GeMe$, 20°C	$Co_2(CO)_6\{\mu\text{-}GeMeCo(CO)_4\}_2$	76
B	$Co_2(CO)_8$ + H_2GePh_2, RT	$Co_2(CO)_6(\mu\text{-}CO)(\mu\text{-}GePh_2)$	75
B	$Co_4(\mu_4\text{-}SiMe)_2(CO)_{11}$ + H_2GeMe_2, 30°C, 7 days	$Co_4(\mu_4\text{-}SiMe)_2(CO)_{10}(\mu\text{-}GeMe_2)$	77
H	$Co(CO)_4(\mu\text{-}GeMe_2)Co(CO)_4$, $h\nu$, 22 h	$Co_2(CO)_6(\mu\text{-}CO)(\mu\text{-}GeMe_2)$	82
C	$Co(CO)_4GeMe_2Cl$, $h\nu$, 18 h	$Co_2(CO)_6(\mu\text{-}GeMe_2)_2$	81,82
H	(3,4-Me₂-germacyclopent-3-ene)$Ge\{Co(CO)_4\}_2$, hν, 15 min or 70 °C, 6 h	(3,4-Me₂-germacyclopent-3-ene)$Ge(\mu)Co_2(CO)_6(\mu\text{-}CO)$ [Me, Me, Ge, $(CO)_3$Co–Co$(CO)_3$, CO]	83
F	$[Ir(\eta\text{-}C_8H_{14})_2(\mu\text{-}Cl)]_2$ + $:Ge\{N(SiMe_3)_2\}_2$, 20 °C	$[Ir_2H_2(\mu\text{-}Cl)_2(\mu\text{-}Ge(NR_2)N(R)SiMe_2CH_2)(CH_2SiMe_2N(R)Ge(NR_2))(Ge(NR_2)_2)]$ ($R = SiMe_3$)	89
J	$Pd[Ge\{N(SiMe_3)_2\}_2]_3$ + CO, RT	$Pd_3(CO)_3[\mu\text{-}Ge\{N(SiMe_3)_2\}_2]_3$	97
J	$Pt[Ge\{N(SiMe_3)_2\}_2]_3$ + CO, RT	$Pt_3(CO)_3[\mu\text{-}Ge\{N(SiMe_3)_2\}_2]_3$	9,97

[a] A germylyne complex.

tri-, or tetrahydrosilanes to coordinatively unsaturated metal species, which are usually generated by the liberation of CO, H_2, $CH_2{=}CH_2$, phosphine, and so forth, or the reductive cleavage of metal–halide bonds. Disilanes containing Si—H bonds are also commonly used as the precursors where not only the Si—H bond but also the Si–Si bond is cleaved for the generation of a μ-silylene unit. The reactions that are less frequently employed involve the oxidative addition of Si–X bonds (X $\neq$ H) to metal complexes, salt elimination reactions, rearrangement reactions, and dimerization of mononuclear silylene complexes. Transformation of one bridged silylene complex to another is also a useful reaction for the synthesis of new bridged silylene complexes, including the elimination of a ligand from **A**-type complexes to give **B**-type complexes or the reverse reaction, and the modification of substituents or ligands on the complexes.

The bridged silylene complexes of types **A**, **B**, and **C**, along with their formation reactions and reaction types (B–K), are listed in Tables I, II, and III, respectively.

B. *Reactions of Hydrosilanes with Carbonyl Complexes*

Under UV irradiation or heating, carbonyl complexes generate coordinatively unsaturated species. In the presence of hydrosilanes having two or more Si—H bonds, these species may insert into the Si—H bonds to give bridged silylene complexes. A typical reaction of this type is the photolysis of $Fe(CO)_5$ in the presence of diphenylsilane [Eq. (1)].[17,18] In this reaction, oxidative addition of the silane at the Si—H bond to photochemically generated $Fe(CO)_4$ is a key step. Use of the less bulky silane $PhMeSiH_2$ allows the formation of triply bridged product $\{Fe(CO)_3\}_2(\mu\text{-CO})(\mu\text{-}SiPhMe)_2$ by the loss of one more CO [Eq. (2)].

$$Fe(CO)_5 + Ph_2SiH_2 \xrightarrow[\text{hexane}]{h\nu} \frac{1}{2}\,(OC)_4Fe(\mu\text{-}SiPh_2)_2Fe(CO)_4 \qquad (1)$$

$$Fe(CO)_5 + PhMeSiH_2 \xrightarrow[\text{hexane}]{h\nu} \frac{1}{2}\,(OC)_3Fe(\mu\text{-CO})(\mu\text{-}SiPhMe)_2Fe(CO)_3 \qquad (2)$$

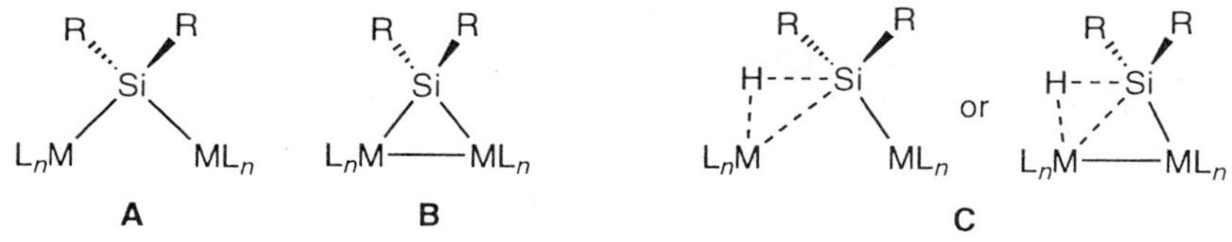

A B C

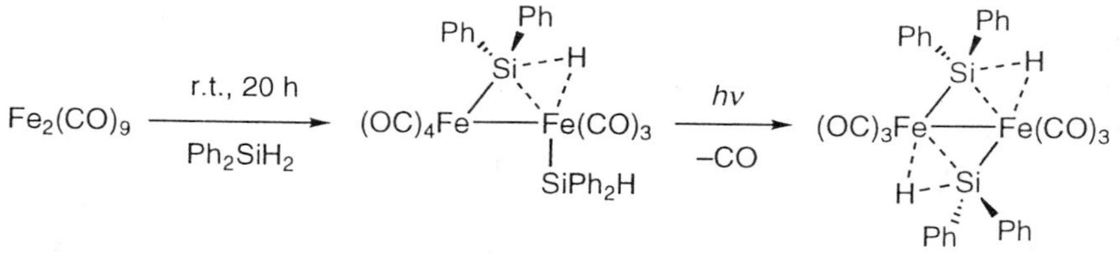

SCHEME 1

Thermolysis of $Fe_2(CO)_9$ or $Fe_3(CO)_{12}$ in the presence of silanes also gives bridged silylene complexes (Schemes 1 and 2).[19,20] Interestingly, the product of the thermal reaction of Ph_2SiH_2 with $Fe_2(CO)_9$, which contains an Fe–H–Si $3c$–$2e$ bond, is different from that of the photochemical reaction in Eq. (1). Furthermore, the photolysis of the thermal reaction product simply promotes the loss of a CO ligand followed by intramolecular coordination of a Si–H σ-bond to give the complex with two Fe–H–Si $3c$–$2e$ bonds (Scheme 1).[19] At present the origin of the difference is not clear. The structure of the product depends on the substituents of silanes, and very bulky substituents stabilize the metal–metal bonded species that is only bridged by a single silylene ligand (Scheme 2),[20] whereas in the case

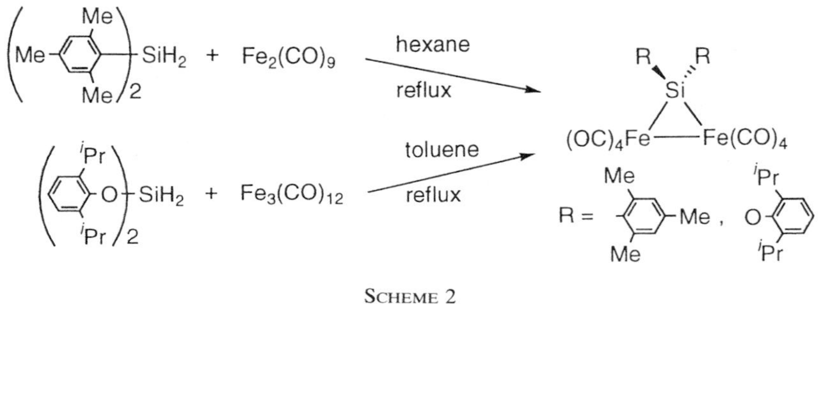

SCHEME 2

of $PhSiH_3$, the product is a triply bridged complex $\{Fe(CO)_3\}_2(\mu\text{-}CO)(\mu\text{-}SiPhH)_2$.[19] These results suggest that the steric bulkiness of the substituents on μ-silylene ligands is an important factor in determining the products.

Photoreactions of cyclopentadienyl complexes of the type $Cp'Fe(CO)_2R$ [Cp' = Cp (η^5-C_5H_5), Cp* (η^5-C_5Me_5); R = $SiMe_3$, Me] in the presence of polyhydrosilanes provide a useful tool for the synthesis of diiron complexes containing a μ-silylene and a μ-carbonyl ligand. Thus, photolysis of $CpFe(CO)_2SiMe_3$ in the presence of *tert*-alkyltrihydrosilane gives a silylene-bridged diiron complex [Eq. (3)].[21,22] An analogous reaction has been also observed in the photolysis of $CpFe(CO)_2Me$ in the presence of $MesSiH_3$ (Mes = 2,4,6-trimethylphenyl).[23] When $CpFe(CO)_2Me$ is photolyzed in the presence of a dihydrosilane with bulky aryl groups, a diiron complex bridged by a silylene ligand and two carbonyl ligands is formed [Eq. (4)].[24] This violet complex is the only known example of a bridged silylene complex having a triplet ground state. Photolysis of a complex with the bulky Cp* ligand, $Cp^*Fe(CO)_2SiMe_3$, in the presence of t-$BuSiH_3$ only gives a mononuclear complex $CpFe(CO)_2SiH_2t$-Bu,[22] whereas the photolysis in the presence of less bulky p-$TolSiH_3$ results in the formation of a *cis–trans* mixture of bridged silylene complexes [Eq. (5)].[25] In general, it has been observed that the photolysis of $CpFe(CO)_2Me$ tends to give a larger amount of by-product $Cp_2Fe_2(CO)_4$ than does $CpFe(CO)_2SiMe_3$, but the Me complex is a better precursor for the synthesis of more sterically congested complexes.

$$2\ CpFe(CO)_2SiMe_3 + RSiH_3 \xrightarrow[-CO,\ -2Me_3SiH]{h\nu} Cp(CO)Fe(\mu\text{-}CO)(\mu\text{-}SiHR)Fe(CO)Cp \quad (3)$$

(R = ^{t}Bu, $(CMe_2)_2H$) *cis(H)*

$$2\ CpFe(CO)_2Me + R_2SiH_2 \xrightarrow[-2MeH,\ -2CO]{h\nu} CpFe(\mu\text{-}SiR_2)(\mu\text{-}CO)_2FeCp \quad (4)$$

R = 2,4,6-^{i}Pr$_3C_6H_2$, 2,6-$Et_2C_6H_3$, 2,4,6-$Me_3C_6H_2$

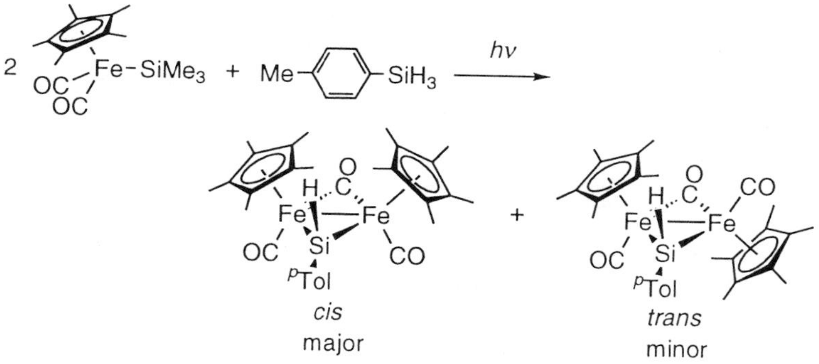

(5)

A proposed mechanism for these reactions is summarized in Scheme 3.[21,22,24] Photochemical CO dissociation and oxidative addition of hydrosilane are the key steps. By controlling the substituents on the silicon and the reaction conditions, complexes corresponding to the intermediates **D**

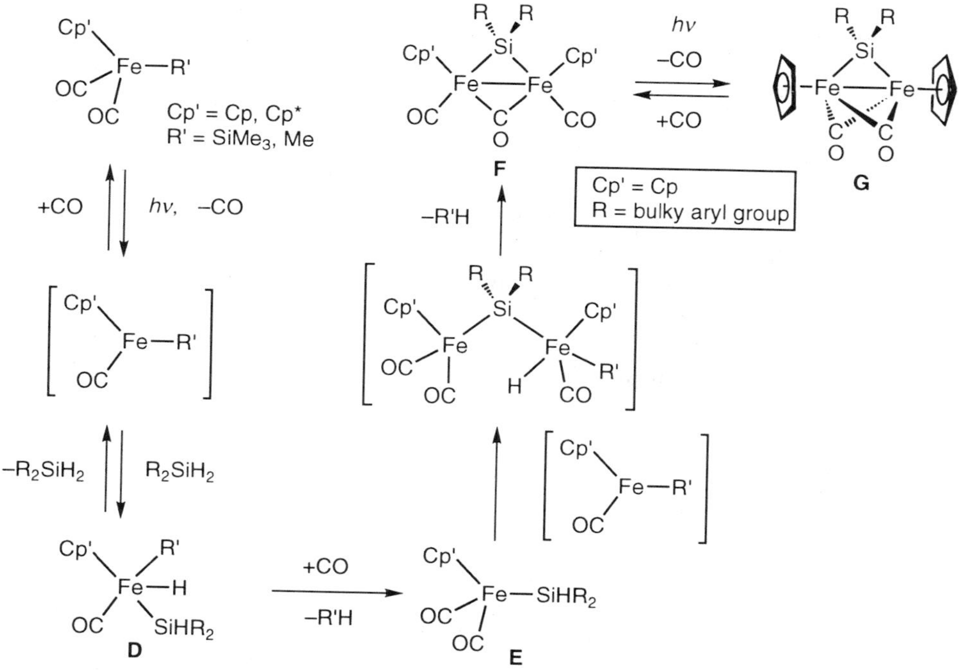

SCHEME 3

and **E** have been isolated.[22] When the silane is a trihydrosilane, a complex with three CO ligands **F** forms. However, when the silane is a dihydrosilane with bulky aryl groups, a triplet complex with two bridging CO ligands **G** forms directly. When R is mesityl or 2,6-diethylphenyl, complex **G** slowly reacts with CO in toluene at room temperature to give *trans*-**F**, whereas irradiation of this *trans*-**F** induces a rapid loss of CO to give **G** quantitatively.[24]

Reaction of a phosphido-bridged unsymmetrical diiron complex $Cp^*Fe_2(CO)_4(\mu\text{-}CO)(\mu\text{-}PPh_2)$ with diarylsilane under gradual elevation of temperature gives a monoarylsilylene-bridged complex and a triarylsilane accompanied by the redistribution of substituents on the silane [Eq. (6)].[26] A proposed mechanism (Scheme 4) involves the initial cleavage of the iron–iron and iron–μ-carbonyl bonds followed by the oxidative addition of two diarylsilanes. Subsequent elimination of H_2 generates a silyl(silylene) complex **H**, on which a substituent on the silyl group migrates to the silylene ligand, and final reductive elimination of triaryl silane and conversion of the terminal silylene ligand to the bridging one gives the products. The 1,3-migration of substituents on silyl(silylene) complexes has been observed in a number of mononuclear systems.[15,27–29]

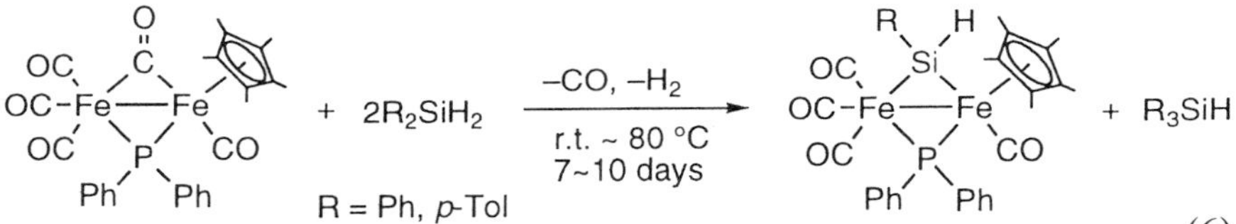

OC, OC–Fe–Fe(CO), OC, C=O, P, Ph, Ph + $2R_2SiH_2$ —(–CO, –H_2; r.t. ~ 80 °C; 7~10 days)→ R, H, Si, OC, OC–Fe–Fe, OC, CO, P, Ph, Ph + R_3SiH

R = Ph, *p*-Tol

(6)

In the reactions of polyhydrosilanes with Group 9 metal complexes, although the mechanism is less clear, the formation of bridged silylene complexes has been reported in the reactions with zerovalent cobalt,[30] rhodium,[31] and iridium[32] carbonyl complexes [Eqs. (7)–(9)]. Interestingly, in the case of rhodium complexes elimination of H_2 occurs, whereas in the case of iridium complexes one of CO ligands dissociates. However, it has been confirmed by ^1H and ^{31}P NMR and IR spectroscopy that in the Rh case the first product is a dihydrido complex analogous to the product in the iridium reaction, but this product easily eliminates H_2 and binds CO to give the final product.

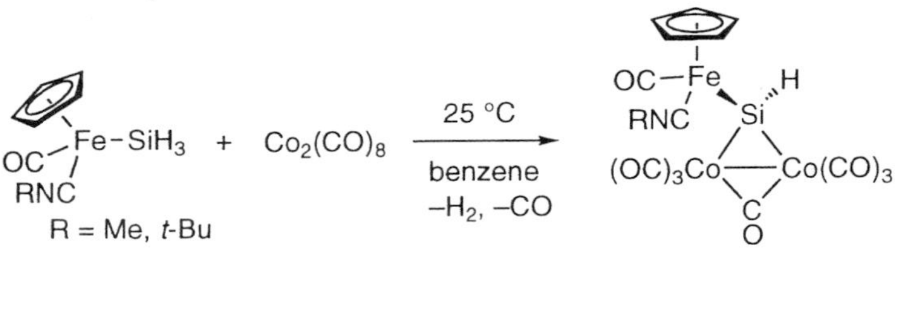

Fe–SiH_3, OC, RNC + $Co_2(CO)_8$ —(25 °C; benzene; –H_2, –CO)→ OC–Fe, RNC, Si, H, $(OC)_3Co$–$Co(CO)_3$, C, O

R = Me, *t*-Bu

(7)

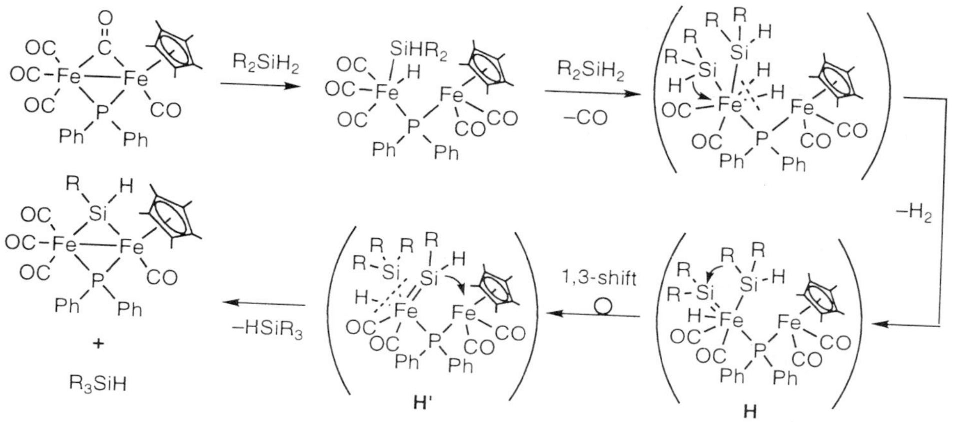

SCHEME 4

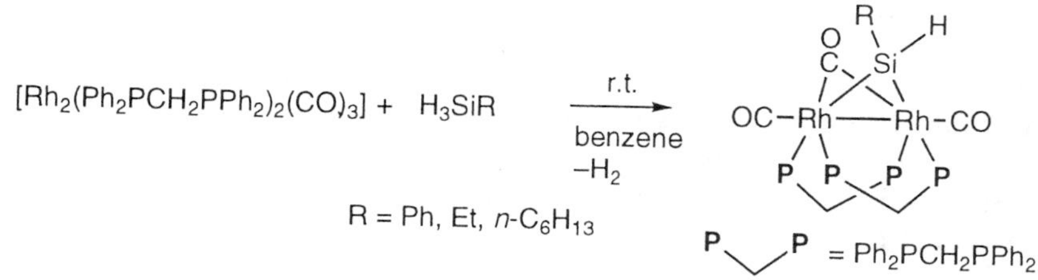

$$[Rh_2(Ph_2PCH_2PPh_2)_2(CO)_3] + H_3SiR \xrightarrow[\text{benzene}, -H_2]{\text{r.t.}} \quad (8)$$

R = Ph, Et, n-C_6H_{13}

P⌒P = $Ph_2PCH_2PPh_2$

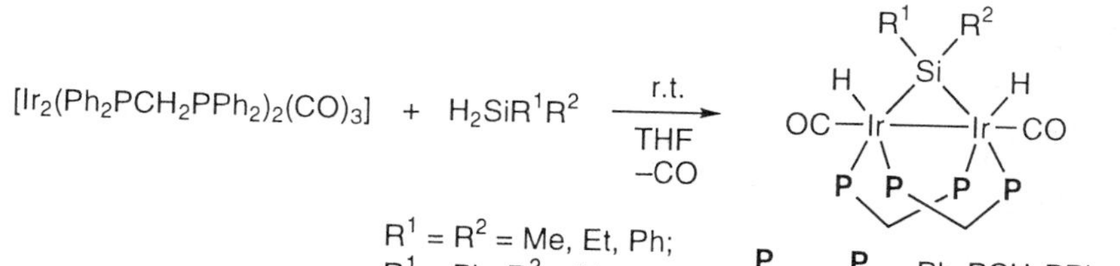

$$[Ir_2(Ph_2PCH_2PPh_2)_2(CO)_3] + H_2SiR^1R^2 \xrightarrow[\text{THF}, -CO]{\text{r.t.}} \quad (9)$$

$R^1 = R^2$ = Me, Et, Ph;
R^1 = Ph, R^2 = H

P⌒P = $Ph_2PCH_2PPh_2$

The reaction of Group 7 metal carbonyl $Re_2(CO)_{10}$ with Ph_2SiH_2 under photolytic conditions was reported to give a bridged silylene complex with hydrides on the Re atoms [Eq. (10)].[33] The X-ray structure analysis of this product and also those of $Re_2(CO)_6H_4(\mu\text{-}SiEt_2)_2$[34] and $Re_2(CO)_7H_2(\mu\text{-}SiEt_2)_2$[35] have been reported, but the bonding mode, that is, whether the Si—H bond is fully oxidatively added or not, is still controversial.[10,11]

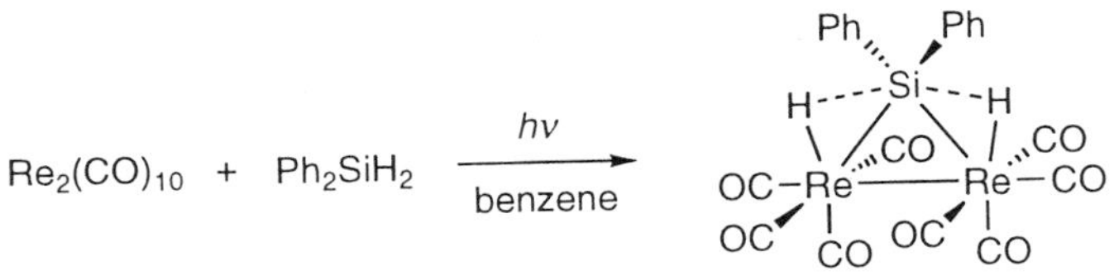

$$Re_2(CO)_{10} + Ph_2SiH_2 \xrightarrow[\text{benzene}]{h\nu} \quad (10)$$

In most cases, bridged silylene complexes are homometallic owing to their synthetic methods, but a heterometallic complex can be prepared by the reaction of a hydrosilyl complex with a metal carbonyl [Eq. (11)].[36]

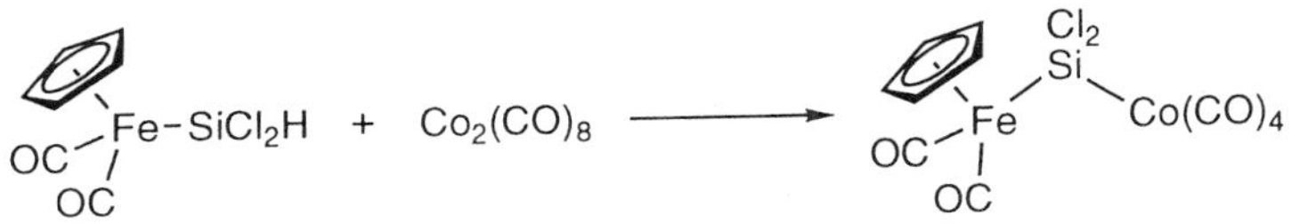

$$CpFe(CO)_2SiCl_2H + Co_2(CO)_8 \longrightarrow \quad (11)$$

C. *Reactions of Hydrosilanes with Hydrido Complexes*

In the reactions of polyhydrosilanes with hydrido complexes, not only **B**-type complexes but also **C**-type complexes containing M–H–Si $3c$–$2e$ bonds are commonly formed. Thus, the reaction of a doubly hydrido-bridged dimanganese complex with Ph_2SiH_2 on heating gives a complex in which the silicon bridges two Mn fragments symmetrically, and two equivalent Mn–H–Si $3c$–$2e$ bonds exist at the positions *trans* to the phosphorus atoms on the basis of the spectroscopic data [Eq. (12)].[37]

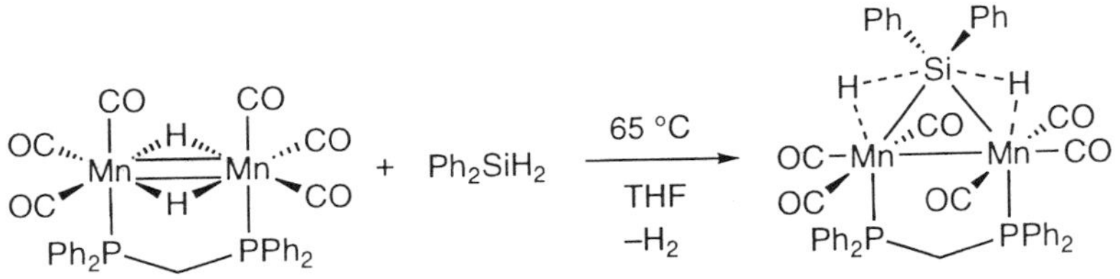

$$+ Ph_2SiH_2 \xrightarrow[\text{THF}, \; -H_2]{65\ °C} \quad (12)$$

A tetrahydride-bridged diruthenium complex $Cp^*_2Ru_2(\mu\text{-}H)_4$ reacts with dihydrosilane below room temperature to give a diruthenium complex with two bridging silyl groups [Eq. (13)].[38] In this product, all the four hydride ligands are equivalent in the ^1H NMR spectrum at 25°C, whereas they split into three signals in 1 : 2 : 1 ratio at −80°C. The location of hydrogen atoms by X-ray structure determination clearly shows the existence of Ru–H–Si $3c$–$2e$ bonds. The thermolysis of this complex in the presence or absence of CO gives other types of bridged silylene complexes that are discussed later.

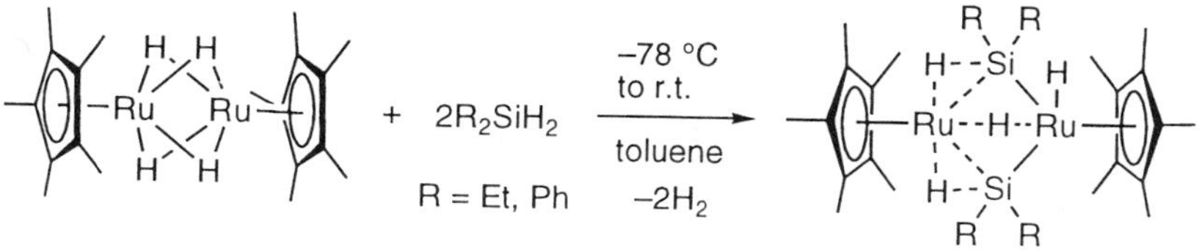

(13)

The reaction of a dirhodium complex bridged by two hydrides and two bis(diphenylphosphino)methane ligands (dppm) reacts with trihydrosilanes at low temperature (−30°C) to give singly silylene-bridged dirhodium complexes, which are then converted to doubly silylene-bridged complexes upon heating or prolonged reaction time (Scheme 5).[31,39] These bridged silylene complexes take a "cradle-like" structure with *cis* dppm ligands, and in the doubly silylene-bridged products, two R groups are in *diaxial* orientation. The analogous complexes are also formed when (diphenylarsino)methane ligands are used instead of dppm ligands.

A dihydrido-bridged dirhodium complex with chelating bis(diisopropylphosphino)ethane (dippe) also reacts with various dihydrosilanes and trihydrosilanes. The reaction with one equivalent of dihydrosilane gives a **C**-type complex containing a μ-η^2-$HSiR_2$ ligand and a μ-H ligand [Eq. (14)].[40,41] Further reaction with one more equivalent of dihydrosilane or the direct reaction of the starting dihydride complex with two equivalents of dihydrosilane gives a doubly silylene-bridged complex of the **B**-type [Eq. (15)].[40] This reaction is reversible, and application of hydrogen pressure (4 atm) to the doubly silylene-bridged complex regenerates the complex with μ-η^2-$HSiR_2$ and μ-H ligands. This reversibility makes possible the catalytic conversion of Ph_2SiH_2 to Ph_2SiD_2 with D_2 using the dihydrido-bridged dirhodium complex.

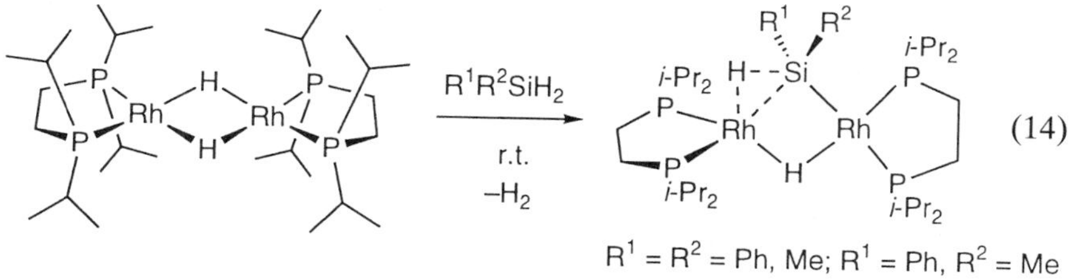

(14)

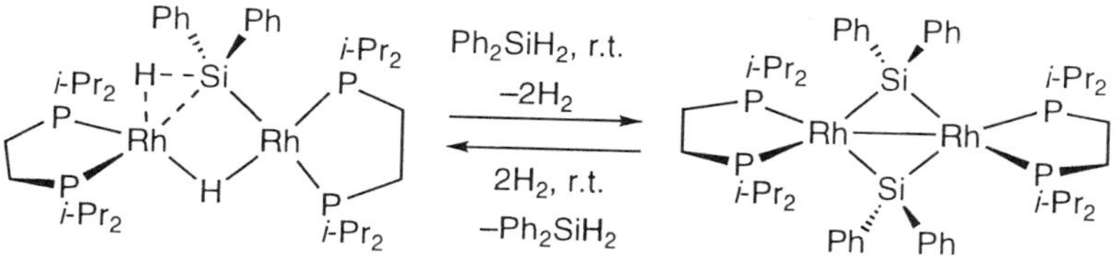

$R^1 = R^2 = Ph, Me; R^1 = Ph, R^2 = Me$

(15)

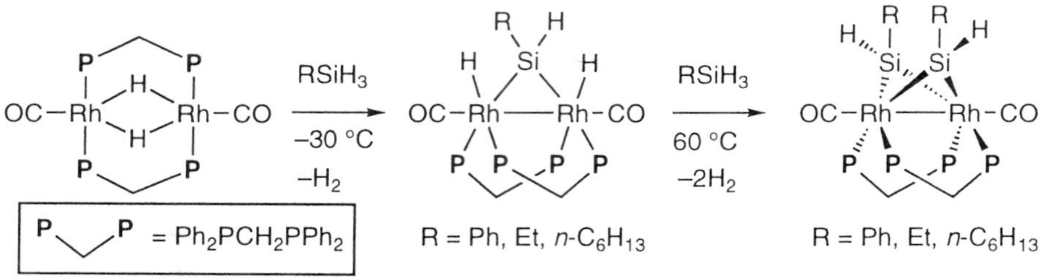

SCHEME 5

In the case of the reactions with trihydrosilanes, employment of 2 equiv $RSiH_3$ gives the doubly silylene-bridged complex as a *cis–trans* mixture, whereas treatment with 4–5 equiv $RSiH_3$ leads the formation of complex with two μ-η^2-$HSiR_2$ ligands and a μ-silylene ligand (Scheme 6).[42]

D. *Reactions of Hydrosilanes with Other Complexes*

Elimination of ligands other than CO or H_2 is sometimes used to generate a coordinatively unsaturated complex that serves as the precursor to bridged silylene complexes. The simplest example is the reaction of Cp*Mn $(CO)_2(THF)$ with SiH_4 giving $\{Cp^*Mn(CO)_2(H)\}_2(\mu\text{-}SiH_2)$ through the dissociation of THF followed by the oxidative addition of SiH_4 to the coordinatively unsaturated species [Eq. (16)].[43]

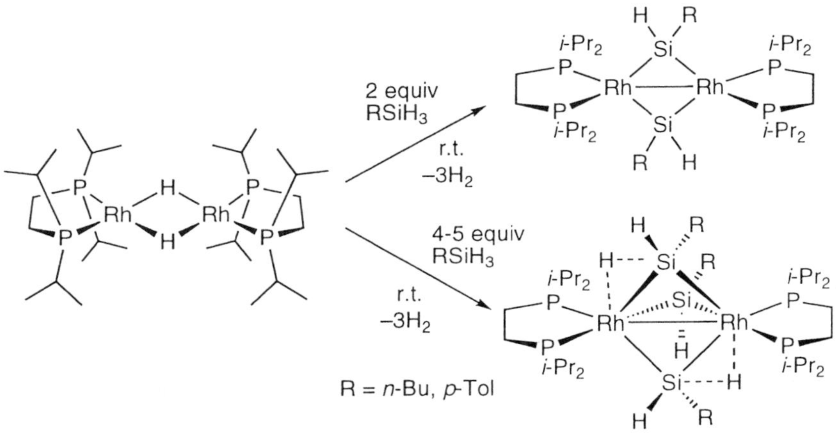

SCHEME 6

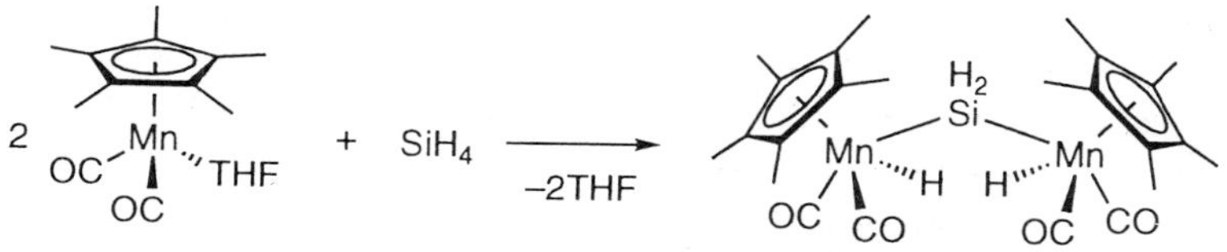

$$2\,(\eta^5\text{-}C_5Me_5)Mn(CO)_2(THF) + SiH_4 \xrightarrow{-2THF} [(\eta^5\text{-}C_5Me_5)Mn(CO)_2H]_2(\mu\text{-}SiH_2) \quad (16)$$

The most typical example of this type of reaction may be provided by the reaction of ethylene–platinum complexes with hydrosilanes. Dihydrosilanes readily react with various bis(ethylene)phosphineplatinum complexes at room temperature to give diplatinum complexes in which two platinum atoms are bridged with two μ-η^2-HSiR$_2$ ligands [Eq. (17)].[44] The same complex with R = Me and PR$'_3$ = PCy$_3$ is also obtained in a low yield by the thermolysis of diplatinum complex {Pt(PCy$_3$)(μ-H)(SiMe$_2$Ph)}$_2$ in refluxing hexane accompanied by the cleavage of Si–Ph bonds.

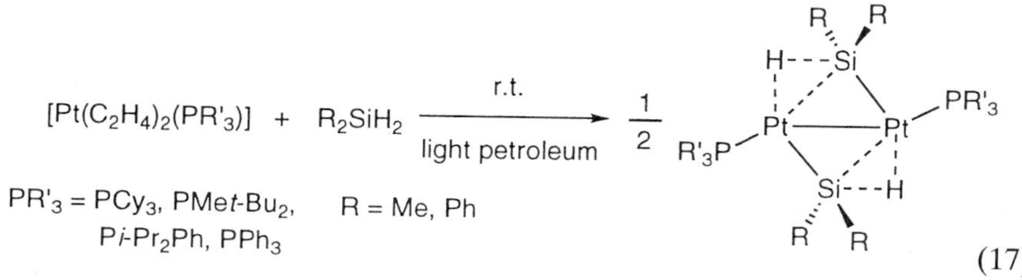

$$[Pt(C_2H_4)_2(PR'_3)] + R_2SiH_2 \xrightarrow[\text{light petroleum}]{\text{r.t.}} \frac{1}{2}\,[(R'_3P)Pt(\mu\text{-}\eta^2\text{-}HSiR_2)]_2 \quad (17)$$

PR'$_3$ = PCy$_3$, PMet-Bu$_2$, Pi-Pr$_2$Ph, PPh$_3$ R = Me, Ph

The reaction of hydrosilylmanganese complexes with ethylenebis(phosphine)platinum or tetrakis(phosphine)platinum offers a convenient method for the synthesis of silylene-bridged heterometallic complexes [Eq. (18)].[45]

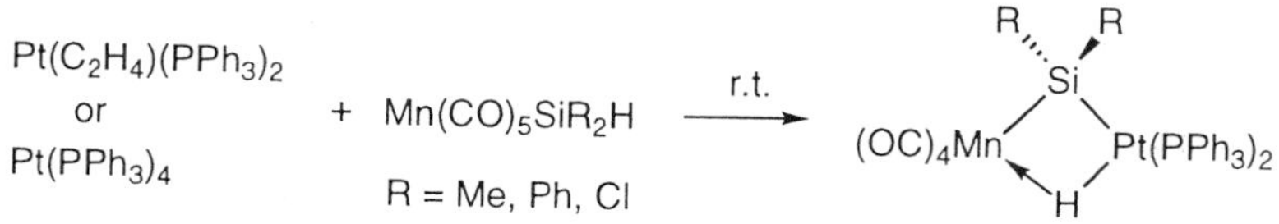

$$Pt(C_2H_4)(PPh_3)_2 \text{ or } Pt(PPh_3)_4 + Mn(CO)_5SiR_2H \xrightarrow{\text{r.t.}} (OC)_4Mn(\mu\text{-}SiR_2)(\mu\text{-}H)Pt(PPh_3)_2 \quad (18)$$

R = Me, Ph, Cl

The reaction of Pt(PEt$_3$)$_3$ with trihydrosilane affords bis(μ-silylene) complexes (Scheme 7).[46] An isolable intermediate is a *cis*-bis(silyl) complex

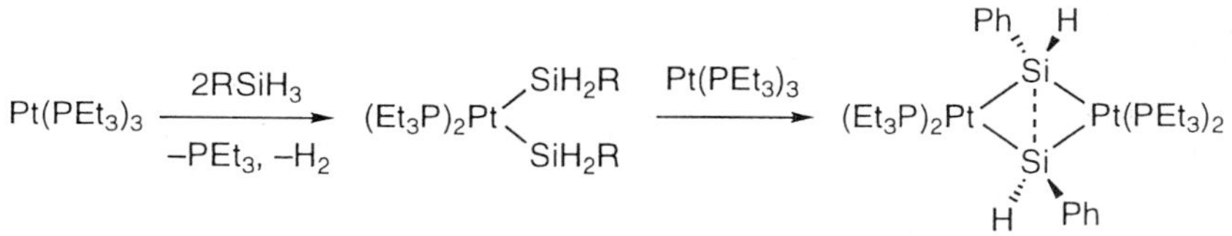

$$Pt(PEt_3)_3 \xrightarrow[-PEt_3,\ -H_2]{2RSiH_3} (Et_3P)_2Pt(SiH_2R)_2 \xrightarrow{Pt(PEt_3)_3} (Et_3P)_2Pt(\mu\text{-}SiHPh)_2Pt(PEt_3)_2$$

Scheme 7

that reacts with additional $Pt(PEt_3)_3$ to give the product. This type of complex was first prepared by the reduction of $(Et_3P)_2PtCl_2$ with sodium in the presence of trihydrosilane or the reaction of $(Et_3P)_2PtCl_2$ with Ph_2SiHLi.[47,48] In these reactions, bis(μ-silylene) complexes in which one or two Si–H bonds are replaced by Si–Cl bonds are also formed.

One rare example of silicon-bridged early transition-metal complex of type **C** has been synthesized by the reaction of dimethyltitanocene with $PhSiH_3$ [Eq. (19)].[49] The structures of these products containing μ-η^2-HSiHPh ligands were determined by X-ray crystal structure analysis.

$$Cp_2TiMe_2 \xrightarrow{PhSiH_3} [Cp_2Ti(\mu\text{-}\eta^2\text{-HSiHPh})_2TiCp_2] + [Cp_2Ti(\mu\text{-H})(\mu\text{-}\eta^2\text{-HSiHPh})TiCp_2] \quad (19)$$

Another interesting reaction is that of $[Cp^*Ru(\mu\text{-OMe})]_2$ with dihydrosilanes. The product depends on the silane substituents: Ph_2SiH_2 reacts with the complex at room temperature to give a phenyl(methoxy)silylene-bridged diruthenium complex, accompanied by the cleavage of the Si–Ph bond [Eq. (20)], whereas the reaction of dialkylsilane $CyMeSiH_2$ gives a complex analogous to the product of the reaction in Eq. (13), $\{Cp^*Ru(\mu\text{-}\eta^2\text{-HSiCyMe})\}_2(\mu\text{-H})(H)$.[50]

$$[Cp^*Ru(\mu\text{-OMe})]_2 + Ph_2SiH_2 \xrightarrow[-MeOH,\ -C_6H_6]{r.t.} Cp^*Ru{=}Ru Cp^*(\mu\text{-Si(OMe)Ph})(\mu\text{-H})(\mu\text{-OMe}) \quad (20)$$

E. *Reactions of Hydrodisilanes with Complexes*

In most cases, the reaction of hydrodisilanes with transition-metal complexes causes cleavage not only of the Si–H bond, but also of the Si–Si bond, to give bridged silylene complexes. Since it is much easier to handle the liquid $HMe_2SiSiMe_3$ and $HMe_2SiSiMe_2H$ than the gaseous Me_2SiH_2 and $MeSiH_3$, hydrodisilanes are conveniently used as precursors of silylene units. Reactions of $HMe_2SiSiMe_2H$ with $Fe_2(CO)_9$ and $Co_2(CO)_8$ at room temperature give doubly silylene-bridged dinuclear complexes $[Fe(CO)_3]_2(\mu\text{-CO})(\mu\text{-SiMe}_2)_2$ and $Co_2(CO)_6(\mu\text{-SiMe}_2)_2$, respectively.[51] In the case of ruthenium and osmium carbonyl complexes, thermolysis and photolysis of $M_3(CO)_{12}$, $[Me_3EM(CO)_4]_2$ (E = Si, Ge), and $(Me_3Si)_2M(CO)_4$ with hydrodisilanes have been studied. In all cases, the products contain two or three bridging silylene ligands, and the structures depend on the precursors [Eq.(21); Scheme 8].[52]

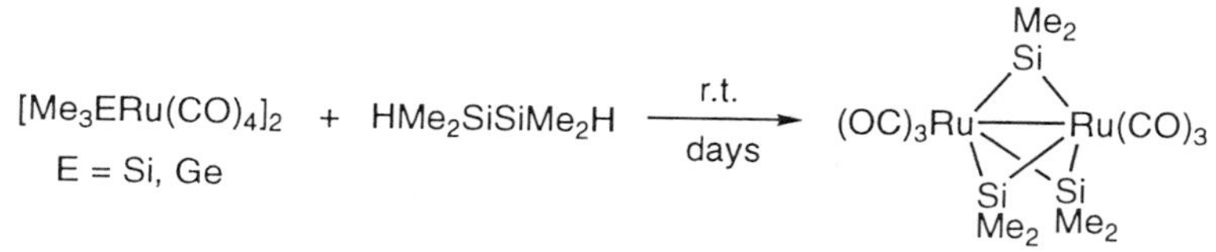

$$[Me_3ERu(CO)_4]_2 + HMe_2SiSiMe_2H \xrightarrow[\text{days}]{\text{r.t.}} (OC)_3Ru(\mu\text{-}SiMe_2)_3Ru(CO)_3 \quad (21)$$

E = Si, Ge

Redistribution of substituents on silanes that is closely related to the reaction of Eq. (6) also occurs when dihydrodisilanes are used instead of dihydrosilanes [Eq. (22)].[26] In this reaction, a silyl(silylene) intermediate that is analogous to **H** shown in Scheme 4 is proposed to form via the oxidative addition of dihydrodisilane followed by migration of the terminal silyl group to the metal center.

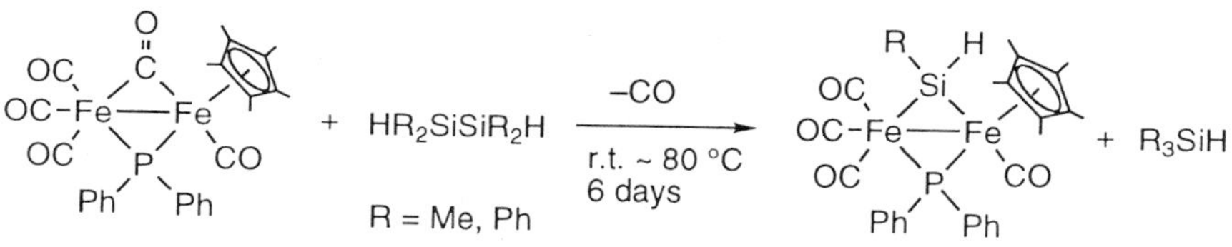

$$(OC)_3Fe(\mu\text{-}CO)(\mu\text{-}PPh_2)Fe(CO)Cp^* + HR_2SiSiR_2H \xrightarrow[\text{r.t.} \sim 80\ ^\circ C,\ 6\ \text{days}]{-CO} (OC)_3Fe(\mu\text{-}SiRH)(\mu\text{-}PPh_2)Fe(CO)Cp^* + R_3SiH \quad (22)$$

R = Me, Ph

F. *Salt Elimination Reactions*

Although this is a potentially widely useful preparative method, especially for the synthesis of **A**-type silylene-bridged complexes, only two types of reactions are reported. One is the reaction of silyl-alkali metal derivatives

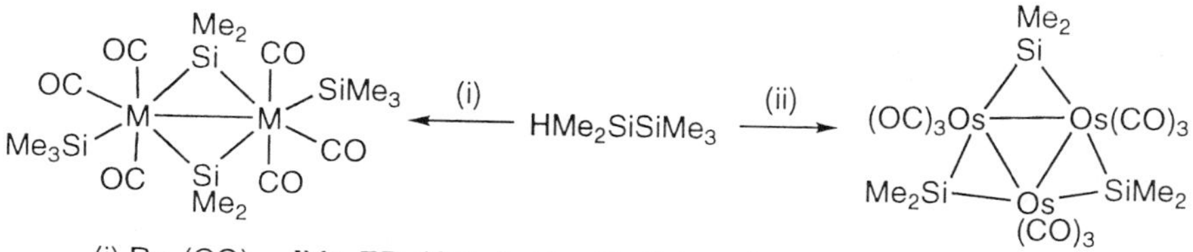

$$(Me_3Si)(OC)_3M(\mu\text{-}SiMe_2)_2M(CO)_3(SiMe_3) \xleftarrow{(i)} HMe_2SiSiMe_3 \xrightarrow{(ii)} Os_3(CO)_9(\mu\text{-}SiMe_2)_3$$

(i) $Ru_3(CO)_{12}$, $[Me_3ERu(CO)_4]_2$ (E = Si, Ge), or $(Me_3Si)_2Ru(CO)_4$, *hv*; $Os_3(CO)_{12}$, 160 °C, 44 h; $[Me_3SiOs(CO)_4]_2$ or $(Me_3Si)_2Os(CO)_4$, *hv*;
(ii) $Os(CO)_4H_2$, 80 °C, 18 h

SCHEME 8

with dichlorotitanocene [Eq. (23)][53] or dichlorobis(triethylphosphine)platinum.[47] The other one is the reaction of anionic transition-metal species with chlorosilyl transition-metal complexes [Eq. (24)].[23,54]

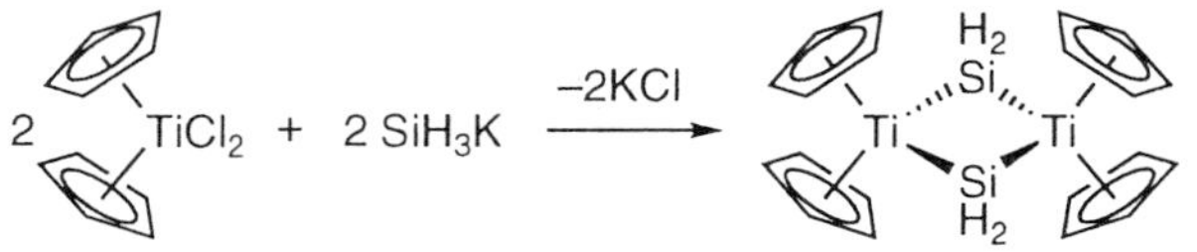

$$2\ Cp_2TiCl_2 + 2\ SiH_3K \xrightarrow{-2KCl} [Cp_2Ti(\mu\text{-}SiH_2)]_2 \quad (23)$$

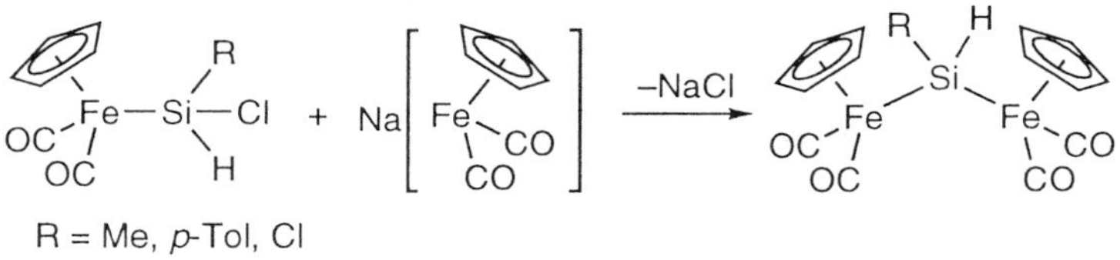

$$Cp(CO)_2Fe\text{-}SiRHCl + Na[Fe(CO)_2Cp] \xrightarrow{-NaCl} Cp(CO)_2Fe\text{-}SiRH\text{-}Fe(CO)_2Cp \quad (24)$$

R = Me, *p*-Tol, Cl

G. *Rearrangement of a Disilanylene Unit in a Complex*

The only example of this type of reaction is the photolysis of $Cp(CO)_2FeMe_2SiSiMe_2Fe(CO)_2Cp$ (Scheme 9).[55,56] In the early stage of this reaction, methyl(trimethylsilyl)silylene-bridged complex **I** mainly forms, but prolonged irradiation leads to the quantitative formation of bis(dimethylsilylene)-bridged complex **J**. Complex **I** is obtained as a mixture of all of the three possible geometric isomers, and complex **J** as a mixture of *cis* and *trans* isomers. Scrambling of substituents occurs extensively in this reaction. A possible mechanism for the reaction is shown in Scheme 10, where after photochemical dissociation of a CO ligand, 1,2-silyl shift and 1,3-methyl shift are suggested to operate as the key steps.

H. *Dimerization of Mononuclear Silylene Complexes*

Removal of the base from a base-stabilized mononuclear silylene complex can induce the dimerization of the generated base–free silylene complex

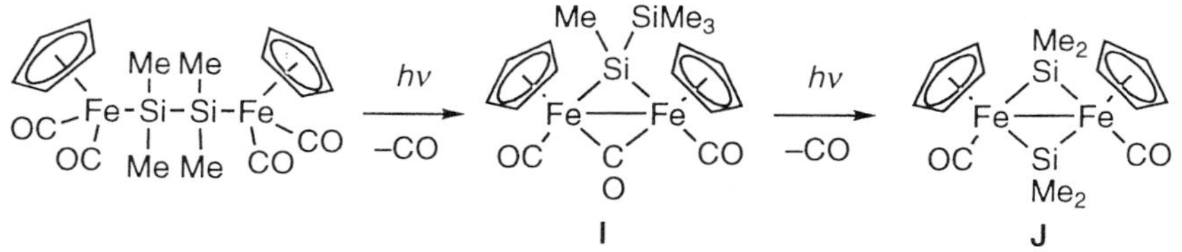

$$Cp(CO)_2Fe\text{-}SiMe_2\text{-}SiMe_2\text{-}Fe(CO)_2Cp \xrightarrow[-CO]{h\nu} \mathbf{I}\ [Cp(CO)Fe]_2(\mu\text{-}Si(Me)SiMe_3)(\mu\text{-}CO) \xrightarrow[-CO]{h\nu} \mathbf{J}\ [Cp(CO)Fe]_2(\mu\text{-}SiMe_2)_2$$

Scheme 9

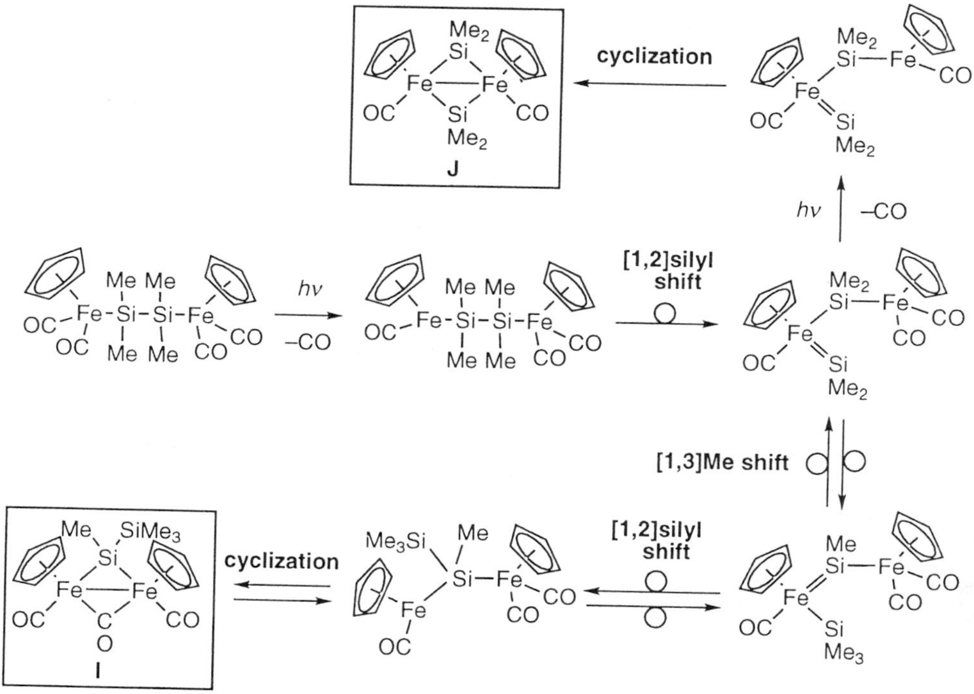

SCHEME 10

[Eq. (25)].[57] This type of reaction is more common to stannylene complexes.[8] The inverse reaction, that is, formation of base-stabilized mononuclear silylene complexes from doubly silylene-bridged dinuclear complexes, is described in Section V,B.

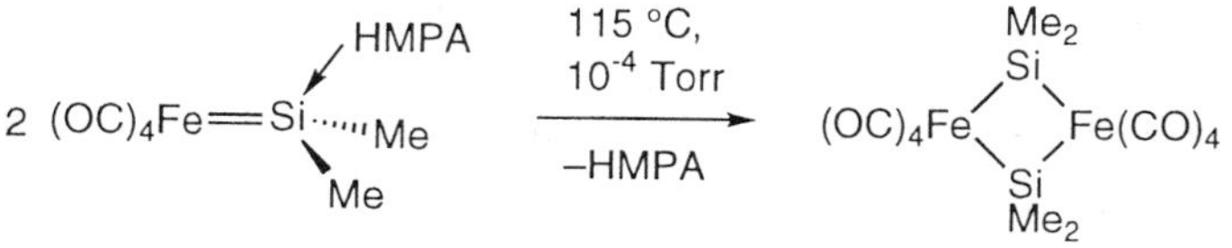

$$2\ (OC)_4Fe{=}Si(HMPA)Me_2 \xrightarrow[-HMPA]{115\ ^\circ C,\ 10^{-4}\ Torr} (OC)_4Fe(\mu\text{-}SiMe_2)_2Fe(CO)_4 \qquad (25)$$

Another reaction in which the dimerization of a mononuclear silylene complex is suggested to play an important role is shown in Scheme 11.[58] After the oxidative addition of disilacyclobutene to each iron center, fluorine migration accompanied by the cleavage of the Fe–Fe bond leads to a silyl(silylene) complex as one of the intermediates, which then dimerizes in head-to-tail fashion to form the product.

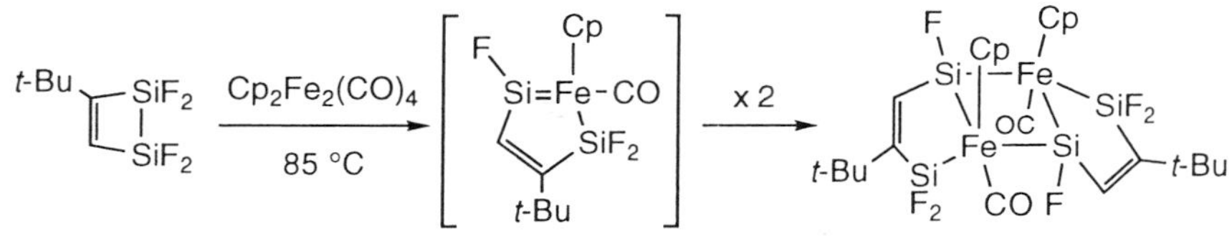

SCHEME 11

I. *Removal or Addition of Ligands on Bridged Silylene Complexes*

Photodissociation of a CO ligand from an **A**-type complex leads to the formation of a metal–metal bond and a CO bridge to give a **B**-type complex [Eq. (26)].[54] The product is a mixture of three geometric isomers (a *trans* and two *cis* isomers) that interconvert slowly to attain equilibrium at room temperature. A similar reaction of $Fe_2(CO)_8(\mu\text{-}SiMe_2)_2$ leads to the formation of $\{Fe(CO)_3\}_2(\mu\text{-}CO)(\mu\text{-}SiMe_2)_2$.[59]

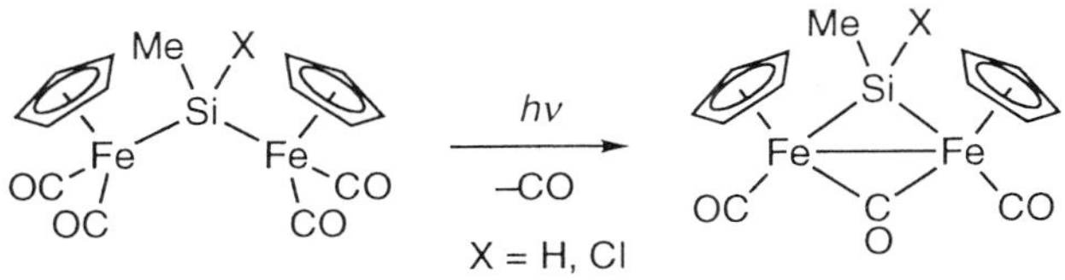

(26)

Both the elimination of H_2 from the **C**-type complex to give the **B**-type complex and the addition of H_2 to the **B**-type complex to give the **C**-type complex have been reported. Heating of a diruthenium complex with two $\mu\text{-}\eta^2\text{-}HSiR_2$ ligands affords a complex in which two ruthenium atoms are bridged by two silylene and two hydrido ligands. Thermolysis of the starting complex or the product in the presence of CO gives rise to the elimination of all the hydrido ligands and addition of CO ligands (Scheme 12).[38]

In contrast, doubly silylene-bridged dirhodium complexes reversibly catch two molecules of H_2 at room temperature to form the complexes with two $\mu\text{-}\eta^2\text{-}HSiR_2$ ligands [Eq. (27)].[42]

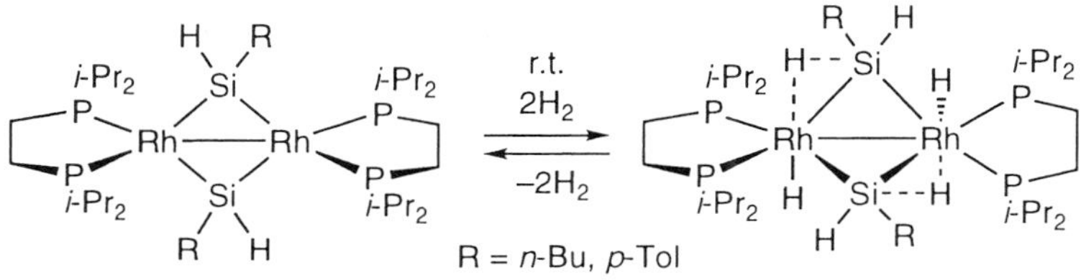

(27)

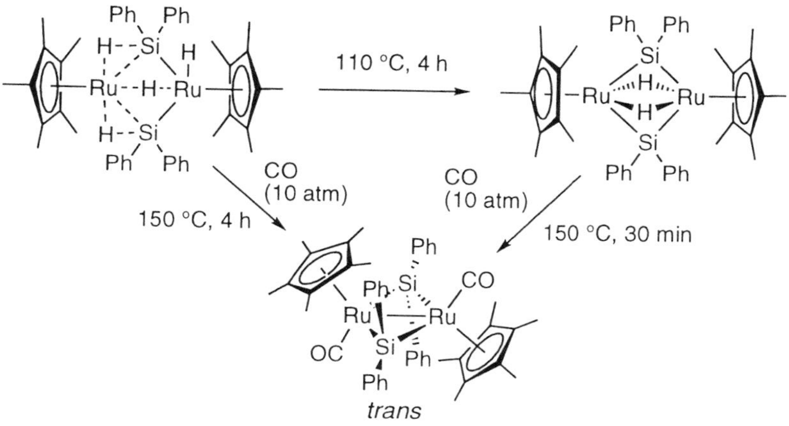

110 °C, 4 h

CO (10 atm) 150 °C, 4 h

CO (10 atm) 150 °C, 30 min

trans

SCHEME 12

J. *Replacement of Substituents or Ligands on Bridged Silylene Complexes*

The Si–H or Si–X bond on bridging silylene ligands is found to be highly activated toward substitution reactions. Malisch *et al.* first observed the rapid exchange of hydrogen for chlorine on the silylene bridge of $Cp_2Fe_2(CO)_2(\mu\text{-}CO)(\mu\text{-}SiMeH)$ by CCl_4 solvent at room temperature.[54] The analogous complex $Cp_2Fe_2(CO)_2(\mu\text{-}CO)(\mu\text{-}Si\textit{t}\text{-}BuH)$ (100% *cis*) reacts with halomethane solvents at room temperature to give halosilylene-bridged complexes [Eq. (28)].[60,61] In halogenated products, the *trans* isomer is observed as a minor component in an equilibrium state.

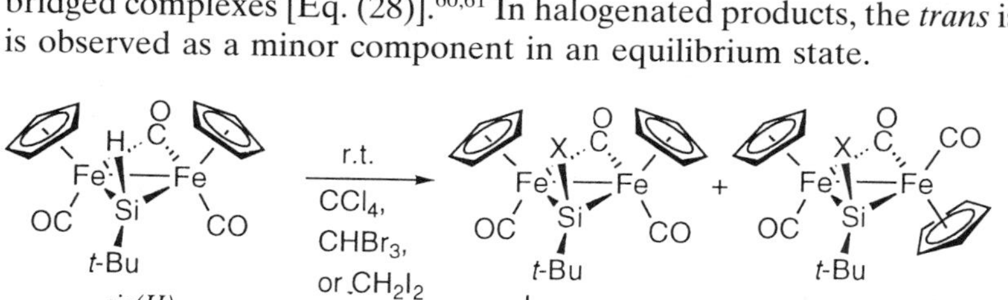

cis(H) —(r.t., CCl_4, $CHBr_3$, or CH_2I_2)→ *cis(X)* + *trans* (28)

	cis(X)		*trans*
X = Cl	93	:	7
Br	93	:	7
I	92	:	8

The hydrogen atom on the silylene bridge of **A**-type complexes is also rapidly replaced by Cl in CCl_4, and the chorine atom(s) on the silylene bridge is further changed to F by treatment with $AgBF_4$ (Scheme 13).[62]

Oxygenation of hydrosilylene-bridged diiron complexes with dimethyldi-

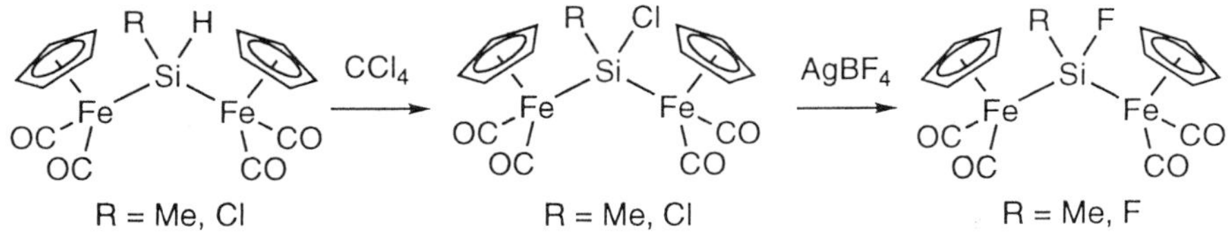

SCHEME 13

oxirane yields the hydroxysilylene-bridged complexes [Eq. (29)].[23] In an analogous manner, the **B**-type complex $Cp_2Fe_2(CO)_2(\mu\text{-}CO)(\mu\text{-}SiMesH)$ can be converted into its hydroxysilylene derivative.

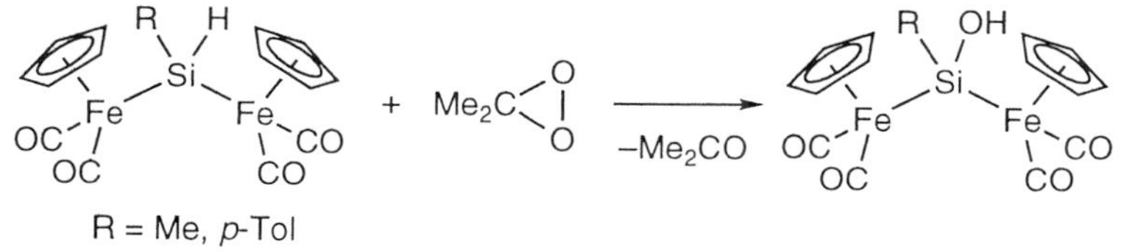

(29)

The first base-stabilized silylene-bridged dinuclear complex has been synthesized by the exchange of an iodide ion on the iodosilylene-bridged complex to a strong Lewis base, *N*-methylimidazole (NMI) or 4-(dimethylamino)pyridine (DMAP) [Eq. (30)].[61,63]

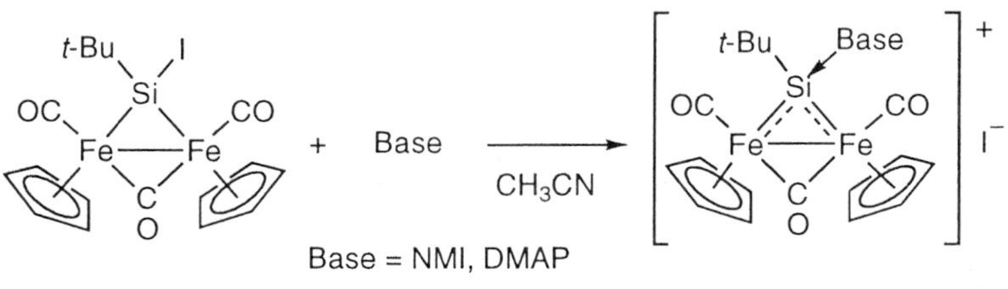

(30)

Ligand substitution reactions have been reported for some bridged silylene complexes. On a manganese–platinum complex, phosphine ligands on the platinum atom are easily replaced by carbonyl or other phosphine ligands (Scheme 14).[45] The displacement of hydrogen by CO in dirhodium complexes containing a $\mu\text{-}\eta^2\text{-}HSiR_2$ ligand and a μ-H ligand generates dirhodium complexes with bridging carbonyl and silylene units [Eq. (31)].[41] This reaction occurs readily at room temperature with addition of ca. 1 equiv CO, and this fact demonstrates that a hydrogen molecule is removed easily through the complete oxidative addition of the second Si–H bond.

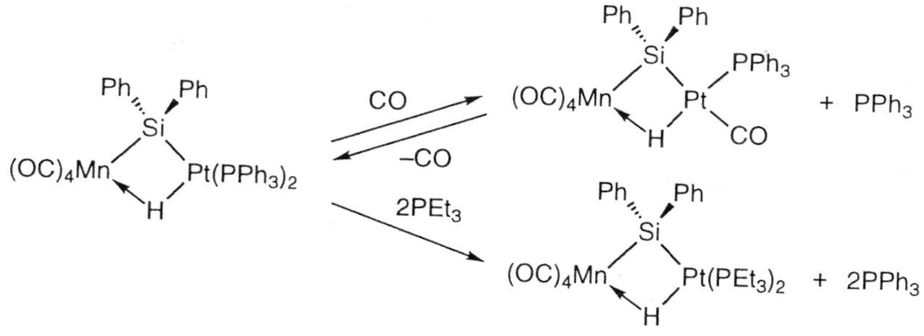

SCHEME 14

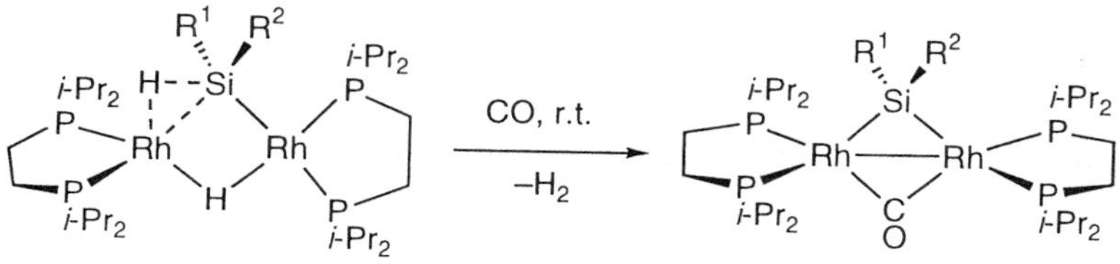

$R^1 = R^2 = Ph, Me; R^1 = Ph, R^2 = Me$ (31)

Reaction of a doubly silylene-bridged diiron complex with a silyliron complex gives a triply silylene-bridged dinuclear complex in a low yield, perhaps through the displacement of carbonyl ligands by a silylene unit [Eq. (32)].[64]

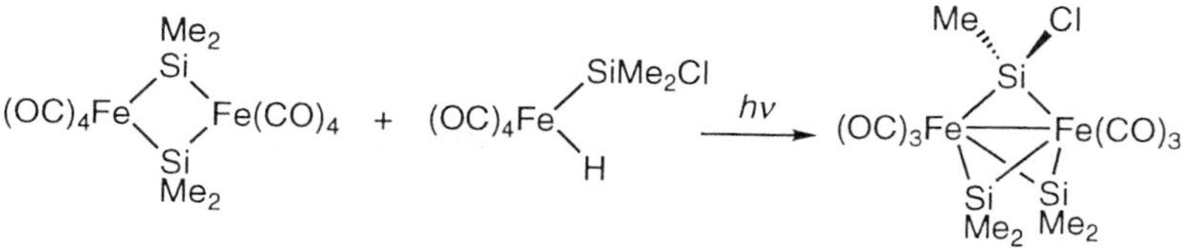

(32)

K. *Other Reactions*

The first complex with a metal–metal bond that is bridged by only a single silylene ligand has been prepared in a low yield by treatment of chloro(hydro)silane with base in the presence of $Fe_2(CO)_9$ [Eq. (33)].[65] However, this product is reported to be inflammable in air and stable only at low temperature.

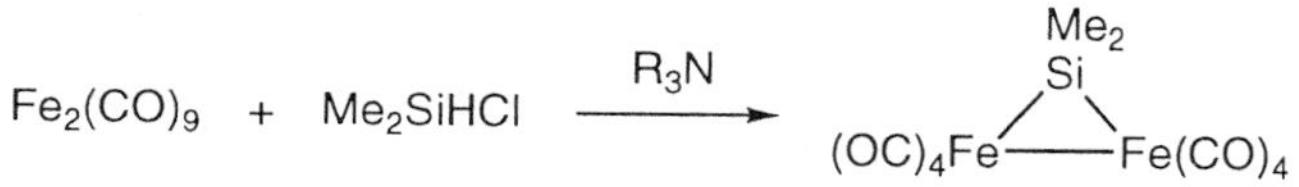

$$Fe_2(CO)_9 + Me_2SiHCl \xrightarrow{R_3N} (OC)_4Fe(\mu\text{-}SiMe_2)Fe(CO)_4 \quad (33)$$

Si–C bond fission in organosilyl complexes sometimes gives silylene-bridged complexes. A typical example giving a **C**-type complex is shown in Eq. (34).[66]

$$\xrightarrow[80\ ^{\circ}\mathrm{C}]{\mathrm{H_2\ (5\ atm)}} \qquad (34)$$

Braunstein *et al.* recently reported an interesting reaction of a base-stabilized mononuclear silylene complex with platinum–ethylene complexes [Eq. (35)]. In this reaction, the Pt-bound ethylene ligand is displaced by the base-free silylene complex, but the products can be also regarded as a dinuclear complex where two metals are bridged by a silylene unit.[67]

$$(\mathrm{OC})_4\mathrm{Fe{=}Si(NHMe_2)(NMe_2)_2} + \mathrm{Pt(C_2H_4)L_2} \xrightarrow[-\mathrm{NHMe_2},\ -\mathrm{C_2H_4}]{\mathrm{r.t.}} (\mathrm{OC})_4\mathrm{Fe}(\mu\text{-}\mathrm{Si(NMe_2)_2})\mathrm{PtL_2} \qquad (35)$$

$$\mathrm{L_2 = (PPh_3)_2,\ (P}p\text{-}\mathrm{Tol_3)_2,\ Ph_2PCH_2CH_2PPh_2}$$

III
FORMATION OF BRIDGED GERMYLENE COMPLEXES

A. *General Considerations*

The variety of reactions forming bridged germylene complexes is comparable to that for bridged silylene complexes. Several reactions are commonly used for the synthesis of bridged silylene and germylene complexes, but some reactions are so far only used for the synthesis of bridged silylene complexes. There are also some reactions that are exclusively used for the synthesis of bridged germylene complexes, such as the reaction of halogermanes with carbonyl complexes, elimination of hydrogermane from bis(germyl) complexes, and reactions of stable germylene with transition-metal complexes. However, when the same type of reaction is used with the aim of getting bridged silylene and germylene complexes, the structures of products are frequently different.

A notable difference in the formation reactions and structures of bridged germylene complexes from those of the silicon analog is that by a simple reaction such as that of polyhydrogermane with metal carbonyl complexes, complexes or clusters with two or more bridging germylene ligands tend to form. By a similar reaction of polyhydrosilane, doubly bridged silylene complexes rarely form and only one triply silylene bridged complex was obtained in a low yield. Another difference is the scarcity of the **C**-type complexes in the germanium system. In the chemistry of bridged silylene complexes, we frequently encountered **C**-type complexes, that is, complexes with hydrogen atoms bridging the metal–silylene bond. In contrast, the corresponding **C**-type bridged germylene complex was not synthesized until 1995.[68,69]

For these reasons, we classified the formation reactions of bridged germylene complexes independently of those of bridged silylene complexes. The bridged germylene and germylyne complexes of types **A** and **B** and their formation reactions and reaction types B–J are listed in Tables IV and V, respectively.

B. *Reactions of Hydrogermanes with Carbonyl Complexes*

Like the synthesis of bridged silylene complexes, the reaction of hydrogermane with coordinatively unsaturated complexes has frequently been used for the synthesis of bridged germylene complexes, but the transition-metal moieties have been limited so far to carbonyl complexes. The earliest example of the synthesis of a bridged germylene complex is the reaction of GeH_4 with $Mn(CO)_5H$ at room temperature reported by Stone *et al.* that gave $\{Mn(CO)_5\}_2(\mu\text{-}GeH_2)$ in a high yield [Eq. (36)].[1] Since no Mn $(CO)_5GeH_3$ or $\{Mn(CO)_5\}_3GeH$ has been detected among the products, the authors proposed a mechanism involving germylene GeH_2, perhaps formed from GeH_4 by the reduction with $Mn(CO)_5H$. The same product is also formed, but in a low yield, by the thermal reaction of GeH_4 with $Mn_2(CO)_{10}$ at 140°C.

$$Mn(CO)_5H \quad + \quad GeH_4 \xrightarrow{\text{r.t., 8 days}} (OC)_5Mn\text{–}Ge(H_2)\text{–}Mn(CO)_5 \tag{36}$$

The first structurally well-characterized germylene-bridged complex is triply germylene-bridged diiron complex, $Fe_2(CO)_6(\mu\text{-}GeMe_2)_3$, formed by the reaction of Me_2GeH_2 and $Fe_2(CO)_9$ at 65°C together with $Fe_2(CO)_8$ $(\mu\text{-}GeMe_2)$ [Eq. (37)].[70,71] In this molecule, the two iron atoms lie on a threefold axis and the three $GeMe_2$ residues lie on the perpendicular mirror plane.

$$Fe_2(CO)_9 \xrightarrow[65\ ^\circ C,\ 5\ h]{Me_2GeH_2} \{Fe(CO)_3\}_2(\mu\text{-}GeMe_2)_3 + (OC)_4Fe(\mu\text{-}GeMe_2)Fe(CO)_4 \quad (37)$$

An interesting Ge–C bond fission reaction is reported on the thermolysis of a mixture of Me_3GeH and $M_3(CO)_{12}$ (M = Ru, Os).[72] By this reaction, a doubly germylene-bridged complex forms in a low yield together with a major product $M(CO)_4(GeMe_3)_2$ [Eq. (38)]. Thermal decomposition of these major products at higher temperatures affords triangular clusters in high yields as well as trace amounts of dinuclear complexes, each having three bridging germylene units [Eq. (39)]. This reaction also proceeds through extensive Ge–C bond fission. In contrast, the Si–C bond fission was not observed in the thermolysis of a mixture of silane Me_3SiH and $M_3(CO)_{12}$ (M = Ru, Os) under analogous conditions, and trimethylsilyl-substituted complexes were obtained as main products.[73,74]

$$\underset{M\,=\,Ru,\ Os}{M_3(CO)_{12}} \xrightarrow[100\ ^\circ C\ (M\,=\,Ru),\ 150\ ^\circ C\ (M\,=\,Os)]{Me_3GeH} \{M(CO)_3(GeMe_3)\}_2(\mu\text{-}GeMe_2)_2 + (OC)_4M(GeMe_3)_2 \quad (38)$$

$$(OC)_4M(GeMe_3)_2 \xrightarrow[160\ ^\circ C\ (M\,=\,Os)]{140\ ^\circ C\ (M\,=\,Ru)\ or} \underset{major}{\{M(CO)_3\}_3(\mu\text{-}GeMe_2)_3} + \underset{minor}{\{M(CO)_3\}_2(\mu\text{-}GeMe_2)_3} \quad (39)$$

Reaction of $Co_2(CO)_8$ with Ph_2GeH_2 proceeds smoothly at room temperature to afford a diphenylgermylene-bridged dicobalt complex in high yield [Eq. (40)].[75] In this reaction, the **A**-type complex $\{Co(CO)_4\}_2(\mu\text{-}GePh_2)$ can be ruled out as an intermediate, because under the same conditions, it does not lose CO to give the product.

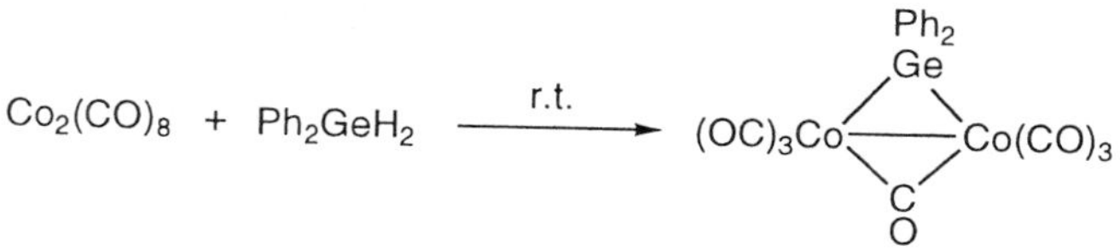

$$Co_2(CO)_8 + Ph_2GeH_2 \xrightarrow{\text{r.t.}} (OC)_3Co(\mu\text{-}GePh_2)(\mu\text{-}CO)Co(CO)_3 \qquad (40)$$

The products of the reaction of trihydrogermane $MeGeH_3$ with $Co_2(CO)_8$ depend mainly on the starting material ratio and, to some extent, on reaction time. Thus, the reaction of $MeGeH_3$ and $Co_2(CO)_8$ in 1:2 or 1:1 ratio at 10–20°C gives a singly germylene-bridged complex, whereas the reaction in 2:1 ratio for a longer reaction time yields a doubly germylene-bridged complex in high yields (Scheme 15).[76] In these reactions, $[CoH(CO)_4]$ is established as a primary product. In both cases, all the Ge–H bonds are replaced by Ge–Co bonds.

A bridging CO ligand on a tetracobalt cluster, which is prepared by the reaction of $MeSiH_3$ and $Co_4(CO)_{12}$, was found to be selectively and quantitatively replaced by a μ-$GeMe_2$ group in the reaction with Me_2GeH_2 [Eq. (41)].[77]

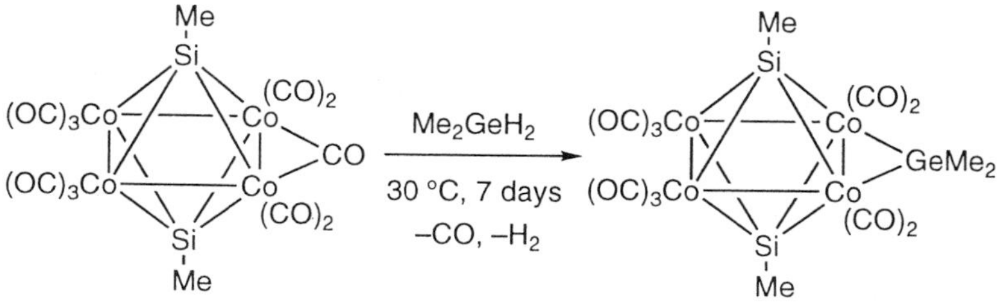

$$[(\mu_4\text{-}SiMe)_2Co_4(CO)_{10}(\mu\text{-}CO)] \xrightarrow[\text{30 °C, 7 days; } -CO, -H_2]{Me_2GeH_2} [(\mu_4\text{-}SiMe)_2Co_4(CO)_{10}(\mu\text{-}GeMe_2)] \qquad (41)$$

As in the cases of silicon analogs, the photolysis of $CpFe(CO)_2SiMe_3$ or $Cp^*Fe(CO)_2SiMe_3$ in the presence of di- or trihydrogermane is a useful

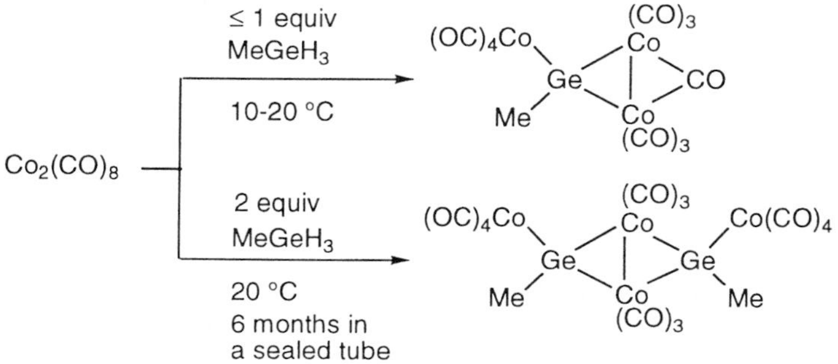

$Co_2(CO)_8$ — $\le$ 1 equiv $MeGeH_3$, 10-20 °C → $(OC)_4Co$–Ge(Me) bridging Co(CO)₃–Co(CO)₃ with μ-CO

$Co_2(CO)_8$ — 2 equiv $MeGeH_3$, 20 °C, 6 months in a sealed tube → $(OC)_4Co$(Me)Ge and Ge(Me)$Co(CO)_4$ bridging $(CO)_3Co$–$Co(CO)_3$

SCHEME 15

method for the synthesis of germylene-bridged diiron complexes. However, two notable points are not observed in the corresponding reactions of polyhydrosilanes: (i) the formation of doubly germylene-bridged diiron complexes, and (ii) the formation of unprecedented complexes containing two μ-η^2-HGeHR bridging ligands.

Thus, the photolysis of $CpFe(CO)_2SiMe_3$ in the presence of t-$BuGeH_3$ or p-Tol_2GeH_2 gives not only a singly germylene-bridged diiron complex, but also a doubly germylene-bridged diiron complex [Eq. (42)].[78,79] Both products form as the mixtures of geometric isomers. When $Cp^*Fe(CO)_2SiMe_3$ is photolyzed in the presence of trihydrogermane, t-$BuGeH_3$ or p-$TolGeH_3$, a complex containing two μ-η^2-HGeHR bridging ligands **K** forms besides the singly and doubly germylene-bridged complexes [Eq. (43)].[68,69] The complex of **K** type is obtained in a higher yield by the photolysis of $Cp^*Fe(CO)_2GeH_2t$-Bu.[68] Complex **K** adopts a *cis* configuration and readily loses H_2 upon UV irradiation or heating to give doubly germylene-bridged complexes.

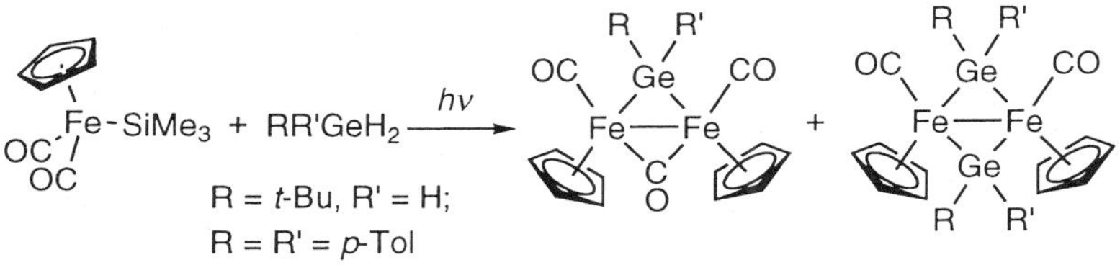

Fe-SiMe3 + RR'GeH2 —hv→ (42)

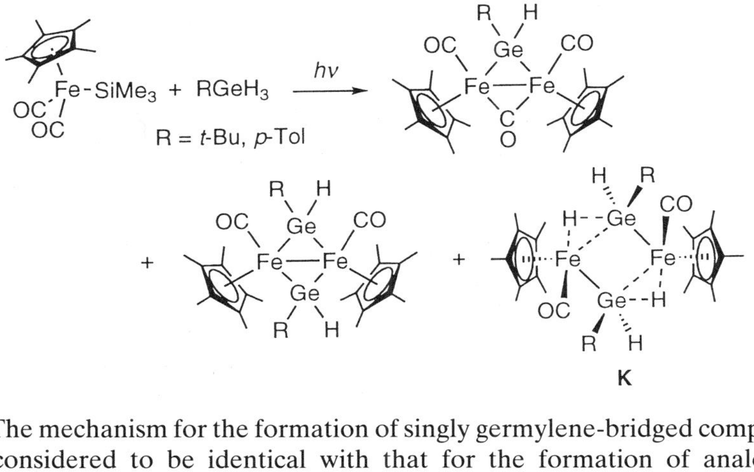

R = t-Bu, R' = H;
R = R' = p-Tol

Fe-SiMe3 + RGeH3 —hv→ (43)

R = t-Bu, p-Tol

K

The mechanism for the formation of singly germylene-bridged complexes is considered to be identical with that for the formation of analogous silylene-bridged complexes shown in Scheme 3. On the other hand, the doubly germylene-bridged diiron complexes in Eqs. (42) and (43) are likely to form via the intermediate formation of **K**-type complexes.

C. *Reactions of Halogermanes with Carbonyl Complexes*

Some reactions are reported in which halogermanes react with neutral carbonyl complexes either thermally or photochemically to give bridged germylene complexes. This type of reaction has not been reported for the silicon analog. The reaction of $Co_2(CO)_8$ with $GeCl_4$ proceeds smoothly in THF at room temperature to give an **A**-type complex [Eq. (44)].[80] This reaction is thought to involve cleavage of the Co–Co bond of $Co_2(CO)_8$ by the Ge–Cl bond, followed by decomposition of the unstable Co $(CO)_4Cl$ product.

$$3Co_2(CO)_8 + 2GeCl_4 \xrightarrow[\text{THF}]{\text{r.t.}} 2\ (OC)_4Co(\mu\text{-}GeCl_2)Co(CO)_4 + 2CoCl_2 + 8CO \qquad (44)$$

Photolysis of chlorodimethylgermyl carbonyl complexes causes dimerization accompanied by loss of a chlorine atom and a CO ligand, and doubly germylene-bridged complexes form when the metal fragment is $Co(CO)_4$ and $Mn(CO)_5$ [Eq. (45)], whereas the complex with a μ-germylene and a μ-CO ligand becomes a main product when the metal fragment is $CpFe(CO)_2$ [Eq. (46)].[81,82]

$$M(CO)_nGeMe_2Cl \xrightarrow[-Cl,\ -CO]{h\nu} (CO)_{n-1}M(\mu\text{-}GeMe_2)_2M(CO)_{n-1} \qquad (45)$$

$M(CO)_n = Co(CO)_4,\ Mn(CO)_5 \qquad M(CO)_{n-1} = Co(CO)_3,\ Mn(CO)_4$

$$CpFe(CO)_2GeMe_2Cl \xrightarrow{h\nu} \textit{cis}\text{-}[CpFe(CO)]_2(\mu\text{-}CO)(\mu\text{-}GeMe_2)\ (\mathbf{4}) + \textit{trans}\text{-}[CpFe(CO)]_2(\mu\text{-}CO)(\mu\text{-}GeMe_2)\ (\mathbf{1}) \qquad (46)$$

cis 4 *trans* 1

D. *Salt Elimination Reactions*

This method has been more widely used for the synthesis of bridged germylene complexes than for their silicon analogs. The reactions of transition-metal carbonyl monoanions with Me_2GeCl_2 give **A**-type bridged germylene complexes [Eq. (47)].[80,82] This reaction has been extensively applied by Lei *et al.* for the synthesis of transition-metal carbonyl-substituted germacyclopent-3-enes [Eq. (48)].[83] They also prepared Co–Mn and Co–Fe heterobimetallic carbonyl derivatives by reactions of 1,1-dibromogermacyclopent-3-ene with two different metal carbonyl anions. Most of these **A**-type complexes have been further used for the synthesis of **B**-type complexes through the removal of a CO ligand (see Section III,H).

$$Me_2GeCl_2 + 2Na[ML_n] \xrightarrow[-2NaCl]{} L_nM{-}GeMe_2{-}ML_n \quad (47)$$

$ML_n = Mn(CO)_5, CpFe(CO)_2, Co(CO)_4$

$$\text{1,1-dibromogermacyclopent-3-ene } (R^1,R^2,R^3,R^4) + 2Na[ML_n] \xrightarrow[-2NaBr]{} (R^1,R^2,R^3,R^4)\text{-germacyclopent-3-ene-1,1-}(ML_n)_2 \quad (48)$$

R^1	R^2	R^3	R^4	ML_n
H	Me	Me	H	$Mn(CO)_5$
H	Me	H	H	$Mn(CO)_5$
H	H	H	H	$Mn(CO)_5$
H	Me	Me	H	$Fe(CO)_2Cp$
H	Me	Me	H	$Co(CO)_4$

The reaction of osmium carbonyl dianion with Me_2GeCl_2 has been reported to give a doubly germylene-bridged diosmium complex [Eq. (49)].[84] This product serves as a good precursor for several higher clusters containing μ-$GeMe_2$ ligands (see Section III,J).

$$Na_2[Os(CO)_4] + Me_2GeCl_2 \xrightarrow[\text{THF-hexane}]{-40\ ^\circ C\text{ - r.t.}} (OC)_4Os(\mu\text{-}GeMe_2)_2Os(CO)_4 \quad (49)$$

E. *Elimination of Hydrogermane from Bis(germyl) Complexes*

This method has been extensively studied by Mackay *et al.* using various combinations of two germyl groups [Eq. (50)].[85–88] The tendency to eliminate a methylgermane (derived from the germyl group with the largest number of Me groups) and condense to four-membered-ring Ge_2Fe_2 complexes of **A** type increases rapidly with increasing numbers of Me groups.[88] Interestingly, no such reaction was observed from complexes with no Me groups such as $[Fe(CO)_4(GeH_3)_2]$, $[Fe(CO)_4(GeH_3)(Ge_2H_5)]$, or $[Fe(CO)_4(Ge_2H_5)_2]$. A proposed mechanism (represented by the reaction of $[Fe(CO)_4(GeH_3)(GeR_3)]$) involves the initial bimolecular condensation accompanied by the elimination of methyl(hydro)germane followed by intramolecular condensation liberating the second methyl(hydro)germane (Scheme 16).[88]

$$2\,(OC)_4Fe(GeMe_mH_{3-m})(GeMe_nH_{3-n}) \xrightarrow[-2GeMe_nH_{4-n}]{r.t.} [(OC)_4Fe(\mu\text{-}GeMe_mH_{2-m})]_2 \quad (50)$$

$n \geq m,$
$n \geq 1,\ m = 0 - 2$

F. *Reactions of Stable Germylene with Complexes*

A number of isolable germylenes are stabilized by special substituents, and two of them have been used for the synthesis of bridged germylene complexes. The first is the reaction of $[Ir(\eta\text{-}C_8H_{14})_2(\mu\text{-}Cl)]_2$ with 4 equiv $Ge\{N(SiMe_3)_2\}_2$ at 20°C, which gives a germylene-bridged diiridium complex [Eq. (51)].[89] In this reaction, oxidative addition of two C–H bonds of the $SiMe_3$ groups to two iridium centers also occurs to form two five-membered chelate rings.

$$[Ir(\eta\text{-}C_8H_{14})_2(\mu\text{-}Cl)]_2 + :Ge(NR_2)_2 \xrightarrow{20\ °C} \quad (51)$$

4 equiv
$R = SiMe_3$

The second example is the reaction of heterobimetallic complexes with a Pt–Fe bond bridged by a bidentate diphosphine and a CO ligand with a cyclic germanium(II) amide. This reaction gives rise to the substitution of the bridging CO ligand by the germanium(II) species under very mild conditions to afford germylene-bridged heterobimetallic complexes [Eq. (52)].[90] Addition of an excess amount of the cyclic germanium(II) amide brings about the reversible substitution of the Ph_3P ligand on the platinum atom by the second germanium(II) unit.

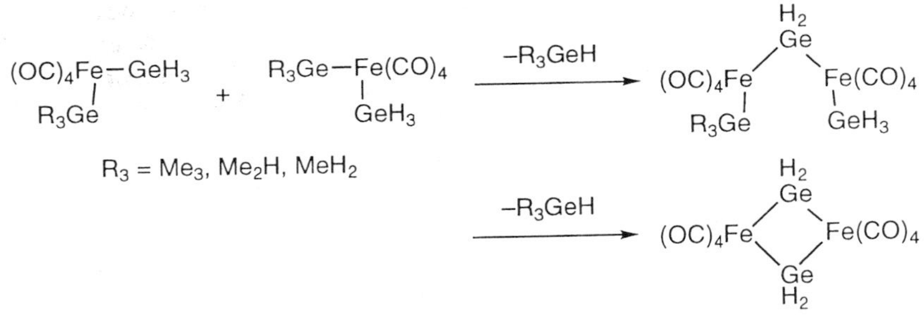

$(OC)_4Fe(GeR_3)GeH_3$ + $R_3Ge{-}Fe(CO)_4GeH_3$ $\xrightarrow{-R_3GeH}$ $(OC)_4Fe(GeR_3){-}GeH_2{-}Fe(CO)_4GeH_3$ $\xrightarrow{-R_3GeH}$ $(OC)_4Fe(\mu{-}GeH_2)_2Fe(CO)_4$

$R_3 = Me_3, Me_2H, MeH_2$

SCHEME 16

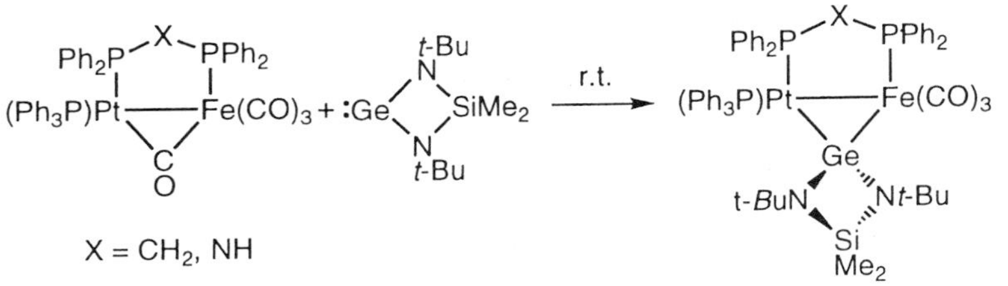

$$(Ph_3P)Pt(\mu{-}CO)(\mu{-}Ph_2PXPPh_2)Fe(CO)_3 + :Ge(\mu{-}Nt{-}Bu)_2SiMe_2 \xrightarrow{r.t.} (Ph_3P)Pt(\mu{-}Ph_2PXPPh_2)(\mu{-}Ge(Nt{-}Bu)_2SiMe_2)Fe(CO)_3 \qquad (52)$$

$X = CH_2, NH$

G. *Rearrangement of a Germylenesilylene Unit in a Complex*

Photolysis of a diiron complex containing a silylene–germylene chain $Cp(CO)_2FeMe_2SiGeMe_2Fe(CO)_2Cp$ selectively gives a $Me_3Si(Me)Ge$-bridged complex as a mixture of three equilibrating geometric isomers (Scheme 17).[91] The lack of formation of the isomeric $Me_3Ge(Me)Si$-bridged complex confirms the general tendency that the germylene complex is more stable than the silylene complex. On further photolysis, this complex transforms slowly to a *cis* and *trans* mixture of a complex doubly bridged by a silylene and a germylene. The proposed reaction mechanism is closely analogous to that of the photochemical transformation of the disilanylene analog shown in Scheme 10.

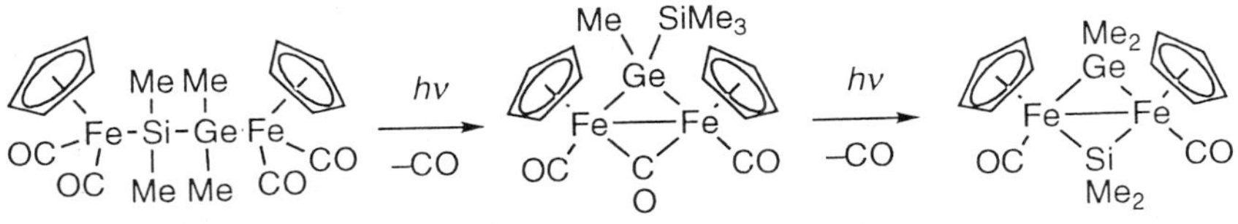

$$Cp(CO)_2Fe{-}SiMe_2{-}GeMe_2{-}Fe(CO)_2Cp \xrightarrow[-CO]{h\nu} Cp(CO)Fe(\mu{-}CO)(\mu{-}Ge(Me)SiMe_3)Fe(CO)Cp \xrightarrow[-CO]{h\nu} Cp(CO)Fe(\mu{-}GeMe_2)(\mu{-}SiMe_2)Fe(CO)Cp$$

SCHEME 17

H. *Removal of Carbonyl Ligands from Bridged Germylene Complexes*

Photodissociation of a carbonyl ligand offers a simple and practical method to convert **A**-type complexes of manganese, iron, and cobalt carbonyls to **B**-type complexes with a bridging CO ligand [Eq. (53)].[82,83,92] In the iron case, the same product has been also obtained by the photolysis of $CpFe(CO)_2GeMe_2(CH{=}CH_2)$, probably through the initial formation of the **A**-type μ-germylene complex.[93] In the manganese case, Curtis and co-workers reported that in solution the product exists as a mixture of rapidly interconverting isomers, and the structure with a bridging CO is not predominant, since the IR spectrum shows only a very weak absorption in the bridging carbonyl region (Scheme 18).[82] However, this compound in the solid state, crystallized from petroleum ether, has the structure with a bridging CO. The X-ray crystal structure and the molecular orbitals for the bridge bond have been discussed in detail.[94] The precursors of these reactions, namely singly germylene-bridged complexes of **A**-type, can be prepared by salt elimination reactions from metal carbonyl anions and dihalogermanes (see Section III,D).

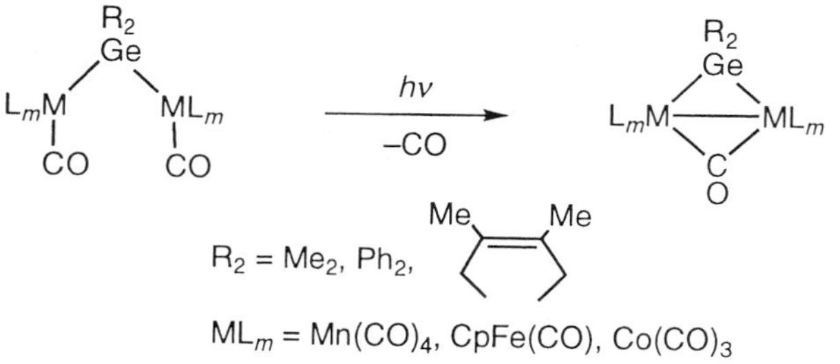

$$L_mM(CO)(\mu\text{-}GeR_2)ML_m(CO) \xrightarrow[-CO]{h\nu} L_mM(\mu\text{-}GeR_2)(\mu\text{-}CO)ML_m \qquad (53)$$

$R_2 = Me_2$, Ph_2, $CH_2C(Me){=}C(Me)CH_2$

$ML_m = Mn(CO)_4$, $CpFe(CO)$, $Co(CO)_3$

A doubly germylene-bridged diiron complex also dissociates a carbon monoxide slowly under UV irradiation to give a triply bridged complex [Eq. (54)].[86] Interestingly, photolysis or thermolysis of the corresponding osmium complex is not so simple and yields a mixture of several germylene-bridged clusters. A mechanism involving terminal germylene complexes **L** and **M** as intermediates is proposed (Scheme 19).[84]

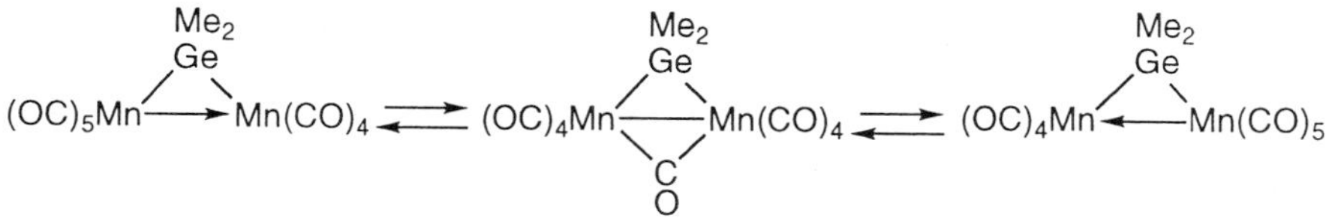

$$(OC)_5Mn\rightarrow Mn(CO)_4(\mu\text{-}GeMe_2) \rightleftharpoons (OC)_4Mn(\mu\text{-}GeMe_2)(\mu\text{-}CO)Mn(CO)_4 \rightleftharpoons (OC)_4Mn(\mu\text{-}GeMe_2)\leftarrow Mn(CO)_5$$

SCHEME 18

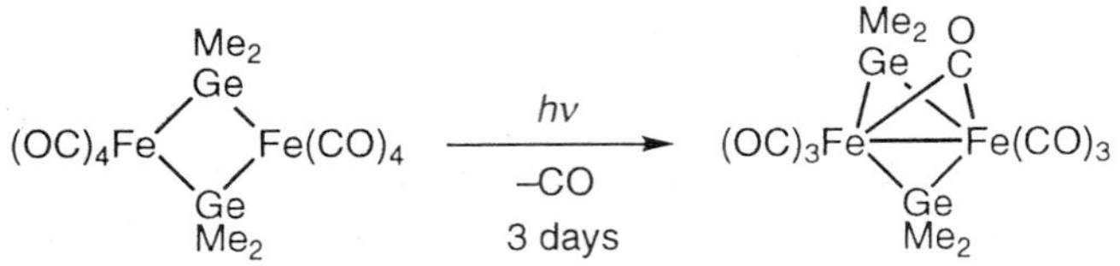

$$(OC)_4Fe(\mu\text{-}GeMe_2)_2Fe(CO)_4 \xrightarrow[-CO,\ 3\ days]{h\nu} (OC)_3Fe(\mu\text{-}GeMe_2)_2(\mu\text{-}CO)Fe(CO)_3 \quad (54)$$

I. *Replacement of Substituents on Bridged Germylene Complexes*

The Ge–H bond on the bridging germylene ligand of $Cp_2Fe_2(CO)_2$ (μ-CO)(μ-Ge*t*-BuH) is highly activated, as is the Si–H bond on the silicon analog (see Section II,J), and is halogenated by even less reactive halomethane solvents (CH_2Cl_2, CH_3I) at room temperature to give $Cp_2Fe_2(CO)_2$ (μ-CO)(μ-Ge*t*-BuX) (X = Cl, I).[78,95] Addition of DMAP to an acetonitrile

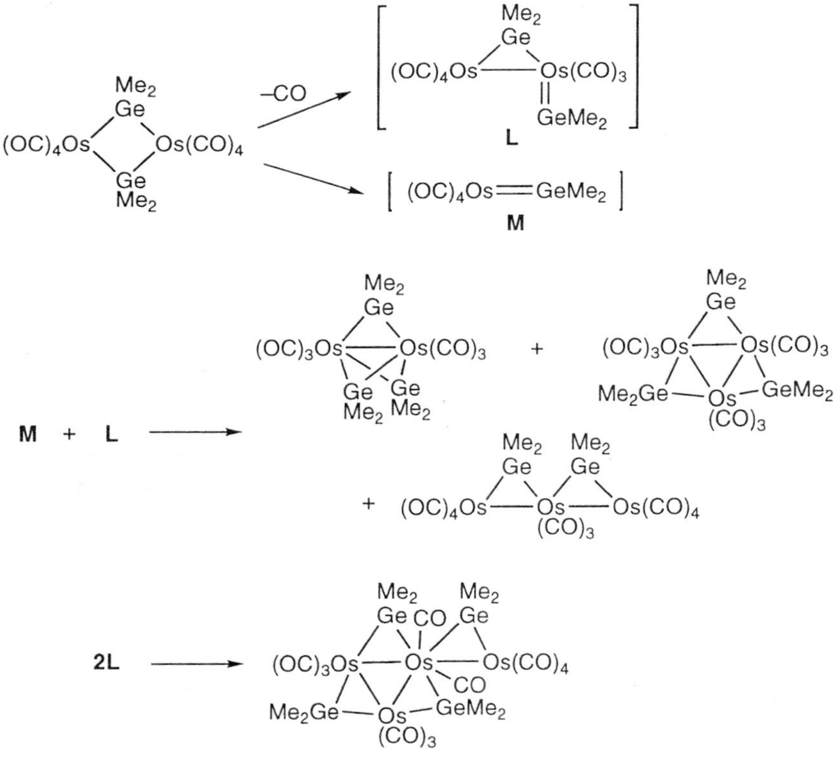

$$(OC)_4Os(\mu\text{-}GeMe_2)_2Os(CO)_4 \xrightarrow{-CO} \left[(OC)_4Os(\mu\text{-}GeMe_2)Os(CO)_3(=GeMe_2)\right]\ \mathbf{L}$$

$$(OC)_4Os(\mu\text{-}GeMe_2)_2Os(CO)_4 \longrightarrow \left[(OC)_4Os{=}GeMe_2\right]\ \mathbf{M}$$

$$\mathbf{M} + \mathbf{L} \longrightarrow (OC)_3Os(\mu\text{-}GeMe_2)_3Os(CO)_3 + Os_3(CO)_9(\mu\text{-}GeMe_2)_3 + (OC)_4Os(\mu\text{-}GeMe_2)Os(CO)_3(\mu\text{-}GeMe_2)Os(CO)_4$$

$$2\mathbf{L} \longrightarrow (OC)_3Os\text{–}Os(CO)_2\text{–}Os(CO)_4 \text{ cluster with } Os(CO)_3 \text{ and four } \mu\text{-}GeMe_2$$

SCHEME 19

solution of the iodogermylene-bridged complex only forms the equilibrium mixture of the precursor and the DMAP-stabilized cationic μ-germylyne complex. This behavior is in contrast to that of the iodosilylene-bridged analog in which the equilibrium lies far to the product side. However, addition of AgOTf (silver trifluoromethanesulfonate) led the reaction to completion to give the first base-stabilized germylyne-bridged dinuclear complex [Eq. (55)].[95]

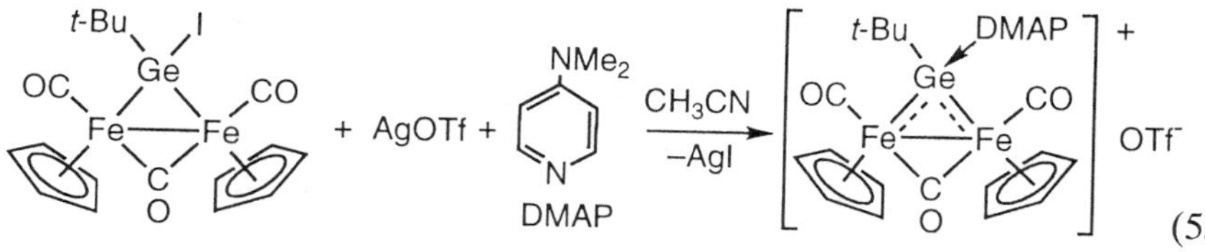

(55)

J. *Other Reactions*

Halogenation of the mesitylthio groups of a mononuclear germylene complex by boron trihalides induces the formation of halogermylene-bridged dinuclear complexes (Scheme 20).[96] In this reaction, one germylene unit is removed from the monogermylene complex and the remaining $W(CO)_5$ unit is stabilized by the coordination of the double bond of the mononuclear halogermylene complex.

Conversion of homoleptic terminal germylene complexes to trinuclear triply germylene-bridged complexes has been reported by Lappert *et al.* [Eq. (56)].[97] This reaction is induced by the action of CO under ambient conditions. The starting homoleptic terminal germylene complexes can be prepared in high yields by the reaction of $Ge\{N(SiMe_3)_2\}_2$ with $[Pd(cod)Cl_2]$ or $[Pt(cod)_2]$. The tin analogs of these complexes have been also prepared in a similar manner. The planar triangular structure of these tin analogs[98] and the germylene–platinum complex[9] have been determined by X-ray crystallography.

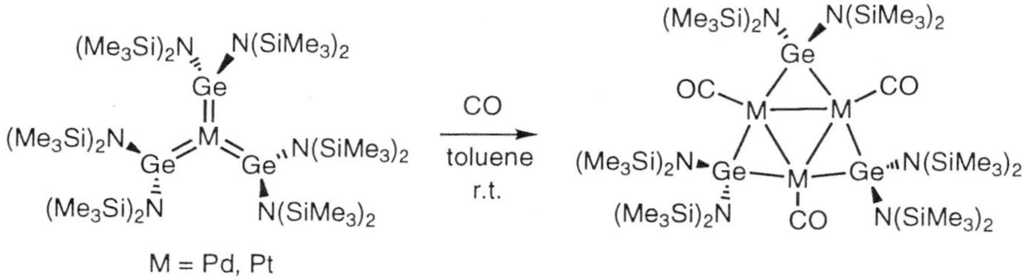

M = Pd, Pt

(56)

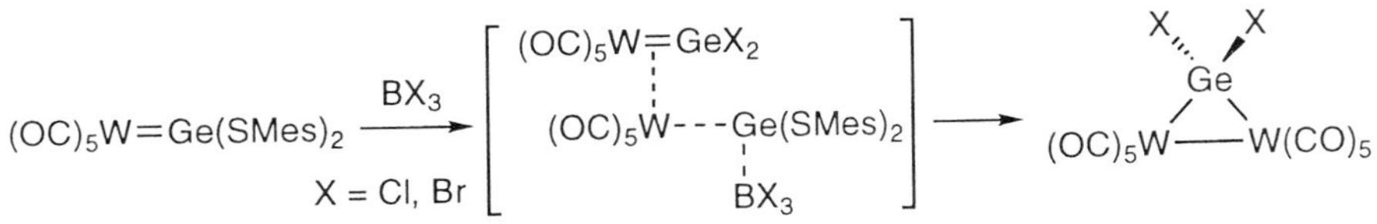

SCHEME 20

The final example of the reaction forming a Ge(II)-bridged complex is the thermolysis of $Cp_2W(SiMe_3)(GeMe_2OTf)$ [Eq. (57)].[99] In this reaction, a methyl group of a possible intermediate, bimetallic complex **O** in Scheme 21, is considered to be finally abstracted by the starting complex. In fact, the complex **O** prepared by treatment of the cationic germylyne complex **N** with MeMgBr reacts rapidly at room temperature with $Cp_2W(SiMe_3)(GeMe_2OTf)$ to regenerate **N** together with $Cp_2W(SiMe_3)(GeMe_3)$ (Scheme 21). Complex **O** may be formed by dimerization and rearrangement of the neutral mononuclear germylene complex $Cp_2W{=}GeMe_2$ generated by 1,2-elimination of Me_3SiOTf from the starting complex.

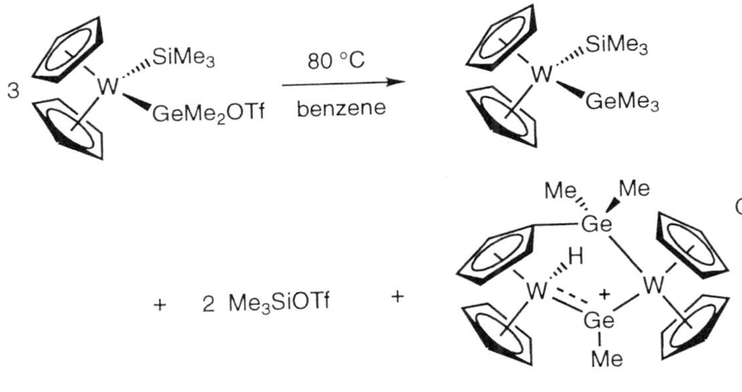

(57)

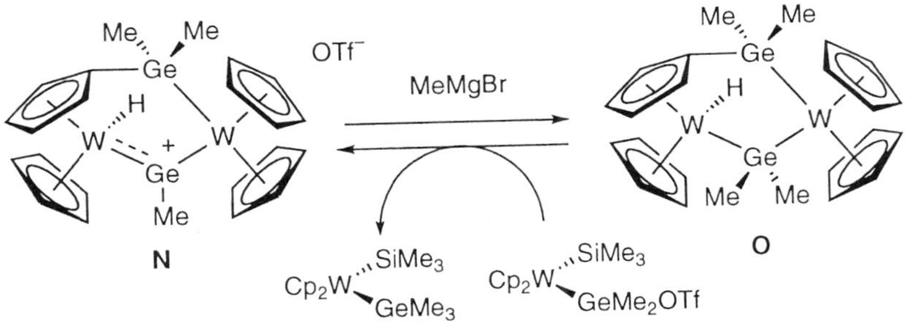

SCHEME 21

IV

PHYSICAL PROPERTIES OF BRIDGED SILYLENE AND GERMYLENE COMPLEXES

A. *Structure and Bonding*

A detailed discussion of the structure and bonding of **C**-type complexes has already appeared in the review articles of Graham[10] and Schubert,[11] and thus this topic is not covered here.

In contrast to the alkylidene complexes, the number of isolated bridged silylene or germylene complexes is much larger than that of the terminal counterparts. This is apparently due to the higher reactivity of the double bond containing silicon or germanium compared to that containing carbon. This general tendency is well demonstrated by studies on silenes ($R_2Si{=}CR_2$),[100] disilenes ($R_2Si{=}SiR_2$),[101,102] and terminal silylene complexes ($R_2Si{=}ML_n$),[103] as well as their germanium analogs.[8,102] These species are only stable when protected by bulky substituents or ligands, and they readily react with various small molecules to give stable tetravalent silicon or germanium species.

The higher stability of the bridged silylene and germylene complexes compared to their terminal counterparts is attributable to the steric and electronic factors. Sterically, the silylene or germylene group is more efficiently protected by the ligands of two metal fragments from the attack of reagents, especially nucleophiles. As for the electronic factors, **A**-, **B**-, and **C**-type complexes are discussed separately, because their orbital interactions between the silylene or germylene unit and the two metal fragments are different.

In **A**- and **C**-type complexes, the silicon or germanium can be regarded as sp^3-hybridized, so that, unlike terminal silylene or germylene complexes, there is no low-lying empty *p*-orbital on the atom that is vulnerable to the attack of nucleophiles. An exception in **A**-type complexes is the doubly bridged complexes, which are discussed later.

However, for the **B**-type complexes, Curtis *et al.* proposed an MO model for germylene-bridged dinuclear complexes[94] based on Teo's MO calculation on the doubly phosphido-bridged manganese and chromium species $[(\mu\text{-}PH_2)_2M_2(CO)_8]^n$ ($n = 0$–$2+$ for M = Mn and $n = 0$–$2-$ for M = Cr),[104] in which the metal–metal bond interacts with the essentially sp^2-hybridized germylene unit. The orbital interaction in this model is a little bit different from that reported for alkylidene-bridged complexes,[7,105,106] but fundamentally they seem to be similar. It consists of a σ-type interaction between the sp^2-hybridized orbital of the bridging ligand and the symmetry adapted orbitals of the bimetallic unit (one is shown in Scheme 22a) and a π-type

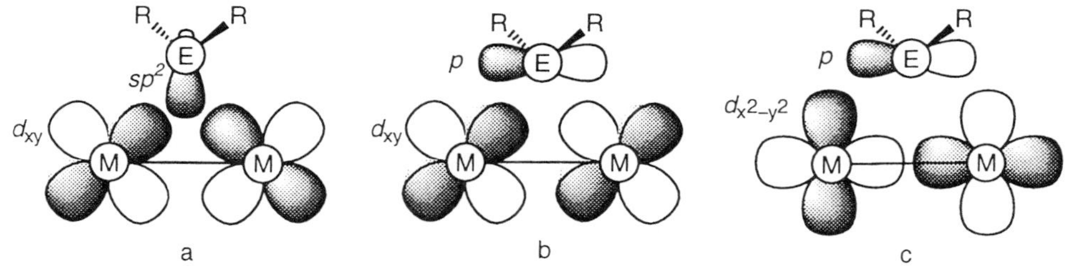

SCHEME 22

interaction between the p-orbital of the bridging ligand and the symmetry-adapted orbitals of the bimetallic unit (one is shown in Scheme 22b). Note that the π-type interaction shown in Scheme 22b reduces the antibonding character of the metal–metal π^* orbital. The nd_{z^2} and nd_{xy} as well as perhaps $(n+1)s$ and $(n+1)p$ orbitals of the metals are mainly used for metal-(bridging ligand) bonding. The interaction of the bridging ligand orbitals with the remaining nd orbitals, $nd_{x^2-y^2}$, nd_{xz}, and nd_{yz}, is weak, and therefore they are primarily metallic in nature. Importantly, the highest orbital among these metallic orbitals has the antibonding σ^*-type combination of the $nd_{x^2-y^2}$ orbitals of two metals (Scheme 22c). Thus, if this σ^*-orbital is occupied by two electrons as well as all other metallic orbitals, the M–M bond order should be zero, whereas if this orbital is vacant, the M–M bond order will become 1.[104] It should be noted that, in the latter case, a direct metal–metal bond does not necessarily exist, or rather it may be meaningless to discuss the existence of a direct metal–metal bond, because the formal bonds among the two metal fragments and the bridging ligand(s) are actually not independent.

The former case is exemplified by doubly bridged **A**-type complexes such as $\{Cp_2Ti\}_2(\mu\text{-}SiH_2)_2$,[53] $Fe_2(CO)_8(\mu\text{-}EMe_2)_2$ (E = Si, Ge),[57,59,86,88] and $\{(Et_3P)_2Pt\}_2(\mu\text{-}SiHR)_2$.[46–48] In these complexes, the M $\cdots$ M distances are much longer than the single bonds and the M-E-M internal angles are usually larger than 100° (see Tables VI and VII).

Most of the **B**-type complexes have the molecular orbitals of the latter type, although there are a few exceptions. The Ru–Ru distances in the diruthenium complexes $Cp^*_2Ru_2(\mu\text{-}H)_2(\mu\text{-}SiPh_2)_2$ [2.665(0) Å][38] and $Cp^*_2Ru_2(\mu\text{-}SiPhOMe)(\mu\text{-}OMe)(\mu\text{-}H)$ [2.569(5) Å][50] are much shorter than the Ru–Ru single bond, which is usually longer than 2.9 Å. Of course, the 18-electron rule predicts that a double bond exists between the metals in these complexes. More accurately, in these complexes, the second highest metallic orbital is also vacant, and thus the formal bond order of the Ru–Ru bonds becomes 2. In a similar manner, the triply bridged diiron complexes $Cp_2Fe_2(\mu\text{-}CO)_2(\mu\text{-}SiR_2)$ are also predicted to have a double bond between

TABLE VI

M ··· M DISTANCES AND M–Si–M BOND ANGLES IN BRIDGED SILYLENE COMPLEXES[a]

Complex	M ··· M distance (Å)	M–Si–M angle (°)	Ref.
A-type complexes			
$\{Cp_2Ti\}_2(\mu\text{-}SiH_2)_2$	3.368(10)	102.8(7)	53
$\{Cp^*Mn(CO)_2(H)\}_2(\mu\text{-}SiH_2)$	4.306	124.4(3)	43
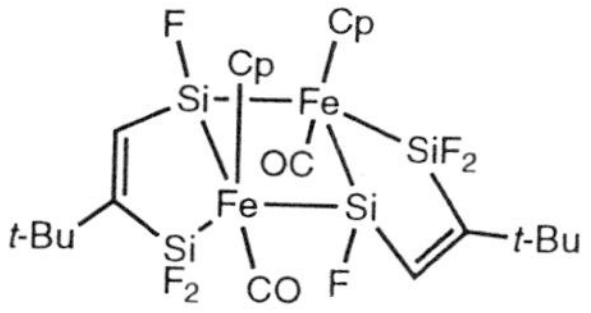	3.939	113.2(1)	58
$(Et_3P)_2Pt(\eta^2,\eta^2\text{-}Ph(H)SiSi(H)Ph)Pt(PtEt_3)_2$	3.973(1)–3.997(2)	113.6(1)–114.2(3)	47
$(Et_3P)_2Pt(\eta^2,\eta^2\text{-}Cy(H)SiSi(H)Cy)Pt(PtEt_3)_2$			
(*trans*)	4.046(1)	115.5(2)	48
(*cis*)	3.661(2)	99.4(2)	
B-type complexes			
$Mn_2(CO)_8(\mu\text{-}SiPh_2)_2$	2.871(2)	73.39(6)	107
$Fe_2(CO)_8(\mu\text{-}SiMes_2)$	2.721	68.95	20
$Fe_2(CO)_6(\mu\text{-}SiMe_2)_2(\mu\text{-}SiMeCl)$	2.705(1)	71.2(1)	64
$Cp_2Fe_2(CO)_2(\mu\text{-}CO)(\mu\text{-}Si\textit{t}\text{-}BuH)$	2.614(1)	70.28(2)	21,22
$[Cp_2Fe_2(CO)_2(\mu\text{-}CO)(\mu\text{-}Si\textit{t}\text{-}Bu\cdot NMI)]I$[b]	2.626(2)	70.96(7)	61,63
$[Cp_2Fe_2(CO)_2(\mu\text{-}CO)(\mu\text{-}Si\textit{t}\text{-}Bu\cdot DMAP)]I$[b]	2.633(2)	71.05(9)	61
$Cp_2Fe_2(\mu\text{-}CO)_2\{\mu\text{-}Si(2,4,6\text{-}\textit{i}\text{-}Pr_3C_6H_2)_2\}$	2.303(2)	58.7(1)	24
$Cp_2Fe_2(CO)_2(\mu\text{-}CO)\{\mu\text{-}SiMe(SiMe_3)\}$	2.622(1)	69.59(3)	55
$Cp^*_2Ru_2(\mu\text{-}H)_2(\mu\text{-}SiPh_2)_2$	2.665(0)	68.70(3)	38
$Cp^*_2Ru_2(\mu\text{-}SiPhOMe)(\mu\text{-}OMe)(\mu\text{-}H)$	2.569(5)	67.8	50
$\{Ru(CO)_3(SiMe_3)\}_2(\mu\text{-}SiMe_2)_2$	2.958(3)	—	108
$Rh_2(Ph_2PCH_2PPh_2)_2(CO)_2(\mu\text{-}SiHPh)_2$	2.813(1)	73.52(4) 73.47(4)	31
$Rh_2(Ph_2PCH_2PPh_2)_2(CO)_2(\mu\text{-}SiHEt)_2$	2.814(1)	73.62(6)	39
$\{(dippe)Rh\}_2(\mu\text{-}SiH\textit{n}\text{-}Bu)_2$[c]	2.8850(7)	76.34(4)	42
$Ir_2(Ph_2PCH_2PPh_2)_2(H)_2(CO)_2(\mu\text{-}SiPh_2)$	2.8361(9)	73.5(1)	32
$Ir_2(Ph_2PCH_2PPh_2)_2(H)_2(CO)_2(\mu\text{-}SiHPh)$	2.879(1)	75.7(2)	32
C-type complexes			
$\{Cp_2Ti(\mu\text{-}\eta^2\text{-}HSiHPh)\}_2$	3.866(2)	—	49
$Cp_2Ti(\mu\text{-}H)(\mu\text{-}\eta^2\text{-}HSiHPh)TiCp_2$	3.461(1)	—	49
$W_2(CO)_8H_2(\mu\text{-}SiEt_2)_2$	3.183(1)	73.97(12)	109
$\{Re(CO)_4(H)\}_2(\mu\text{-}SiPh_2)$	3.121(3)	76.1(3)	33
$Re_2(CO)_6H_4(\mu\text{-}SiEt_2)_2$	3.084(1)	74.95(7)	34
$Re_2(CO)_7H_2(\mu\text{-}SiEt_2)_2$	3.052(1)	73.7(1) 73.5(1)	35
$Fe_2(CO)_7(\mu\text{-}\eta^2\text{-}HSiPh_2)(SiPh_2H)$	2.759(1)	72.1(1)	19
$Fe_2(CO)_6(\mu\text{-}\eta^2\text{-}HSiPh_2)_2$	2.759(6)	70.8(2)	19
$\{Cp^*Ru(\mu\text{-}\eta^2\text{-}HSiEt_2)\}_2(\mu\text{-}H)(H)$	2.971(1)	74.84(4) 74.76(4)	38

(*continues*)

TABLE VI (*continued*)

Complex	M ··· M distance (Å)	M–Si–M angle (°)	Ref.
$\{(dippe)Rh\}_2(\mu\text{-H})(\mu\text{-}\eta^2\text{-HSiPh}_2)^c$	2.937(1)	74.29(3)	40
$\{(dippe)Rh\}_2(\mu\text{-}\eta^2\text{-HSiH}p\text{-Tol})_2(\mu\text{-SiH}p\text{-Tol})^c$	2.8499(9)	74.79(6)(μ-SiHR) 72.34(5) 72.92(6)	42
$\{Pt(PCy_3)(\mu\text{-}\eta^2\text{-HSiMe}_2)\}_2$	2.708(1)	69.5(3)	44

[a] Average value is written when there are two or more independent molecules in the crystal.
[b] A silylyne complex.
[c] dippe = $i\text{-Pr}_2PCH_2CH_2Pi\text{-Pr}_2$.

two iron atoms according to the 18-electron rule. The Fe–Fe distance [2.303(2) Å when R = 2,4,6-i-$Pr_3C_6H_2$], which is significantly shorter than the usual Fe–Fe single bond (2.5–2.7 Å) is consistent with the prediction. However, these complexes actually have a triplet ground state, which was confirmed by the measurement of the effective magnetic moment (μ_{eff} = 2.8–2.9 B.M. from 300 to 10 K when R = 2,4,6-i-$Pr_3C_6H_2$).[24] This means that each of the second and third highest metallic orbitals of these complexes is singly occupied.

An important result of the π-type interaction shown in Scheme 22b is the transfer of substantial electron density from the metal–metal bond into the ER_2 unit. This effect greatly reduces the electrophilicity of highly electropositive silylene and germylene ligands to make them stable enough for isolation, and also activates the cleavage of the E–R bond to form the cationic silylene- or germylene-bridged dinuclear complexes (see Section II,J).[61,63,95]

There are some general tendencies concerning the structure of bridged silylene and germylene complexes. The M ··· M distances and M–E–M angles are summarized in Tables VI and VII. The M–E–M internal angles of complexes without M–M bonds (**A**-type complexes) range rather widely from 99° to 130°. The M ···M distances are longer than 3.3 Å, which indicates that there is essentially no direct interaction between two metals. In contrast, the M–E–M angles of complexes with formal M–M bonds (**B**-type complexes) is more acute: They are between 67° and 77° in μ-silylene complexes with only one exception (58.7° in $Cp_2Fe_2(\mu\text{-CO})_2\{\mu\text{-Si}(2,4,6\text{-}i\text{-Pr}_3C_6H_2)_2\}$)[24] and between 65° and 73° in μ-germylene complexes. These values are smaller than those for alkylidene-bridged **B**-type complexes, which mostly range from 75° to 82°.[7,105] Since most of the M–E bond lengths and M–M bond lengths of the **B**-type complexes are in normal ranges of the single bonds, the magnitude of the M–E–M angles (E = C, Si, and Ge) apparently reflects the M–E bond lengths or the size of the bridging

TABLE VII

M ··· M DISTANCES AND M–Ge–M BOND ANGLES IN BRIDGED GERMYLENE COMPLEXES

Complex	M ··· M distance (Å)	M–Ge–M angle (°)	Ref.
A-type complexes			
Me, Me, Ge, H, W, +, W, Ge, Me, OTf^- [a]	—	129.9(1)	99
Me, Me, Ge, $Co(CO)_4$, $Co(CO)_4$	—	116.24(4)	83
$Os_2(CO)_8(\mu\text{-}GeMe_2)_2$	—	104.80(4)	84
B-type complexes			
$Mn_2(CO)_8(\mu\text{-}CO)(\mu\text{-}GeMe_2)$	2.854(2)	71.13(5)	94
$Fe_2(CO)_6(\mu\text{-}GeMe_2)_3$	2.750(11)	70.0(2)	71
$Cp_2Fe_2(CO)_2(\mu\text{-}CO)(\mu\text{-}GeMe_2)$	2.628(1)	68.15(3)	111
$Cp_2Fe_2(CO)_2(\mu\text{-}CO)(\mu\text{-}Ge\textit{t}\text{-}BuH)$	2.641(1)	68.56(2)	78
$[Cp_2Fe_2(CO)_2(\mu\text{-}CO)(\mu\text{-}Ge\textit{t}\text{-}Bu \cdot DMAP)]OTf$[a]	2.665(3)	70.05(7)	95
$Ru_3(CO)_9(\mu\text{-}GeMe_2)_3$	2.926(9)	71.9(3)	110
$Os_3(CO)_9(\mu\text{-}GeMe_2)_3$	2.920(1)	70.82(4)	84
$Os_2(CO)_6(\mu\text{-}GeMe_2)_3$	2.944(1)	70.68(4)	84
$Os_3(CO)_{11}(\mu\text{-}GeMe_2)_2$	2.981(1)	72.49(2)	84
$Os_4(CO)_{12}(\mu\text{-}GeMe_2)_4$	2.860(1)–3.069(1)	69.40(6)–75.05(7)	84
$Co_2(CO)_6(\mu\text{-}CO)\{\mu\text{-}GeMeCo(CO)_4\}$	2.587(2)	65.8(1)	76
$Co_4(\mu_4\text{-}SiMe)_2(CO)_{10}(\mu\text{-}GeMe_2)$	2.692(1)	70.4(1)	77
R_2N, R, N, $SiMe_2$, Me_2Si, H, Ge, RN, Ir, Ir, H, Cl, Ge, NR_2, Cl, $Ge(NR_2)_2$ ($R = SiMe_3$)	2.723(1)	66.96(2)	89
$Pt_3(CO)_3[\mu\text{-}Ge\{N(SiMe_3)_2\}_2]_3$	2.723(2)–2.735(2)	—	9

[a] A germylyne complex.

atom. Interestingly, the M–Si–M angles of complexes with $\mu\text{-}\eta^2\text{-}HSiR_2$ ligands (**C**-type complexes) are in the same range as those of the **B**-type complexes.

The molecular structures of some **B**-type silylene- and silylyne-bridged diiron complexes are shown in Fig. 1. All of (a), (b), and (c) have a bridging

a

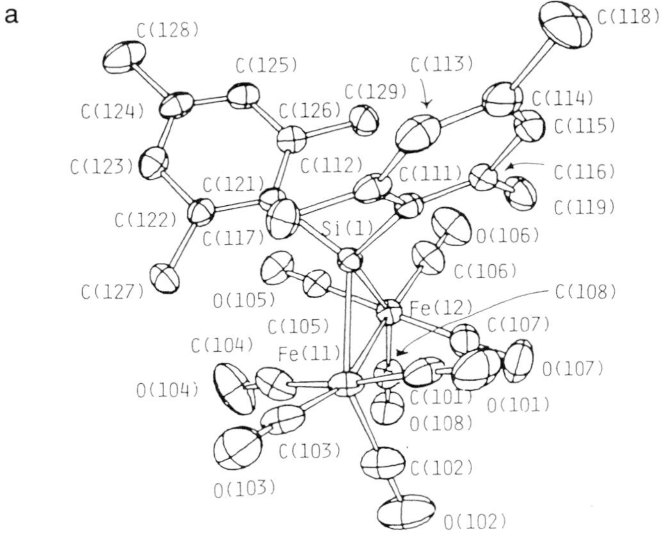

b

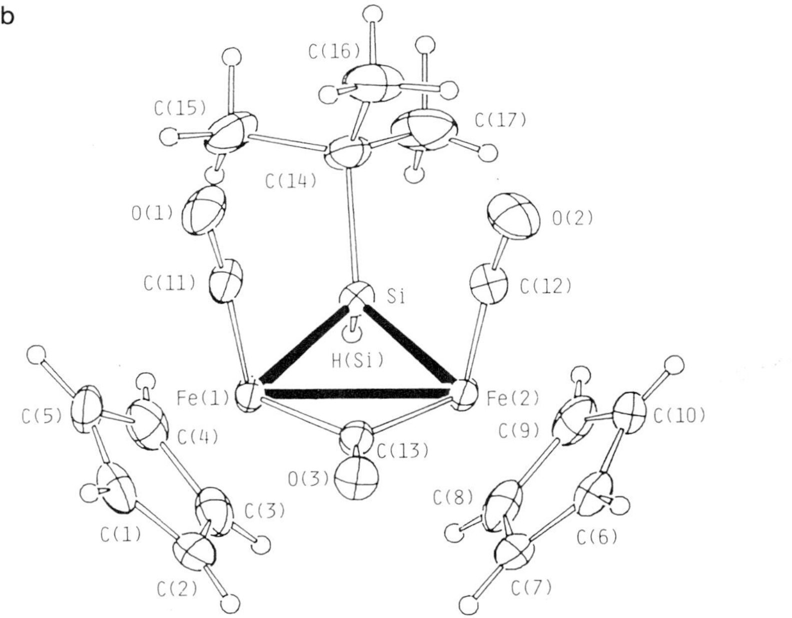

FIG. 1. Molecular structures of silylene- and silylyne-bridged diiron complexes. (a) $Fe_2(CO)_8(\mu\text{-}SiMes_2)$ (Ref. 20), (b) $Cp_2Fe_2(CO)_2(\mu\text{-}CO)(\mu\text{-}Si\textit{t}\text{-}BuH)$ (Ref. 22).

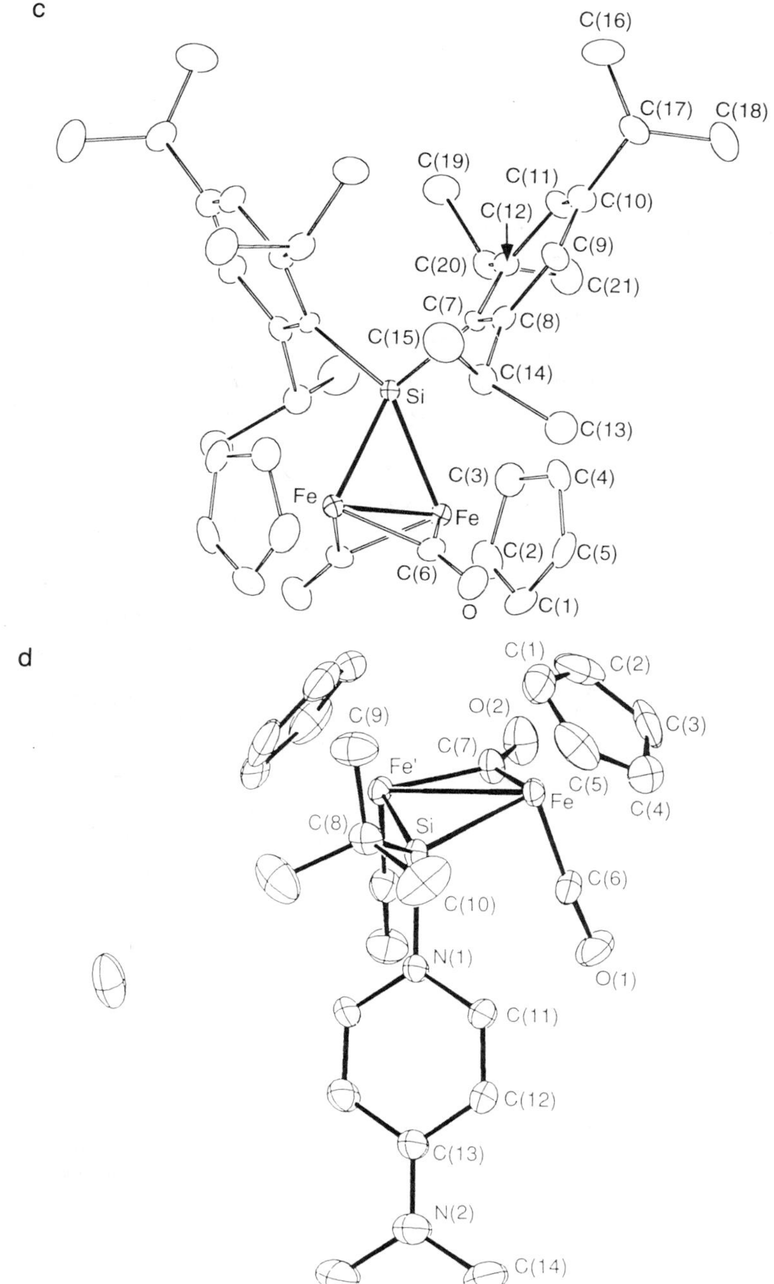

FIG. 1. (*continued*) (c) $Cp_2Fe_2(\mu\text{-}CO)_2\{\mu\text{-}Si(2,4,6\text{-}i\text{-}Pr_3C_6H_2)_2\}$ (Ref. 24), and (d) $[Cp_2Fe_2(CO)_2(\mu\text{-}CO)(\mu\text{-}Si\mathit{t}\text{-}Bu{\cdot}DMAP)]I$ (Ref. 61).

silylene ligand, and in addition, (b) has a bridging CO ligand and (c) has two bridging CO ligands. (d) is a cationic silylyne-bridged complex in which the silylyne ligand is coordinated by a strong Lewis base DMAP. The Fe–Fe distance in complex (b) [2.614(1) Å] is shorter than that in complex (a) (2.721 Å), perhaps because of the additional CO bridge in (b). The Fe–Fe distance in complex (c) [2.303(2) Å] is much shorter than that in complex (b) mainly because of the difference of the bond order as mentioned previously, but the triple bridges in (c) can be another factor that shortens the Fe–Fe distance. Three geometric isomers (two *cis* and one *trans*) are possible for complexes (b) and (d), but only the *cis*(*X*) isomer exists in each case because of the steric interaction between the cyclopentadienyl ligands and substituents on the silylene ligand (for isomerization between the geometric isomers, see Section V,A).

The Fe–Si bond lengths in the silylyne-bridged complex (d) [2.266(3) Å] and its NMI derivative [2.262(2) Å] are the shortest among the structurally determined silicon-bridged diiron complexes despite the existence of very bulky groups attached to the silicon atom.[61,63] The same tendency is also observed for the Fe–Ge bond lengths in the germylyne derivative of (d).[95] This bond shortening is obviously due to the enhanced π-type donation of electrons (see Scheme 22b) from the metal–metal bond to the positively charged silylyne or germylyne unit.

Triangular clusters with three briding germylene ligands are structurally intriguing. The molecular structure of $Os_3(CO)_9(\mu\text{-}GeMe_2)_3$[84] is shown in Fig. 2. The ruthenium analog of this complex[110] and the structurally closely related palladium and platinum complexes $M_3(CO)_3[\mu\text{-}Ge\{N(SiMe_3)_2\}_2]_3$ (M = Pd, Pt) [see Eq. (56)][97] are also known. These complexes have a triangular planar M_3Ge_3 framework. So far, the only one known example of the silicon analog of this type of complex is $Os_3(CO)_9(\mu\text{-}SiMe_2)_3$.[52]

B. *Spectroscopic Properties*

The most characteristic aspect of the spectroscopic properties of bridged silylene complexes is undoubtedly the chemical shift of the bridging silicon atom in their ^{29}Si NMR spectra. Unfortunately, the data for most of the Group 9 metal complexes and the complexes prepared before 1980 are not reported, so that almost 80% of the data are those of iron complexes. The ^{29}Si NMR chemical shifts for bridging silylene groups are listed in Table VIII.

The ^{29}Si NMR signals of bridging silylene groups in **A**-type complexes appear approximately between 60 and 160 ppm, and those of bridging silyl

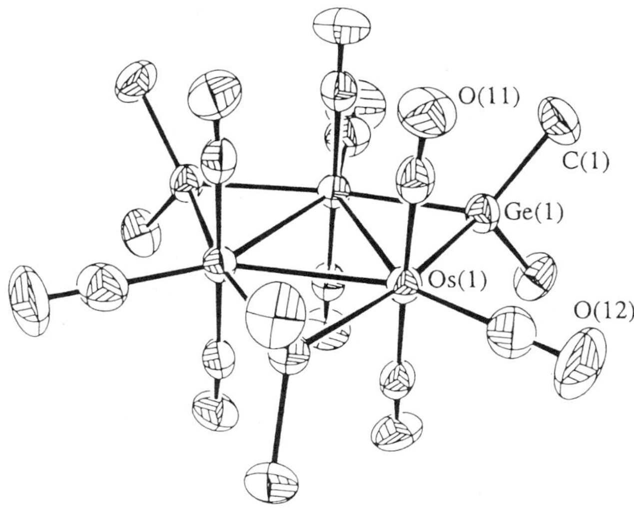

FIG. 2. Molecular structure of $Os_3(CO)_9(\mu\text{-}GeMe_2)_3$.[84]

groups in **C**-type complexes appear between 85 and 165 ppm. At least a part of this downfield shift of the signals is attributable to the diamagnetic deshielding effect originated from the polarization in $M^{\delta-}$–$Si^{\delta+}$ fashion. As to the **B**-type complexes, the complexes bridged by one silylene ligand and one carbonyl ligand with the generalized structure of $Cp_2Fe_2(CO)_2(\mu\text{-}CO)(\mu\text{-}SiR_2)$ show their signals at extremely low field (230–290 ppm). The complexes with a silylene bridge and a phosphido bridge also show their signals at the field lower than 200 ppm. These large downfield shifts can be explained by considering the paramagnetic shielding effect. This effect is well known to dominate the chemical shift of heavy atoms including ^{29}Si, and according to Ramsey's expression,[113] the chemical shift δ is inversely proportional to the average electronic excitation energy ΔE localized on the silicon atom. Thus, a decrease of ΔE leads to a downfield shift of the signal. The ΔE value of the bridging silylene ligand in the previously mentioned complexes is distinctly smaller than those of usual σ-bonded silicon compounds because the MOs of the $SiM_2(CO)$ four-membered ring involve π-type interaction. This ΔE value is supposed to correlate with the HOMO–LUMO gap of the complex, so that the complexes with downfield-shifted signals and thus small ΔE values should show the λ_{max}'s of the UV-vis spectra at the long-wavelength region. In fact, all the complexes with the structure $Cp_2Fe_2(CO)_2(\mu\text{-}CO)(\mu\text{-}SiR_2)$ are dark red because of the

TABLE VIII

^{29}Si NMR Chemical Shifts For Bridging Silylene Groups

Complex	δ(ppm)	Solvent	Ref.
A-type complexes			
$Fe_2(CO)_8(\mu\text{-}SiMe_2)_2$	161.0	C_6D_6	57
$Fe_2(CO)_8(\mu\text{-}SiPh_2)_2$	111.0	C_6D_6	112
$CpFe(CO)_2(\mu\text{-}SiMeH)Fe(CO)_2Cp$	62.85	C_6D_6	54
$CpFe(CO)_2(\mu\text{-}Si\textit{p}\text{-}TolH)Fe(CO)_2Cp$	59.52	C_6D_6	23
$CpFe(CO)_2(\mu\text{-}SiClH)Fe(CO)_2Cp$	110.72	C_6D_6	62
$CpFe(CO)_2(\mu\text{-}SiMeCl)Fe(CO)_2Cp$	142.00	C_6D_6	62
$CpFe(CO)_2(\mu\text{-}SiCl_2)Fe(CO)_2Cp$	146.65	C_6D_6	62
$CpFe(CO)_2(\mu\text{-}SiMeOH)Fe(CO)_2Cp$	120.85	C_6D_6	23
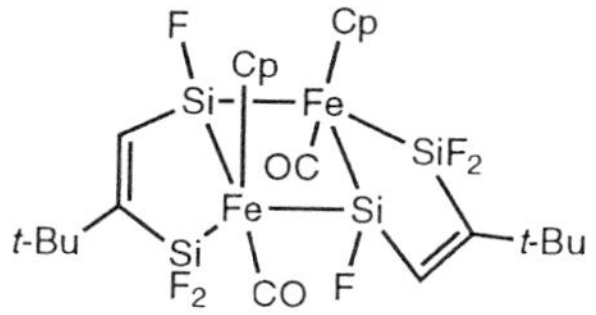F Cp Cp Si Fe SiF2 OC Fe Si t-Bu Si t-Bu F2 CO F	89.74	C_6D_6	58
B-type complexes			
$Fe_2(CO)_8(\mu\text{-}SiMe_2)$	173.0	—	65
$Fe_2(CO)_8(\mu\text{-}SiMes_2)$	145.1	C_6D_6	20
$Fe_2(CO)_8[\mu\text{-}Si\{O(2,6\text{-}\textit{i}\text{-}Pr_2C_6H_3)\}_2]$	88.4	C_6D_6	20
$Fe(CO)_4\{\mu\text{-}Si(NMe_2)_2\}Pt(PPh_3)_2$	113.9	C_6D_6	67
$\{Fe(CO)_3\}_2(\mu\text{-}CO)(\mu\text{-}SiPhH)_2$	161.71, 161.67	C_6D_6	19
$Cp_2Fe_2(CO)_2(\mu\text{-}CO)(\mu\text{-}Si\textit{t}\text{-}BuH)$	254.4	CD_2Cl_2	22
$Cp_2Fe_2(CO)_2(\mu\text{-}CO)[\mu\text{-}Si\{(CMe_2)_2H\}H]$	255.1	CD_2Cl_2	22
$Cp_2Fe_2(CO)_2(\mu\text{-}CO)(\mu\text{-}Si\textit{t}\text{-}BuCl)$	276.3	CD_2Cl_2	61
$Cp_2Fe_2(CO)_2(\mu\text{-}CO)(\mu\text{-}Si\textit{t}\text{-}BuBr)$	284.8	CD_2Cl_2	61
$Cp_2Fe_2(CO)_2(\mu\text{-}CO)(\mu\text{-}Si\textit{t}\text{-}BuI)$	289.1	CD_2Cl_2	61
$Cp_2Fe_2(CO)_2(\mu\text{-}CO)(\mu\text{-}Si\textit{t}\text{-}BuMe)$	267.4	CD_2Cl_2	61
$[Cp_2Fe_2(CO)_2(\mu\text{-}CO)(\mu\text{-}Si\textit{t}\text{-}Bu \cdot NMI)]I$[a]	251.5	CD_3CN	61, 63
$[Cp_2Fe_2(CO)_2(\mu\text{-}CO)(\mu\text{-}Si\textit{t}\text{-}Bu \cdot DMAP)]I$[a]	264.6	CD_3CN	61
$Cp_2Fe_2(CO)_2(\mu\text{-}SiMe(SiMe_3)\}$ (three isomers)	245.5, 242.7, 232.1	CD_2Cl_2	55
$Cp_2Fe_2(CO)_2(\mu\text{-}SiMe_2)_2$ (two isomers)	243.8, 229.5	CD_2Cl_2	55
$Cp^*_2Fe_2(CO)_2(\mu\text{-}CO)(\mu\text{-}Si\textit{p}\text{-}TolH)$ (two isomers) (*cis*)	250.4	C_6D_6	25
(*trans*)	235.5		
$Cp^*Fe_2(CO)_4(\mu\text{-}SiPhH)(\mu\text{-}PPh_2)$	209.1	C_6H_6	26
$Cp^*Fe_2(CO)_4(\mu\text{-}Si\textit{p}\text{-}TolH)(\mu\text{-}PPh_2)$	209.1	C_6D_6	26
$Cp^*Fe_2(CO)_4(\mu\text{-}SiMeH)(\mu\text{-}PPh_2)$	202.8	C_6D_6	26
$Cp^*_2Ru_2(\mu\text{-}SiPhOMe)(\mu\text{-}OMe)(\mu\text{-}H)$ (1st isomer)	211.12	C_6D_6	50
(2nd isomer)	193.50		
C-type complexes			
$Cp_2Ti(\mu\text{-}H)(\mu\text{-}\eta^2\text{-}HSiHPh)TiCp_2$	86.7	—	49
$Fe_2(CO)_7(\mu\text{-}\eta^2\text{-}HSiPh_2)(SiPh_2H)$	142.1	C_6D_6	19
$Fe_2(CO)_6(\mu\text{-}\eta^2\text{-}HSiPh_2)_2$	109.2	C_6D_6	19
$Cp^*_2Ru_2(\mu\text{-}H)(H)(\mu\text{-}SiPh_2CH{=}CH_2)(\mu\text{-}\eta^2\text{-}HSiPh_2)$	121.9	C_6D_6	66

TABLE VIII (*continued*)

Complex	δ(ppm)	Solvent	Ref.
$\{Cp^*Ru(\mu\text{-}\eta^2\text{-}HSiCyMe)\}_2(\mu\text{-}H)(H)$	115.17	C_7D_8	50
$\{(dippe)Rh\}_2(\mu\text{-}H)(\mu\text{-}\eta^2\text{-}HSiPh_2)^b$	137–141	C_7D_8	41
$\{(dippe)Rh\}_2(\mu\text{-}H)(\mu\text{-}\eta^2\text{-}HSiMe_2)^b$	163	C_7D_8	41
$\{(dippe)Rh\}_2(\mu\text{-}\eta^2\text{-}HSiH\textit{n}\text{-}Bu)_2(\mu\text{-}SiH\textit{n}\text{-}Bu)^b$	125.5, 90.2	C_6D_6	42
$\{(dippe)Rh\}_2(\mu\text{-}\eta^2\text{-}HSiH\textit{p}\text{-}Tol)_2(\mu\text{-}SiH\textit{p}\text{-}Tol)^b$	142-100	C_6D_6	42

[a] A silylyne complex.
[b] dippe = $i\text{-}Pr_2PCH_2CH_2Pi\text{-}Pr_2$.

existence of a band above 500 nm that can be assigned to the excitation of the $SiFe_2(CO)$ core.[61]

C. *Fluxional Behaviors*

Fluxional behavior of a bridged germylene complex was first reported by Adams and Cotton in 1970 for $Co_2(CO)_6(\mu\text{-}GeMe_2)_2$.[108] Because of the folded Co_2Ge_2 framework, there are methyl groups in two different environments in this complex, and they appear inequivalently in the ^1H NMR spectrum at −89°C. Upon raising the temperature, these two signals coalesce and become a sharp signal at room temperature. A similar rapid exchange of two inequivalent methyl groups at room temperature has also been reported for a monogermylene-bridged complex $Co_2(CO)_6(\mu\text{-}CO)(\mu\text{-}GeMe_2)$ with an analogous folded framework.[82] A converse Berry pseudorotation has been proposed for the former complex to account for the observed site exchange (Scheme 23). In this mechanism, the cobalt atoms with square-pyramidal coordination undergo pseudorotation through a trigonal–bipyramidal geometry. An alternative pathway involving the transition state or intermediate having terminal germylene ligands **P** has also been mentioned as an unlikely mechanism.[111] Although the mechanism of this behavior is still controversial, recent study on the isomerization of doubly germylene-bridged diiron complexes seems to give an evidence against the latter mechanism: The *cis* isomer of $Cp_2Fe_2(CO)_2(\mu\text{-}Ge\textit{p}\text{-}Tol_2)_2$ does not show any exchange between two inequivalent *p*-tolyl groups on germanium nor thermal *cis–trans* isomerization (see Section V,A).[79] This result demonstrates that this complex does not open to give a species having two terminal germylene ligands. Since this complex is isolobal with $Co_2(CO)_6(\mu\text{-}GeMe_2)_2$, this cobalt complex also seems unlikely to give a transition state or intermediate having terminal germylene ligands faster than the NMR time scale.

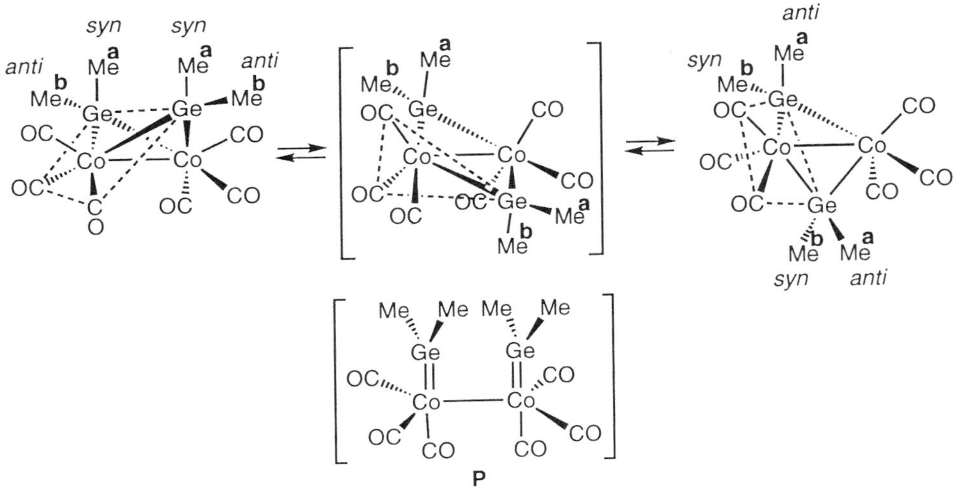

SCHEME 23

A new type of fluxional behavior of the **B**- and **C**-type rhodium complexes with hydrido ligands has been recently reported. The fluxional process involving exchange of the hydrogen atoms bound to rhodium and silicon has been observed in silylene-bridged dirhodium complexes with a cradle-like structure[31,39] (**Q** in Scheme 24). Variable-temperature ^{1}H NMR study on this system revealed that the fluxionality is due to rapid reductive elimination and oxidative addition of Si–H bonds via a silylrhodium intermediate (Scheme 25). Interestingly, the analogous diiridium complex with a monosubstituted silylene bridge **R** is not fluxional, whereas those with a disubstituted silylene bridge **S** are fluxional (Scheme 24).[32] The X-ray crystal struc-

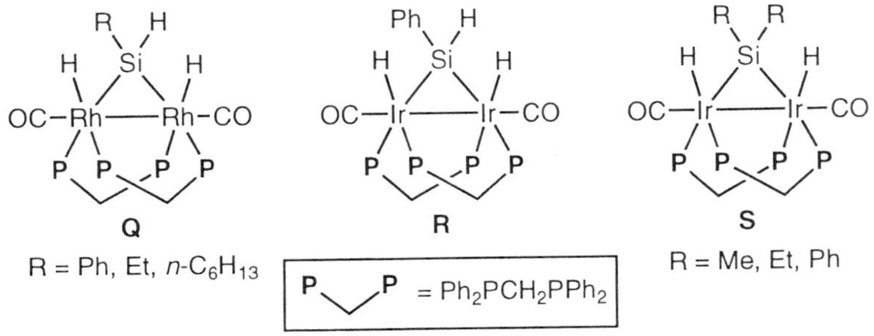

Q: R = Ph, Et, n-C_6H_{13}

R

S: R = Me, Et, Ph

P⌒P = $Ph_2PCH_2PPh_2$

SCHEME 24

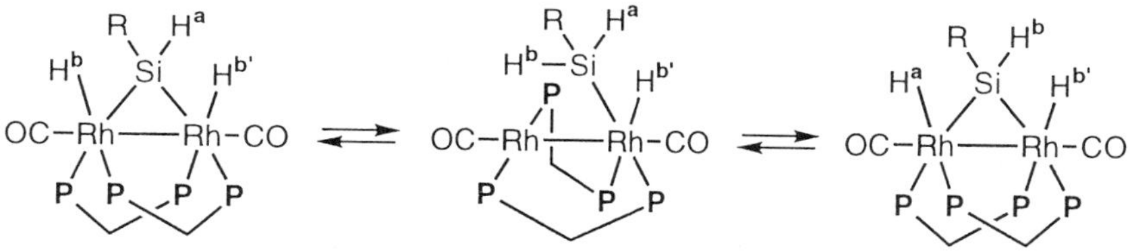

SCHEME 25

ture of **S** (R = Ph) displays a twisted configuration showing substantial steric interaction between the phenyl groups of μ-$SiPh_2$ and dppm. This steric interaction is suggested to be the origin of the fluxionality.

Variable-temperature ^1H and ^{31}P NMR spectra of **C**-type dirhodium complexes $\{(i\text{-}Pr_2PCH_2CH_2Pi\text{-}Pr_2)Rh\}_2(\mu\text{-}H)(\mu\text{-}\eta^2\text{-}HSiR_2)$ ($R_2 = Ph_2, Me_2$, PhMe) revealed the fluxionality of these complexes involving exchange of a bridging hydride and a hydrogen on silicon.[40,41] In the ^1H NMR spectrum at room temperature, a bridging hydride and a hydrogen on silicon appear equivalently at −6.17 ppm as a septet coupled with two Rh and four phosphorus atoms. At −96°C, it separates to two broad signals. The ^{31}P NMR spectrum at room temperature shows only one doublet, but it separates to four doublets at −96°C. A proposed mechanism for this exchange is shown in Scheme 26. It involves the complete oxidative addition of Si–H^a and the swinging of the resulting terminal hydride H^a into a bridging position, yielding a symmetric intermediate **T**. Subsequently, the reverse of these steps with H^b swinging out of the terminal position instead of H^a exchanges the two hydrogens.

Dirhodium complexes with two silyl bridges $\{(i\text{-}Pr_2PCH_2CH_2Pi\text{-}Pr_2)Rh(H)\}_2(\mu\text{-}\eta^2\text{-}HSiHR)_2$ (R = *n*-Bu, *p*-Tol) are also fluxional, and terminal hydrides and hydrogens bridging Si and Rh exchange rapidly at room temperature.[42] For this exchange, a mechanism involving simultaneous "twisting" of the two rhodium coordination spheres was proposed (Scheme 27). There are *cis* (major) and *trans* (minor) isomers with respect to the substituents on two silylene bridges of these complexes, and the ^1H NMR spectrum shows two Si–H signals for the *trans* isomer at room temperature. This result is inconsistent with the alternative mechanism with a "flapping" process (Scheme 28), where the Si–H in *syn* and *anti* sites would exchange and should give a single, average signal. It is also found, by introducing deuterides instead of hydrides in the complex (R = *n*-Bu), that the terminal Si–*H* and the Rh–*H* (terminal) or Rh–*H*–Si is exchanging slowly, perhaps through the dissociation of the agostic Si–H to give terminal $SiRH_2$ and recoordination of another Si–H.[42]

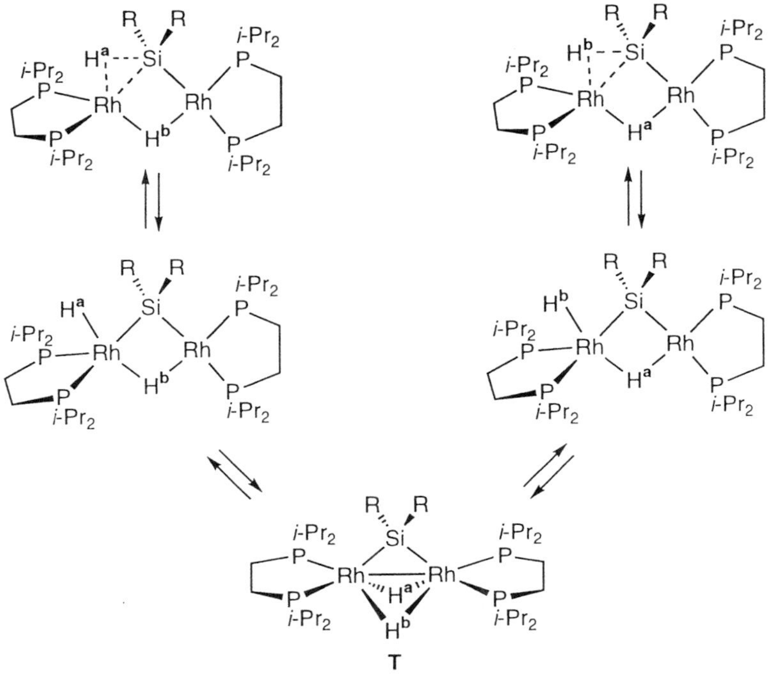

SCHEME 26

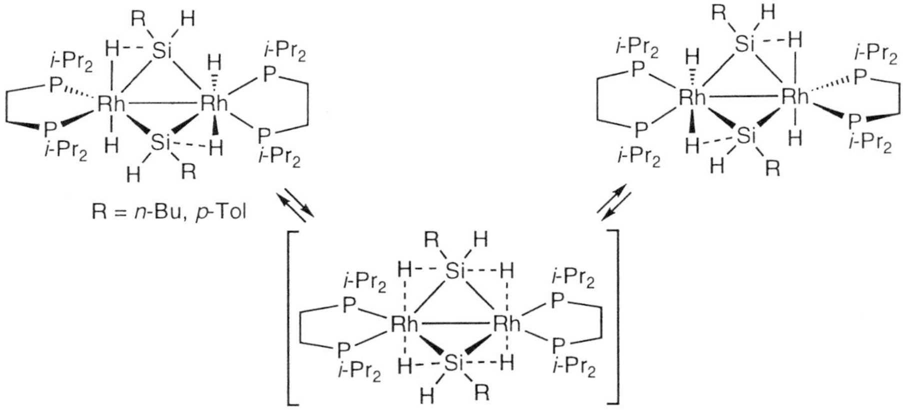

SCHEME 27

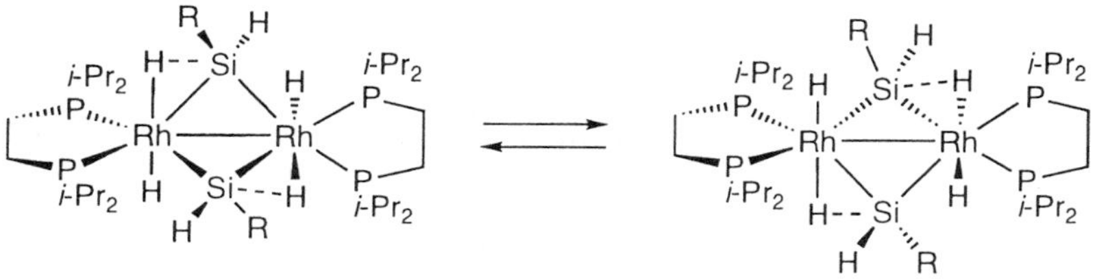

SCHEME 28

A similar rapid exchange of bridging and terminal hydride ligands and the hydrogen atoms in Ru–H–Si $3c$–$2e$ bonds is observed in the **C**-type diruthenium complexes $\{Cp^*Ru(\mu\text{-}\eta^2\text{-}HSiR_2)\}_2(\mu\text{-}H)(H)$ (R = Et, Ph) [see Eq. (13)].[38] In these complexes, all the four hydride ligands are equivalent in the ^1H NMR spectrum at 25°C, whereas they separate to three signals at −80°C.

V

REACTIONS OF BRIDGED SILYLENE AND GERMYLENE COMPLEXES

A. *Geometric Isomerizations*

Geometric isomerization is certainly one of the most intriguing reactions of bridged silylene and germylene complexes. This is also regarded as a dynamic behavior of such molecules that is closely related to the fluxional behavior discussed in Section IV,C, but the reaction rate of the former is much slower than that of the latter. The only example that has been mechanistically thoroughly investigated is the *cis–trans* isomerization of $Cp'_2Fe_2(CO)_2(\mu\text{-}CO)(\mu\text{-}ER_2)$ (Cp′ = Cp, Cp*; E = Si, Ge).

$Cp_2Fe_2(CO)_2(\mu\text{-}CO)(\mu\text{-}GeMe_2)$ is the first complex for which the geometric isomerization was reported.[82,111] The dynamic NMR as well as the structure of this complex was investigated by Cotton *et al.*[111] In solution, this complex exists as a mixture of *cis* and *trans* isomers in a ratio of *cis*:*trans* = 8:1. At room temperature, the ^1H NMR signals of Cp and Me groups of these isomers appear separately and sharply, whereas at higher temperatures (above ca. 90°C), these signals broaden and coalesce. From the line shape analysis of the temperature-dependent NMR spectra, the activation parameters, $E_a = 85.8 \pm 2.9$ kJ mol^{-1}, $\Delta H^\ddagger = 82.4 \pm 2.9$ kJ mol^{-1}, $\Delta S^\ddagger = -20.9 \pm 8.4$ J K^{-1} mol^{-1}, $\Delta G^\ddagger_{298} = 88.3 \pm 3.8$ kJ mol^{-1}, were determined. The best results were obtained when the methyl resonance of the *trans* isomer was allowed to exchange with each of the methyl resonances of the *cis* isomer and vice versa. This means that *direct exchange between the two* cis *methyl groups is not permitted.*

A proposed mechanism based on these data involves nonbridged intermediates with a terminal germylene ligand (Scheme 29) and is analogous to the Adams–Cotton mechanism originally conceived for the fluxional behavior of $Cp_2Fe_2(CO)_4$. The E_a of the *cis–trans* isomerization of $Cp_2Fe_2(CO)_4$ is about 50 kJ mol^{-1}, which is much smaller than the E_a of the process of Scheme 29. This was explained by the lower stability of nonbridged species containing a terminal germylene ligand instead of a terminal CO. A debatable assumption in the mechanism in Scheme 29 is that rotation about the Fe–Ge bond in the nonbridged intermediate is hindered because of its double-bond character. However, MO calculations[114] and also the behavior of isolated germylene complexes[115] suggest that the rotation barrier of a M=Ge bond is not very high. Actually, the mechanism proposed recently for the geometric isomerization of silylene-bridged complexes seems to be also applicable to this system (see later discussion).

For the silylene-bridged diiron complexes of the type $Cp_2Fe_2(CO)_2(\mu\text{-CO})(\mu\text{-SiRX})$, three geometric isomers are possible, as illustrated in Scheme 30. Malisch and Ries reported that the NMR spectra immediately after dissolution of crystalline $Cp_2Fe_2(CO)_2(\mu\text{-CO})(\mu\text{-SiMeH})$ show only the signals of two *cis* isomers, whereas the equilibrium containing all three

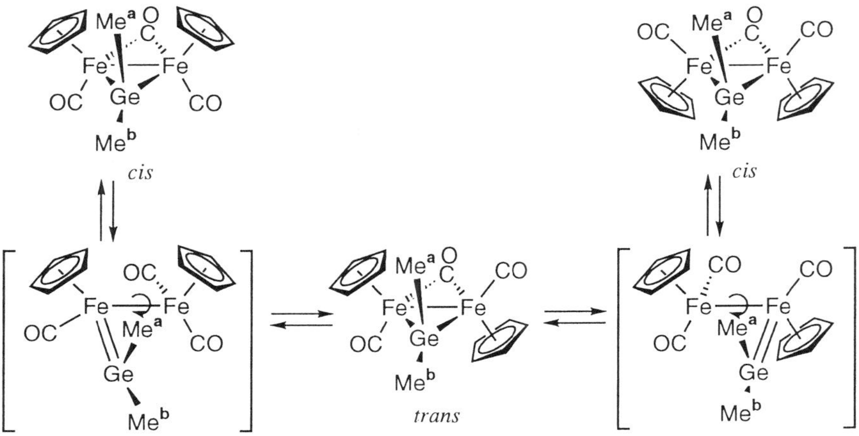

Scheme 29

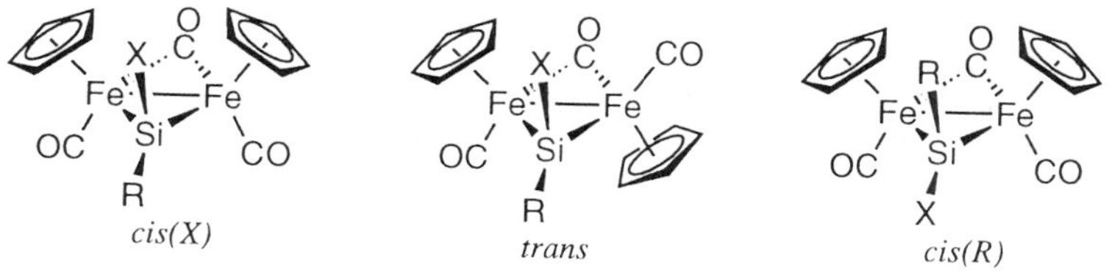

cis(X) *trans* *cis(R)*

SCHEME 30

isomers is attained after 20 min at room temperature. This result clearly shows the existence of slow thermal interconversion between these isomers.

The ratio of geometric isomers in equilibrium depends on the combination of substituents on the silylene bridge. Thus, when X is H and R is bulky *t*-Bu, only *cis(H)* isomer is observed,[21,22] but when the bulkiness of R and X becomes similar, such as R = Me and X = H[54] or R = *t*-Bu and X = Cl, Br, I, or Me,[60,61] other isomers begin to be observed [see Eq. (28)]. Furthermore, when both R and X are very bulky, such as R = Mes or 2,6-$Et_2C_6H_3$, the geometry becomes entirely *trans*.[24] The ratio is also controlled by the bulkiness of cyclopentadienyl ligands: In pentamethylcyclopentadienyl-substituted complexes $Cp^*_2Fe_2(\mu\text{-CO})(\mu\text{-E}p\text{-TolH})$ (E = Si, Ge), the ratio of the *trans* isomer exceeds 90% in thermal equilibrium.[25,69] Steric interaction between two cyclopentadienyl ligands and substituents on silylene ligands obviously plays an important role in determining these ratios.

Kinetic studies on the geometrical isomerization of two types of complexes provide insight into the mechanism of these geometric isomerizations.

The *cis(SiMe₃)* isomer of $Cp_2Fe_2(CO)_2(\mu\text{-CO})\{\mu\text{-SiMe}(SiMe_3)\}$ obtained by fractional crystallization slowly isomerizes at 296 K and gives the equilibrium mixture with the ratio of $cis(SiMe_3):cis(Me):trans = 44.0:29.1:26.9$ after 450 min.[116] Kinetic studies using ^1H NMR spectroscopy revealed that direct interconversion between the two *cis* isomers is forbidden or is very slow compared to the other two isomerizations. A proposed mechanism that can reasonably explain these experimental data is shown in Scheme 31. The basic processes in this mechanism, that is, cleavage of an Fe–(μ-CO) bond and the Fe–Fe bond to form a 18*e* and a 16*e* Fe center and coordination of solvent to the 16*e* Fe center, have been proposed to explain the results of the kinetic study on the *cis–trans* isomerization of a methylene-bridged diiron complex $Cp_2Fe_2(CO)_2(\mu\text{-CO})(\mu\text{-CH}_2)$.[117] In this mechanism, the isomerization of one *cis* isomer to the other inevitably goes through the *trans* isomer.

The alternative Adams–Cotton mechanism was ruled out because all known rotation barriers of $M{=}ER_2$ bonds (E = C, Si, Ge)[115,118] are much

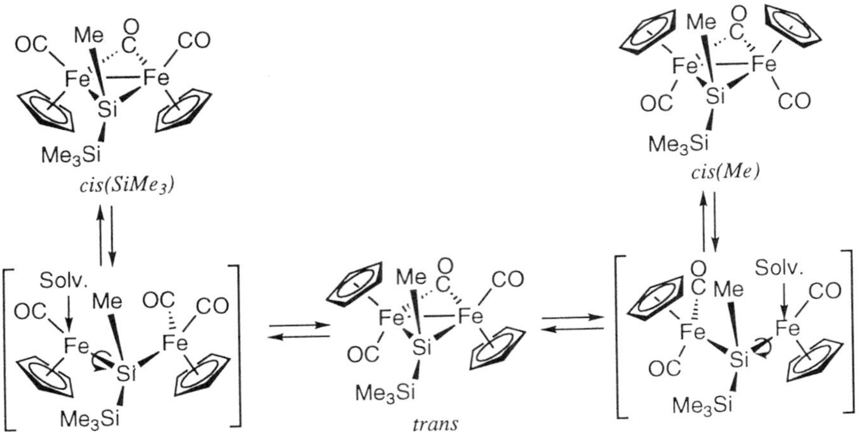

SCHEME 31

smaller than the barrier between *cis* and *trans* isomers of $Cp_2Fe_2(CO)_2(\mu\text{-}CO)\{\mu\text{-}SiMe(SiMe_3)\}$ ($\Delta G^{\ddagger}_{296}$ = 93–95 kJ mol^{-1}). If, then, the Fe=Si(Me)$SiMe_3$ bond rotates rapidly in the nonbridged intermediate, the Adams–Cotton mechanism should permit direct interconversion between the two *cis* isomers, which is actually forbidden.

Another example is the thermal and photochemical *cis–trans* isomerization of $Cp^*_2Fe_2(CO)_2(\mu\text{-}CO)(\mu\text{-}Si$*p*-TolH).[25] In this case, both *cis*(*H*) and *trans* isomers can be isolated at full purity by flash chromatography. Interconversion between these isomers occurs both thermally and photochemically in cyclohexane-d_{12}, and the composition in the thermal equilibrium state (*cis*(*H*) : *trans* = 2 : 98 at 25°C) is extremely different from that in the photostationary state (*cis*(*H*) : *trans* = 70 : 30). Kinetics of the thermal isomerization in decalin afforded the activation parameters shown in Eq. (58). The large negative activation entropies imply that this reaction also

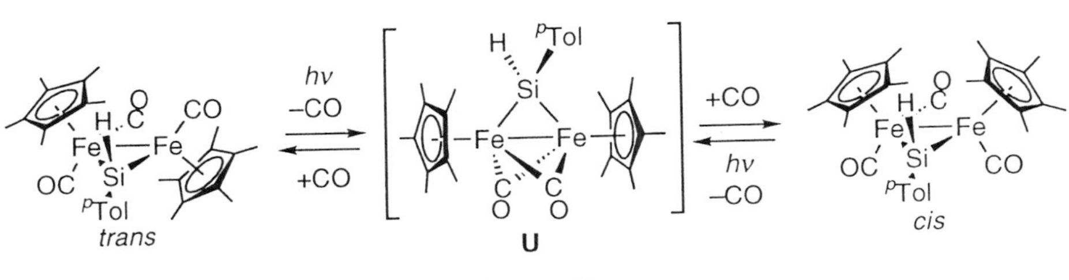

SCHEME 32

proceeds via cleavage of Fe–(μ-CO) and Fe–Fe bonds, followed by coordination of solvent to the coordinatively unsaturated Fe center.

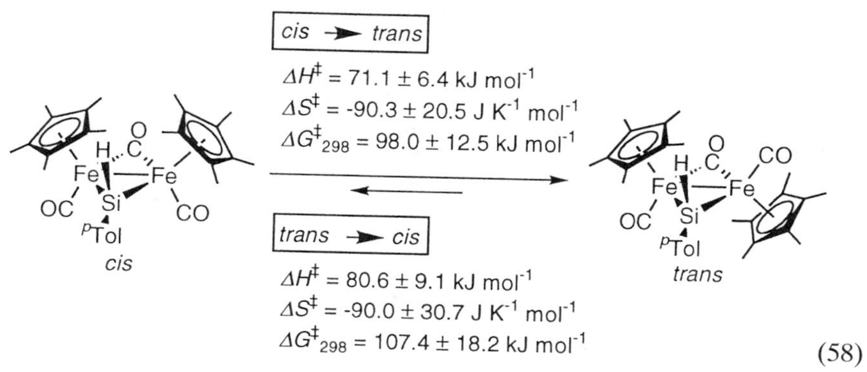

(58)

However, the photochemical isomerization seems to proceed through the dissociation of a terminal CO ligand (Scheme 32). The triply bridged complex $Cp^*_2(Fe_2(\mu\text{-}CO)_2(\mu\text{-}Si\textit{p}\text{-}TolH)$ (**U**) may be either the intermediate itself or its stabilized isomer. As mentioned in Section II,B (Scheme 3), when the substituents on a bridging silylene ligand are bulky, the triply bridged complex readily forms via photochemical dissociation of CO and can be isolated as a complex having a triplet ground state. An analogous triply bridged diiron complex $Cp^*_2(Fe_2(\mu\text{-}CO)_3$ was previously prepared by the photolysis of $Cp^*_2Fe_2(CO)_4$ and structurally determined by Wrighton *et al.*[119] The formation of $Cp_2Fe_2(\mu\text{-}CO)_3$[120,121] and carbene-bridged complex $Cp_2Fe_2(\mu\text{-}CO)_2(\mu\text{-}CHMe)$[122] in low-temperature matrices has also been established spectroscopically. These facts support the mechanism shown in Scheme 32. Furthermore, the authors have succeeded in the isolation of **U** from the solution of the photolysis of $Cp^*_2Fe_2(CO)_2(\mu\text{-}CO)(\mu\text{-}Si\textit{p}\text{-}TolH)$ under an Ar flow and determined its structure by X-ray crystallography.[123] However, the reason for the dominance of *cis*(*H*) over *trans* in the photostationary state is not yet clear. Analogous thermal and photochemical *cis–trans* isomerization was also observed for the germylene analog $Cp^*_2Fe_2(CO)_2(\mu\text{-}CO)(\mu\text{-}Ge\textit{p}\text{-}TolH)$ in C_6D_6, for which the composition in the thermal equilibrium state was *cis*(*H*) : *trans* = 5 : 95 at 25°C, whereas the photostationary state was *cis*(*H*) : *trans* = 84 : 16.[69]

Unlike the singly silylene- or germylene-bridged complexes described previously, the doubly silylene- or germylene-bridged diiron complexes $Cp_2Fe_2(CO)_2(\mu\text{-}ER_2)_2$ (ER_2 = $SiMe_2$, $Ge\textit{p}\text{-}Tol_2$) do not undergo thermal *cis–trans* isomerization.[79,124] However, irradiation of these complexes induces the isomerization. Thus, photolysis of *trans*-$Cp_2Fe_2(CO)_2(\mu\text{-}Ge\textit{p}\text{-}Tol_2)_2$ in C_6D_6 for 40 min resulted in a photostationary state with the ratio of *cis*:*trans* = 85:15.[79] It is conceivable from these results that the M–E bonds (E = Si, Ge) in bridged silylene and germylene complexes are quite sturdy and the bridging silylene and germylene ligands do not readily convert to the terminal form.

Slow isomerization with respect to the substituents on the bridging silylene ligand was reported for a diruthenium complex $Cp^*_2Ru_2(\mu\text{-Si PhOMe})(\mu\text{-OMe})(\mu\text{-H})$ [Eq. (59)].[50] This reaction proceeds at room temperature and the conversion reaches 60% after 24 h.

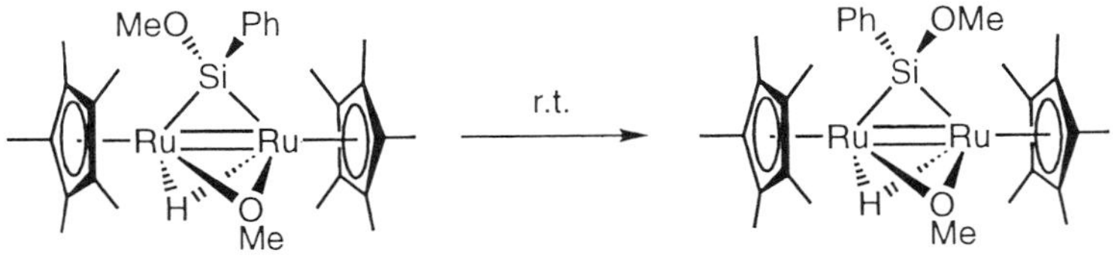

MeO Ph Si Ru Ru H O Me —r.t.→ Ph OMe Si Ru Ru H O Me (59)

B. *Miscellaneous Reactions*

Reactions with alkynes are reported for two bridged silylene complexes. Thus, the reaction of $Fe_2(CO)_8(\mu\text{-SiPh}_2)_2$ with disubstituted acetylenes on heating in toluene gives the disilametallacyclopentenes in moderate yields (Scheme 33).[17,18] Interestingly, the photolysis of these disilametallacyclopentenes in the presence of nitriles $R'CH_2CN$ leads to the formation of disilanylated enamines in high yields, which can be subsequently converted to the corresponding aldehydes $R'CH_2CHO$. This is one of very few reactions in which the silylene-bridged complex has been used as a reagent for organic synthesis.

Another example is the reaction of diruthenium complex $Cp^*_2Ru_2$

$(OC)_4Fe(\mu\text{-SiPh}_2)_2Fe(CO)_4$ + RC≡CR —Δ, toluene→ $(OC)_4Fe$–$SiPh_2$–CR=CR–$SiPh_2$ (ring)

R = Ph, Et, Me

R = Et, Me | hν, R'CH$_2$C≡N (R' = H, Ph)

R'-CH=CH–N(–$SiPh_2$–CR=CR–$SiPh_2$–) (ring)

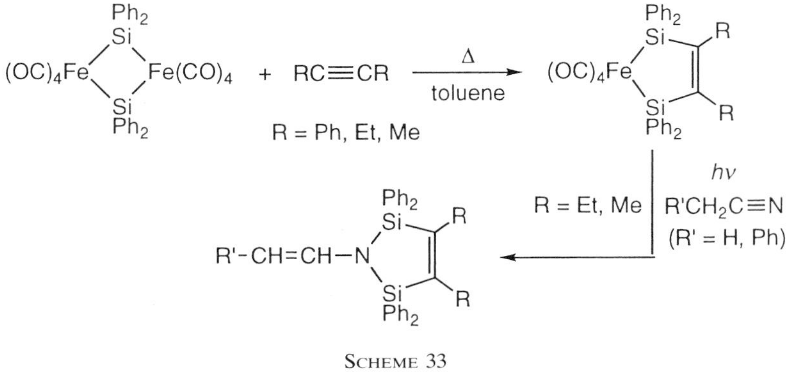

SCHEME 33

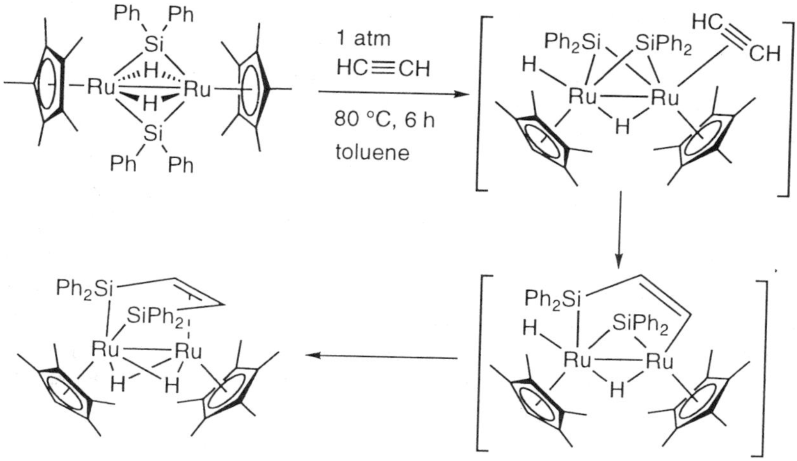

SCHEME 34

$(\mu\text{-H})_2(\mu\text{-SiPh}_2)_2$ with acetylene, which also gives a disilametallacyclopentene while maintaining the dinuclear structure (Scheme 34).[66] In the proposed mechanism, isomerization of the starting bis(μ-silylene) complex to the *cis* configuration allows an acetylene molecule to coordinate onto a Ru center, and subsequently insertion of the coordinated acetylene into the Ru–Si bond occurs. Finally, reductive coupling between two bridging ligands gives the product.

Doubly silylene-bridged **A**-type complexes are sometimes cleaved by treatment with base to the corresponding base-stabilized mononuclear silylene complexes [Eqs. (60) and (61)].[58,112] This corresponds to the inverse reaction of those described in Section II,H and provides a good method for synthesizing base-stabilized silylene complexes.

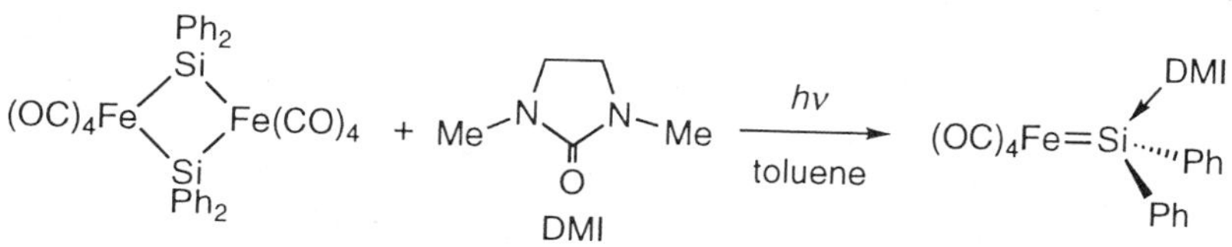

(60)

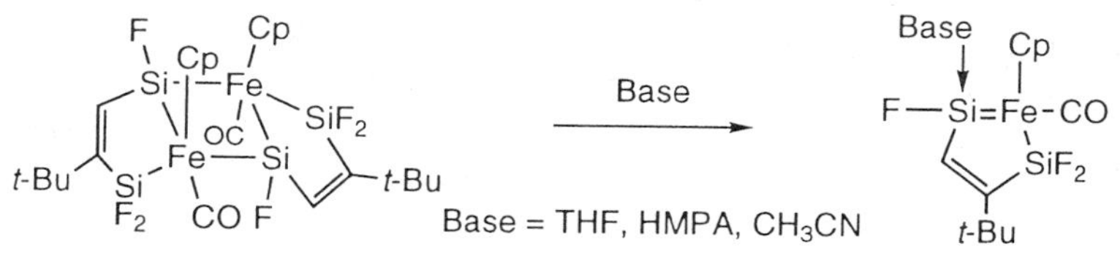

Base = THF, HMPA, CH_3CN (61)

Exchange reactions of the silylene units between primary and secondary silanes and some silylene-bridged dirhodium complexes were found to occur at room temperature [Eq. (62)].[115] For instance, the exchange reaction between $PhSiH_3$ and the complex with μ-Si(n-C_6H_{13})H occurs within a few minutes. In contrast, no exchange is observed between (n-C_6H_{13})SiH_3 and the complex with μ-SiPhH. This means that the equilibrium of Eq. (62) lies entirely on one side. Based on the experiments with various combinations, a qualitative order of the stability of the silylene bridge has been established as μ-SiPhH > μ-Si(n-C_6H_{13})H > μ-$SiPh_2$ > μ-SiMePh > μ-$SiEt_2$. Clearly, the monosubstituted μ-SiRH dihydride complexes are more stable than the disubstituted μ-SiRR′ dihydride complexes, and the reduced steric interaction between the μ-SiRH ligand and dppm phenyl groups has been proposed as the reason. Both the associative and dissociative mechanisms have been proposed for this exchange reaction.

$$[(\mathrm{CO})\mathrm{HRh}(\mu\text{-}\mathrm{SiR^1R^2})(\mu\text{-dppm})_2\mathrm{RhH}(\mathrm{CO})] \underset{\mathrm{R^1R^2SiH_2}}{\overset{\mathrm{R^3R^4SiH_2}}{\rightleftharpoons}} [(\mathrm{CO})\mathrm{HRh}(\mu\text{-}\mathrm{SiR^3R^4})(\mu\text{-dppm})_2\mathrm{RhH}(\mathrm{CO})] \quad (62)$$

P⌒P = $Ph_2PCH_2PPh_2$

VI

CONCLUDING REMARKS

In the past three decades, the chemistry of bridged silylene and germylene complexes has been developing steadily. Many complexes of this type as well as various synthetic methods for new silylene- and germylene-bridged complexes are now available. Nevertheless, it is only very recently that the reactivities of these complexes as catalysts or reagents began to be studied.[41,42,125,126] Nowadays, the cooperative reactivity of two or more metals supported in close proximity is receiving considerable attention of chemists who search for unprecedented reactivities. Since many bridged silylene and germylene complexes have relatively sturdy metal–silicon or –germanium bonds, as described in this chapter, these complexes may serve as excellent dinuclear catalysts as well as model systems for study of bimetallic reactivity. However, as theoretically expected and implied by some experiments, the bridging atom and the bonds in the bridging silylene and germylene ligands are considered to be activated for various chemical reactions. The extensive studies in this respect may open a door to fruitful chemistry involving not only the new transformation reactions of organosilicon and organogermanium compounds, but also unprecedented silicon and germanium species bridging two or more metals.

Acknowledgments

We are grateful to our various co-workers cited in the references for their effective collaboration. We are also indebted to Dr. L.-S. Luh for critical reading of the manuscript. We thank the Ministry of Education, Science, Sports and Culture of Japan, Shin-Etsu Chemical Co., Ltd., Dow Corning Toray Silicone Co., Ltd., Japan Electronic Metals Co., and Asai Germanium Research Institute for support.

References

(1) Massey, A. G.; Park, A. J.; Stone, F. G. A. *J. Am. Chem. Soc.* **1963**, *85*, 2021.
(2) Brooks, E. H.; Cross, R. J. *Organomet. Chem. Rev. A* **1970**, *6*, 227.
(3) Cundy, C. S.; Kingston, B. M.; Lappert, M. F. *Adv. Organomet. Chem.* **1973**, *11*, 253.
(4) Bonny, A. *Coord. Chem. Rev.* **1978**, *25*, 229.
(5) Nicholson, B. K.; Mackay, K. M.; Gerlach, R. F. *Rev. Silicon, Germanium, Tin, Lead Compd.* **1981**, *5*, 67.
(6) Aylett, B. J. *Adv. Inorg. Chem. Radiochem.* **1982**, *25*, 2.
(7) Herrmann, W. A. *Adv. Organomet. Chem.* **1982**, *20*, 159.
(8) Petz, W. *Chem. Rev.* **1986**, *86*, 1019.
(9) Lappert, M.; Rowe, R. S. *Coord. Chem. Rev.* **1990**, *100*, 267.
(10) Graham, W. A. G. *J. Organomet. Chem.* **1986**, *300*, 81.
(11) Schubert, U. *Adv. Organomet. Chem.* **1990**, *30*, 151.
(12) Tilley, T. D. In *The Chemistry of Organic Silicon Compounds;* Patai, S., Rappoport, Z., Eds., Wiley: New York, 1989, Chapter 24; Tilley, T. D. In *Silicon–Heteroatom Bond;* Patai, S., Rappoport, Z., Eds.; Wiley: New York, 1991, Chapters 9 and 10.
(13) Lickiss, P. D. *Chem. Soc. Rev.* **1992**, 271.
(14) Zybill, C. *Top. Curr. Chem.* **1991**, *160*, 1; Zybill, C.; Handwerker, H.; Friedrich, H. *Adv. Organomet. Chem.* **1994**, *36*, 229.
(15) Sharma, H. K.; Pannell, K. H. *Chem. Rev.* **1995**, *95*, 1351.
(16) Braunstein, P.; Knorr, M. *J. Organomet. Chem.* **1995**, *500*, 21.
(17) Corriu, R. J. P.; Moreau, J. J. E. *J. Chem. Soc., Chem. Commun.* **1980**, 278.
(18) Carré, F. H.; Moreau, J. J. E. *Inorg. Chem.* **1982**, *21*, 3099.
(19) Simons, R.; Tessier, C. A. *Organometallics* **1996**, *15*, 2604.
(20) Tobita, H.; Shinagawa, I.; Ohnuki, S.; Abe, M.; Izumi, H.; Ogino, H. *J. Organomet. Chem.* **1994**, *473*, 187.
(21) Tobita, H.; Kawano, Y.; Shimoi, M.; Ogino, H. *Chem. Lett.* **1987**, 2247.
(22) Kawano, Y.; Tobita, H.; Ogino, H. *J. Organomet. Chem.* **1992**, *428*, 125.
(23) Malisch, W.; Hindahl, K.; Käb, H.; Reising, J.; Adam, W.; Prechtl, F. *Chem. Ber.* **1995**, *128*, 963.
(24) Tobita, H.; Izumi,H.; Ohnuki, S.; Ellerby, M. C.; Kikuchi, M.; Inomata, S.; Ogino, H. *J. Am. Chem. Soc.* **1995**, *117*, 7013.
(25) Kawano, Y.; Tobita, H.; Ogino, H. *Organometallics* **1992**, *11*, 499.
(26) Hashimoto, H.; Tobita, H.; Ogino, H. *J. Organomet. Chem.* **1995**, *499*, 205.
(27) Pannell, K. H.; Cervantes, J.; Hernandez, C.; Cassias, H.; Vincenti, S. *Organometallics* **1986**, *5*, 1056.

(28) Tobita, H.; Ueno, K.; Ogino, H. *Chem. Lett.* **1986**, 1777.
(29) Ueno, K.; Nakano, K.; Ogino, H. *Chem. Lett.* **1996**, 459, and references cited therein.
(30) Wekel, H.-U.; Malisch, W. *J. Organomet. Chem.* **1984**, *264*, C10.
(31) Wang, W.-D.; Eisenberg, R. *J. Am. Chem. Soc.* **1990**, *112*, 1833.
(32) McDonald, R.; Cowie, M. *Organometallics* **1990**, *9*, 2468.
(33) Hoyano, J. K.; Elder, M.; Graham, W. A. G. *J. Am. Chem. Soc.* **1969**, *91*, 4568.
(34) Cowie, M.; Bennett, M. J. *Inorg. Chem.* **1977**, *16*, 2321.
(35) Cowie, M.; Bennett, M. J. *Inorg. Chem.* **1977**, *16*, 2325.
(36) Malisch, W.; Wekel, H.-U.; Grob, I.; Köhler, F. H. *Z. Naturforsch., B* **1982**, *37*, 601.
(37) Carreño, R.; Riera, V.; Ruiz, M. A.; Jeannin, Y.; Philoche-Levisalles, M. *J. Chem. Soc., Chem. Commun.* **1990**, 15.
(38) Suzuki, H.; Takao, T.; Tanaka, M.; Moro-oka, Y. *J. Chem. Soc., Chem. Commun.* **1992**, 476.
(39) Wang, W.-D., Hommeltoft, S. I.; Eisenberg, R. *Organometallics* **1988**, *7*, 2417.
(40) Fryzuk, M. D.; Rosenberg, L.; Rettig, S. J. *Organometallics* **1991**, *10*, 2537.
(41) Fryzuk, M. D.; Rosenberg, L.; Rettig, S. J. *Organometallics* **1996**, *15*, 2871.
(42) Fryzuk, M. D.; Rosenberg, L.; Rettig, S. J. *Inorg. Chim. Acta* **1994**, *222*, 345.
(43) Herrmann, W. A.; Voss, E.; Guggolz, E.; Ziegler, M. L. *J. Organomet. Chem.* **1985**, *284*, 47.
(44) Auburn, M.; Ciriano, M.; Howard, J. A. K.; Murray, M.; Pugh, N. J.; Spencer, J. L.; Stone, F. G. A.; Woodward, P. *J. Chem. Soc., Dalton Trans.* **1980**, 659.
(45) Powell, J.; Sawyer, J. F.; Shiralian, M. *Organometallics* **1989**, *8*, 577.
(46) Heyn, R. H.; Tilley, T. D. *J. Am. Chem. Soc.* **1992**, *114*, 1917.
(47) Zarate, E. A.; Tessier-Youngs, C. A.; Youngs, W. J. *J. Am. Chem. Soc.* **1988**, *110*, 4068.
(48) Zarate, E. A.; Tessier-Youngs, C. A.; Youngs, W. J. *J. Chem. Soc., Chem. Commun.* **1989**, 577.
(49) Aitken, C. T.; Harrod, J. F.; Samuel, E. *J. Am. Chem. Soc.* **1986**, *108*, 4059.
(50) Campion, B. K.; Heyn, R. H.; Tilley, T. D. *Organometallics* **1992**, *11*, 3918.
(51) Kerber, R. C.; Pakkanen, T. *Inorg. Chim. Acta* **1979**, *37*, 61.
(52) Brookes, A.; Knox, S. A. R.; Stone, F. G. A. *J. Chem. Soc.* (*A*) **1971**, 3469.
(53) Hencken, G.; Weiss, E. *Chem. Ber.* **1973**, *106*, 1747.
(54) Malisch, W.; Reis, W. *Angew. Chem., Int. Ed. Engl.* **1978**, *17*, 120.
(55) Ueno, K.; Hamashima, N.; Shimoi, M.; Ogino, H. *Organometallics* **1991**, *10*, 959.
(56) Pannell, K. H.; Sharma, H. *Organometallics* **1991**, *10*, 954.
(57) Leis, C.; Wilkinson, D. L.; Handwerker, H.; Zybill, C.; Müller, G. *Organometallics* **1992**, *11*, 514.
(58) Horng, K. M.; Wang, S. L.; Liu, C. S. *Organometallics* **1991**, *10*, 631.
(59) Schmid, G.; Welz, E. *Z. Naturforsch., B* **1979**, *34B*, 929.
(60) Tobita, H.; Kawano, Y.; Ogino, H. *Chem. Lett.* **1989**, 2155.
(61) Kawano, Y.; Tobita, H.; Shimoi, M.; Ogino, H. *J. Am. Chem. Soc.* **1994**, *116*, 8575.
(62) Malisch, W.; Ries, W. *Chem. Ber.* **1979**, *112*, 1304.
(63) Kawano, Y.; Tobita, H.; Ogino, H. *Angew. Chem., Int. Ed. Engl.* **1991**, *30*, 843.
(64) Simons, R. S.; Tessier, C. A. *Acta Cryst.* **1995**, *C51*, 1997.
(65) Bikovetz, A. L.; Kuzmin, O. V.; Vdovin, V. M.; Krapivin, A. M. *J. Organomet. Chem.* **1980**, *194*, C33.
(66) Takao, T.; Suzuki, H.; Tanaka, M. *Organometallics* **1994**, *13*, 2554.
(67) Bodensieck, U.; Braunstein, P.; Deck, W.; Faure, T.; Knorr, M.; Stern, C. *Angew. Chem., Int. Ed. Engl.* **1994**, *33*, 2440.
(68) Tobita, H.; Fujita, J.; Kawano, Y.; El-Maradny, A.; Ogino, H. Presented at the 8th International Conference on the Organometallic Chemistry of Germanium, Tin, and Lead, Sendai, Japan, Sept. 1995.

(69) El-Maradny, A.; Tobita, H.; Ogino, H. *Organometallics* **1996**, *15*, 4954.
(70) Brooks, E. H.; Elder, M.; Graham, W. A. G.; Hall, D. *J. Am. Chem. Soc.* **1968**, *90*, 3587.
(71) Elder, M.; Hall, D. *Inorg. Chem.* **1969**, *8*, 1424.
(72) Knox, S. A. R.; Stone, F. G. A. *J. Chem. Soc. (A)* **1971**, 2874.
(73) Knox, S. A. R.; Stone, F. G. A. *J. Chem. Soc. (A)* **1969**, 2559.
(74) Knox, S. A. R.; Stone, F. G. A. *J. Chem. Soc. (A)* **1970**, 3147.
(75) Fieldhouse, S. A.; Freeland, B. H.; O'Brien, R. J. *J. Chem. Soc., Chem. Commun.* **1969**, 1297.
(76) Anema, S. G.; Lee, S. K.; Mackay, K. M.; Nicholson, B. K.; Service, M. *J. Chem. Soc., Dalton Trans.* **1991**, 1201.
(77) Anema, S. G.; Lee, S. K.; Mackay, K. M., Nicholson, B. K. *J. Organomet. Chem.* **1993**, *444*, 211.
(78) Kawano, Y.; Sugawara, K.; Tobita, H.; Ogino, H. *Chem. Lett.* **1994**, 293.
(79) El-Maradny, A.; Tobita, H.; Ogino, H. *Chem. Lett.* **1996**, 83.
(80) Patmore, D. J.; Graham, W. A. G. *Inorg. Chem.* **1967**, *6*, 981.
(81) Curtis, M. D.; Job, R. C. *J. Am. Chem. Soc.* **1972**, *94*, 2153.
(82) Job, R. C.; Curtis, M. D. *Inorg. Chem.* **1973**, *12*, 2514.
(83) Lei, D.; Hampden-Smith, M. J.; Garvey, J. W.; Huffman, J. C. *J. Chem. Soc., Dalton Trans.* **1991**, 2449.
(84) Leong, W. K.; Einstein, W. B.; Pomeroy, R. K. *Organometallics* **1996**, *15*, 1589.
(85) Bonny, A.; Mackay, K. M. *J. Chem. Soc., Dalton Trans.* **1978**, 506.
(86) Bonny, A.; Mackay, K. M. *J. Chem. Soc., Dalton Trans.* **1978**, 722.
(87) Anema, S. G.; Audett (née Christie), J. A.; Mackay, K. M.; Nicholson, B. K. *J. Chem. Soc., Dalton Trans.* **1988**, 2629.
(88) Audett (née Christie), J. A.; Mackay, K. M. *J. Chem. Soc., Dalton Trans.* **1988**, 2635.
(89) Hawkins, S. M.; Hitchcock, P. B.; Lappert, M. F.; Rai, A. K. *J. Chem. Soc., Chem. Commun.* **1989**, 1689.
(90) Knorr, M.; Hallauer, E.; Huch, V.; Veith, M.; Braunstein, P. *Organometallics* **1996**, *15*, 3868.
(91) Sharma, H.; Pannell, K. H. *Organometallics* **1994**, *13*, 4946.
(92) Cleland, A. J.; Fieldhouse, S. A.; Freeland, B. H.; O'Brien, R. J. *J. Organomet. Chem.* **1971**, *32*, C15.
(93) Job, R. C.; Curtis, M. D. *Inorg. Chem.* **1973**, *12*, 2510.
(94) Triplett, K.; Curtis, M. D. *J. Am. Chem. Soc.* **1975**, *97*, 5747.
(95) Fujita, J.; Kawano, Y.; Tobita, H.; Ogino, H. *Chem. Lett.* **1994**, 1353.
(96) Jutzi, P.; Stroppel, K. *Chem. Ber.* **1980**, *113*, 3366.
(97) Hitchcock, P. B.; Lappert, M. F.; Misra, M. C. *J. Chem. Soc., Chem. Commun.* **1985**, 863.
(98) Campbell, G. K.; Hitchcock, P. B.; Lappert, M. F.; Misra, M. C. *J. Organomet. Chem.* **1985**, *289*, C1.
(99) Figge, L. K.; Carroll, P. J.; Berry, D. H. *Angew. Chem., Int. Ed. Engl.* **1996**, *35*, 435.
(100) Brook, A. G.; Baines, K. M. *Adv. Organomet. Chem.* **1985**, *24*, 179; Brook, A. G.; Brook, M. A. *Adv. Organomet. Chem.* **1996**, *39*, 71.
(101) West, R. *Angew. Chem., Int. Ed. Engl.* **1987**, *26*, 1201; Okazaki, R.; West, R. *Adv. Organomet. Chem.* **1996**, *39*, 231.
(102) Tsumuraya, T.; Batcheller, S. A.; Masamune, S. *Angew. Chem., Int. Ed. Engl.* **1991**, *30*, 902.
(103) Grumbine, S. D.; Tilley, T. D.; Arnold, F. P.; Rheingold, A. L. *J. Am. Chem. Soc.* **1993**, *115*, 7884, and references cited therein.
(104) Teo, B. K.; Hall, M. B.; Fenske, R. F.; Dahl, L. F. *J. Organomet. Chem.* **1974**, *70*, 413.
(105) Hahn, J. E. *Prog. Inorg. Chem.* **1984**, *31*, 205.

(106) Hofmann, P. *Angew. Chem., Int. Ed. Engl.* **1979**, *18*, 554.
(107) Simon, G. L.; Dahl, L. F. *J. Am. Chem. Soc.* **1973**, *95*, 783.
(108) Crozat, M. M.; Watkins, S. F. *J. Chem. Soc., Dalton Trans.* **1972**, 2512.
(109) Bennett, M. J.; Simpson, K. A. *J. Am. Chem. Soc.* **1971**, *93*, 7156.
(110) Howard, J.; Woodward, P. *J. Chem. Soc. (A)* **1971**, 3648.
(111) Adams, R. D.; Brice, M. D.; Cotton, F. A. *Inorg. Chem.* **1974**, *13*, 1080.
(112) Corriu, R. J. P.; Lanneau, G. F.; Chauhan, B. P. S. *Organometallics* **1993**, *12*, 2001.
(113) Ramsey, N. F. *Phys. Rev.* **1950**, *78*, 699; Jameson, C. J.; Gutowsky, H. S. *J. Chem. Phys.* **1964**, *40*, 1714; Mason, J. *Multinuclear NMR;* Plenum Press: New York, 1987.
(114) Márquez, A.; Sanz, J. F. *J. Am. Chem. Soc.* **1992**, *114*, 2903.
(115) Lee, K. E.; Arif, A. M.; Gladysz, J. A. *Organometallics* **1991**, *10*, 751; Koe, J. R.; Tobita, H.; Suzuki, T.; Ogino, H. *Organometallics* **1992**, *11*, 150; Koe, J. R.; Tobita, H.; Ogino, H. *Organometallics* **1992**, *11*, 2479.
(116) Ueno, K.;Hamashima, N.; Ogino, H. *Organometallics* **1992**, *11*, 1435.
(117) Korswagen, R. P.; Alt, R.; Speth, D.; Ziegler, M. L. *Angew. Chem., Int. Ed. Engl.* **1981**, *20*, 1049; Altbach, M. I.; Muedas, C. A.; Korswagen, R. P.; Ziegler, M. L. *J. Organomet. Chem.* **1986**, *306*, 375.
(118) Brookhart, M.; Tucker, J. K.; Flood, T. C.; Jensen, J. *J. Am. Chem. Soc.* **1980**, *102*, 1203.
(119) Blaha, J. P.; Bursten, B. E.; Dewan, J. C.; Frankel, R. B.; Randolph, C. L.; Wilson, B. A.; Wrighton, M. S. *J. Am. Chem. Soc.* **1985**, *107*, 4561.
(120) Hepp, A. F.; Blaha, J. P.; Lewis, C.; Wrighton, M. S. *Organometallics* **1984**, *3*, 174.
(121) Hooker, R. H.; Mahmoud, K. A.; Rest, A. J. *J. Chem. Soc., Chem. Commun.* **1983**, 1022.
(122) McKee, S. D.; Bursten, B. E. *J. Am. Chem. Soc.* **1991**, *113*, 1210.
(123) Kawano, Y.; Tobita, H.; Ogino, H. Presented at the 40th Symposium on Organometallic Chemistry, Japan, Hokkaido, Oct. 1993.
(124) Ueno, K.; Kanehashi, K.; Ogino, H. Presented at the 43rd Symposium on Organometallic Chemistry, Japan, Osaka, Oct.–Nov. 1996.
(125) Wang, W.-D.; Eisenberg, R. *Organometallics* **1992**, *11*, 908.
(126) Wang, W.-D.; Eisenberg, R. *Organometallics* **1991**, *10*, 2222.

ADVANCES IN ORGANOMETALLIC CHEMISTRY, VOL. 42

Organometallic Complexes in Nonlinear Optics I: Second-Order Nonlinearities

IAN R. WHITTALL, ANDREW M. McDONAGH, and MARK G. HUMPHREY

Department of Chemistry
Australian National University, Canberra, ACT 0200, Australia

MAREK SAMOC

Australian Photonics Cooperative Research Centre
Laser Physics Centre, Research School of Physical Sciences and Engineering
Australian National University, Canberra, ACT 0200, Australia

I

INTRODUCTION

Nonlinear optics arises from interactions of electromagnetic fields (light) with matter. Interactions in materials possessing nonlinear optical (NLO) properties change the nature of the incident light such that new field components with differing phase, frequency, amplitude, polarization, path, or other propagation characteristics are produced. These materials are of technological importance in areas that use optical devices, with potential applications as optical signal processors, switches, and frequency generators (exploiting such processes as harmonic generation, frequency mixing, optical parametric oscillation, and Raman shifting). Such materials may also contribute to areas such as optical data storage, optical communication, optical switching, image processing, and, ultimately, optical computing.

Materials currently employed for their NLO properties are mostly inor-

0065-3055/98 $25.00

ganics: Salts such as $LiNbO_3$ and KH_2PO_4 (KDP) are used for frequency mixing and electrooptic modulation, whereas glasses such as silica are the materials of choice in applications where third-order nonlinear processes, such as self-phase modulation, are needed. In these materials, the NLO effects of a purely electronic nature are often accompanied by lattice distortions. For example, loosely bound positive ions of lithium or potassium within crystals of the salts provide an asymmetric field, giving rise to NLO effects with response times in the order of nanoseconds. This is of importance for relatively slow NLO processes such as the electro-optic effect but is not ideal for frequency conversions that require a purely electronic NLO response. Advantages of inorganic salts include a large transparency range (UV to IR), high optical damage thresholds, high perfection of single crystals, and very low optical losses. In many applications, such as doubling the frequency of lasers or optical parametric amplification, there is a need for synchronization of the phases of the interacting optical fields (phase matching). This requirement is not always easy to satisfy, which severely limits the applicability of the nonlinearity of some materials. Quasi-phase-matched second-harmonic generation in crystals such as $LiNbO_3$, $LiTaO_3$, and $KTiOPO_4$ (KTP) has, however, been achieved, with efficiencies up to 10%.[1]

Many organic molecules (in solution or as crystals, in polymer, or in guest–host systems) have been probed for their NLO responses (for examples, see Refs. 2–14). In these compounds, the main source of the NLO response is usually the electronic nonlinearities. Organic materials possess a number of advantages for NLO applications. Some have a higher optical damage threshold than inorganic crystals, and they are generally cheaper and easier to synthesize and fabricate than inorganic materials. They possess other advantages stemming from their structural diversity and architectural flexibility, which allow for molecular design and engineering. It is possible to grow single crystals of some organics, but many of them can also be used in other forms: For example, thin films of polymers can be electric field poled to introduce the asymmetry needed for the appearance of second-order NLO effects. Low-lying electronic transitions in the UV–visible region improve the NLO efficiency in organic molecules, but a disadvantage of such molecules is a trade-off between nonlinear efficiency and optical transparency. Other disadvantages of organics may include lower thermal stability and facile relaxation of the chromophores to random orientation in poled guest–host systems.

During the evolution of the study of organic molecules for nonlinear optics, experimental observation has enabled certain structure/NLO property relationships to be developed, from which useful insights may be gained. Computational investigations using quantum theory have also afforded general qualitative rules. High polarizability of any order is associated with the existence of low-energy molecular excited states that, because they are close in energy to the ground state, mix easily when the molecule is perturbed. If an excited state is to make a large contribution to the second-order nonlinearity, there must also be a substantial difference between the excited- and ground-state dipole moments, and the transition to the excited state should have a large oscillator strength.[15]

Organic molecules containing conjugated π systems with charge asymmetry have been shown to exhibit extremely large second-order NLO properties. Donor–acceptor substituted azo dyes, Schiff bases, and stilbenes are examples of molecules with an extensive cloud of π-electrons that are easily polarized, and that show large second-order nonlinearities. The nonlinearity of such molecules can be enhanced by either increasing the conjugation length, thereby increasing electron delocalization, or increasing the strength of donor or acceptor groups to improve electron asymmetry within a molecule.[16] NLO properties of polyene-like molecules can be increased further by controlling bond-length alternation (the difference in length between C–C single and double bonds in a π-conjugated system). Changing the alternation may alter the charge distribution in the electronic ground state. For optimized alternation, enhancements of up to a factor of 5 may be observed.[17,18]

Like organic molecules, organometallic complexes can possess large NLO responses, fast response times, and ease of fabrication and integration into composites, as well as having the advantage over organic systems of much greater flexibility at the design stage. Variation in metal, oxidation state, ligand environment, and geometry can in principle permit NLO responses to be tuned in ways not possible for purely organic molecules. Studying the nonlinear properties of a variety of metals, oxidation states, and ligands in systematic series of "families" of organometallic compounds can lead to an understanding of structure–property relationships.

As described previously, large differences in excited- and ground-state dipole moments and large oscillator strengths are associated with large second-order nonlinearities. Organometallic complexes possessing metal-to-ligand or ligand-to-metal charge transfer bands in the UV–visible region of the spectrum satisfy these criteria. Novel bonding patterns and coordination geometries allow for spatial arrangements of atoms that may not be easily accessible in other systems. Organometallic compounds are often strong oxidizing or reducing agents, since metal centers may be electron rich or poor depending on their oxidation state and ligand environment. Thus, the metal center may be an extremely strong donor or acceptor, which is a requirement for electron asymmetry and hence second-order nonlinearity within a molecule. Unusual and/or unstable organic fragments (e.g., carbenes) may be stabilized on metals, allowing the NLO properties of these species to be assessed. Organometallics can also form polymers or be included in polymers, either as side chains or in the polymer backbone, which affords the possibility of introducing more polarizable atoms in a polymer chain than may be accessible in purely organic systems.

The NLO properties of organometallic compounds have been reviewed previously together with those of some related coordination complexes.[19–21] This article reviews the field through mid-1996, although some more recent results have been incorporated when possible. The first part of this article covers the theoretical background and some of the practical aspects of nonlinear optics, including a description of the origins of nonlinearities, systems of units that are encountered, experimental techniques that have been used or may be used to probe the NLO properties of organometallic complexes, and computational methods that have or could be used to calculate NLO properties. The second part of the review collects comprehensive data of organometallic complexes in tables categorized by complex type, discussions of the results of second-order NLO measurements, and the results of calculations performed on organometallic complexes. Although it is intended that coverage of transition-metal complexes be comprehensive in scope, data for metalloidal systems (such as silicon compounds) are indicative only.

II

BACKGROUND

A. *Theory of Nonlinear Optics*

Optical nonlinearities can be explained by considering the interaction of strong electric fields with matter. If the fields have optical frequencies, the phenomena resulting from the nonlinear interactions are called nonlinear optical phenomena. Most texts on nonlinear optics (e.g., Refs. 22–25) begin the discussion of this area from considerations of macroscopic relations between the vector quantities $\mathbf{P}$ (the polarization vector), $\mathbf{D}$ (the displacement vector), and $\mathbf{E}$ (the electric field vector). Chemists, however, consider the molecular origin of physical phenomena, so the description of NLO phenomena that follows starts from consideration of the behavior of a single molecule in a strong electric field.

An electric field $\mathbf{E}_{loc}$ acting on a molecule is termed a *local* field since it may differ substantially from the macroscopic field outside the medium (because of the influence of neighboring molecules). The field will, in general, distort the electron density distribution $\rho(\mathbf{r})$ in a molecule. Such a distortion may be described in terms of changes in the electron distribution moments. The first moment of the electron distribution, the dipole moment μ, is the most important quantity from the aspect of optical properties (hence, one often talks about a so-called dipolar approximation). The changes in the dipole moment induced by a relatively weak field can be expected to be linear with the magnitude of the field. However, this will not be the case for the field $\mathbf{E}_{loc}$, comparable in strength to the internal electric fields within the molecule. In these circumstances, the distortion and the induced dipole moment have to be treated as nonlinear functions of the field strength, usually being presented in terms of a power series:

$$\mu = \mu_0 + \alpha\mathbf{E}_{loc} + \beta\mathbf{E}_{loc}\mathbf{E}_{loc} + \gamma\mathbf{E}_{loc}\mathbf{E}_{loc}\mathbf{E}_{loc} + \ldots \qquad (1)$$

In many texts (e.g., some of those relating to quantum chemical calculations), Eq. (1) is treated as a Taylor series and therefore has $1/n!$ multipliers in front of the consecutive $(\mathbf{E}_{loc})^n$ terms.

The tensorial α, β, and γ quantities defined by the preceding equation are called the linear polarizability, the second-order or quadratic hyperpolarizability (or, sometimes, the first hyperpolarizability), and the third-order or cubic hyperpolarizability (the second hyperpolarizability), respectively. As both μ and $\mathbf{E}_{loc}$ are vectors, the relation between the three Cartesian components of μ and the three Cartesian components of $\mathbf{E}_{loc}$ needs nine proportionality factors, and so α is a second-rank tensor (or a 3×3 matrix). A full description of the second- and third-order interactions involves assessing the effect on the dipole moment of *combinations* of Cartesian components of the field, so β is a third-rank tensor (or a $3 \times 3 \times 3$ matrix) and γ is a fourth-rank tensor (or a $3 \times 3 \times 3 \times 3$ matrix). Fortunately, many of the tensor components of α, β, and γ are equivalent by various symmetry rules or equal to zero. The most straightforward simplification comes from permutation symmetry (the products of the Cartesian components of the field can be freely permuted), which results in some indices in the tensor elements of the polarizabilities being permuted, too.[24] Additional simplification comes from polarizabilities being invariant with respect to all point group symmetry operations. The latter rule is especially important when the second-order hyperpolarizability β is considered: In the same way that a vectorial property must be absent in an object that has a center of symmetry (the only vector that stays invariant after being inverted through a center of symmetry is a vector of zero length), all the components of β (and any third-rank tensor) must vanish in centrosymmetric point groups.

The electric field of a light wave can be expressed as

$$\mathbf{E}(t) = \mathbf{E}_0 \cos(\omega t) = \frac{\mathbf{E}_0}{2}(\exp(i\omega t) + \exp(-i\omega t))$$

Therefore, for an arbitrary point in space, Eq. (1) can be written as

$$\begin{aligned}\mu(t) &= \mu_0 + \alpha\mathbf{E}_0\cos(\omega t) + \beta\mathbf{E}_0^2\cos^2(\omega t) + \gamma\mathbf{E}_0^3\cos^3(\omega t) + \dots \\ &= \mu_0 + \tfrac{1}{2}\alpha\mathbf{E}_0\exp(i\omega t) + \tfrac{1}{2}\beta\mathbf{E}_0^2 + \tfrac{1}{4}\beta\mathbf{E}_0^2\exp(2i\omega t) \\ &\quad + \tfrac{3}{8}\gamma\mathbf{E}_0^3\exp(i\omega t) + \tfrac{1}{8}\gamma\mathbf{E}_0^3\exp(3i\omega t) + \text{c.c.} + \dots\end{aligned}$$

where c.c. stands for complex conjugate terms. It is easily seen from the preceding expansions in terms of exponential factors or, equivalently, trigonometric relations such as $\cos^2(\omega t) = \frac{1}{2} + \frac{1}{2}\cos(2\omega t)$ that the effect of the nonlinear terms in the dipole moment expansion has been to introduce contributions at different frequencies: the second-order (β) term has introduced a time-independent (dc) contribution (optical rectification) as well as a term oscillating at the frequency of 2ω (the second harmonic generation component). It can readily be verified that the quadratic term also provides a frequency mixing phenomenon if the input field is a sum of two components with different frequencies, and that a constant (dc) field may influence an oscillating field if the two are combined in a medium containing second-order nonlinear molecules [the linear electro-optic (Pockels) effect]. In a similar way, the cubic term in Eq. (1) leads to various nonlinear optical effects, one being oscillation of the induced dipoles at 3ω (third harmonic generation).

Equation (1) is, strictly speaking, not suitable for optical fields, which are rapidly varying in time. Even for linear polarization, the oscillation of the induced dipole moment may be damped (by material resonances) and thereby phase-shifted with respect to the oscillation of the external electric field. The usual way of expressing this phase shift is by considering the relationship between the Fourier components of the induced effect (oscillation of the induced dipole) and the stimulus (the electric field), with the damping and phase shift conveniently expressed by treating the terms involved as complex. Thus, the linear polarizability can be written as

$$\Delta\mu^{(1)}(\omega) = \alpha(\omega)E(\omega)$$

where $\alpha(\omega)$ is complex, $E(\omega)$ is the Fourier amplitude of the field at frequency ω, and $\Delta\mu^{(1)}(\omega)$ is the linear component of the oscillation of the dipole at the same frequency. Dispersion of α (the frequency dependence of the linear polarizability) will show characteristic rapid changes of the real part of α and enhanced values of the imaginary part of α near to the resonance frequencies of the molecule.

In the same way, frequency-dependent hyperpolarizabilities can be defined as complex quantities by considering the relations between the nonlinear (quadratic and cubic) components of the induced dipole moment oscillations at particular frequencies. A complication is that more than a single field frequency is, in general, involved. The usual notation is

$$\Delta\mu^{(2)}(\omega_3) = \beta(-\omega_3;\, \omega_1,\, \omega_2)E(\omega_1)E(\omega_2)$$

and

$$\Delta\mu^{(3)}(\omega_4) = \gamma(-\omega_4;\, \omega_1,\, \omega_2,\, \omega_3)E(\omega_1)E(\omega_2)E(\omega_3)$$

for the quadratic and cubic nonlinear optical effects, respectively. The first frequency in the brackets describing the frequency dependence of the hyperpolarizability refers to the output frequency, and the remaining frequencies are those of the input fields. Positive and negative signs of the frequencies can occur, depending on the type of interaction; for example, the β responsible for second harmonic generation is represented as $\beta(-2\omega;\, \omega, \omega)$, whereas β for optical rectification is written as $\beta(0;\, -\omega, \omega)$. Dispersion of β and γ is therefore quite complicated, the dispersion of β being a function in two-variable space (the frequency ω_3 is always $\omega_3 = \omega_1 + \omega_2$) and the dispersion of γ needing three-variable space for a full description. It should be noted that resonant behavior of the hyperpolarizabilities (a rapidly changing real part and enhanced imaginary part) is expected not only when any of the frequencies in $\beta(-\omega_3;\, \omega_1,\, \omega_2)$ or $\gamma(-\omega_4;\, \omega_1,\, \omega_2,\, \omega_3)$ approaches a resonance, but also when some combination of the input frequencies is close to a resonance. One of the best-known examples of such behavior is that the so-called degenerate third-order hyperpolarizability $\gamma(-\omega;\, \omega,\, -\omega,\, \omega)$ can be expected to exhibit resonant behavior when 2ω approaches resonance.

Macroscopic description of the NLO phenomena is very similar to the microscopic approach presented previously. The macroscopic quantities of interest are the susceptibilities of various orders defined by

$$\mathbf{P} = \varepsilon_0(\chi^{(1)}\mathbf{E} + \chi^{(2)}\mathbf{E}^2 + \chi^{(3)}\mathbf{E}^3 + \ldots) \text{ in the SI system}$$

or

$$\mathbf{P} = \chi^{(1)}\mathbf{E} + \chi^{(2)}\mathbf{E}^2 + \chi^{(3)}\mathbf{E}^3 + \ldots \text{ in the cgs system}$$

$\chi^{(i)}$ are tensors of the same ranks as their corresponding molecular analogues and, again, the equation relating the polarization to the macroscopic optical field is rewritten in terms of the Fourier components of the polarization and of the input fields. The important problem is, therefore, expressing the macroscopic NLO properties in terms of the molecular ones.

The usual way of treating the optical properties of systems containing organic molecules is in terms of the oriented gas model, that is, the macroscopic property is treated as the sum of molecular contributions, allowing for orientation of the molecules and for differences between the local field and the macroscopic electrical field. By following the usual approach of transforming tensor properties from one coordinate system to another using matrices of orientational cosines, expressions for macroscopic nonlinear susceptibilities can be derived. For example, the second-order susceptibility $\chi^{(2)}$ of a crystal composed of organic molecules with second-order hyperpolarizability β will be equal to[26]

$$\chi^{(2)}_{IJK}(-\omega_3; \omega_1, \omega_2) = L_I(\omega_3)L_J(\omega_1)L_K(\omega_2) \sum_{t=1}^{p} N_t b^t_{IJK}(-\omega_3; \omega_1, \omega_2)$$

where the L factors are the local field factors (often approximated by the Lorentz expression $L = (n^2 + 2)/3$, where n = refractive index), and

$$b^t_{IJK}(-\omega_3; \omega_1, \omega_2) = \frac{1}{N_g} \sum_{ijk} \sum_{s=1}^{N_g} \cos \theta^{(s)}_{Ii_t} \cos \theta^{(s)}_{Jj_t} \cos \theta^{(s)}_{Kk_t} \beta_{ijk}(-\omega_3, \omega_1, \omega_2)$$

In the above, ijk denote the Cartesian coordinates of a molecule, IJK those of a crystal (a unit cell), N_t is the number of molecules in a unit volume occupying each particular inequivalent site in the unit cell, p is the number of inequivalent positions of a molecule in a unit cell, and N_g is the number of equivalent positions in a unit cell. The directional cosines are used to transform each of the molecular β components to those of the new coordinate system (b_{IJK}) and the contributions are summed.

As mentioned previously, the second-order nonlinearities are mostly used for various frequency mixing schemes. Among the possible processes, several have specific technological applications and are therefore of significant interest: (i) second harmonic generation, that is, the $\omega + \omega \rightarrow 2\omega$ mixing process, which doubles the energy of photons (e.g., to convert infrared into visible); (ii) the linear electronoptic (Pockels) effect, that is, the $\omega + 0 \rightarrow \omega$ process, which is often used to modulate the phase or amplitude of a light wave (to make it carry information); and (iii) parametric generation, that is, the $\omega \rightarrow \omega_1 + \omega_2$ process, which involves splitting an energetic photon into a sum of two less energetic ones (a popular way of generating laser beams at tunable wavelengths).

B. *Systems of Units*

The two common unit systems employed for the description of nonlinear optical properties are the SI (or MKS) and Gaussian (or cgs) systems (Boyd[24] mentions an alternative system of SI units that are not discussed further in this article, as it has not been used with organometallic complexes). In the Gaussian system, properties are described in units of esu.

The main source of confusion arises from the fact that not only do the units vary between each system, but the dimensions of the properties also vary: for example, the polarizability α has dimensions of length cubed in the Gaussian system (units: cm^3) but dimensions of charge times length squared times inverse potential (units: $C\ m^2\ V^{-1}$) in the SI system. Furthermore, vacuum permittivity ε_0 exists in the SI system (having units of $F\ m^{-1}$) but has no equivalent in the Gaussian system (i.e., $\varepsilon_0 = 1$). It is important to be able to convert between the two systems, but care must be taken to ensure not only that the units are converted correctly but also that the quantities of interest are treated according to their different definitions in different systems.

The dimensions of the first-, second-, and third-order susceptibilities in both systems are simply derived from the polarizability power series equation.[24] In the Gaussian system, polarization **P** and electric field strength **E** have equivalent dimensions [units: $statV\ cm^{-1} = statC\ cm^{-2} = (erg\ cm^{-3})^{1/2}$] and are related by

$$\mathbf{P} = \chi^{(1)}\mathbf{E} + \chi^{(2)}\mathbf{E}^2 + \chi^{(3)}\mathbf{E}^3 + \ldots$$

so that the electrostatic units of the susceptibilities are as follows:

$\chi^{(1)}$ dimensionless

Units of $\chi^{(2)}$ = units of $1/E$ = $cm\ statV^{-1} = (erg\ cm^{-3})^{-1/2}$

Units of $\chi^{(3)}$ = units of $1/E^2$ = $cm^2\ statV^{-2} = (erg\ cm^{-3})^{-1}$

The preceding units are traditionally written as esu for any of the preceding quantities.

In the SI system, **P** and **E** have different dimensions (units of **P**: $C\ m^{-2}$, and units of **E**: $V\ m^{-1}$), and **P** is related to **E** by

$$\mathbf{P} = \varepsilon_0[\chi^{(1)}\mathbf{E} + \chi^{(2)}\mathbf{E}^2 + \chi^{(3)}\mathbf{E}^3 + \ldots]$$

where $\varepsilon_0 = 8.85 \times 10^{-12}\ F\ m^{-1}$. The units of the susceptibilities in the SI system are

$\chi^{(1)}$ dimensionless

Units of $\chi^{(2)}$ = units of $1/E$ = $m\ V^{-1}$

Units of $\chi^{(3)}$ = units of $1/E^2$ = $m^2\ V^{-2}$

Similarly, the units for α, β, and γ in both SI and Gaussian systems can be derived from the equation describing polarization on the molecular scale (noting units μ_{SI}: C m, and units μ_{cgs}: $statV\ cm^2$) and are given in Table I.

To convert $\chi^{(1)}$, $\chi^{(2)}$, and $\chi^{(3)}$ between the systems of units, it is also necessary to include a factor of 4π, as the displacement vector **D** is defined differently for the two systems. In the Gaussian system,

$$\mathbf{D}_{cgs} = \mathbf{E}_{cgs} + 4\pi\mathbf{P}_{cgs} = \mathbf{E}_{cgs}(1 + 4\pi\chi^{(1)}_{cgs})$$

and in the SI system,

$$\mathbf{D}_{SI} = \varepsilon_0\mathbf{E}_{SI} + \mathbf{P}_{SI} = \varepsilon_0\mathbf{E}_{SI}(1 + \chi^{(1)}_{SI})$$

so that

$$4\pi\chi^{(1)}_{cgs} = \chi^{(1)}_{SI}$$

To convert α, β, and γ between the systems of units, we use the fact that μ in SI (units: C m) and μ in cgs (units: StatC cm) are defined the same way and, since 1 C = 3×10^9 StatC, the conversion factor is $\mu_{SI} = \frac{1}{3} \times 10^{-11}\,\mu_{cgs}$. The unit system conversions for the polarizability and hyperpolarizabilities can be deduced by analogy to the derivation of the susceptibilities (i.e., rearranging the equations and equating $\alpha_{SI}E_{SI}/\mu_{0SI} = \alpha_{cgs}E_{cgs}/\mu_{0cgs}$, etc.). A summary of units and conversion factors for important properties is shown in Table I.

C. *Experimental Techniques*

A variety of experimental techniques have been used to obtain both qualitative and quantitative information about the second-order optical

TABLE I

UNITS AND CONVERSION FACTORS FOR IMPORTANT PROPERTIES

Property		System of units		Conversion factor
		SI	cgs	
Dipole moment	μ	C m	statC cm = statV cm^2	$\mu_{SI} = 1/3 \times 10^{-11}\,\mu_{cgs}$
Electric field	E	V m^{-1}	statV cm^{-1} = (erg cm^{-2})$^{1/2}$	$E_{SI} = 3 \times 10^4\,E_{cgs}$
Linear polarizability	α	C m^2 V^{-1}	cm^3	$\alpha_{SI} = (1/3)^2 \times 10^{-15}\,\alpha_{cgs}$
Linear susceptibility	$\chi^{(1)}$	Dimensionless	Dimensionless	$\chi^{(1)}_{SI} = 4\pi\,\chi^{(1)}_{cgs}$
Hyperpolarizability	β	C m^3 V^{-2}	cm^2 statV^{-1} = esu	$\beta_{SI} = (1/3)^3 \times 10^{-19}\,\beta_{cgs}$
Second-order susceptibility	$\chi^{(2)}$	m V^{-1}	cm statV^{-1} = (cm^3 erg^{-1})$^{1/2}$ = esu	$\chi^{(2)}_{SI} = (4\pi/3) \times 10^{-4}\,\chi^{(2)}_{cgs}$
Second hyperpolarizability	γ	C m^4 V^{-3}	cm^5 statV^{-2} = esu	$\gamma_{SI} = (1/3)^4 \times 10^{-23}\,\gamma_{cgs}$
Third-order susceptibility	$\chi^{(3)}$	m^2 V^{-2}	cm^2 statV^{-2} = cm^3 erg^{-1} = esu	$\chi^{(3)}_{SI} = (4\pi/3)^2 \times 10^{-8}\,\chi^{(3)}_{cgs}$

nonlinearities of materials. This section includes descriptions of those techniques that have been used, or have potential use, for the measurements of second-order nonlinear optical properties of organometallics. For an excellent source of information about other techniques, the interested reader is directed to Ref. 25.

1. *Kurtz Powder Technique*

In this technique,[27] a laser beam is directed onto a solid microcrystalline ("powder") sample that is sometimes immersed in an index matching liquid. The emitted second harmonic light is collected, filtered, detected, and compared with a standard (usually a urea powder for the organometallic complexes measured to date). This technique is crude: The magnitude of the response depends on particle size, and care must be taken in preparation of samples (e.g., sieving samples to ensure a narrow particle size range).

Materials can generally be classed as phase matchable or nonphase matchable (Fig. 1). For a nonphase matchable material, the second harmonic is only generated effectively over distances smaller than the coherence length (the coherence length L_c for a second harmonic process is related to the fundamental wavelength λ_ω and the refractive indices of the material at the fundamental n_ω and second harmonic $n_{2\omega}$ by $L_c = [\lambda_\omega/4(n_{2\omega} - n_\omega)]$). For light paths smaller than the coherence length, the intensity of the second harmonic increases with the square of the interaction distance, so for small crystal sizes there is an increase of the second harmonic intensity

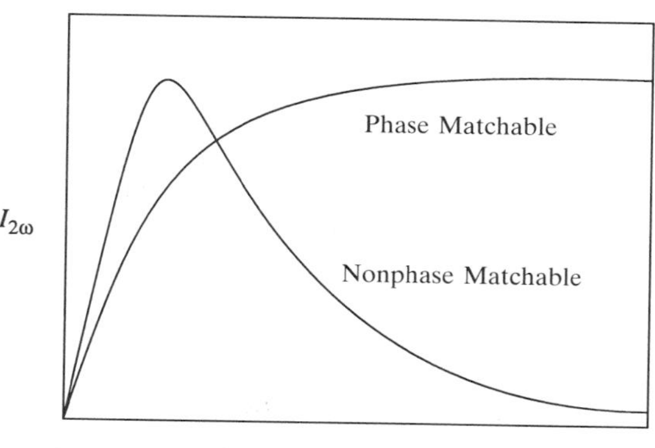

FIG. 1. Dependence of the second-harmonic intensity on particle size for phase-matchable and nonphase matchable materials. (Adapted with permission from S. K. Kurtz and T. T. Perry, *J. Appl. Phys.* **1968**, *39*, 3798. Copyright © 1968 American Institute of Physics.)

with the crystal size. However, when the crystal sizes are of the order of the coherence length (averaged over crystal orientations: The coherence length also depends on the direction of propagation of light in the crystal and on the polarization of the interacting fields), there is no further increase of the light intensity with the propagation distance and the powder second harmonic generation (SHG) signal actually decreases, which can be understood as due to the fact that the number of crystals decreases as their sizes grow. For materials that are phase matchable, there is a direction of propagation for which the second harmonic intensity grows quadratically with the propagation path without a limit (the coherence length is formally infinite in such a direction). It follows that the SHG intensity should not show a falling-off behavior in such a case because the decrease in the number of crystallites as they become bigger will now be compensated by the contribution from phase-matched interactions. Figure 1 shows the difference between the non-phase-matched and phase-matched cases as predicted by Kurtz and Perry.[27]

Results from Kurtz powder SHG studies do not have quantitative significance as the light intensities measured in the powder SHG technique depend on several factors. The magnitude of the tensor components of the molecular hyperpolarizability β is only one of these factors. A very important issue is how the molecules are oriented in the unit cell of the crystal: In a centrosymmetric arrangement, the unit cell hyperpolarizability tensor components are all identically equal to zero, but even in noncentrosymmetric groups there may be very substantial differences in the nonlinear coefficients arising from packing the nonlinear molecules in different ways. Unit cell hyperpolarizability is transformed into macroscopic second-order susceptibility with the contribution of local field factors; this can modify the properties also. The efficiency for second harmonic generation depends critically on the coherence lengths, which depend in turn on crystal optics. The measured second harmonic intensities also depend on factors such as reflection coefficients at the crystal–air interfaces (introduction of an index matching fluid is therefore important) and absorption and scattering of fundamental and second harmonic light. Given these complicating factors, the powder technique cannot be considered to give quantitative information about the molecular properties of molecules in the crystals being investigated. Observation of high powder SHG is a good indication of large β for a compound, whereas the absence of strong powder SHG does not necessarily preclude high molecular nonlinearities.

Materials that crystallize in centrosymmetric space groups theoretically cannot exhibit second harmonic generation, so the applicability of this technique is limited as about three-quarters of organometallic complexes crystallize centrosymmetrically. Despite the need for noncentrosymmetric crystal packing and the reservations concerning the quantitative nature of the results, the Kurtz powder technique has been used extensively to characterize organometallics because it can be used to screen a large number of samples quickly and is a convenient method for studying second harmonic generation of materials without the need to grow large single crystals.

2. *Solvatochromism*

Solvatochromism[28,29] is the shift of the absorption spectrum of a molecule with varying solvent. The use of this phenomenon for the determination of β is based on the two-level microscopic model of the first hyperpolarizability, in which the infinite sum-over-states quantum perturbation expression for β is reduced to two states, the ground and excited states. This model allows β_{CT} (β in the direction of the charge-transfer axis) to be determined in terms of other measurable microscopic quantities, namely the ground to excited-state transition energy ω_{eg} (from λ_{max} in the electronic absorption spectrum), the transition dipole moment μ_{eg} (from integration of the absorption band in the UV–vis spectrum), the ground-state dipole moment μ_g (measured separately), and the excited-state dipole moment μ_e. The last quantity is found by measuring the solvatochromic shift of λ_{max} of the solute, in solvents of varying polarity for which the dielectric constant and refractive index are known (or can be measured). The value for β_{CT} is then obtained from

$$\beta_{CT} = \frac{3}{2}\hbar^2\left[\frac{\mu_{eg}^2(\mu_e - \mu_g)}{(\omega_{eg}^2 - \omega^2)(\omega_{eg}^2 - 4\omega^2)}\right]$$

This technique has been applied to organic compounds where charge transfer is dominated by one transition; this is not often the case for organometallics. The applicability of this technique to organometallics thus far has not been tested; there are no reports where solvatochromism has been used to examine the second-order nonlinearities of transition-metal organometallics, and only one report of its application to two organoboron compounds.[30]

3. *SHG Measurements of Single Crystals and Films*

Detailed information about the components of the second-order susceptibility $\chi^{(2)}(-2\omega; \omega, \omega)$ can be obtained from second harmonic measurements on well-defined samples such as single crystals or oriented thin films, the latter obtained by procedures such as the asymmetric Langmuir-Blodgett deposition technique or electric-field poling of NLO chromophore-doped polymers.[31] In the case of single-crystal samples, the second harmonic is generated by a crystal with plane parallel surfaces and the distance of interaction of the fundamental and second harmonic radiation is varied by rotation of the sample about an axis perpendicular to the laser beam axis. This produces minima and maxima of the second harmonic that are called Maker fringes. Alternatively, a wedge made out of a crystal may be used to vary the interaction length. The amplitude of the second harmonic is proportional to $[\chi^{(2)}]^2[L_c \sin(L/L_c)]^2$, L being the variable interaction distance and L_c the coherence length. Comparison of measurements with a well-characterized material, such as quartz, allows $\chi^{(2)}$ to be determined. The coherence length is obtained from the periodicity of the signal and can be compared to that obtained from measurements of crystal refractive indices. The major disadvantage of the technique is the requirement for large (of the order of cubic millimeters) single crystals. Large crystals are difficult to obtain for organometallic complexes, and this is probably the main reason this technique has never been applied to purely organometallic materials.

A second harmonic signal can also be obtained from thin films of some compounds if the films are made noncentrosymmetric. A technique similar to that of Maker fringe measurements can then be used to determine the second-order nonlinearity. Rotating a second-order optically nonlinear thin film deposited on a second-order inactive substrate (e.g., glass) will result in an angularly dependent second harmonic with no fringes, whereas a glass plate covered on both sides with an NLO film shows an angular dependence of the second harmonic with fringes similar to those obtainable from a crystal. For both films and crystals, additional information about individual tensor components of $\chi^{(2)}$ can be obtained by varying the polarization of the fundamental radiation. An important issue is that these measurements determine the macroscopic nonlinearity, and obtaining information about the molecular hyperpolarizabilities usually requires certain assumptions, such as about the nature of local field corrections.

4. *Electric-Field-Induced Second Harmonic Generation*

In the electric-field-induced second harmonic generation (EFISH) technique, a liquid or solution sample is subjected to a high voltage dc pulse to align molecules, the pulse being synchronized with the laser beam pulse. $\chi^{(2)}$ can then be observed in what was previously an isotropic medium. All materials will produce an EFISH signal as it is formally a third-order nonlinear process described by the susceptibility $\chi^{(3)}(-2\omega; \omega, \omega, 0)$. There are two contributions to this susceptibility, one of them arising from the sum of the orientationally averaged third-order hyperpolarizabilities $\gamma(-2\omega; \omega,$

$\omega, 0)$ of the medium, and another due to the vectorial sum of the components of the second-order hyperpolarizabilities. Molecules that possess a permanent dipole μ partially align with the dc field, the degree of the alignment usually described in terms of the Langevin function. The net second-order effect can be shown to depend on the $\mu \cdot \beta_{vec}$ product, where μ is the dipole moment of the molecule and β_{vec} is the vectorial component of the second-order hyperpolarizability (the hyperpolarizability β is a symmetric third-rank tensor that can be treated as being composed of a vector part and a septor part).[32] In general, the directions of β_{vec} and of μ do not coincide. The effective hyperpolarizability measured by the EFISH technique can be defined as β_{EFISH} and is given by $\mu \cdot \beta_{vec} = \mu\beta_{EFISH}$. For molecules with strong electron donor and acceptor groups (at opposing ends of the molecule), β_{CT} (the hyperpolarizability along the charge-transfer axis) usually accounts for most of β_{EFISH}.

A wedge-shaped cell is used to hold a solution of the sample. This is translated in a direction perpendicular to the incident laser beam, creating Maker fringes whose periodicity is related to the wedge design and to the coherence length, which can therefore be determined. An analogous measurement can also be made on a reference wedge such as quartz, but most often a measurement on a pure solvent of well-known properties, e.g., chloroform, is used to calibrate the system.

The EFISH third-order macroscopic susceptibility defined as $\Gamma = 3\chi^{(3)}(-2\omega; \omega, \omega, 0)$ is related to the microscopic second hyperpolarizability γ' by local field factors and the molecule number density. In turn, β can be obtained from $\gamma' = \gamma + \mu\beta_{EFISH}/(5k_bT)$, where γ' is the effective second hyperpolarizability (third-order hyperpolarizability), γ is the intrinsic second hyperpolarizability consisting of electronic and vibrational parts, k_b is Boltzmann's constant, and T is the temperature in K. In the experiment, comparison against a reference enables Γ values to be determined. To determine the $\mu\beta_{EFISH}$ product of an unknown substance, one usually performs EFISH measurements as a function of concentration in a well-characterized solvent; this concentration dependence study is necessary to resolve ambiguities occurring because the $\mu\beta_{EFISH}$ products for the solvent and the solute may be of the same or of opposite signs, and the SHG signal is proportional to the square of the EFISH susceptibility. Other quantities that may be required for the interpretation of the results are the dielectric constant, the permanent dipole moment, and the intrinsic second hyperpolarizability of the solute (found from a separate experiment or ignored).

The application of EFISH to organometallics is limited to neutral complexes. The presence of ionic species makes it impossible to apply high electric fields to a solution. It is also not possible to use EFISH when the complex has no net dipole moment.

5. *Hyper-Rayleigh Scattering*

The hyper-Rayleigh scattering (HRS)[33,34] technique involves detecting the incoherently scattered second harmonic light generated from an isotropic solution to determine the first hyperpolarizability. HRS is due to orientational fluctuations of asymmetric molecules in solution that give rise to local asymmetry, on a microscopic scale, in an isotropic liquid.[34] The light scattered from such a system can have a component at the second harmonic that depends only on the first hyperpolarizability of the solute molecules and varies quadratically with the incident intensity; an example of the data obtained from an HRS experiment is displayed in Fig. 2. The solute concentration has a linear relationship with the square of the nonlinearity of all of the molecules in the system (Fig. 3), so measurements of different concentrations of solute allow β^2 to be extracted. The experimental setup for HRS is shown in Fig. 4. A seed injected, Q-switched laser is used to pump the HRS cell. The incident intensity and polarization are controlled by a half-wave plate polarizer/combination and monitored by a photodiode or energy meter. The incident beam is focused into the sample solution. A concave mirror, with its focus at the interaction focal volume, and a lens are used to collect the scattered light that is filtered to isolate the second

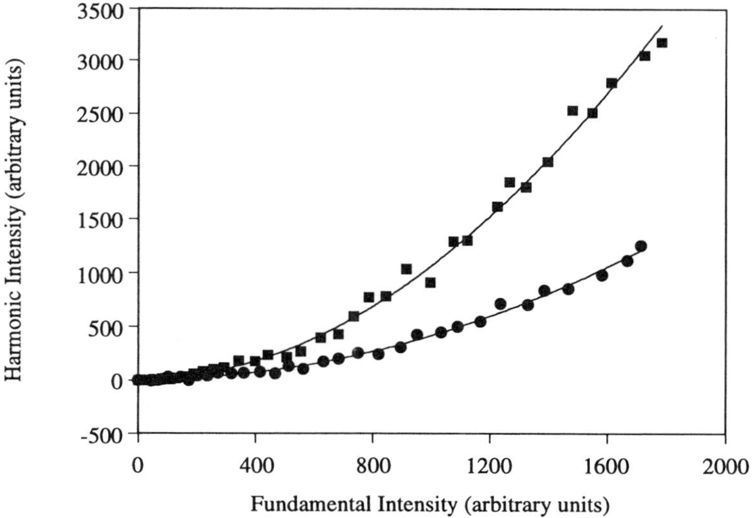

FIG. 2. Quadratic dependence of $I_{2\omega}$ vs I_{ω} for complex $[Ru(4,4'\text{-}C{\equiv}CC_6H_4C{\equiv}CC_6H_4NO_2)(PPh_3)_2(\eta^5\text{-}C_5H_5)]$.

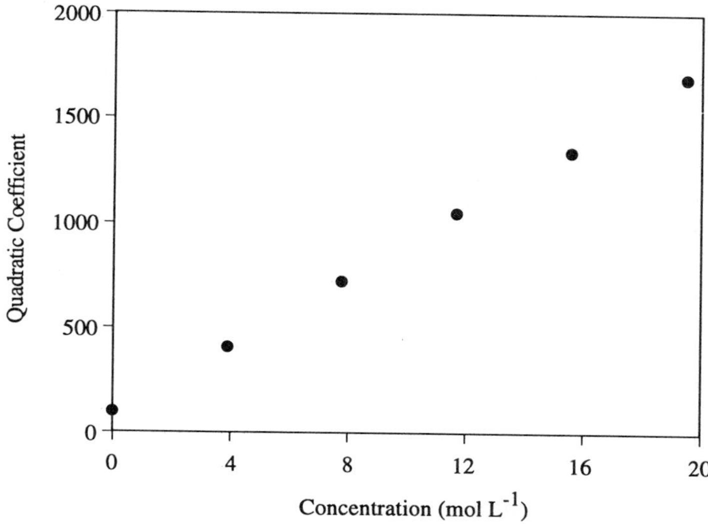

FIG. 3. Quadratic coefficient $GB^2 = G[N_{solvent}\beta^2_{solvent} + N_{solute}\beta^2_{solute}]$, obtained from the curves in Fig. 2, vs N_{solute} for complex $[Ru(4,4'\text{-}C{\equiv}CC_6H_4C{\equiv}CC_6H_4NO_2)(PPh_3)_2(\eta^5\text{-}C_5H_5)]$.

harmonic light, detected by a photomultiplier tube, and averaged by a gated integrator.

Advantages of HRS are (i) its simplicity when compared to EFISH (there is no need for a dc field to be applied, and it does not need complementary

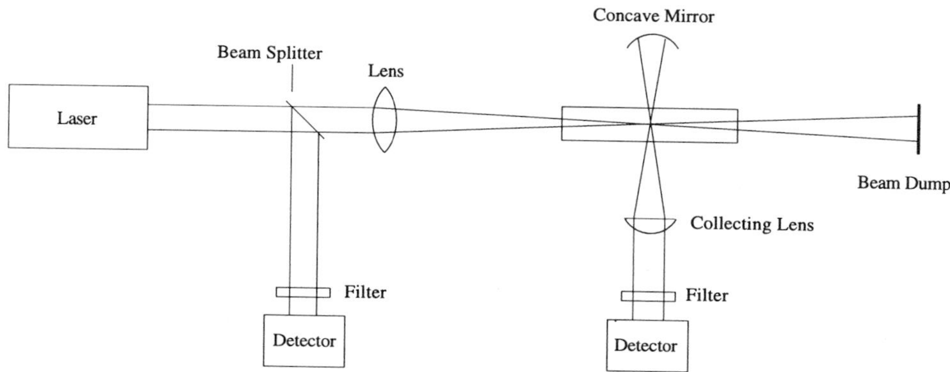

FIG. 4. Schematic diagram of the HRS experiment. (Adapted with permission from K. Clays and A. Persoons, *Rev. Sci. Instrum.* **1992**, *63*, 3286. Copyright © 1992 American Institute of Physics.)

measurements of μ or γ); (ii) its sensitivity to nonvector components of the β tensor; and (iii) the fact that, unlike EFISH, it can be used to measure octupolar molecules and ionic molecules, the latter having important implications for organometallics where a particular system may have a range of accessible oxidation states. Disadvantages of HRS include (i) that it needs sensitive detection and high intensity of the fundamental, because of the low intensity of the second harmonic light (high intensity of the fundamental may be detrimental to the experiment due to stimulated Raman or Brillouin scattering, self-focusing, or dielectric breakdown)[25]; (ii) that it is only possible to find the magnitude of β, because of the quadratic dependence on the HRS signal; and (iii) that HRS can give unreliable results when the complex fluoresces at the frequency-doubled wavelength.[35]

D. *Comparisons between Results*

There are several problems in comparing results obtained by different groups using the many experimental techniques that are available to investigate the NLO properties of molecules; the most important of these include dispersion effects, the measurement of different tensorial components, different physical processes contributing to nonlinearity, and solubility problems.

The dispersion of NLO properties is a major source of problems. Measurements are frequently available at one wavelength only and the degree to which the results are influenced by material resonances close to the measurement wavelength is often difficult to quantify. It is possible to compensate for some of the dispersion effects in certain cases. For example, off-resonant dispersion of the second-order hyperpolarizability for linear intramolecular charge-transfer type molecules is reasonably well described by a two-state model in which parameters of the dominant excited state are assumed. However, a two-state model is probably not sufficient for more complicated second-order molecules, such as those with significant contribution to the nonlinearity from octupolar origins. The tensorial character of the nonlinear polarizabilities is another experimental complication. Experimental techniques only provide access to specific tensorial components or combinations thereof, such as the vector part of β. Last, a trivial, but important, issue in NLO measurements on many organometallics is that the most convenient technique for investigating the nonlinearity (as solutions in common solvents) is of little value if the solubility of the compounds is not sufficiently high. For example, if the nonlinearity of a solution can be measured with a 10% accuracy and the compound being investigated can only be dissolved at 1% concentration, the practical limit of detection is that the nonlinearity of the organometallic has to be at least 10 times higher than that of the solvent.

E. *Computational Methodologies*

Computational techniques can afford useful insights into structure–property relationships of molecules and materials. It is possible to use on-line database search procedures to identify classes of organometallic complexes that have potentially high NLO susceptibility. Searches have been promulgated that use crystal space group, dipole moment, specific substructure, and oxidation potential as indicator properties that have been related to NLO response; these have suggested classes of candidate materials.[36] More commonly, computational procedures have been used to calculate responses. Time-consuming syntheses may be avoided by predicting responses computationally. Otherwise inaccessible structural variations may be probed, such as the effect of biphenyl dihedral angle rotation.[37] A number of structural variables may be analyzed in an efficient manner, permitting trends to be uncovered that may lead to the development and refinement of molecular design rules. One limitation of computational procedures is the problem of direct comparison with experimental results, specifically, the difficulty of modeling environmental interactions (intermolecular or solvent/molecule).[38–40] It is therefore not surprising that the vast majority of calculated NLO responses are those of individual molecules rather than of bulk materials. This section reviews some of the computational methods that have been or may be applied to afford molecular optical nonlinearities of organometallic compounds. Although the more popular methods are mentioned here, this is not intended to be a comprehensive treatment of computational approaches, more detailed treatments being found in the literature.[38]

1. *Fundamentals*

From molecular orbital theory, a many-electron wavefunction, Ψ, may be defined by a determinant of molecular wavefunctions, ψ_i. The ψ_i may in turn be expressed as a linear combination of one-electron functions, that is, $\psi_i = \sum_{\mu=1}^{N} c_{\mu i}\phi_\mu$, where $c_{\mu i}$ are the molecular orbital expansion coefficients and ϕ_μ are one-electron basis functions. Several mathematical descriptions of basis functions exist, two of the more common types used in calculations being Slater-type orbitals (STO) and Gaussian-type atomic functions; a linear combination of Gaussian-type atomic functions may also be used to construct a basis function. The set of basis functions required to describe a given system is the basis set. The reader is directed elsewhere for a more comprehensive treatment of the concepts presented here.[41] It is important to note that selection of basis set size is often a trade-off between accuracy and computational expense, and that a satisfactory basis set size depends on the property being examined; significant changes in electron density as in some NLO systems require an extended basis set to generate accurate results.[42,43]

To assign values to the molecular orbital coefficients, $c_{\mu i}$, many computational methods apply Hartree–Fock theory (which is based on the variational method).[44] This uses the result that the calculated energy of a system with an approximate, normalized, antisymmetric wavefunction will be higher than the exact energy, so to obtain the optimal wavefunction (of the single determinant type), the coefficients $c_{\mu i}$ should be chosen such that they minimize the energy E, i.e., $\partial E/\partial c_{\mu i} = 0$. This leads to a set of equations to be solved for $c_{\mu i}$ known as the Roothaan–Hall equations. For the closed shell case, the equations are

$$\sum_{\nu=1}^{N} (F_{\mu\nu} - \varepsilon_i S_{\mu\nu}) c_{\nu i} = 0, \qquad \mu = 1, 2, \ldots, N \tag{2}$$

where ε_i is the energy of orbital ψ_i, and $S_{\mu\nu}$ are the overlap matrix elements:

$$S_{\mu\nu} = \int \phi_\mu^*(1)\,\phi_\nu\,dx_1\,dy_1\,dz_1$$

and the elements of the Fock matrix, $F_{\mu\nu}$, are

$$F_{\mu\nu} = H_{\mu\nu}^{\text{core}} + \sum_{\lambda=1}^{N}\sum_{\sigma=1}^{N} P_{\lambda\sigma}\left[(\mu\nu|\lambda\sigma) - \frac{1}{2}(\mu\lambda|\nu\sigma)\right]$$

where $H_{\mu\nu}^{\text{core}}$ is the matrix of single electron energies for electrons experiencing the field of the nucleus in the absence of other electrons:

$$H_{\mu\nu}^{\text{core}} = \int \phi_\mu^*(1)\hat{H}^{\text{core}}(1)\,\phi_\nu(1)\,dx_1\,\mathrm{d}y_1\,dz_1$$

where

$$\hat{H}^{\text{core}}(1) = -\frac{1}{2}\left(\frac{\partial^2}{\partial x_i^2} + \frac{\partial^2}{\partial y_i^2} + \frac{\partial^2}{\partial z_i^2}\right) - \sum_{A=1}^{M}\frac{Z_{\text{A}}}{r_{1\text{A}}}$$

for M atoms with atomic number Z_{A}. The integrals $(\mu\nu|\lambda\sigma)$ are the two-electron repulsion terms:

$$(\mu\nu|\lambda\sigma) = \int\int \phi_\mu^*(1)\,\phi_\nu(1)\left(\frac{1}{r_{12}}\right)\phi_\lambda^*(2)\,\phi_\sigma(2)\,dx_1\,dy_1\,dz_1\,dx_2\,dy_2\,dz_2$$

and $P_{\lambda\sigma}$ are elements of the density matrix, $P_{\lambda\sigma} = 2\sum_{i=1}^{occ} c_{\lambda i}^{*} c_{\sigma i}$. The total energy is then given as

$$E = \frac{1}{2}\sum_{\mu=1}^{N}\sum_{\nu=1}^{N} P_{\mu\nu}(F_{\mu\nu} + H_{\mu\nu}^{\text{core}}) + \sum_{A<B}\sum^{M} \frac{Z_A Z_B}{R_{AB}}$$

In solving Eq. (2), an iterative process is used to adjust the $c_{\mu i}$ until the best wavefunction is found [self-consistent field (SCF) theory]. For the open shell case where incompletely filled orbitals exist, spin-restricted Hartree-Fock (RHF) methods or unrestricted Hartree-Fock (UHF) methods may be used to calculate the energies.[41] The extent of calculation, approximation, or neglect of the two-electron integral terms largely defines the computation method.

Ab initio methods calculate each of the S, $(\mu\nu|\lambda\sigma)$, and $H_{\mu\nu}^{\text{core}}$ integrals exactly. This does not correspond to exact calculation of the energy, as the technique is still limited by factors such as the Born–Oppenheimer approximation, linear combination of orbitals, and choice of basis sets. Although *ab initio* methods give quantitatively accurate results, calculations performed on large systems (e.g., those containing transition metals) are computationally expensive and limited by the availability of suitable basis sets. No *ab initio* calculations of nonlinear optical properties have been promulgated for organometallic complexes.

Semiempirical models still assign values to the S, $(\mu\nu|\lambda\sigma)$, and $H_{\mu\nu}^{\text{core}}$ integrals but rather than calculating them all exactly, some are set by choosing values that give agreement with empirical data.[45–50] The ZINDO package, an intermediate neglect of differential overlap (INDO)-based routine, has been used to calculate the first hyperpolarizabilities of a number of organometallic complexes.[45–49] This method is discussed next.

Hückel methods use variational theory to derive the set of equations $\sum_{\nu=1}^{N}(H_{\mu\nu} - \varepsilon_i S_{\mu\nu})c_{\nu i} = 0$. In the basic Hückel method,[51] the overlap integrals $S_{\mu\nu}$ are set to $S_{\mu\nu} = \delta_{\mu\nu}$ (δ is the Kronecker delta) and the $H_{\mu\nu}$ are empirically parametrized. The symbol α is generally used to represent $H_{\mu\mu}$ and β to represent $H_{\mu\nu}$ if the centers μ and ν are bonded. If μ and ν are nonbonded, $H_{\mu\nu}$ is set to zero. In extended Hückel methods,[52] extended basis sets are employed, generally using STO, overlap integrals $S_{\mu\nu}$ are retained and calculated, and all of the $H_{\mu\nu}$ are retained but parametrized. Although Hückel methods have been used to calculate the NLO properties of some organic compounds,[38] there are no reports thus far of calculations on organometallic complexes.

Density functional methods (DFMs) have also been used to estimate NLO properties,[43,53] although this approach has not yet been applied to organometallic complexes. Hartree–Fock theory expresses the energy in terms of wavefunction Ψ and the relative positions of the atoms, that is, $E = E(\Psi, r)$. In contrast, density functional theory expresses the energy in terms of the electron density ρ and the relative atomic positions, that is, $E = E(\rho, r)$. The total density ρ is expressed as a combination of occupied single-electron wavefunctions ψ_i, that is, $\rho = \sum_{\text{occ}} |\psi_i(r)|^2$. Variation of the total electron density leads to an energy minimum, $\partial E/\partial\rho = 0$, from which a series of equations similar to the Hartree–Fock equations may be written. A more detailed description of DFMs may be found in Ref. 54, but it is worth noting that the scaling of the computational effort with molecular size is significantly less than with traditional correlated molecular orbital methods. DFMs have been used successfully for calculations involving organotransition metal complexes[55] and, although no nonlinear optical calculations of organometallic complexes have been reported thus far, it is likely to become an important technique in the future.

2. *Calculating Nonlinear Optical Coefficients*

There exist two main methods for implementing nonlinear optical calculations into a given computational technique: coupled methods and uncoupled methods. Coupled methods [sometimes called finite field (FF) or coupled–perturbed Hartree–Fock (CPHF) methods] include the effect of the perturbing field into the Hamiltonian. The energy (e) of the system in the field **E**

$$e(\mathbf{E}) = \int \Psi(\mathbf{E})H(\mathbf{E})\Psi(\mathbf{E})\,d\tau \tag{3}$$

and the dipole moment

$$\mu(\mathbf{E}) = \int \Psi(\mathbf{E}) \sum_i q_i(\mathbf{E}) \cdot r_i(\mathbf{E})\Psi(\mathbf{E})\,d\tau \tag{4}$$

where q_i is the charge and r_i the position vector of particle i, may then be calculated. Equation (2) may be written (using tensor notation and the Einstein convention of summing over repeated indices) as

$$\mu_i(E) = \mu_i^0 + \alpha_{ij}E_j + \beta_{ijk}E_jE_k + \gamma_{ijkl}E_jE_kE_l + \ldots \tag{5}$$

and

$$e(E) = e(0) - \mu_iE_i - \alpha_{ij}E_iE_j - \beta_{ijk}E_iE_jE_k - \gamma_{ijkl}E_iE_jE_kE_l - \ldots \tag{6}$$

Then, equating Eq. (4) with Eq. (5) gives

$$\alpha_{ij} = \left.\frac{\partial\mu_i}{\partial E_j}\right|_{E=0}$$

$$\beta_{ijk} = \left. \frac{\partial^2 \mu_i}{\partial E_j \partial E_k} \right|_{E=0}$$

$$\gamma_{ijkl} = \left. \frac{\partial^3 \mu_i}{\partial E_j \partial E_k \partial E_l} \right|_{E=0}$$

and Eq. (3) with Eq. (6) gives

$$\alpha_{ij} = \left. \frac{\partial^2 e}{\partial E_i \partial E_j} \right|_{E=0}$$

$$\beta_{ijk} = \left. \frac{\partial^3 e}{\partial E_i \partial E_j \partial E_k} \right|_{E=0}$$

$$\gamma_{ijkl} = \left. \frac{\partial^4 e}{\partial E_i \partial E_j \partial E_k \partial E_l} \right|_{E=0}$$

The preceding derivatives are evaluated at zero field, and there are several ways of evaluating them that are discussed in Ref. 38. If frequency dependence is to be included, that is, the field is $\mathbf{E}(\omega) = \mathbf{E}_0 + \mathbf{E}\sin(\omega t)$, then the calculations become somewhat more complex. A summary of methods used to calculate frequency dependent responses is given in Ref. 38.

Uncoupled methods [sometimes called the sum over states (SOS) methods] do not include the field in the Hamiltonian but use a time-dependent perturbation theory approach.[38,56] A sum over all excited states is used that requires values for dipole moments in ground and excited states and excitation energies to be evaluated. One must choose the number of states at which to terminate the series. It has been shown in several studies of second-order nonlinearities[38] that the β values converge after a finite number of states are chosen. Furthermore, this approach intrinsically accounts for frequency dependence.

3. *Other Factors*

Using a single determinant to derive the Hartree–Fock wavefunction does not take into account the correlation between electrons of different spins. If the wavefunction is described by a linear combination of determinants, then configuration interaction is incorporated. Another method to model correlation energy involves Möller–Plesset perturbation theory. A more detailed description of this may be found in Ref. 41.

Geometry selection is another important consideration when calculating NLO properties. The NLO responses have been shown to be sensitive to geometry alterations.[37,46] Three common sources of molecular geometries are used: (i) crystallographically derived coordinates for atoms, (ii) scanning the crystallographic literature and using average geometries for moieties, and (iii) geometry optimization computational routines (where no empirical data exists).

4. *Methods Applied to Organometallics*

ZINDO[57,58] is a semiempirical intermediate neglect of differential overlap/spectroscopy (INDO/S) based routine. It can be combined with an SOS method to calculate second-order nonlinear optical coefficients. ZINDO is parametrized to accommodate transition-metal calculations and is therefore suited for calculation on organometallic compounds. To achieve computational efficiency, some of the terms in Eq. (2) are replaced by empirical data or neglected. To see how the INDO/S does this, the closed-shell case will be examined.[57,58] It is useful to introduce the following:

$$(\phi_\mu^A \phi_\nu^B | \phi_\sigma^C \phi_\lambda^D) \equiv \delta_{AB}\delta_{CD} \int\int \phi_\mu^A(1)\,\phi_\nu^B(1)\,\frac{1}{r_{12}}\,\phi_\sigma^C(2)\,\phi_\lambda^D(2)\,dx_1\,dy_1\,dz_1\,dx_2\,dy_2\,dz_2$$

$$= (\phi_\alpha^A \phi_\beta^A | \phi_\sigma^A \phi_\lambda^A) \qquad A = C$$

$$= (\overline{\phi}_\alpha^A \overline{\phi}_\alpha^A | \overline{\phi}_\sigma^C \overline{\phi}_\alpha^C) = \gamma_{\alpha\sigma} \quad A \neq C$$

where ϕ_μ^A is the atomic orbital centered on atom "A," $\overline{\phi}_\mu^A$ is an *s*-symmetric atomic orbital with the same exponents and expansion coefficients as ϕ_μ^A, and δ_{AB} is the Kronecker delta. The Fock matrix may be examined in three parts, one part describing the one-center, same function terms, another part describing the one-center, different function terms, and the third part the two-center, two-function terms:

$$F_{\mu\mu}^{AA} = U_{\mu\mu}^{AA} - \sum_{B\neq A} Z_B\gamma_{AB} + \sum_{\sigma,\lambda}^{A} P_{\sigma\lambda}\left[(\sigma\lambda|\mu\mu) - \frac{(\sigma\mu|\mu\lambda)}{2}\right] + \sum_{\sigma\notin A} P_{\sigma\sigma}\gamma_{\sigma\mu}$$

$$F_{\mu\nu}^{AA} = \sum_{\sigma,\lambda}^{A} P_{\sigma\lambda}\left[(\mu\nu|\sigma\lambda) - \frac{(\mu\sigma|\nu\lambda)}{2}\right]$$

$$F_{\mu\nu}^{AB} = \frac{(\beta_\mu^A + \beta_\nu^B)S'}{2} - \frac{P_{\mu\nu} - \gamma_{\mu\nu}}{2}$$

In the $F_{\mu\mu}^{AA}$ component, the core integral term $U_{\mu\mu}^{AA}$ is assigned values derived from experimental atomic spectroscopy.[57,58] All one-center two-electron terms are retained, and the two-center two-electron terms are obtained from the following:

$$\gamma_{\mu\nu} = \frac{f_\gamma}{2f_\gamma/(F^0(\mu\mu) + F^0(\nu\nu)) + R_{AB}}$$

where the F^0 are Slater–Condon one-center two-electron integrals, values for which can be found in the literature.[57] In the $F_{\mu\nu}^{AA}$ component, the core terms are set to zero and the remaining terms calculated. In the $F_{\mu\nu}^{AB}$ component, the core terms are approximated by the parameter β and the S' is a weighted overlap integral ensuring correct positioning of π molecular orbitals relative to the σ orbitals.[57] The details of the implementation of these equations may be found in Refs. 58 and 59.

The SOS treatment is then used with the mono-excited-state configuration interaction (MECI) approximation.[60,61] The values reported are β_{EFISH} (denoted in the literature as β_{vec} and defined as the value sampled by the EFISH technique) and β_{tot}, the total quadratic hyperpolarizability [defined as $\beta_{\mathrm{tot}} = \sqrt{(\beta_x^2 + \beta_y^2 + \beta_z^2)} = |\boldsymbol{\beta}_{\mathbf{vec}}|$, where β_x, β_y, and β_z are the Cartesian components of the vector part of the β tensor]. The number of transitions to undergo CI mixing needs to be chosen and then the sum needs to be truncated. Several studies have reported values for each of these factors after which calculated β on the given system converged.[38,62] Of the large number of reports with a comparison between experimental and calculated nonlinear coefficients, those of organometallic complexes are presented in Section III,C.

III

ORGANOMETALLICS FOR SECOND-ORDER NONLINEAR OPTICS

A. *Bulk Second-Order Measurements*

The SHG efficiencies of more than 300 organometallic complexes measured using the Kurtz powder method have been reported, but most responses are either very low or zero. Table II contains data for samples with reported Kurtz powder efficiencies greater than or equal to twice that of urea, with Fig. 5 containing the structural formulas of complexes with very large ($\geq$50 $\times$ urea) responses.

Ferrocenyl-type complexes are prominent in Table II, with many having SHG efficiencies comparable to those of efficient organic compounds (although these data are significantly smaller than the largest reported powder measurement for an organic: 1000 $\times$ urea for [*N*-MeC_5H_4N-4-(*E*)-$CH{=}CHC_6H_4$-4′-NMe_2][$MeC_6H_4SO_3$]).[86] Complexes with ferrocenyl groups as donors and either organic or metalorganic acceptors are responsible for all of the very large powder responses for organometallics reported

TABLE II

COMPLEXES WITH KURTZ POWDER EFFICIENCIES ≥2 × UREA

Complex	SHG Urea = 1	Fund. (μm)	Ref.
[Fc-(*E*)-CH=CH-4-C_5H_4N-1-Me)][I]	125–240	1.91	63–66
[Fc-(*E*)-CH=CH-4-C_5H_4N-1-Me)][Br]	165–170	1.91	63–66
[Fc-(*E*)-CH=CH-4-C_5H_4N-1-Me)][NO_3]	75–120	1.91	63–66
[Fc-(*E*)-CH=CH-4-C_5H_4N-1-Me)][BF_4]	50	1.91	63–66
[Fc-(*E*)-CH=CH-4-C_5H_4N-1-Me)][BPh_4]	13	1.91	63–66
[Fc-(*E*)-CH=CH-4-C_5H_4N-1-Me)][MeC_6H_4-4-SO_3]	13	1.91	63–66
[Fc-(*Z*)-CH=CHC_6H_4-4-NO_2]	62	1.91	67–70
[Fc-(*E*)-CH=CHC_6H_3-2,4-$(NO_2)_2$]	21	1.91	71
[(*S*)-Fe(η^5-C_5H_5)(η^5-C_5H_3-2-Me-(*E*)-CH=CHC=CHCH=(NO_2)O]	17–17.5	1.91	68–70
[(*S*)-Fe(η^5-C_5H_5)(η^5-C_5H_3-2-Me-(*E*)-CH=CHC_6H_4-4-NO_2)]	8.0–8.4	1.91	68–70
[(*S*)-Fe(η^5-C_5H_5)(η^5-C_5H_3-2-Me-(*E*)-CH=CH-(*E*)-CH=CHC_6H_4-4-NO_2)]	6.4	1.91	68–70
[(*S*)-Fe(η^5-C_5H_5)(η^5-C_5H_3-2-Me-(*E*)-CH=CHC_6H_4-4-CHO)]	1.8–2.5	1.91	68–70
[(η^5-C_5H_5)$_2$$Fe_2$$(CO)_2$($\mu$-CO)($\mu$-C-(*E*)-CH=CH$C_6H_4$-4-$NMe_2$)][$CF_3SO_3$]	3.6	1.91	72
[(*S*)-Ru(η^5-C_5H_5)(η^5-C_5H_4CHMeSC_6H_4-4-NO_2)]	27	1.06	73
[(*S*)-Os(η^5-C_5H_5)(η^5-C_5H_4CHMeSC_6H_4-4-NO_2)]	21	1.06	73
[MoCl(NO)(NHC_6H_3-3-Me-4-N=NC_6H_4-4-Fc)(HB$(dmpz)_3$)]	50–123	1.91	74–76
[MoI(NO)(NHC_6H_4-4-(*E*)-CH=CHC_6H_4-4-Fc)(HB$(dmpz)_3$)]	85	1.91	77
[MoCl(NO)(NHC_6H_4-4-N=NC_6H_4-4-Fc)(HB$(dmpz)_3$)]	30–59	1.91	75, 76
[WCl(NO)(NHC_6H_3-3-Me-4-N=NC_6H_4-4-Fc)(HB$(dmpz)_3$)]	8–53	1.91	75, 77
[MoCl(NO)(NHC_6H_4-4-(*E*)-CH=CHC_6H_4-4-Fc)(HB$(dmpz)_3$)]	35	1.91	77
[MoBr(NO)(NHC_6H_3-3-Me-4-N=NC_6H_4-4-Fc)(HB$(dmpz)_3$)]	25	1.91	76
[WCl(NO)(NHC_6H_4-4-N=NC_6H_4-4-Fc)(HB$(dmpz)_3$)]	12	1.91	76
[WCl(NO)(NHC_6H_4-4-(*E*)-CH=CHC_6H_4-4-Fc)(HB$(dmpz)_3$)]	8	1.91	77
[$Cr(CO)_3$(η^6-C_6H_6)] : thiourea, 1 : 3	2.3	1.06	78–81
[$Cr(CO)_3$(η^6-C_6H_5F)] : thiourea, 1 : 3	2.0	1.06	79–81
[PtBr(C_6H_3-2-Me-4-NO_2)$(PEt_3)_2$]	2.0–14.0	1.06	82
[PdBr(C_6H_3-2-Me-4-NO_2)$(PEt_3)_2$]	5.0–10.0	1.06	82
[PtI(C_6H_4-4-NO_2)$(PEt_3)_2$]	5.0–8.0	1.06	82, 83
[Pt(NCO)(C_6H_4-4-NO_2)$(PEt_3)_2$]	6.7–9.2	1.06	83
[PdBr(C_6H_4-4-NO_2)$(PEt_3)_2$]	2.1–7.2	1.06	82
[PtBr(C_6H_4-4-NO_2)$(PEt_3)_2$]	4.2	1.06	82, 83
[PtBr(C_6H_4-4-CHO)$(PEt_3)_2$]	1.5–3.4	1.06	82
[PtCl(C_6H_4-4-NO_2)$(PEt_3)_2$]	3.0	1.06	83
[PdI(C_6H_4-4-NO_2)$(PEt_3)_2$]	2.3	1.06	82
[PdBr(C_6H_4-4-CHO)$(PEt_3)_2$]	2.3	1.06	82, 83
[Pt(CN)(C_6H_4-4-NO_2)$(PEt_3)_2$]	2.0	1.06	83
[Fe(N≡CC_6H_4-4-Ph)((+)-diop)(η^5-C_5H_5)][PF_6]	4.2	1.91	84
[Fe(N≡C-(*E*)-CH=CHC_6H_4-4-NMe_2)((+)-diop)(η^5-C_5H_5)][PF_6]	3.3	1.91	84
[Fe(N≡CC_6H_4-4-C_6H_4-4-NO_2)((+)-diop)(η^5-C_5H_5)][PF_6]	38.0	1.91	84
[Ru(N≡CC_6H_4-4-NO_2)((+)-diop)(η^5-C_5H_5)][CF_3SO_3]	10	1.06	85
[Ru(N≡CC_6H_4-4-NO_2)((+)-diop)(η^5-C_5H_5)][ClO_4]	2.9	1.06	85
[Ru(N≡CC_6H_4-4-NO_2)((+)-diop)(η^5-C_5H_5)][PF_6]	2.7	1.06	85

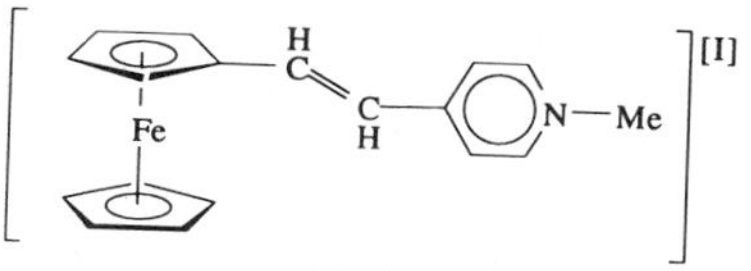

125-240

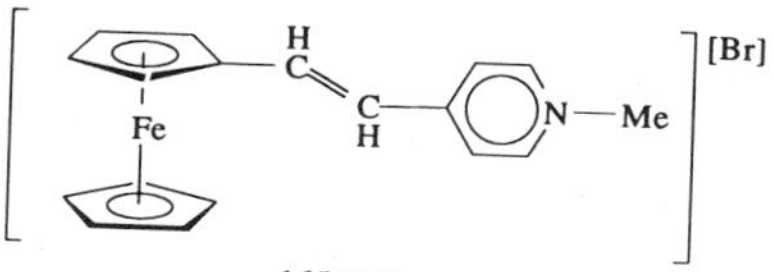

165-170

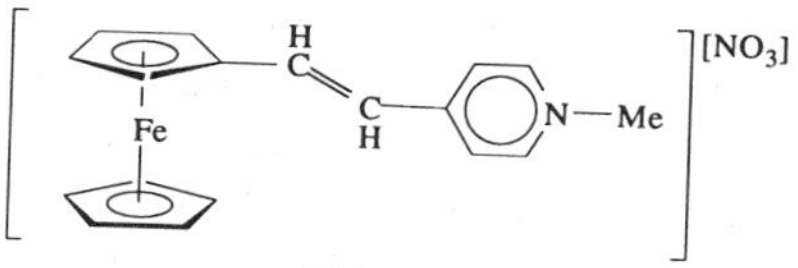

75-120

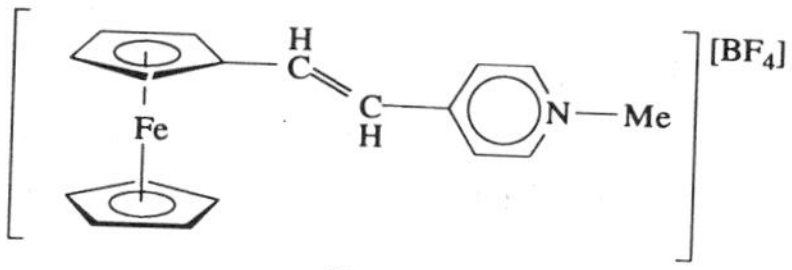

50

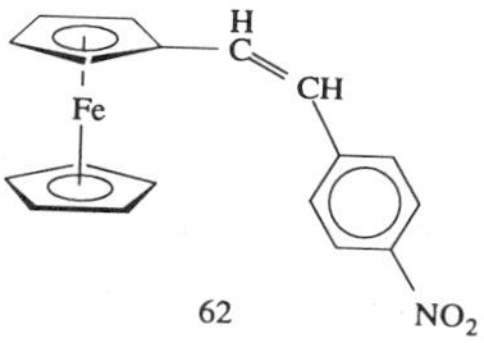

62

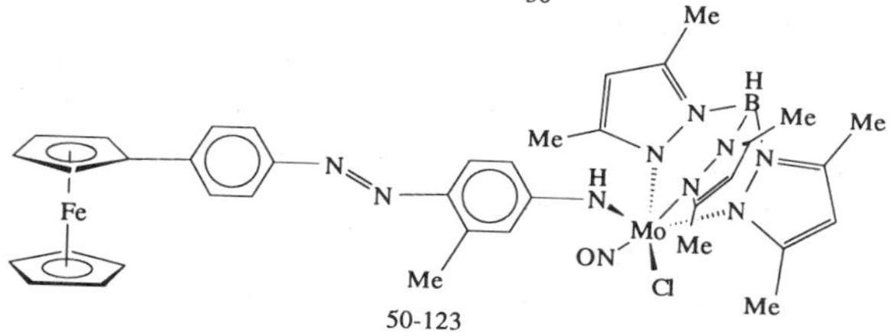

50-123

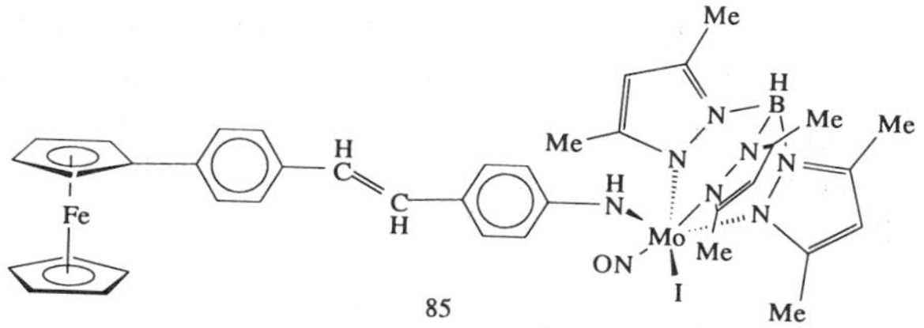

85

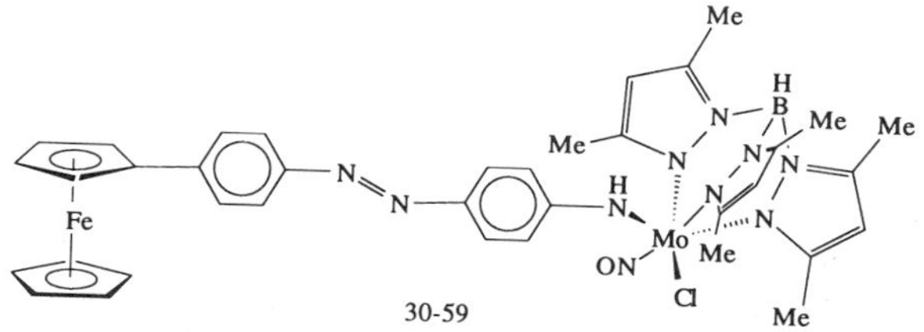

30-59

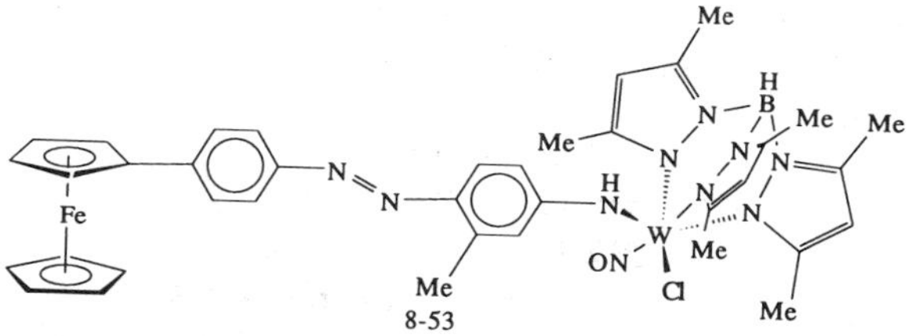

8-53

FIG. 5. Organometallic complexes with large $\chi^{(2)}$ measured by the Kurtz powder technique.

thus far, despite all ferrocenyl data being obtained at 1.91 μm, which should minimize resonance enhancement. Large bulk nonlinearities are a function of both significant molecular nonlinearities and favorable crystal packing. As many types of organometallic complexes have been the subject of powder measurements and have comparable molecular nonlinearities to ferrocenyl complexes (see Section III,B), the implication is that ferrocenyl-based complexes have a greater proclivity for optimal lattice alignment. Two ruthenocenyl and osmocenyl complexes having identical cyclopentadienyl substituents are featured in Table II; these complexes show little difference in their SHG efficiencies, but comment on the significance of intramolecular structural changes (metal variation in this case) upon bulk NLO merit is not warranted (this shortcoming of the powder technique was discussed in Section II,C). The iron analogue of these compounds has also been prepared and measured,[73] but has an SHG efficiency only 0.8 times that of urea, in contrast to molecular measurements of the effect of metal variation where ferrocenyl complexes have been shown to be significantly more efficient than ruthenocenyl analogues (Section III,B). Phosphine-containing complexes exhibit low to moderate SHG efficiencies, although they mostly have the donor–π-conjugated bridge–acceptor structure, which should favor enhanced molecular nonlinearity. Comparison of the significant number of data for the Group 10 $MX(C_6H_4\text{-}4\text{-}A)(PEt_3)_2$ complexes reveals little difference in observed efficiency for systematic structural modifications. As was discussed in Section II,C, the usefulness of the Kurtz technique is to rapidly identify SHG activity; structure–NLO property correlations are not justified.

The powder technique can establish that a basic structural type is SHG active; once NLO activity of this type is confirmed, molecular modification has most impact when it affects packing in the solid state. Variation of counterions can have a significant effect on the SHG efficiencies (see, e.g., the first six entries in Table II). Marder and co-workers have shown that variation of counterions is a highly successful and very straightforward method to rapidly sample different lattice arrangements and thus engineer materials with large nonlinearities.[65] Meredith[87] has explained these results by proposing that Coulombic interactions in salts may negate deleterious dipolar interactions, enhancing the possibility of favorable noncentrosymmetric packing.

A design strategy of this type for optimizing bulk nonlinearity is clearly not applicable to neutral dipolar molecules. Possibilities for organizing favorable crystal alignment in neutral complexes exist, though, including the formation of guest–host inclusion complexes, and incorporation of chiral ligands. Results obtained thus far with guest–host inclusion complexes are unimpressive, but this may be a function of the included guest molecule; guest complexes with high molecular nonlinearities have not thus far been assayed. Table III contains results for complexes with chiral ligands, the incorporation of which should enforce the noncentrosymmetric packing required for observable bulk nonlinearity. Although this methodology has afforded materials with higher nonlinearities than those obtained by employing the guest–host inclusion complex strategy, the SHG efficiencies observed are still mostly quite low, showing that chiral ligands do not ensure large SHG. One major problem is that a noncentrosymmetric crystal lattice may be consistent with opposed molecular dipoles (see Fig. 6); observable bulk nonlinearity derives then from charge transfer to co-ligands, which involve much less electron polarization.

Optimizing molecular alignment is important for enhancing SHG efficiencies in noncentrosymmetric space groups. As discussed previously, the introduction of chiral ligands theoretically ensures noncentrosymmetry (although it does not preclude pseudocentrosymmetric arrangements), but it does not ensure the most favorable molecular arrangement in the lattice. Table IV contains results for complexes with nonzero SHG for which the crystal structures have been reported, emphasizing the range of SHG intensities observed for complexes that crystallize in noncentrosymmetric space groups (and, incidentally, the range of values obtained *for a single complex* and the consequent lack of quantitative significance of Kurtz powder results). The complexes $[(\eta^5\text{-}C_5H_5)_2Fe_2(CO)_2(\mu\text{-}CO)(\mu\text{-}C\text{-}(E)\text{-}CH{=}CHC_6H_4\text{-}4\text{-}NMe_2)][BF_4]$, $[MoCl(p\text{-}O\text{-tpph})(NO)\{HB(dmpz)_3\}]$, and

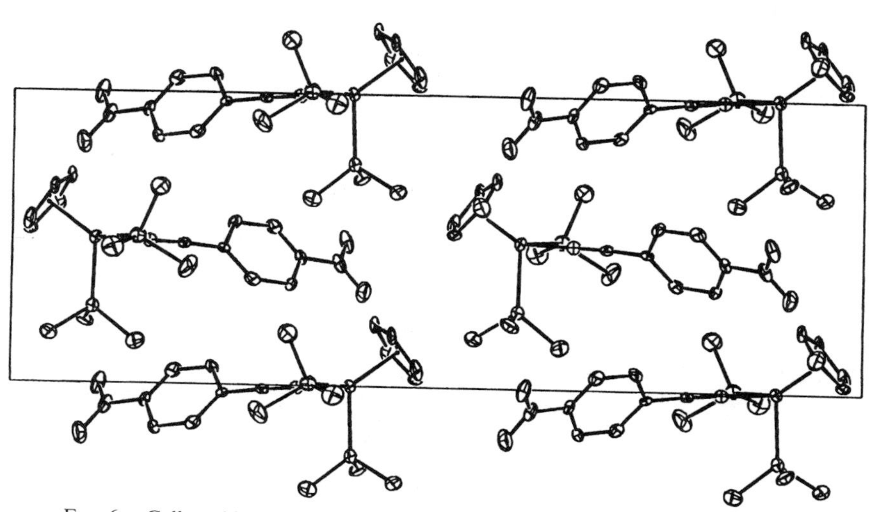

FIG. 6. Cell packing diagram for $[Ru(C \equiv CC_6H_4\text{-}4\text{-}NO_2)(PMe_3)_2(\eta^5\text{-}C_5H_5)]$.

TABLE III

KURTZ POWDER MEASUREMENTS OF COMPLEXES CONTAINING CHIRAL LIGANDS

Complex	SHG urea = 1	Fund. (μm)	Ref.
[(*S*)-Fe(η^5-C_5H_5)(η^5-C_5H_3-2-Me-(*E*)-CH=CHC_6H_4-4-NO_2)]	8.0–8.4	1.91	68–70
[(*S*)-Fe(η^5-C_5H_5)(η^5-C_5H_3-2-Me-(*E*)-CH=CHC_6H_4-4-CN)]	1.2–1.6	1.91	68–70
[(*S*)-Fe(η^5-C_5H_5)(η^5-C_5H_3-2-Me-(*E*)-CH=CHC_6H_4-4-CHO)]	1.8–2.5	1.91	68–70
[(*S*)-Fe(η^5-C_5H_5)(η^5-C_5H_3-2-Me-(*E*)-CH=CH-(*E*)-CH=CHC_6H_4-4-NO_2)]	6.4	1.91	68–70
[(*S*)-Fe(η^5-C_5H_5)(η^5-C_5H_3-2-Me-(*E*)-CH=CHC=CHCH=(NO_2)O)]	17.0–17.5	1.91	68–70
[(*R*)-FcCHMeSC_6H_4-4-NO_2]	0.8	1.06	73
[(*S*)-Ru(η^5-C_5H_5)(η^5-C_5H_4CHMeSC_6H_4-4-NO_2)]	27	1.06	73
[(*S*)-Os(η^5-C_5H_5)(η^5-C_5H_4CHMeSC_6H_4-4-NO_2)]	21	1.06	73
[(*R*)-FcCHMe-2-SC_6H_3N-5-NO_2]	~0	1.06	73
[(*S*)-Ru(η^5-C_5H_5)(η^5-C_5H_4CHMe(2-SC_6H_3N-5-NO_2)]	1.9	1.06	73
[(*R*)-FcCHMeNHC_6H_4-4-NO_2]	~0	1.06	73
[(*S*)-Ru(η^5-C_5H_5)(η^5-C_5H_4CHMeNHC_6H_4-4-NO_2)]	~0	1.06	73
[(*S*)-Os(η^5-C_5H_5)(η^5-C_5H_4CHMeNHC_6H_4-4-NO_2)]	~0	1.06	73
[(−)-FcCHMeOH]	0	1.06	73
[(−)-Fe(η^5-C_5H_5)(η^5-OCHC_5H_3-2-CH_2OH)]	0	1.06	73
[(−)-Fe(η^5-$C_5H_4CH_2$SOMe$)_2$]	0	1.06	73
[MoI(NO)(2-(−)-OCH_2CHMe$C_{10}H_6$-6-OMe)(HB(dmpz$)_3$)]	0.007	1.91	75
[MoCl(NO)(2-(−)-OCH_2CHMe$C_{10}H_6$-6-OMe)(HB(dmpz$)_3$)]	0.004	1.91	75
[MoI(NO)(3-*O*-1,2:5,6-($CMe_2)_2$-α-D-glucofuranose)(HB(dmpz$)_3$)]	0.245	1.91	75
[MoI(NO)((1*S*,2*R*,5*S*)-(+)-mentholate)(HB(dmpz$)_3$)]	0.074	1.91	75
[WCl((*R*)-(+)-NHCHMePh)(NO)(HB(dmpz$)_3$)]	0	1.91	75
[Mo((+)-NHCHMePh)(*p*-*O*-tpph)(NO)(HB(dmpz$)_3$)]	0	1.91	75
[Cr$(CO)_3$(η^6-(−)-α-ethylphenethylalcohol)]	0.003[a]	1.06	88
[Cr$(CO)_3$(η^6-(*S*)-CH_3C(O)C_6H_4-2-CH(OH)COOH)]	0[a]	1.06	88
[Cr$(CO)_3$(η^6-(*S*)-PhCH(OH)COOH)]	0[a]	1.06	88
[Cr$(CO)_3$(η^6-(*R*)-PhCH(OH)COOH)]	0[a]	1.06	88
[Cr$(CO)_3$(η^6-(*S*)-PhCH_2CHMeEt)]	0.5[a]	1.06	88
[Cr$(CO)_3$(η^6-nopolbenzyl ether)]	4[a]	1.06	88
[Cr$(CO)_3$(η^6-L-phenylaniline ethyl ester hydrochloride)]	0.2[a]	1.06	88
[Cr$(CO)_3$(η^6-PhCHMeCH_2OH)]	0[a]	1.06	88
[(+)-Cr$(CO)_3$(η^6-MeC_6H_4-2-CHO)]	0	1.06	73
[(+)-Cr$(CO)_3$(η^6-C_6H_4-2-CH_2OHCH_2OC(O)$C_{15}H_{31}$)]	0	1.06	73
[(+)-Cr$(CO)_3$(η^6-C_6H_4-2-Me-COOH)]	0	1.06	73
[Cr$(CO)_2$(nmdpp)(η^6-C_6H_4-1,4-Me_2)]	0.003[a]	1.06	88
[Fe$(CO)_4$(nmdpp)]	0.003[a]	1.06	88
[Mo$(CO)_5$(nmdpp)]	<0.003[a]	1.06	88
[W$(CO)_5$(nmdpp)]	0.02[a]	1.06	88
[Fe(N≡CC_6H_4-4-NH_2)((+)-diop)(η^5-C_5H_5)][PF_6]	0.0	1.06	84
[Fe(N≡CC_6H_4-4-NMe_2)((+)-diop)(η^5-C_5H_5)][PF_6]	0.2	1.91	84
[Fe(N≡CC_6H_4-4-Ph)((+)-diop)(η^5-C_5H_5)][PF_6]	4.2	1.91	84
[Fe(N≡C-(*E*)-CH=CHC_6H_4-4-NMe_2)((+)-diop)(η^5-C_5H_5)][PF_6]	3.3	1.91	84
[Fe(N≡CC_6H_4-4-NO_2)((+)-diop)(η^5-C_5H_5)][PF_6]	38.0	1.91	84
[Fe(N≡C-(*E*)-CH=CHC_6H_4-4-NO_2)((+)-diop)(η^5-C_5H_5)][PF_6]	0.0	1.06	84

TABLE III (*continued*)

Complex	SHG urea = 1	Fund. (μm)	Ref.
$[Fe(N{\equiv}CC_6H_4\text{-}4\text{-}C_6H_4\text{-}NO_2)((+)\text{-diop})(\eta^5\text{-}C_5H_5)][PF_6]$	0.0	1.91	84
$[Ru(N{\equiv}CC_6H_4\text{-}4\text{-}NO_2)((+)\text{-diop})(\eta^5\text{-}C_5H_5)][MeC_6H_4\text{-}4\text{-}SO_3]$	0.4	1.06	85
$[Ru(N{\equiv}CC_6H_4\text{-}4\text{-}NO_2)((+)\text{-diop})(\eta^5\text{-}C_5H_5)][Cl]$	0.5	1.06	85
$[Ru(N{\equiv}CC_6H_4\text{-}4\text{-}NO_2)((+)\text{-diop})(\eta^5\text{-}C_5H_5)][NO_3]$	1.2	1.06	85
$[Ru(N{\equiv}CC_6H_4\text{-}4\text{-}NO_2)((+)\text{-diop})(\eta^5\text{-}C_5H_5)][BF_4]$	1.9	1.06	85
$[Ru(N{\equiv}CC_6H_4\text{-}4\text{-}NO_2)((+)\text{-diop})(\eta^5\text{-}C_5H_5)][PF_6]$	2.7	1.06	85
$[Ru(N{\equiv}CC_6H_4\text{-}4\text{-}NO_2)((+)\text{-diop})(\eta^5\text{-}C_5H_5)][ClO_4]$	2.9	1.06	85
$[Ru(N{\equiv}CC_6H_4\text{-}4\text{-}NO_2)((+)\text{-diop})(\eta^5\text{-}C_5H_5)][CF_3SO_3]$	10	1.06	85

[a] Measured relative to ADP, conversion of ADP = 3.0 × U was used.[89]

$[Ru(C{=}CPhN{=}NC_6H_4\text{-}4\text{-}OMe)(PPh_3)_2(\eta^5\text{-}C_5H_5)][BF_4]$ crystallize in centric space groups, but surprisingly give nonzero responses.[72,74,90,91] A variety of reasons have been proposed to explain these results, including SHG by the particle surfaces (which must of necessity be noncentrosymmetric), crystal defects (i.e., the existence of a noncentrosymmetric phase), decomposition in the laser beam due to the high power of the laser, or fluorescence.

Second-order susceptibilities of a further 300 complexes have been reported,[63–65,68–70,72–76,78–83,85,88,90–93,95–104] all of which are <2 × urea, with about 200 examples giving a zero response. About 100 complexes have been reported relative to ADP[88,104] and 7 relative to quartz,[64,99] but, assuming conversion factors ADP = 0.3 × urea and quartz = 0.1 × urea,[89] all of their nonlinearities are low. The complexes listed in Table II are not representative of all complex types that have been assayed; all examples investigated of $[M(CO)_3(\eta^6\text{-arene})]$ (M = Cr, Mo, W), $[M(CO)_4(L)(\eta^1\text{-}N\text{-pyridine})]$ (M = Cr, Mo, W; L = CO, PR_3),[88] $[M(CO)_4(bipy)]$ (M = Cr, Mo, W),[93] $[Ru(\eta^5\text{-}C_5H_5)(\eta^6\text{-arene})]^+$ salts,[97] $[Ru(C{=}CPhN{=}NR)(PPh_3)_2(\eta^5\text{-}C_5H_5)]^+$ salts,[91] $[Pt(C{\equiv}CR)(C{\equiv}CR')(PR''_3)_2]$,[99,100,105] $[WI(\eta^2\text{-}RC{\equiv}CR')(CO)\{HB(dmpz)_3\}]$,[102] [(pentane-2,4-dionato)ML_2] (M = Pd, Rh, Ir; L = CO, PPh_3),[103] and $[HgX(C_6H_3(2\text{-}R)\text{-}4\text{-}Y)]$ complexes gave low (<2 × urea) responses.

Although most bulk material second-order measurements have been made on powders, other material forms have also been examined. Table V contains data of bulk second-order measurements performed on films and composites containing organometallics. The molecular alignment achieved in Langmuir–Blodgett monolayers leads, perhaps not surprisingly, to little anion dependence compared with that observed in Kurtz powder measurements; the anions do not affect film-forming properties to the same

TABLE IV

COMPLEXES WITH NONZERO SHG FOR WHICH CRYSTAL STRUCTURES ARE KNOWN

Complex	SHG urea = 1	Fund. (μm)	Space group	Ref.
[Fc-(E)-CH=CH-4-C_5H_4N-1-Me)][NO_3]	75–120	1.91	Cc	63–65
[Fc-(Z)-CH=CHC_6H_4-4-NO_2]	62	1.91	Cc	67, 69, 70
[FcCH≡NC_6H_4-4-NO_2]	0.33	1.91	$P2_12_12_1$	92
[MoCl(NO)(NHC_6H_3-3-Me-4-N=NC_6H_4-4-(η^5-C_5H_4)Fe(η^5-C_5H_5))(HB$(dmpz)_3$)]	50–123	1.91	$P2_1$	74, 75
[WCl(NO)(NHC_6H_3-3-Me-4-N=NC_6H_4-4-(η^5-C_5H_4)Fe(η^5-C_5H_5))(HB$(dmpz)_3$)]	8–53	1.91	$P2_1$	74, 75
[Re(CF_3SO_3)$(CO)_3$(bipy)]	1.7–2	1.06	$P2_1$	93
[$PtCl_4$(bipy)]	1.2	1.06	Pn	93
[$PdCl_4$(bipy)]	0.1	1.06	Pn	93
[$Cr(CO)_3$(η^6-C_6H_6)] : thiourea, 1 : 3	2.3	1.06	$R3c$	78–81
[$Fe(CO)_3$(η^4-1,3-cyclohexadienyl)] : thiourea, 1 : 3	0.4	1.06	$Pna2_1$	78–81
[$Mn(CO)_3$(η^4-1,3-cyclohexadienyl)] : thiourea, 1 : 3	0.4	1.06	$R3c$	78–81
[$Fe(CO)_3$(trimethylenemethane)] : thiourea, 1 : 3	0.3	1.06	$R3c$	78–81
[$Cr(CO)_3$(η^6-tetrahydronaphthalene) : tris(o-thymotide), 1 : 1	0.1	1.06	$Pca2_1$	79, 81, 94
[PtI(C_6H_4-4-NO_2)$(PEt_3)_2$]	5.0–8.0	1.06	$Cmc2_1$	82, 83
[Pt($SnCl_3$)(C_6H_4-4-NO_2)$(PEt_3)_2$]	0.8	1.06	$P2_1$	83
[Pt(SPh)(C_6H_4-4-NO_2)$(PEt_3)_2$]	0.6	1.06	$Pca2_1$	83
[(Fe(η^5-C_5H_5)(η^5-$C_5H_4CH_2$))$_2$-2-(=CCN_2)-1,3-diazolidine]	0.67	1.06	$P3_221$	95
[(η^5-C_5H_5)$_2$$Fe_2$$(CO)_2$($\mu$-CO)($\mu$-C-($E$)-CH=$CHC_6H_4$-4-$NMe_2$)][$BF_4$]	0.77	1.91	$P2_1/n$	72, 90
[MoCl(p-O-tpph)(NO)(HB$(dmpz)_3$)]	1.93	1.91	$P\bar{1}$	74, 75
[Ru(C=CPhN=NC_6H_4-4-OMe)$(PPh_3)_2$(η^5-C_5H_5)][BF_4]	1.05	1.06	$P\bar{1}$	91

TABLE V

RESULTS OF BULK SECOND-ORDER MEASUREMENTS ON FILMS AND COMPOSITES

Material	$\chi^{(2)}$ (10^{-7} esu)	β (10^{-30} esu)	λ_{max}	Technique	Fund. (μm)	Ref.
[Fc-(E)-CH = CH-4-C_6H_4N-1-$C_{16}H_{33}$][4,4,4-F_3-1-(2-naphthyl)-1,3-butanedionato)$_4$Dy][a]	2.8	150	560, 330, 286, 256	SHG	1.06	106
[Fc-(E)-CH = CHCH$_2$-4-C_6H_4N-1-$C_{16}H_{33}$][La(dibenzoylmethanato)$_4$][a]	2.3	158	~349, ~250	SHG	1.06	107
[Fc-(E)-CH = CHCH$_2$-4-C_6H_4N-1-$C_{16}H_{33}$][Nd(dibenzoylmethanato)$_4$][a]	2.2	137	~349, ~250	SHG	1.06	107
[Fc-(E)-CH = CHCH$_2$-4-C_6H_4N-1-$C_{16}H_{33}$][Dy(dibenzoylmethanato)$_4$][a]	2.4	150	~349, ~250	SHG	1.06	107
[Fc-(E)-CH = CHCH$_2$-4-C_6H_4N-1-$C_{16}H_{33}$][Yb(dibenzoylmethanato)$_4$][a]	2.9	144	~349, ~250	SHG	1.06	107
[Fc-(E)-CH = CHCH$_2$-4-C_6H_4N-1-$C_{16}H_{33}$]Br[a]	2.9	110	240, 360	SHG	1.06	108
[Fc-(E)-CH = CHCH$_2$-4-C_6H_4N-1-$C_{16}H_{33}$]I[a]	2.8	120	240, 360	SHG	1.06	108
[Ru(N$\equiv$CC_6H_4-4-C_6H_4-4-C_6H_4-4-C_5H_{11})(PPh_3)$_2$(η^5-C_5H_5)][PF_6][a]		40×10^{-50} C^3 m^3 J^{-2}	332	SHG	1.06	109
[Fe(η^5-C_5H_4CH = C(CN)CO_2Et)(η^5-$C_5H_4$$CH_2$OC(O)-CMe$CH_2$-)$_{0.05}$][b]		$d = 1.72$ pm V^{-1} $= \sim 4 \times Q$		SHG	1.06	110

[a] Langmuir–Blodgett monolayer.

[b] Poly(methyl methacrylate) copolymer.

degree as they affect crystal packing. SHG in poled polymers containing organometallic chromophores has also been studied. Decay of SHG signals has been used to monitor the long-term stability of ferrocenyl chromophores copolymerized with methylmethacrylate[111,112] or 1,6-di-isocyanatohexane,[112] and pentacarbonyltungsten attached to polyvinylpyridine or styrene–vinylpyridine copolymer.[113] The long-term stability of the electro-optic response in polymers has also been probed; the electro-optic coefficient r_{33}[12] for *p*-dimethylaminonitrobenzene in a polycarbosilane was invariant after 16 days.[114]

B. *Molecular Second-Order Measurements*

As can be readily ascertained from the tables in this section, researchers have mostly employed the EFISH or HRS techniques to measure second-order nonlinearities of organometallics. Although the initial interest in this field was in metal carbonyl-based complexes, the majority of the measurements are now of ferrocenyl-based complexes or metal acetylides. One reason for this may be pragmatic; despite the vast panoply of possible complex types that exists, ferrocenyl or acetylide complexes can be synthesized in high yield by well-established methodologies, both revealed large molecular nonlinearities in initial studies, and both are (comparatively) oxidatively and thermally stable, the latter an important consideration for (putative) longer-term device applications.

Table VI contains molecular second-order data for ferrocenyl and ruthenocenyl complexes. Ferrocenyl-type compounds had been shown to exhibit large bulk material SHG efficiencies (see Section III,A) and, as a consequence, their molecular nonlinearities were investigated. Some of the observed nonlinearities are large, with the highest values corresponding to bimetallic species with metal-based acceptors as well as ferrocenyl donors (Fig. 7). The data are consistent with the ferrocenyl moiety acting as a reasonable donor in these donor–bridge–acceptor complexes, with observed nonlinearities comparable to those of analogous methoxyphenyl organics.[115] The donor characteristic of the ferrocenyl group has been attributed to the low binding energy of the metal electrons,[116] oxidation from the ferrocene to the analogous ferrocenium being a facile process. The overall effectiveness of the metal as a donor has been questioned, though, with the poor coupling between the metal center and the substituent cited as a shortcoming of this class of complex[117]; it is believed that the orthogonal relationship of the metal-to-ligand charge transfer (MLCT) axis and organic chromophore is unfavorable.

The UV–visible spectra of many of these complexes exhibit absorptions near the second harmonic region for the irradiating frequencies being em-

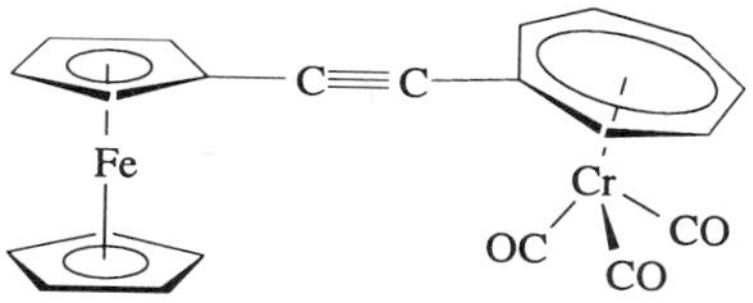

$\beta = 570$; $\beta_0 = 105 \times 10^{-30}$ esu

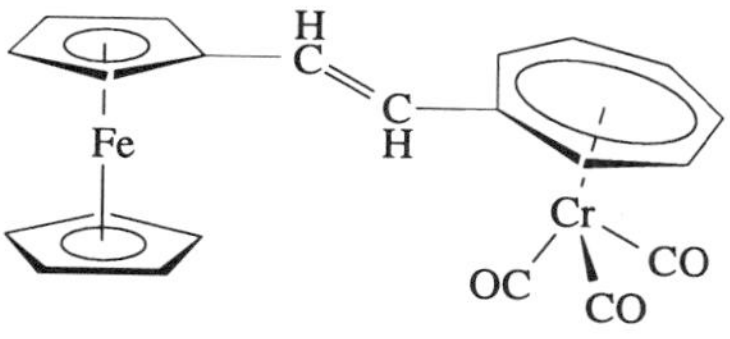

$\beta = 320$; $\beta_0 = 113 \times 10^{-30}$ esu

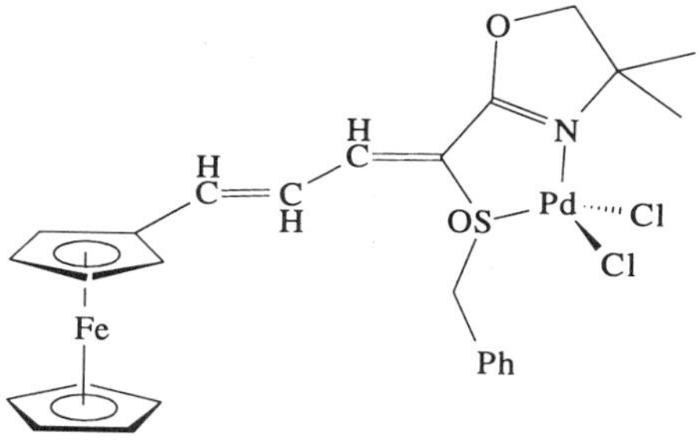

$\beta = 123.5$; $\beta_0 = 19 \times 10^{-30}$ esu

FIG. 7. Ferrocenyl complexes with large β.

ployed; many of the reported β values are therefore resonance enhanced. It is possible to apply the two-level correction to compute frequency-independent nonlinearities. However, both the π–π^* and MLCT transitions are reported by Marder *et al.*[117] to be significant, and the validity of a two-level approximation has therefore been questioned. In contrast to Marder's comments, it has been suggested by Loucif-Saïbi *et al.*[122] that the MLCT is dominant for monometallic ferrocenyl derivatives. The latter group have used a two-level correction to calculate the static hyperpolarizability β_0 for some bimetallic ferrocenyl derivatives[122]; based on the good correlation of β_0 values obtained at two different frequencies, they argue that the MLCT is dominant and that the π–π^* transition is not an important contributor. The two-level correction has also been applied to some bimetallic sesquifulvalene complexes,[119] where the ferrocenyl group is the donor and a (cycloheptatrienyl)tricarbonylchromium moiety is the acceptor. The evidence thus far is consistent with the two-level correction having some utility with organometallics when applied to a closely related and systematically varied class of complexes, but that it must be applied with caution; examples exist of structural variations affecting both the MLCT and π–π^* transitions significantly,[115] and in these instances the two-level model is not justified.

The effect of metal replacement on β has been investigated using analogous ferrocenyl and ruthenocenyl complexes. Replacement of iron by ruthenium in $[M(\eta^5\text{-}C_5H_5)(\eta^5\text{-}C_5H_4\text{-}(E)\text{-}CH{=}CHC_6H_4\text{-}4\text{-}NO_2]$ leads to a reduction in β, perhaps not surprising as ruthenium has a higher ionization potential than iron. Bimetallic complexes coupling $[FcCH{=}C(SCH_2Ph)$

TABLE VI

MOLECULAR NLO RESULTS FOR FERROCENYL AND RUTHENOCENYL COMPLEXES

Complex	Solvent	β^{a} (10^{-30} esu)	λ_{max} (nm)	β_0 (10^{-30} esu)	Fund. (μm)	Ref.
[$FcCOCH_3$]	*p*-dioxane	0.3	—	—	1.91	64, 115, 116
[Fc-(*E*)-CH$=$CHC_6H_3-2,4-$(NO_2)_2$]	—	20	536, 545	—	1.91	71
	p-dioxane	23	366, 536	—	1.91	116
[Fc-(*E*)-CH$=$CHC_6H_4-4-NO_2]	—	34	500	—	1.91	71
	p-dioxane	31–34	356, 496	—	1.91	64, 115–118
[Fc-(*Z*)-CH$=$CHC_6H_4-4-NO_2]	*p*-dioxane	13–14	325, 480	—	1.91	64, 115–117
[Fc*-(*E*)-CH$=$CHC_6H_4-4-NO_2]	*p*-dioxane	40	366, 533	—	1.91	64, 115–117
[Fc-(*E*)-CH$=$CHC_6H_4-4-CN]	*p*-dioxane	10–11	324, 466	—	1.91	115, 117
[Fc-(*Z*)-CH$=$CHC_6H_4-4-CN]	—	4.0	308, 460	—	1.91	115
[Fc-(*E*)-CH$=$C(CN)C_6H_4-4-NO_2]	*p*-dioxane	21	348, 526	—	1.91	116
[Fc*-(*E*)-CH$=$C(CN)C_6H_4-4-NO_2]	*p*-dioxane	35	366, 560	—	1.91	116
[Fc-(*E*)-CH$=$CHC_6H_4-4-CHO]	*p*-dioxane	12–34	338, 474	—	1.91	64, 115, 117
[Ru(η^5-C_5Me_5)(η^5-$C_5H_4NO_2$)]	CH_2Cl_2	0.6	—	—	1.91	64, 115, 116
[Ru(η^5-C_5H_5)(η^5-C_5H_4-(*E*)-CH$=$CHC_6H_4-4-NO_2)]	*p*-dioxane	12–16	350, 390	—	1.91	64, 115–117
[Ru(η^5-C_5Me_5)(η^5-C_5H_4-(*E*)-CH$=$CHC_6H_4-4-NO_2)]	*p*-dioxane	24	370, 424	—	1.91	64, 115–117
[Ru(η^5-C_5Me_5)(η^5-C_5H_4-(*E*)-CH$=$C(CN)C_6H_4-4-NO_2)]	*p*-dioxane	24	370, 443	—	1.91	116
[Fc-(*E*)-CH$=$CH-(*E*)-CH$=$CHC_6H_4-4-NO_2]	*p*-dioxane	66	382, 500	—	1.91	64, 115–117
[FcC$\equiv$C(η^7-C_7H_6)Cr$(CO)_3$]	CH_2Cl_2	570[b]	600	105	1.06	119
[Fc-(*E*)-CH$=$CH(η^7-C_7H_6)Cr$(CO)_3$]	CH_2Cl_2	320[b]	670	113	1.06	119
[Fc(η^6-BC_5H_5)Co(η^5-C_5H_5)][PF_6]	CH_2Cl_2		370, 410(sh), 650	90 ± 30[b]	1.06	132
[Fc-(*E*)-CH$=$CHB$(mes)_2$]	$CHCl_3$	−24	336	—	1.91	120
[FcC$\equiv$CB$(mes)_2$]	$CHCl_3$	4.4	336	—	1.91	120
[FcCH$=$C(SCH_2Ph)$\overline{C=NCMe_2CH_2O}$]	$CHCl_3$	4.5	472	2.0	1.34	121–123
	$CHCl_3$	2.9	472	2.07	1.91	122, 123

[FcCH=C(S(O)CH_2Ph)C=NC$Me_2$$CH_2$O]	$CHCl_3$	8.4	470	3.7	1.34	121–123
[FcCH=C(S(O)$_2$$CH_2$Ph)C=NC$Me_2$$CH_2$O]	$CHCl_3$	6.3	477	1.6	1.34	122, 123
[Fc-(*E*)-CH=CHCH=C(SCH_2Ph)C=NC$Me_2$$CH_2$O]	$CHCl_3$	13.3	482	5.6	1.34	122, 123
	$CHCl_3$	8.8	482	6.14	1.91	122, 123
[Fc-(*E*)-CH=CHCH=C(S(O)CH_2Ph)C=NC$Me_2$$CH_2$O]	$CHCl_3$	11.0	494	4.6	1.34	122, 123
[FcC≡CCH=C(SCH_2Ph)C=NC$Me_2$$CH_2$O]	$CHCl_3$	2.6	460	1.2	1.34	122, 123
[FcCH=C(SCH_2Ph)C=NC$Me_2$$CH_2$O (*N*,*S*-Ni$Cl_2$)]	$CHCl_3$	10.3	500	3.92	1.34	122, 123
[FcCH=C(SCH_2Ph)C=NC$Me_2$$CH_2$O (*N*,*S*-Ni$Br_2$)]	$CHCl_3$	7.14	511	2.55	1.34	122, 123
[FcCH=C(SCH_2Ph)C=NC$Me_2$$CH_2$O (*N*,*S*-Pd$Cl_2$)]	$CHCl_3$	12.2	530	4.13	1.34	121–123
	$CHCl_3$	9	530	5.8	1.91	122, 123
[FcCH=C(SCH_2Ph)C=NC$Me_2$$CH_2$O (*N*,*S*-Pt$Cl_2$)]	$CHCl_3$	16	523	5.2	1.34	122, 123
[FcCH=C(S(O)CH_2Ph)C=NC$Me_2$$CH_2$O (*N*,*S*-Ni$Cl_2$)]	$CHCl_3$	15.8	498	6.0	1.34	122, 123
[FcCH=C(S(O)CH_2Ph)C=NC$Me_2$$CH_2$O (*N*,*S*-Ni$Br_2$)]	$CHCl_3$	5.85	494	2.3	1.34	122, 123
[FcCH=C(S(O)CH_2Ph)C=NC$Me_2$$CH_2$O (*N*,*S*-Pd$Cl_2$)]	$CHCl_3$	30.2	540	7.6	1.34	121–123
[FcCH=C(S(O)CH_2Ph)C=NC$Me_2$$CH_2$O (*N*,*S*-Pt$Cl_2$)]	$CHCl_3$	28.0	548	7.7	1.34	122, 123
[Fc-(*E*)-CH=CHCH=C(SCH_2Ph)C=NC$Me_2$$CH_2$O (*N*,*S*-Pd$Cl_2$)]	$CHCl_3$	58	567	13.5	1.34	122, 123
	$CHCl_3$	18.2	567	10.8	1.91	122, 123
[Fc-(*E*)-CH=CHCH=C(S(O)CH_2Ph)C=NC$Me_2$$CH_2$O (*N*,*S*-Pd$Cl_2$)]	$CHCl_3$	123.5	602	19	1.34	122, 123
[FcC≡CCH=C(SCH_2Ph)C=NC$Me_2$$CH_2$O (*N*,*S*-Pd$Cl_2$)]	$CHCl_3$	33.6	558	8.5	1.34	122, 123

[a] Measured by EFISH. The uncertainties are quoted as ±10%. For lower responses (~1×10^{-30} esu), the uncertainty may be higher.
[b] Measured by HRS.
Fc* ≡ Fc(η^5-C_5Me_5)(η^5-C_5Me_4-)

$\overline{C=NCMe_2CH_2O}]$ to ligated Group 10 metals have been investigated,[122] with the second metal in almost all cases enhancing β compared to the monometallic ferrocenyl precursor; for the acceptor ligated metals, the efficiency series $PtCl_2 \approx PdCl_2 > NiCl_2 > NiBr_2$ was observed. Significantly, coordination of the functionalized ferrocenyl group to a ligated metal results in a larger two-level corrected β_0 than coordination of the dimethylanilino group to the same metal, emphasizing once again that the ferrocenyl group is comparable in strength to the best organic donors. A correlation between $\Delta\mu$ (the difference in dipole moment between ground and excited states) and β_0 has been found with the monometallic ferrocenyl derivatives but not with the bimetallic derivatives.[123] Examination of nonamethylferrocenyl analogues has permitted comment on the effect of making the iron more electron rich; the iron is the donor in these donor–bridge–acceptor complexes, and permethylating both rings results in an increase in nonlinearity over that of the related ferrocenyl complexes as expected. An interesting comparison that was performed for $[Ru(\eta^5\text{-}C_5R_5)(\eta^5\text{-}C_5H_4\text{-}(E)\text{-}CH{=}CHC_6H_4\text{-}4\text{-}NO_2)]$ (R = H, Me) is the effect of permethylating only one of the rings; modifying the ring remote from the chromophore resulted in a 50 to 100% increase in nonlinearity. The borabenzene derivative $[Fc(\eta^6\text{-}BC_5H_5)Co(\eta^5\text{-}C_5H_5)][PF_6]$ has a surprisingly large β_0 value despite the absence of linking units between the ferrocenyl and (borabenzene)cobalt moieties; related sesquifulvalene complexes lacking bridging units do not show measurable β.

Related nitro $[Fc\text{-}(E)\text{-}CH{=}CHC_6H_4\text{-}4\text{-}NO_2]$ and dinitro $[Fc\text{-}(E)\text{-}CH{=}CHC_6H_3\text{-}2,4\text{-}(NO_2)_2]$ ferrocene derivatives have been assessed,[71] with the 4-nitro complex having a larger β than the 2,4-dinitro complex. Two reasons have been suggested to rationalize this surprising observation. First, EFISH measures the β component in the dipolar direction and the two complexes have different dipole moments, with the dipole of the dinitro complex less aligned with β_{CT} than is the dipole of the 4-nitro complex. Second, introduction of the second nitro group may saturate the polarizability, with the second nitro group competing for oscillator strength in the excited state.[71] Table VI also permits comment on relative efficiencies of the bridging groups. The *trans*-complex $[Fc\text{-}(E)\text{-}CH{=}CHC_6H_4\text{-}4\text{-}NO_2]$ has a higher nonlinearity than its *cis*-isomer, and $[Fc\text{-}(E)\text{-}CH{=}CHB(mes)_2]$ has a greater β value than $[FcC{\equiv}CB(mes)_2]$; the greater efficiencies of *E* vs *Z* stereochemistry and double bond vs triple bond in linking units have been established previously with organic compounds.

Table VII contains data for a series of ferrocenyl complexes for which $\mu \cdot \beta$ values only are extant; in the absence of reported dipole moments, only an internal comparison is possible. The results are resonance enhanced and, as the applicability of the two-level correction is questionable (see

TABLE VII

MOLECULAR NLO RESULTS FOR FERROCENYL COMPLEXES FOR WHICH $\mu \cdot \beta$ VALUES HAVE BEEN REPORTED[a]

Complex	$\mu\beta$ (10^{-48} esu)	λ_{max} (nm)	$\mu\beta_0$[c] (10^{-48} esu)	Ref.
[Fc-(*E*)-CH=CHCHO]	60	367,480	42–49 [45]	124
[Fc-(*E*)-CH=CH-(*E*)-CH=CHCHO]	215	330,494	147–184 [165]	124
[Fc-(*E*)-CH=CH-(*E*)-CMe=CH-(*E*)-CH=CH-(*E*)-CH=CMeCHO]	560	398[b]	440	124
[Fc-(*E*)-CH=CH-(*E*)-CH=CH-(*E*)-CMe=CH-(*E*)-CH=CH-(*E*)-CH=CMe-(*E*)-CH=CHCHO]	1150	430[b]	870	124
[FcCH=C(CN)$_2$]	92	320,526	60–80 [70]	124
[Fc-(*E*)-CH=CHCH=C(CN)$_2$]	420	370,556	250–340 [300]	124
[Fc-(*E*)-CH=CH-(*E*)-CH=CHCH=C(CN)$_2$]	1120	406,568	660–875 [770]	124
[Fc-(*E*)-CH=CH-(*E*)-CMe=CH-(*E*)-CH=CH-(*E*)-CH=CMeCH=C(CN)$_2$]	4600	458[b]	3300	124

[a] Measured by EFISH in acetone at 1.91 μm. Uncertainty of 5–20%.

[b] Lower energy band obscured by band to higher energy.

[c] Since the UV–vis spectra show two distinct bands, the two-level model was deemed unsuitable and not applied. The range of values (and average in square brackets) derived from the two-level model is shown.

earlier discussion), a range of values was calculated; approximate static values were obtained by taking the average of this range. Two other assumptions were made to draw conclusions about β in the absence of experimental μ: (i) that the major β component is nearly parallel to the dipolar direction, and (ii) that μ values exhibit a weak length dependence, both of which appear reasonable. The results from these studies suggest a $\mu \cdot \beta_0 = kn^a$ relationship for n ene-linkages, with a values of 1.6 and 2.4 for the formyl and dicyanovinyl acceptor groups, respectively; larger $\mu \cdot \beta_0$ values and larger increases upon chain lengthening were obtained with the stronger acceptor. Coupling of the metal center to the polyene chain was also examined.[124] The two absorption bands assigned to MLCT and π–π^* transitions approach one another with increasing chain length; for sufficiently long chain length, the bands overlap, suggesting mixing of these two bands, resulting in increased coupling between the metal center and the acceptor.

Table VIII contains results from metal carbonyl complexes, principally by Marder, Cheng, Eaton, and their co-workers. In all cases, the dominant contribution to the nonlinearity is presumed to derive from the other ligands present, namely pyridines, arenes, carbenes, acetylides, and nitriles. More than half the examples (the pyridines and arene complexes) have negative β values, indicative of a decrease or even change in direction of dipole moment between ground and excited state. The data show that introduction of an electron-accepting group onto the *N*-ligand leads to larger β[125,126]; the acceptor substituent facilitates back-transfer of charge upon MLCT excitation, as well as providing lower-energy π^* orbitals. The expected increase in β values with increased conjugation length is also seen. Varying the metal has minimal impact on nonlinearity in *N*-ligand complexes, in contrast to its effects on β values in *C*-ligand compounds. Metal substitution across the series [$M(CO)_4$(1,10-phen)] (M = Cr, Mo, W)[126] has little effect on β values, but metal replacement for the carbene complexes [$M(CO)_5$ (C(OMe)-(*E*)-CH=$CHNMe_2$)] (M = Cr, W) results in a significant difference in β. Both metal and halogen replacement in [$MX(CO)_2(NC_5H_4$-4-(*E*)-CH=CHC_6H_4-4-OCH_2CH(Me)Et] had minimal impact on β.

Three types of *C*-ligands ligated to metal carbonyl fragments have been investigated. The chromium arene complexes have much lower second-order responses than the chromium carbene complexes (Fig. 8); unlike the arene complexes, the carbene complexes satisfy a suggested organometallic NLO chromophore design strategy (the metal positioned along the chromophore axis, and the presence of some M–C multiple bonding).[115] The sesquifulvalene complexes have significantly larger nonlinearities, illustrating one important advantage of organometallics, namely, stabilization of reactive organic species. Sesquifulvalenes were predicted computationally to have large nonlinearities, but their high reactivity has prevented experimental

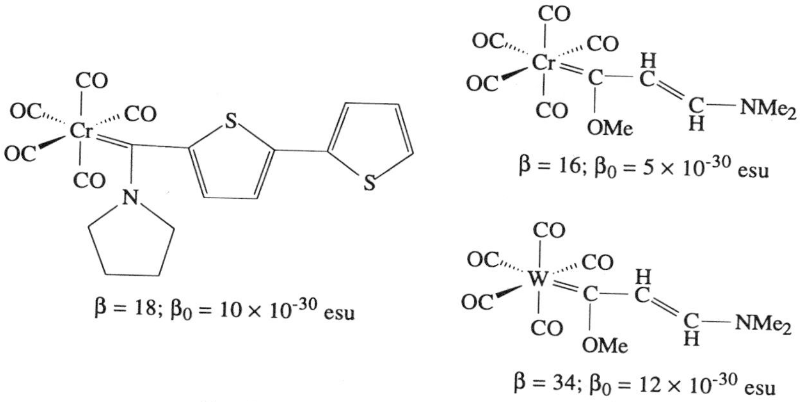

$\beta = 18$; $\beta_0 = 10 \times 10^{-30}$ esu

$\beta = 16$; $\beta_0 = 5 \times 10^{-30}$ esu

$\beta = 34$; $\beta_0 = 12 \times 10^{-30}$ esu

FIG. 8. Carbonyl complexes with large β.

confirmation. Stabilization by coordination to transition metal(s) permits assessment experimentally.

The ruthenium indenyl complexes have second-order data that are strongly resonance enhanced; there is little difference between the two-level corrected β_0 values of the related chromium- and tungsten-containing complexes, the lack of significance of metal replacement being consistent with the observations in the $[M(CO)_4(1,10\text{-phen})]$ series earlier. The β_0 values are much larger than those of the other carbonyl-containing complexes, and among the largest values reported thus far for organometallic complexes.

Table IX contains data from acetylide complexes, with Fig. 9 showing representative examples with extremely large nonlinearities (the largest thus far for organometallic complexes). Humphrey, Gimeno, Persoons, and their co-workers have prepared and measured systematically varied series of ruthenium, nickel, and gold acetylides. Dispersion-enhanced experimentally obtained and two-level-corrected quadratic nonlinearities for the ruthenium examples increase on chain lengthening from one ring to biphenyl and yne-linked two-ring acetylide ligand, with the ene-linked two-ring acetylide the most efficient. Although the data for ruthenium acetylide complexes are substantially resonance enhanced, their absolute values (both experimentally observed and two-level-corrected) are much larger than those for the gold complexes, consistent with the 18-valence-electron, more easily oxidizable ruthenium(II) being a better donor than the 14-valence-electron, less readily oxidizable gold(I). The $Ph_3PAu(C{\equiv}C)$ unit has comparable efficiency to the strongest organic donors, suggesting that the more

TABLE VIII

MOLECULAR NLO MEASUREMENTS FOR METAL CARBONYL COMPLEXES

Complex	Solvent	β^a (10^{-30} esu)	λ_{max} (nm)	β_0 (10^{-30} esu)	Technique	Fund. (μm)	Ref.
$[W(CO)_5(NC_5H_4\text{-}4\text{-}NH_2)]$	dmso	−2.1 ± 0.3	290	—	EFISH	1.91	116, 125, 126
$[W(CO)_5(NC_5H_4\text{-}4\text{-}Bu^n)]$	*p*-dioxane	−3.4	328	—	EFISH	1.91	116, 125, 126
$[W(CO)_5(NC_5H_5)]$	Toluene	−4.4	332	—	EFISH	1.91	116, 125, 126
$[W(CO)_5(NC_5H_4\text{-}4\text{-}Ph)]$	$CHCl_3$	−4.5	330–340	—	EFISH	1.91	116, 125, 126
$[W(CO)_5(NC_5H_4\text{-}4\text{-}COMe)]$	$CHCl_3$	−9.3	420–440	—	EFISH	1.91	116, 125, 126
$[W(CO)_5(NC_5H_4\text{-}4\text{-}CHO)]$	$CHCl_3$	−12	420–440	—	EFISH	1.91	116, 125, 126
$[W(CO)_5(NC_5H_4\text{-}4\text{-}(E)\text{-}CH{=}CHPh)]$	$CHCl_3$	−5.7	302, 440	—	EFISH	1.91	126
$[W(CO)_5(NC_5H_4\text{-}4\text{-}(E)\text{-}CH{=}CHC_6H_4\text{-}4'\text{-}CHO)]$	$CHCl_3$	−17	320, 420		EFISH	1.91	126
$[W(CO)_5(NC_5H_4\text{-}4\text{-}(E)\text{-}CH{=}CHC_6H_4\text{-}4'\text{-}NO_2)]$	$CHCl_3$	−20	322, 425	—	EFISH	1.91	118, 126
$[W(CO)_5(pyrazine)]$	$CHCl_3$	6.0[b]	—	—	EFISH	1.91	126
$[Cr(CO)_4(1,10\text{-}phen)]$	CH_2Cl_2	−13	500	—	EFISH	1.91	126
$[Mo(CO)_4(1,10\text{-}phen)]$	CH_2Cl_2	−13	496	—	EFISH	1.91	126
$[W(CO)_4(1,10\text{-}phen)]$	CH_2Cl_2	−13	492	—	EFISH	1.91	126
$[W(CO)_4(5\text{-}NO_2\text{-}1,10\text{-}phen)]$	CH_2Cl_2	−18	—	—	EFISH	1.91	126
$[Cr(CO)_3(\eta^6\text{-}C_6H_6)]$	Toluene	−0.8 ± 0.08	310	—	EFISH	1.91	116
$[Cr(CO)_3(\eta^6\text{-}C_6H_5OMe)]$	Toluene	−0.9 ± 0.09	310	—	EFISH	1.91	116, 125, 126
$[Cr(CO)_3(\eta^6\text{-}C_6H_5NH_2)]$	*p*-dioxane	−0.6 ± 0.06	313	—	EFISH	1.91	116, 125, 126
$[Cr(CO)_3(\eta^6\text{-}C_6H_5NMe_2)]$	Toluene	−0.4 ± 0.04	318	—	EFISH	1.91	116, 125, 126
$[Cr(CO)_3(\eta^6\text{-}C_6H_5CO_2Me)]$	Toluene	−0.7 ± 0.07	318	—	EFISH	1.91	116, 125, 126
$[Cr(CO)_3(\eta^6\text{-}C_6H_5\text{-}(E)\text{-}CH{=}CHPh)]$	*p*-dioxane	−2.2 ± 0.22	410	—	EFISH	1.91	116, 125, 126
$[Cr(CO)_5(C(N\text{-}pyrrole)\text{-}5\text{-}(2,2'\text{-}bithiophene))]$	$CHCl_3$	18	335	10	HRS	1.06	127
$[Cr(CO)_5(C(OMe)\text{-}(E)\text{-}CH{=}CHNMe_2)]$	$CHCl_3$	16	420	5	HRS	1.06	127
$[W(CO)_5(C(OMe)\text{-}(E)\text{-}CH{=}CHNMe_2)]$	$CHCl_3$	34	411	12	HRS	1.06	127

$[IrCl(CO)_2(NC_5H_4\text{-}4\text{-}(E)\text{-}CH{=}CHC_6H_4\text{-}4\text{-}OCH_2CH(Me)Et)]$	$CHCl_3$	24.4	364		EFISH	1.91	128
$[RhCl(CO)_2(NC_5H_4\text{-}4\text{-}(E)\text{-}CH{=}CHC_6H_4\text{-}4\text{-}OCH_2CH(Me)Et)]$	$CHCl_3$	20.1	355		EFISH	1.91	128
$[RhBr(CO)_2(NC_5H_4\text{-}4\text{-}(E)\text{-}CH{=}CHC_6H_4\text{-}4\text{-}OCH_2CH(Me)Et)]$	$CHCl_3$	23.9	357		EFISH	1.91	128
$[Ru(C{\equiv}C\text{-}(E)\text{-}CH{=}CH\text{-}4\text{-}C_5H_4N\text{-}1\text{-}Cr(CO)_5)(PPh_3)_2(\eta^5\text{-indenyl})]$	CH_2Cl_2	260 ± 26	451	60	HRS	1.06	129, 130
$[Ru(C{\equiv}C\text{-}(E)\text{-}CH{=}CH\text{-}4\text{-}C_5H_4N\text{-}1\text{-}W(CO)_5)(PPh_3)_2(\eta^5\text{-indenyl})]$	CH_2Cl_2	535 ± 54	462	71	HRS	1.06	129, 130
$[Ru(C{\equiv}C\text{-}(E)\text{-}CH{=}CHC_6H_4\text{-}4\text{-}C{\equiv}NCr(CO)_5)(PPh_3)_2(\eta^5\text{-indenyl})]$	CH_2Cl_2	465 ± 47	442	119	HRS	1.06	129, 130
$[Ru(C{\equiv}C\text{-}(E)\text{-}CH{=}CHC_6H_4\text{-}4\text{-}C{\equiv}NW(CO)_5)(PPh_3)_2(\eta^5\text{-indenyl})]$	CH_2Cl_2	700 ± 70	456	150	HRS	1.06	129, 130
$[Ru(C{\equiv}NCr(CO)_5)(PPh_3)_2(\eta^5\text{-indenyl})]$	CH_2Cl_2	25 ± 2.5	392	10	HRS	1.06	129, 130
$[Ru(C{\equiv}NW(CO)_5)(PPh_3)_2(\eta^5\text{-indenyl})]$	CH_2Cl_2	40 ± 4	392	15	HRS	1.06	129, 130
$[Mn(CO)_3(\eta^5\text{-}C_5H_4C_7H_6)][BF_4]$	CH_2Cl_2	45	536	—	HRS	1.06	131
	MeCN	38	506	—	HRS	1.06	131
$[Mn(CO)_2(P(OMe)_3)(\eta^5\text{-}C_5H_4C_7H_6)][BF_4]$	CH_2Cl_2	54	612	—	HRS	1.06	131
	MeCN	53	581	—	HRS	1.06	131
$[Mn(CO)_2(PPh_3)(\eta^5\text{-}C_5H_4C_7H_6)][BF_4]$	CH_2Cl_2	29	643	—	HRS	1.06	131
	MeCN	39	630	—	HRS	1.06	131
$[Mn(CO)_3(\eta^5\text{-}C_5H_4C{\equiv}CC_7H_6)][BF_4]$	CH_2Cl_2	40	537	—	HRS	1.06	131
	MeCN	94	491	—	HRS	1.06	131
$[Mn(CO)_2(P(OMe)_3)(\eta^5\text{-}C_5H_4C{\equiv}CC_7H_6)][BF_4]$	CH_2Cl_2	50	649	—	HRS	1.06	131
	MeCN	151	583	—	HRS	1.06	131
$[Mn(CO)_2(PPh_3)(\eta^5\text{-}C_5H_4C{\equiv}CC_7H_6)][BF_4]$	CH_2Cl_2	44	740	—	HRS	1.06	131
	MeCN	73	650	—	HRS	1.06	131
$[FcC{\equiv}C(\eta^7\text{-}C_7H_6)Cr(CO)_3]$	CH_2Cl_2	570	600	105	HRS	1.06	119
$[Fc\text{-}(E)\text{-}CH{=}CH(\eta^7\text{-}C_7H_6)Cr(CO)_3]$	CH_2Cl_2	320	670	113	HRS	1.06	119
$[Fc(\eta^6\text{-}BC_6H_6)Co(\eta^5\text{-}C_5H_5)]$	CH_2Cl_2	—	370, 410(sh), 650	90 ± 30	HRS	1.06	132

[a] Uncertainties not shown have not been quoted.
[b] Absolute value of β.

TABLE IX

MOLECULAR NLO MEASUREMENTS FOR METAL ACETYLIDE COMPLEXES

Complex	Solvent	β^a (10^{-30} esu)	λ_{max}	β_0 (10^{-30} esu)	Ref.
$[Ru(C{\equiv}CPh)(PPh_3)_2(\eta^5\text{-}C_5H_5)]$	thf	89	310	45	49, 129
$[Ru(C{\equiv}CC_6H_4\text{-}4\text{-}NO_2)(PPh_3)_2(\eta^5\text{-}C_5H_5)]$	thf	468	382, 460	96	47, 129
$[Ru(C{\equiv}CC_6H_4\text{-}4\text{-}NO_2)(PMe_3)_2(\eta^5\text{-}C_5H_5)]$	thf	248	279, 477	39	47, 129
$[Ru(C{\equiv}CC_6H_4\text{-}4\text{-}C_6H_4\text{-}4\text{-}NO_2)(PPh_3)_2(\eta^5\text{-}C_5H_5)]$	thf	560	310, 448	134	49, 129
$[Ru(C{\equiv}CC_6H_4\text{-}4\text{-}(E)\text{-}CH{=}CHC_6H_4\text{-}4\text{-}NO_2)(PPh_3)_2(\eta^5\text{-}C_5H_5)]$	thf	1455	341, 476	232	47, 129
	thf	1464[c]	341, 476	234	47, 129
$[Ru(C{\equiv}CC_6H_4\text{-}4\text{-}C{\equiv}CC_6H_4\text{-}4\text{-}NO_2)(PPh_3)_2(\eta^5\text{-}C_5H_5)]$	thf	865	340, 446	212	49, 129
$[Ru(C{\equiv}CC_6H_4\text{-}4\text{-}N{=}CHC_6H_4\text{-}4\text{-}NO_2)(PPh_3)_2(\eta^5\text{-}C_5H_5)]$	thf	840	298, 496	86	47, 129
	thf	760[c]	298, 496	78	47, 129
$[Ru(C{\equiv}CC_6H_4\text{-}4\text{-}NO_2)(PPh_3)_2(\eta^5\text{-indenyl})]$	CH_2Cl_2	746	476	119	129, 130
$[Ru(C{\equiv}C\text{-}(E)\text{-}CH{=}CHC_6H_4\text{-}4\text{-}NO_2)(PPh_3)_2(\eta^5\text{-indenyl})]$	CH_2Cl_2	1257	507	89	129, 130
$[Ru(C{\equiv}C\text{-}(E)\text{-}CH{=}CHC_6H_4\text{-}4\text{-}CN)(PPh_3)_2(\eta^5\text{-indenyl})]$	CH_2Cl_2	238	427	71	129, 130
$[Ru(C{\equiv}C\text{-}(E)\text{-}CH{=}CH\text{-}(E)\text{-}CH{=}CHC_6H_4\text{-}4\text{-}NO_2)(PPh_3)_2(\eta^5\text{-indenyl})]$	CH_2Cl_2	1320	523	34	129, 130
$[Ru(C{\equiv}C\text{-}(E)\text{-}CH{=}CH\text{-}4\text{-}C_5H_4N)(PPh_3)_2(\eta^5\text{-indenyl})]$	CH_2Cl_2	100	399	37	129, 130
$[Ru(C{\equiv}C\text{-}(E)\text{-}CH{=}CH\text{-}4\text{-}C_5H_4N\text{-}1\text{-}Cr(CO)_5)(PPh_3)_2(\eta^5\text{-indenyl})]$	CH_2Cl_2	260	451	60	129, 130

$[Ru(C{\equiv}C\text{-}(E)\text{-}CH{=}CH\text{-}4\text{-}C_5H_4N\text{-}1\text{-}W(CO)_5)(PPh_3)_2(\eta^5\text{-indenyl})]$	CH_2Cl_2	535	462	71	129, 130
$[Ru(C{\equiv}C\text{-}(E)\text{-}CH{=}CHC_6H_4\text{-}4\text{-}C{\equiv}NCr(CO)_5)(PPh_3)_2(\eta^5\text{-indenyl})]$	CH_2Cl_2	465	442	119	129, 130
$[Ru(C{\equiv}C\text{-}(E)\text{-}CH{=}CHC_6H_4\text{-}4\text{-}C{\equiv}NW(CO)_5)(PPh_3)_2(\eta^5\text{-indenyl})]$	CH_2Cl_2	700	456	150	129, 130
$[Ru(C{\equiv}C\text{-}(E)\text{-}CH{=}CHC_6H_4\text{-}4\text{-}C{\equiv}NRu(NH_3)_5)(PPh_3)_2(\eta^5\text{-indenyl})]^{3+}$	Acetone	315	442	80	129, 130
$[Au(C{\equiv}CPh)(PPh_3)]$	thf	6	268, 282, 296	4	129, 133
$[Au(C{\equiv}CC_6H_4\text{-}4\text{-}NO_2)(PPh_3)]$	thf	22	338	12	129, 133
$[Au(C{\equiv}CC_6H_4\text{-}4\text{-}C_6H_4\text{-}4\text{-}NO_2)(PPh_3)]$	thf	39	274, 287, 350	20	129, 133
$[Au(C{\equiv}CC_6H_4\text{-}4\text{-}(E)\text{-}CH{=}CHC_6H_4\text{-}4\text{-}NO_2)(PPh_3)]$	thf	120	303, 386	49	129, 133
$[Au(C{\equiv}CC_6H_4\text{-}4\text{-}(Z)\text{-}CH{=}CHC_6H_4\text{-}4\text{-}NO_2)(PPh_3)]$	thf	58	298, 362	28	129, 133
$[Au(C{\equiv}CC_6H_4\text{-}4\text{-}C{\equiv}CC_6H_4\text{-}4\text{-}NO_2)(PPh_3)]$	thf	59	301, 362	28	129, 133
$[Au(C{\equiv}CC_6H_4\text{-}4\text{-}N{=}CHC_6H_4\text{-}4\text{-}NO_2)(PPh_3)]$	thf	85	297, 392	34	129, 133
$[Ni(C{\equiv}CPh)(PPh_3)(\eta^5\text{-}C_5H_5)]$	thf	24[b]	307	15	134
$[Ni(C{\equiv}CC_6H_4\text{-}4\text{-}NO_2)(PPh_3)(\eta^5\text{-}C_5H_5)]$	thf	221	368, 439	59	134
$[Ni(C{\equiv}CC_6H_4\text{-}4\text{-}C_6H_4\text{-}4\text{-}NO_2)(PPh_3)(\eta^5\text{-}C_5H_5)]$	thf	193	263, 310, 413	65	134
$[Ni(C{\equiv}CC_6H_4\text{-}4\text{-}(E)\text{-}CH{=}CHC_6H_4\text{-}4\text{-}NO_2)(PPh_3)(\eta^5\text{-}C_5H_5)]$	thf	445	313, 437	120	134
$[Ni(C{\equiv}CC_6H_4\text{-}4\text{-}(Z)\text{-}CH{=}CHC_6H_4\text{-}4\text{-}NO_2)(PPh_3)(\eta^5\text{-}C_5H_5)]$	thf	145	307, 417	47	134
$[Ni(C{\equiv}CC_6H_4\text{-}4\text{-}C{\equiv}CC_6H_4\text{-}4\text{-}NO_2)(PPh_3)(\eta^5\text{-}C_5H_5)]$	thf	326	313, 417	106	134
$[Ni(C{\equiv}CC_6H_4\text{-}4\text{-}N{=}CHC_6H_4\text{-}4\text{-}NO_2)(PPh_3)(\eta^5\text{-}C_5H_5)]$	thf	387	282, 448	93	134

[a] Uncertainty of ±10%. Measured by HRS at 1.06 μm.
[b] Uncertainty of ±20%.
[c] Measured by EFISH at 1.06 μm.

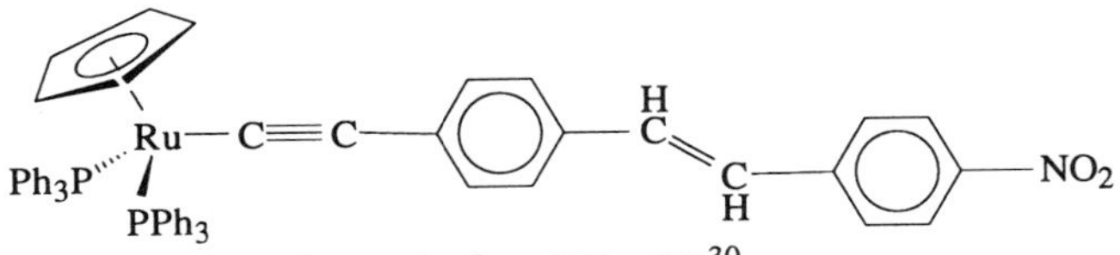

$\beta = 1464;\ \beta_0 = 234 \times 10^{-30}$ esu

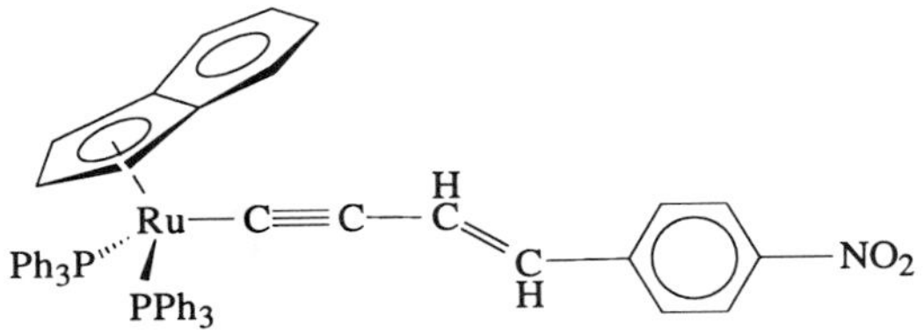

$\beta = 1257;\ \beta_0 = 89 \times 10^{-30}$ esu

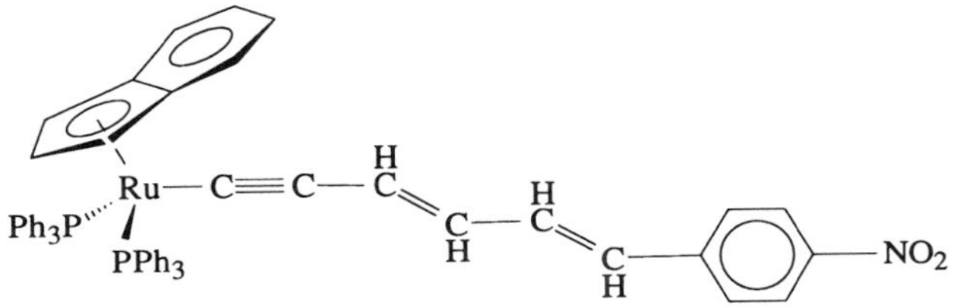

$\beta = 1320;\ \beta_0 = 34 \times 10^{-30}$ esu

FIG. 9. Acetylide complexes with large β.

efficient oxidizable 18-valence-electron organometallic complexes can provide access to stronger donors than are possible in organic systems. Data from the (cyclopentadienyl)(triphenylphosphine)nickel acetylides with an 18-electron, but less easily oxidizable metal are also substantially resonance enhanced, although the relative orderings were maintained with two-level-corrected values. Trends observed in chromophore variation are consistent across the ruthenium, gold, and nickel complexes. Figure 10 shows the relative merit of these varying ligated metal centers, with the $Ru(PPh_3)_2(\eta^5\text{-}C_5H_5)$ unit 10 and 3 times as efficient as the $Au(PPh_3)$ and $Ni(PPh_3)(\eta^5\text{-}C_5H_5)$ moieties, respectively (experimental data).

These structural changes involve the donor group of a donor–bridge–acceptor system; it was therefore expected that a more electron-donating (electron rich) ligand at the donor metal would increase the molecular hyperpolarizability. This was confirmed in the substitution of cyclopentadienyl by the more electron-rich indenyl group for $[Ru(C{\equiv}CC_6H_4\text{-}4\text{-}$

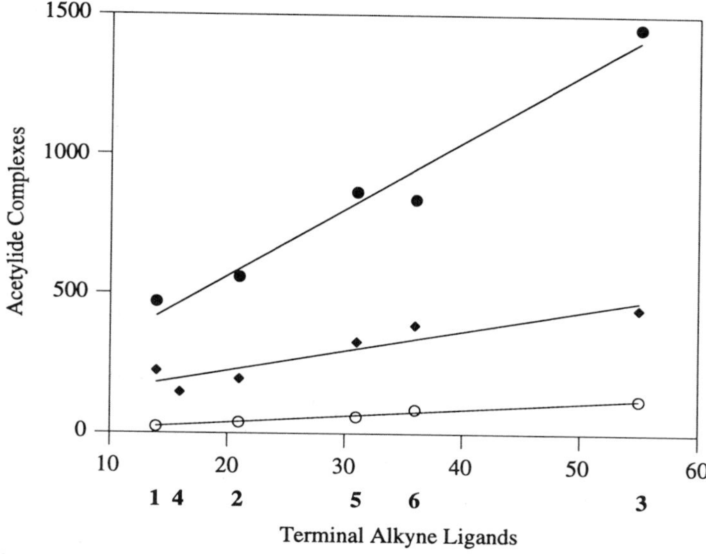

FIG. 10. Correlation of experimentally determined quadratic nonlinearities of metal acetylide complexes with their precursor acetylenes; ○ $[Au(C{\equiv}CR)(PPh_3)]$, ◆ $[Ni(C{\equiv}CR)(PPh_3)(\eta^5\text{-}C_5H_5)]$, ● $[Ru(C{\equiv}CR)(PPh_3)_2(\eta^5\text{-}C_5H_5)]$ (R = 4-$C_6H_4NO_2$, **1**; 4,4′-$C_6H_4C_6H_4NO_2$, **2**; (*E*)-4,4′-$C_6H_4CH{=}CHC_6H_4NO_2$, **3**; (*Z*)-4,4′-$C_6H_4CH{=}CHC_6H_4NO_2$, **4**; 4,4′-$C_6H_4C{\equiv}CC_6H_4NO_2$, **5**; 4,4′-$C_6H_4N{=}CHC_6H_4NO_2$, **6**).

$NO_2)(PPh_3)_2(\eta^5\text{-}C_5H_4R)]$ (R = H, C_4H_7). However, the replacement of triphenylphosphine by the stronger base trimethylphosphine $[Ru(C{\equiv}CC_6H_4\text{-}4\text{-}NO_2)L_2(\eta^5\text{-}C_5H_5)]$ (L = PPh_3, PMe_3) resulted in a decrease of the observed hyperpolarizability; this result is consistent with the report that *N*-phenyl substitution of 4-nitroaniline produces a greater increase in β than *N*-methyl substitution,[135] but further data assessing ligand replacement are required to confirm this observation. Chain lengthening in proceeding from $[Ru(C{\equiv}CC_6H_4\text{-}4\text{-}NO_2)(PPh_3)_2(\eta^5\text{-indenyl})]$ to $[Ru(C{\equiv}C\text{-}(E)\text{-}CH{=}CHC_6H_4\text{-}4\text{-}NO_2)(PPh_3)_2(\eta^5\text{-indenyl})]$ and then $[Ru(C{\equiv}C\text{-}(E)\text{-}CH{=}CH\text{-}(E)\text{-}CH{=}CHC_6H_4\text{-}4\text{-}NO_2)(PPh_3)_2(\eta^5\text{-indenyl})]$ leads to a dramatic increase in observed nonlinearity upon introduction of the first ene-linkage, but results in no significant difference upon introduction of the second alkene group. Surprisingly, examination of the two-level corrected data reveals a dramatic *decrease* in nonlinearity upon stepwise introduction of the ene-linkers. Coordination of the free pyridine in $[Ru(C{\equiv}C\text{-}(E)\text{-}CH{=}CH\text{-}4\text{-}C_5H_4N)(PPh_3)_2(\eta^5\text{-indenyl})]$ to $M(CO)_5$ (M = Cr, W) centers leads to a substantial increase in nonlinearity. Coordination of $[Ru(C{\equiv}C\text{-}(E)\text{-}CH{=}CHC_6H_4\text{-}4\text{-}C{\equiv}N)(PPh_3)_2(\eta^5\text{-indenyl})]$ to the same metal carbonyl fragments was also investigated and, for both bimetallic systems, β(tungsten-containing complex) > β(chromium-containing complex).

Given the increased nonlinearities obtained by incorporating one metal into organic chromophores, it is not surprising that introduction of two metals has been assayed. Examples have been met in earlier tables organized by various metal–ligand groupings, but Table X contains collected data for all bimetallic complexes for which β values have been reported. Figure 11 displays representative examples with large nonlinearities.

The first complex in Fig. 11 is a mixed valence Ru^{II}/Ru^{III} compound. The very large β value reported in 1993 has now been revised down, but the nonlinearity is still large. This β value may be attributed to the intervalence charge transfer excited states as well as some resonance enhancement.[136] Results with the sesquifulvalene complexes in Fig. 11 have been mentioned before and are similarly strongly resonance enhanced. The difference in measured β values for the two tabulated examples suggests that the triple bond linkage leads to a higher nonlinearity than the double bond linkage; however, the two-level corrected β_0 values do not support this conclusion. The two ferrocene derivatives containing the dimesitylboranyl acceptor group show that a marked effect on nonlinearity is observed when a triple bond is replaced by a double bond, with a significantly smaller $|\beta|$ for the triple-bond-containing complex. As both complexes were measured far from resonance and have very similar λ_{max} values, an ene vs yne linkage comparison seems valid in this instance. The series of ferrocenyl bimetallic complexes with Group 10 metals has been discussed previously; whereas ligated nickel is a poor acceptor, replacing $PdCl_2$ by $PtCl_2$ has only a weak influence on the measured nonlinearity. Some of the ruthenium indenyl complexes have also been discussed previously; their nonlinearities are among the largest observed for bimetallic complexes. The differences observed between the bimetallic compounds and their precursor monometallic complexes can be explained by the differences in the effectiveness of the metal σ-acceptor groups; the metal may usefully tune and in many instances enhance the NLO response. For the ruthenium indenyl complexes, the tungsten-containing complexes have larger hyperpolarizabilities than their chromium analogues, probably because the electron density at tungsten is better reduced by back donation to carbonyl ligands than it is at chromium.

Table XI contains data for silicon- and tin-containing complexes, chiefly from Mignani, Simon, and their co-workers. Figure 12 displays representative examples with large nonlinearities. It has been suggested from studies with the silicon-containing compounds that the trimethylsilyl group can function as both a donor and an acceptor, being a weak acceptor in [$SiMe_3(C_6H_4\text{-}4\text{-}NMe_2)$].[139] As this complex has a similar β value to

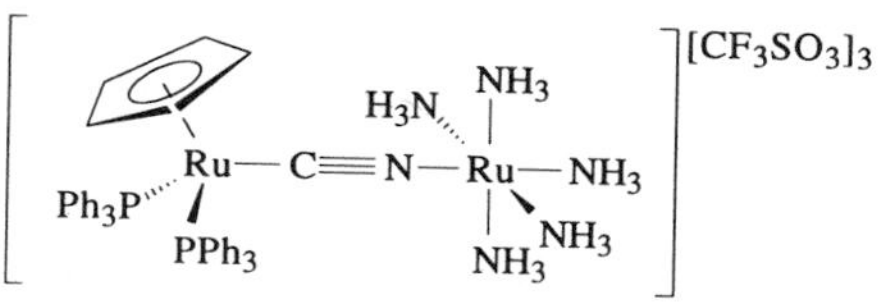

β = 157; 69 × 10^-30 esu

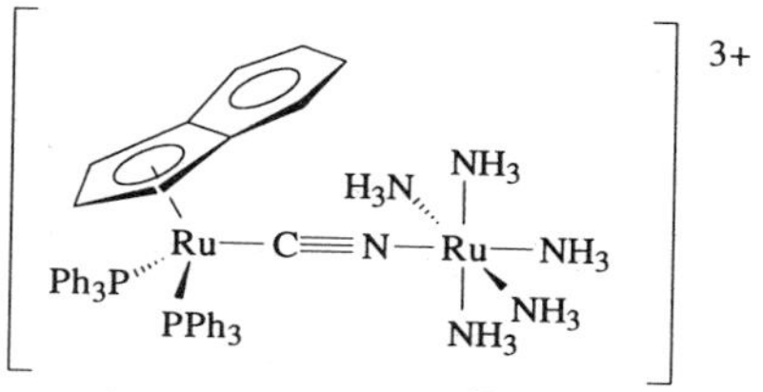

β = 108; β0 = 26 × 10^-30 esu

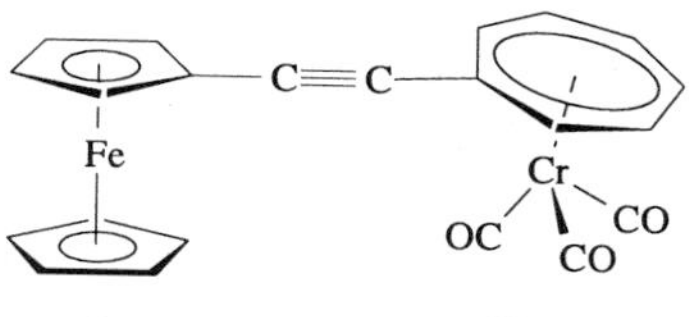

β = 570; β0 = 105 × 10^-30 esu

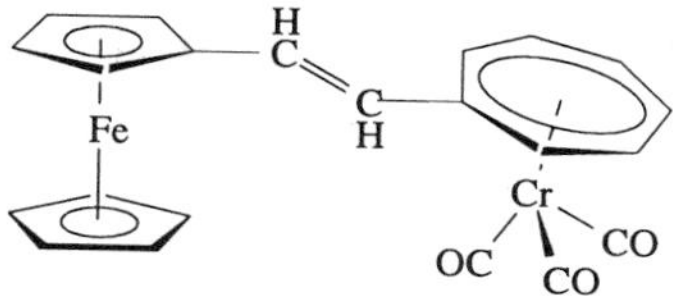

β = 320; β0 = 113 × 10^-30 esu

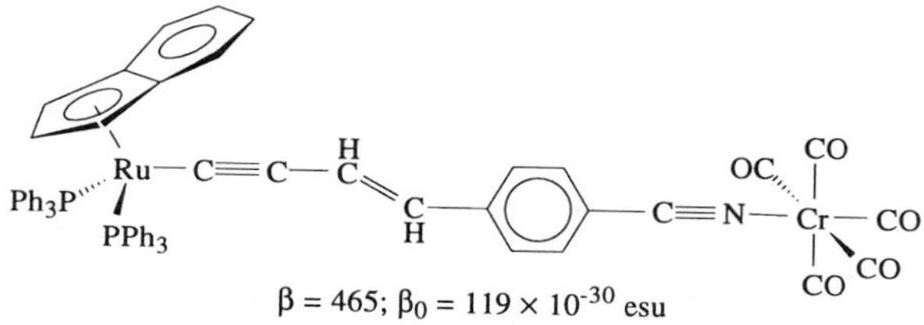

β = 465; β0 = 119 × 10^-30 esu

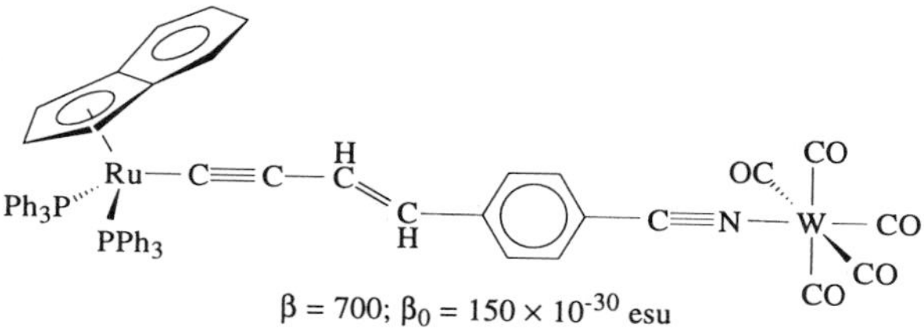

β = 700; β0 = 150 × 10^-30 esu

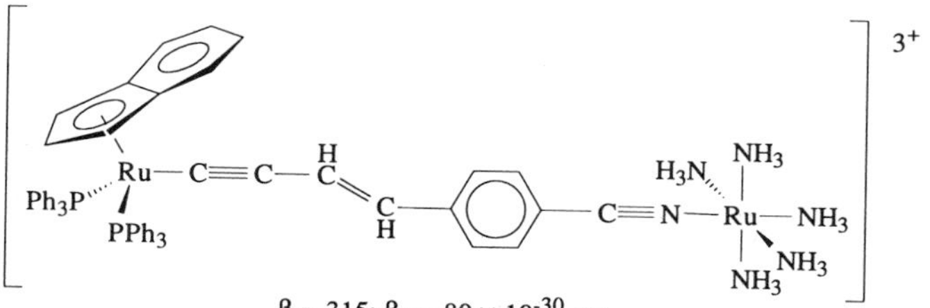

β = 315; β0 = 80 × 10^-30 esu

FIG. 11. Bimetallic complexes with large β.

TABLE X

MOLECULAR NLO MEASUREMENTS FOR BIMETALLIC COMPLEXES

Complex	Solvent	β^a (10^{-30} esu)	λ_{max} (nm)	β_0 (10^{-30} esu)	Technique	Fund. (μm)	Ref.
[FcC≡C(η^7-C_7H_6)Cr$(CO)_3$]	CH_2Cl_2	570	600	105	HRS	1.06	119
[Fc-(*E*)-CH=CH(η^7-C_7H_6)Cr$(CO)_3$]	CH_2Cl_2	320	670	113	HRS	1.06	119
[Fc(η^6-BC_6H_6)Co(η^5-C_5H_5)]	CH_2Cl_2	—	370, 410(sh), 650	90 ± 30	HRS	1.06	132
[Fc-(*E*)-CH=CHB$(mes)_2$]	$CHCl_3$	−24	336	—	EFISH	1.91	120
[FcC≡CB$(mes)_2$]	$CHCl_3$	4.4	336	—	EFISH	1.91	120
[FcCH=C(SCH_2Ph)C=$NCMe_2CH_2O$ (*N*,*S*-$NiCl_2$)]	$CHCl_3$	10.3	500	3.92	EFISH	1.34	122, 123
[FcCH=C(SCH_2Ph)C=$NCMe_2CH_2O$ (*N*,*S*-$NiBr_2$)]	$CHCl_3$	7.14	511	2.55	EFISH	1.34	122, 123
[FcCH=C(SCH_2Ph)C=$NCMe_2CH_2O$ (*N*,*S*-$PdCl_2$)]	$CHCl_3$	12.2	530	4.13	EFISH	1.34	121–123
	$CHCl_3$	9	530	5.8	EFISH	1.91	122, 123
[FcCH=C(SCH_2Ph)C=$NCMe_2CH_2O$ (*N*,*S*-$PtCl_2$)]	$CHCl_3$	16	523	5.2	EFISH	1.34	122, 123
[FcCH=C(S(O)CH_2Ph)C=$NCMe_2CH_2O$ (*N*,*S*-$NiCl_2$)]	$CHCl_3$	15.8	498	6.0	EFISH	1.34	122. 123
[FcCH=C(S(O)CH_2Ph)C=$NCMe_2CH_2O$ (*N*,*S*-$NiBr_2$)]	$CHCl_3$	5.85	494	2.3	EFISH	1.34	122, 123

$[FcCH{=}C(S(O)CH_2Ph)C{=}NCMe_2CH_2O$ $(N,S\text{-}PdCl_2)]$	$CHCl_3$	30.2	540	7.6	EFISH	1.34	121–123
$[FcCH{=}C(S(O)CH_2Ph)C{=}NCMe_2CH_2O$ $(N,S\text{-}PtCl_2)]$	$CHCl_3$	28.0	548	7.7	EFISH	1.34	122, 123
$[Fc\text{-}(E)\text{-}CH{=}CHCH{=}C(SCH_2Ph)C{=}NCMe_2CH_2O$ $(N,S\text{-}PdCl_2)]$	$CHCl_3$	58	567	13.5	EFISH	1.34	122, 123
	$CHCl_3$	18.2	567	10.8	EFISH	1.91	122, 123
$[Fc\text{-}(E)\text{-}CH{=}CHCH{=}C(S(O)CH_2Ph)C{=}NCMe_2CH_2O$ $(N,S\text{-}PdCl_2)]$	$CHCl_3$	123.5	602	19	EFISH	1.34	122, 123
$[FcC{\equiv}CCH{=}C(SCH_2Ph)C{=}NCMe_2CH_2O$ $(N,S\text{-}PdCl_2)]$	$CHCl_3$	33.6	558	8.5	EFISH	1.34	122, 123
$[Ru(C{\equiv}NRu(NH_3)_5)(PPh_3)_2(\eta^5\text{-}C_5H_5)][CF_3SO_3]_3$	H_2O	1080[b]	~700	—	HRS	1.06	136
	$MeNO_2$	157	716	69	HRS	1.06	137
$[Ru(C{\equiv}NOs(NH_3)_5)(PPh_3)_2(\eta^5\text{-}C_5H_5)]^{3+}$	dmso	65	440	16	HRS	1.06	137
$[Mo(C{\equiv}NRu(NH_3)_5)(CO)_5]^{2+}$	Acetone	225	693	90	HRS	1.06	137
$[W(C{\equiv}NRu(NH_3)_5)(CO)_5]^{2+}$	Acetone	130	708	56	HRS	1.06	137
$[Ru(C{\equiv}NCr(CO)_5)(PPh_3)_2(\eta^5\text{-indenyl})]$	CH_2Cl_2	25	392	10	HRS	1.06	129, 130
$[Ru(C{\equiv}NW(CO)_5)(PPh_3)_2(\eta^5\text{-indenyl})]$	CH_2Cl_2	40	392	15	HRS	1.06	129, 130
$[Ru(C{\equiv}NRu(NH_3)_5)(PPh_3)_2(\eta^5\text{-indenyl})]^{3+}$	Acetone	108	621	26	HRS	1.06	129, 130
$[Ru(C{\equiv}C\text{-}(E)\text{-}CH{=}CHC_6H_4\text{-}4\text{-}C{\equiv}NCr(CO)_5)(PPh_3)_2(\eta^5\text{-indenyl})]$	CH_2Cl_2	465	442	119	HRS	1.06	129, 130
$[Ru(C{\equiv}C\text{-}(E)\text{-}CH{=}CHC_6H_4\text{-}4\text{-}C{\equiv}NW(CO)_5)(PPh_3)_2(\eta^5\text{-indenyl})]$	CH_2Cl_2	700	456	150	HRS	1.06	129, 130
$[Ru(C{\equiv}C\text{-}(E)\text{-}CH{=}CHC_6H_4\text{-}4\text{-}C{\equiv}NRu(NH_3)_5)(PPh_3)_2(\eta^5\text{-indenyl})]^{3+}$	Acetone	315	442	80	HRS	1.06	129, 130

[a] Uncertainty of ±10% except where indicated.
[b] No uncertainty reported.

TABLE XI

MOLECULAR NLO MEASUREMENTS FOR SILICON AND TIN COMPLEXES

Complex	Solvent	β (10^{-30} esu)	λ_{max} (nm)	β_0 (10^{-30} esu)	Technique	Fund. (μm)	Ref.
$[SiMe_2(C_6H_4\text{-}4\text{-}NMe_2)(C_6H_4\text{-}4\text{-}CH{=}C(CN)_2)]$	Acetone	16 ± 4	320	11	EFISH	1.34	123, 138, 139
$[C_6H_4(4\text{-}NMe_2)(SiMe_2)_2C_6H_4\text{-}4\text{-}CH{=}C(CN)_2]$	Acetone	22 ± 5	334	16	EFISH	1.34	123, 138, 139
$[C_6H_4(4\text{-}NMe_2)(SiMe_2)_6C_6H_4\text{-}4\text{-}CH{=}C(CN)_2]$	Acetone	38 ± 4	320, 385	26	EFISH	1.34	123, 138, 139
$[SiMe_3C_6H_4\text{-}4\text{-}NMe_2]$	Acetone	3.8 ± 1	266[b]	4	EFISH	1.34	139
$[SiMe_3C_6H_4\text{-}4\text{-}CH{=}C(CN)_2]$	Acetone	12 ± 3	322[b]	9	EFISH	1.34	139
$[SiMe(C_6H_4\text{-}4\text{-}NMe_2)(C_6H_4\text{-}4\text{-}CH{=}C(CN)_2)_2]$	Acetone	22.5 ± 5	317[b]	13	EFISH	1.34	139
$[SiMe(C_6H_4\text{-}4\text{-}NMe_2)_2(C_6H_4\text{-}4\text{-}CH{=}C(CN)_2)]$	Acetone	18 ± 4	319[b]	16	EFISH	1.34	139
$[Sn(C_6H_4\text{-}4\text{-}NMe_2)_4]$	CH_2Cl_2	13.0 ± 1[a]	274	9.0	HRS	1.06	140
$[Sn(C_6H_4\text{-}4\text{-}C_6H_4\text{-}4\text{-}NMe_2)_4]$	CH_2Cl_2	24.0 ± 2[a]	319	14.0	HRS	1.06	140
$[Sn(C_6H_4\text{-}4\text{-}N{=}NC_6H_4\text{-}4\text{-}NMe_2)_4]$	CH_2Cl_2	159.0 ± 16[a]	427	47.5	HRS	1.06	140
$[SnPh_3C_6H_4\text{-}4\text{-}N{=}NC_6H_4\text{-}4\text{-}NMe_2]$	CH_2Cl_2	181 ± 18	421	57	HRS	1.06	140

[a] β_{xyz}.

[b] In $CHCl_3$.

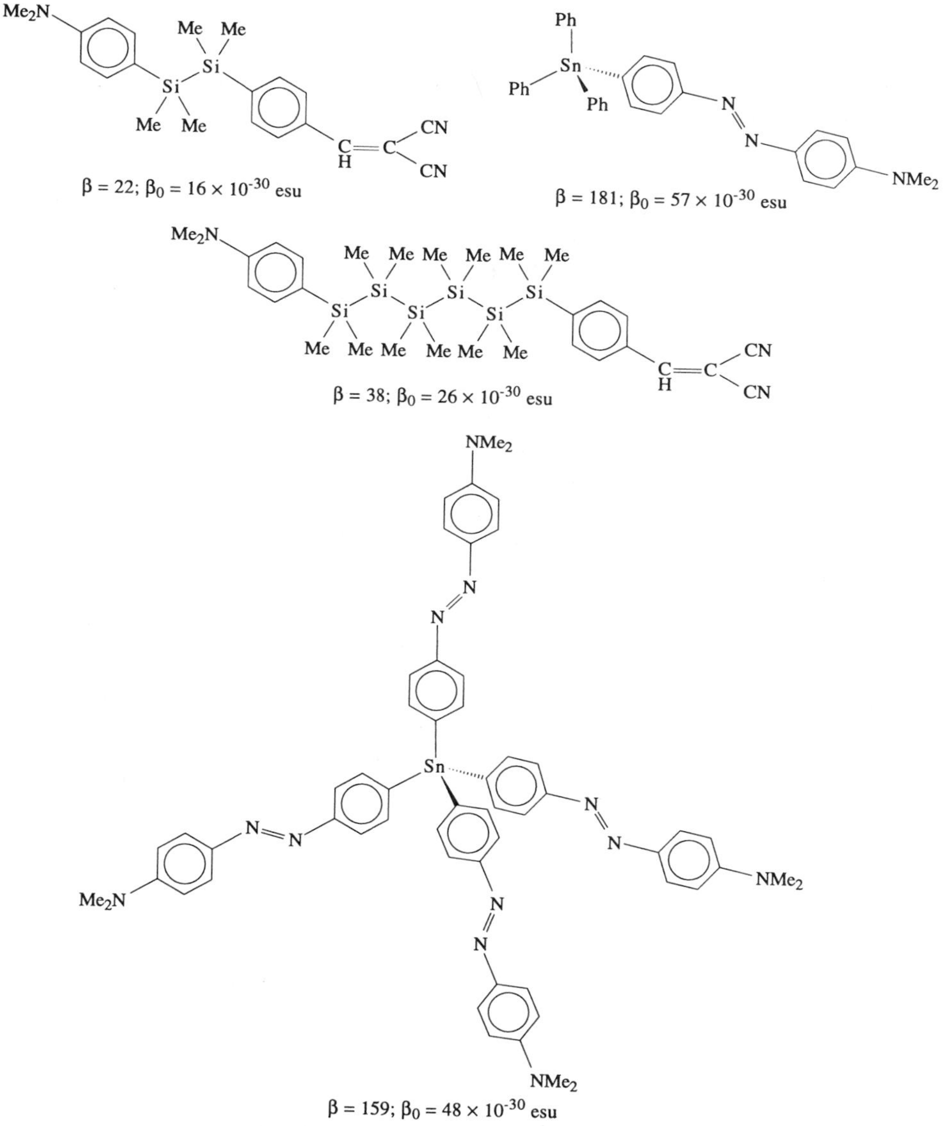

$\beta = 22$; $\beta_0 = 16 \times 10^{-30}$ esu

$\beta = 181$; $\beta_0 = 57 \times 10^{-30}$ esu

$\beta = 38$; $\beta_0 = 26 \times 10^{-30}$ esu

$\beta = 159$; $\beta_0 = 48 \times 10^{-30}$ esu

FIG. 12. Silicon and tin complexes with large β.

N,*N*-dimethylaniline, the trimethylsilyl group is behaving similarly to hydrogen, and its description as an acceptor seems inappropriate. Mignani *et al.*[139] suggest that the trimethylsilyl group is more effective as a donor, with its efficiency ascribed to σ–π interaction. However, it is certainly less effective as a donor than the dimethylamino group, as the compound 1-(4-dimethylamino)phenyl-2,2-dicyanoethene has a β value approximately three times larger than [$SiMe_3(C_6H_4$-4-CH=C(CN)$_2$)]. The efficiency of silicon as part of a conjugation pathway is believed to be small because of the nonplanar silicon–phenyl geometry. In addition to its lack of effectiveness in donor or acceptor groups, therefore, silicon is also an inefficient bridge connecting conjugated systems; it is certainly much less efficient than, for example, the ethenyl linkage. Its advantage as a bridge lies in improved transparency for polysilyl vs polyene units; chromophores of similar length in the two systems have λ_{max} to higher energy in the silyl system. All but one of the tin compounds in Table XI have tetrahedral symmetry and are nonpolar, which should improve the chance of noncentrosymmetric packing in the solid state, as there are no dipoles to interact. The nonlinearities for these examples result from the octupolar contribution to β, as the vectorial component must be zero.[141] As expected, the measured β values increase with increased conjugation length, with the azo derivatives leading to higher β values than their biphenyl analogues.

Table XII contains data for boron-containing compounds. Figure 13 displays representative examples with large nonlinearities. Results, principally from Marder and co-workers, suggest that dimesitylboranyl is an effective π-acceptor. Although only moderate nonlinearities have been observed, a feature of the boranes presented here, as with other main group examples (the silanes and stannanes in Table XI) is their transparency at wavelengths >450 nm. The experimental data show the trend [4-RC_6H_4B(mes)$_2$] < [4-$RC_6H_4C{\equiv}CB$(mes)$_2$] < [4-RC_6H_4-(*E*)-CH=CHB(mes)$_2$] for R = SMe, OMe, and NMe_2, and changing the R group in 4-RC_6H_4XYB(mes)$_2$ reveals an efficiency series NMe_2 > OMe > SMe > Ph > Ph_2P, suggesting that diphenylphosphino is a very poor donor.

The high degree of delocalization in icosahedral carboranes has led Mingos and co-workers to examine their nonlinearities. Although the observed β values are comparatively low, the full potential of these materials has not yet been assessed; the compounds that were examined contained either a donor or an acceptor substituent, and studies of a 1-donor–12-acceptor substituted carborane that may reveal the extent of their NLO merit have not yet been promulgated.

Table XIII contains data for organometallic complexes that do not fall into one of the preceding categories. The nonlinearities of the chloro complexes were used in conjunction with those of terminal acetylenes to demonstrate the importance of electronic communication between ligated metal and acetylide ligand in derivative metal acetylide complexes formed from combining these precursors.[49,133,134] Evaluation of vinylidene, as with sesquifulvalene and carbene, is only experimentally straightforward following metal complexation. The vinylidene cation [Ru(C=CHC_6H_4-4-NO_2)

TABLE XII

Molecular NLO Measurements for Boron Complexes

Complex	Solvent	β (10^{-30} esu)	λ_{max} (nm)	β_0 (10^{-30} esu)	Technique	Fund. (μm)	Ref.
$[BrC_6H_4\text{-}4\text{-}B(mes)_2]$	$CHCl_3$	1.6 ± 0.2	269, 314	—	EFISH	1.91	142
$[MeSC_6H_4\text{-}4\text{-}B(mes)_2]$	$CHCl_3$	2.5 ± 0.5	249, 335	—	EFISH	1.91	142
$[MeOC_6H_4\text{-}4\text{-}B(mes)_2]$	$CHCl_3$	3.3 ± 0.3	274, 317	—	EFISH	1.91	142
$[Me_2NC_6H_4\text{-}4\text{-}B(mes)_2]$	$CHCl_3$	11 ± 1	309, 359	—	EFISH	1.91	142
$[Ph_2P\text{-}(E)\text{-}CH{=}CHB(mes)_2]$	$CHCl_3$	2.6 ± 1	244, 342	—	EFISH	1.91	142, 143
$[Ph\text{-}(E)\text{-}CH{=}CHB(mes)_2]$	$CHCl_3$	5.1 ± 0.5	290, 330	—	EFISH	1.91	142, 143
$[MeSC_6H_4\text{-}4\text{-}(E)\text{-}CH{=}CHB(mes)_2]$	$CHCl_3$	8.6–9.3 ± 0.9	358	—	EFISH	1.91	142, 143
$[MeOC_6H_4\text{-}4\text{-}(E)\text{-}CH{=}CHB(mes)_2]$	$CHCl_3$	8.6–9.3 ± 0.9	347	—	EFISH	1.91	142, 143
$[H_2NC_6H_4\text{-}4\text{-}(E)\text{-}CH{=}CHB(mes)_2]$	$CHCl_3$	18 ± 2	368	—	EFISH	1.91	142, 143
$[Me_2NC_6H_4\text{-}4\text{-}(E)\text{-}CH{=}CHB(mes)_2]$	$CHCl_3$	18–33 ± 3	401	—	EFISH	1.91	142
$[Fc\text{-}(E)\text{-}CH{=}CHB(mes)_2]$	$CHCl_3$	−24 ± 2	336	—	EFISH	1.91	142, 143
$[Ph_2PC{\equiv}CB(mes)_2]$	$CHCl_3$	3.3 ± 0.3	338	—	EFISH	1.91	120
$[PhC{\equiv}CB(mes)_2]$	$CHCl_3$	3.4 ± 0.3	336	—	EFISH	1.91	142
$[FcC{\equiv}CB(mes)_2]$	$CHCl_3$	4.4 ± 0.4	336	—	EFISH	1.91	142
$[MeSC_6H_4\text{-}4\text{-}C{\equiv}CB(mes)_2]$	$CHCl_3$	7.1 ± 1.4	368	—	EFISH	1.91	120
$[MeOC_6H_4\text{-}4\text{-}C{\equiv}CB(mes)_2]$	$CHCl_3$	9.3 ± 0.9	356	—	EFISH	1.91	142
$[Me_2NC_6H_4\text{-}4\text{-}C{\equiv}CB(mes)_2]$	$CHCl_3$	25 ± 2.5	398	—	EFISH	1.91	142
$[Me_2NC_6H_4\text{-}4\text{-}C_6H_4\text{-}4\text{-}B(mes)_2]$	CH_2Cl_2	—	376	42 ± 10	EFISH	1.91	144
	Various	37	~376	16.5	Solvatochromism	—	30
$[Me_2NC_6H_4\text{-}4\text{-}N{=}NC_6H_4\text{-}4\text{-}B(mes)_2]$	CH_2Cl_2	—	446	72 ± 18	EFISH	1.91	144
	Various	210	~446	50	Solvatochromism	—	30
$[1\text{-}(4\text{-}MeSC_6H_4)\text{-}1,2\text{-}C_2B_{10}H_{11}]$	—	~5	—	—	EFISH	—	145
$[1\text{-}(4\text{-}O_2NC_6H_4)\text{-}1,2\text{-}C_2B_{10}H_{11}]$	—	~2.7	—	—	EFISH	—	145
$[1\text{-}(4\text{-}H_2NC_6H_4)\text{-}1,2\text{-}C_2B_{10}H_{11}]$	—	~6.3	—	—	EFISH	—	145
$[9\text{-}(2,4\text{-}(NO_2)_2C_6H_3)\text{-}1,2\text{-}C_2B_{10}H_{11}]$	—	~3.5	—	—	EFISH	—	145
$[1,2\text{-}C_2B_{10}H_{11}]$	—	~3.4	—	—	EFISH	—	145

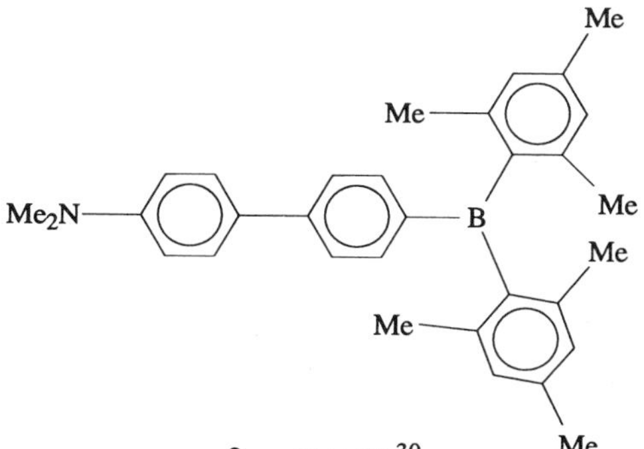

$\beta_0 = 42 \times 10^{-30}$ esu

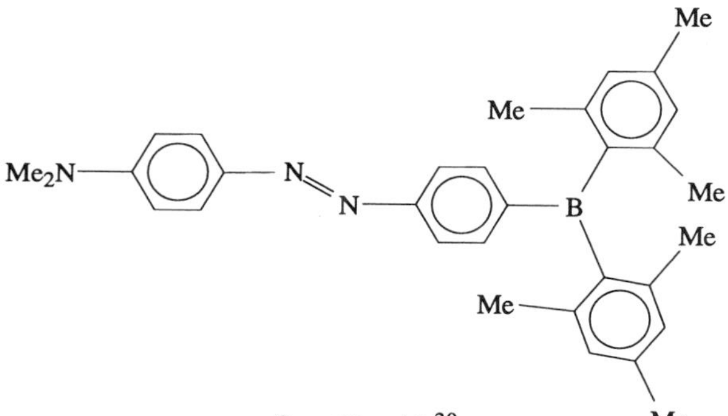

$\beta_0 = 72 \times 10^{-30}$ esu

FIG. 13. Boron complexes with large β_0.

$(PPh_3)_2(\eta^5\text{-indenyl})]^+$ can be directly compared to its parent acetylide complex $[Ru(C{\equiv}CC_6H_4\text{-}4\text{-}NO_2)(PPh_3)_2(\eta^5\text{-indenyl})]$; the corrected nonlinearity of the vinylidene is large, but less than half that of the acetylide. Although the ligated metal center in the acetylide complex is the donor in a classical donor–bridge–acceptor composition, the vinylidene metal-bound carbon is electron deficient, and the complex cation therefore has two potential acceptors. To assess the merit of vinylidene complexes as NLO materials, evaluation of a complex bearing an electron-donating phenylvinylidene ligand would be of interest.

The 16-electron square planar palladium and platinum complexes show low responses. The reported figures suggest that (i) the platinum complex has a larger nonlinearity than its palladium analogue, (ii) the triphenylphosphine ligand leads to higher nonlinearities than triethylphosphine (consistent with results above for ruthenium acetylides bearing PPh_3 or PMe_3

TABLE XIII

MOLECULAR NLO MEASUREMENTS FOR OTHER COMPLEXES

Complex	Solvent	β^a (10^{-30} esu)	λ_{max}	β_0 (10^{-30} esu)	Technique	Fund. (μm)	Ref.
$[NiCl(PPh_3)(\eta^5\text{-}C_5H_5)]$	thf	89	263, 330	45	HRS	1.06	134
$[RuCl(PPh_3)_2(\eta^5\text{-}C_5H_5)]$	thf	<7	357	<4	HRS	1.06	47
$[Ru(C{\equiv}N)(PPh_3)_2(\eta^5\text{-indenyl})]$	CH_2Cl_2	13	396	5	HRS	1.06	129, 130
$[Ru({=}C{=}CHC_6H_4\text{-4-}NO_2)(PPh_3)_2(\eta^5\text{-indenyl})]^+$	CH_2Cl_2	116	379	50	HRS	1.06	129, 130
trans-$[PdI(C_6H_4\text{-4-}NO_2)(PEt_3)_2]$	$CHCl_3$	0.5	—	—	EFISH	1.91	116
trans-$[PdI(C_6H_4\text{-4-}NO_2)(PPh_3)_2]$	$CHCl_3$	1.5	—	—	EFISH	1.91	116
trans-$[PtBr(C_6H_4\text{-4-}CHO)(PEt_3)_2]$	$CHCl_3$	2.1	—	—	EFISH	1.91	116
trans-$[PtBr(C_6H_4\text{-4-}NO_2)(PEt_3)_2]$	$CHCl_3$	3.8	—	—	EFISH	1.91	116
trans-$[PtI(C_6H_4\text{-4-}NO_2)(PEt_3)_2]$	$CHCl_3$	1.7	—	—	EFISH	1.91	116

[a] Uncertainty of ±10%.

ligands), (iii) the nitro group, not surprisingly, leads to higher nonlinearities than the formyl group, which is a much poorer acceptor, and (iv) the iodo ligand leads to lower nonlinearity than the bromo ligand. As the measured nonlinearities are all low, conclusions from this series of complexes should be viewed as tentative.

C. *Molecular Second-Order Computational Results*

Table XIV contains data for computationally derived second-order hyperpolarizabilities, principally by Marder, Ratner, and their co-workers, and more recently by Humphrey *et al.* The focus of these studies has been similar to that of experimental investigations, namely ferrocenyl, metal carbonyl, and metal acetylide complexes, and very recently carboranes. Calculated values were all obtained using ZINDO, the methodology for which is described in Section II,E.

A comparison of the computationally derived values with experimental data shows that ZINDO gives reasonable results. It correctly predicts the sign of the measured values, it generally predicts the order of magnitude, and in most cases the trends in β across a given series are predicted. For the ferrocenyl complexes, the predicted nonlinearities are close to experimentally observed values. Inclusion of acceptor substituents in the plane of the substituted cyclopentadienyl ring causes only slight deviation of calculated β_{EFISH} from β_{tot}.[62] The dipole moment along the cyclopentadienyl–iron–cyclopentadienyl axis is very small and the majority of the ground-state dipole moment is in the plane perpendicular to this axis. The complex [Fc-(*E*)-CH=CHC_6H_4-2,4-$(NO_2)_2$] is anomalous in affording a result that does not agree with experiment. However, the molecular geometry input into ZINDO was based on crystallographic data, and the experimental techniques employ solutions of sample. The degree of rotation of the nitro-containing aromatic ring out of the ethylenic plane in the solid state is almost certainly much greater than that in the solution phase. Results from a geometry calculation suggest that rotation in solution is of the order of 10–20°, significantly less than that in the solid state. Using the smaller rotation, a calculated β_{EFISH} value of 24.3×10^{-30} esu was obtained,[62] in much better agreement with experiment, which highlights the usefulness of computational techniques in probing geometry/optical nonlinearity relationships.

Kanis *et al.*[62] suggest that the qualitative structure/NLO property relationships for organic molecules may be applied to the ferrocenyl derivatives here. A greater electronic asymmetry arising from stronger donor and acceptor moieties leads to larger β values; the strongly donating permethylated ferrocenyl complexes have larger nonlinearities than their

TABLE XIV

Computationally Derived Optical Nonlinearities (all by ZINDO at 1.91 μm)

Complex	β_{EFISH} (10^{-30} esu)	β_{tot} (10^{-30} esu)	Ref.	Experimental			
				Effect	β (10^{-30} esu)	Freq. (μm)	Ref.
[Fc-(E)-CH=CH-4-C_5H_4N-1-Me]$^+$	67.6	68.9	62				
[Fc-(Z)-CH=CHC_6H_4-4-NO_2]	20.7–29	21.0–29.1	62, 146	β_{EFISH}	13–14	1.91	115
[Fc-(E)-CH=CHC_6H_4-4-NO_2]	30.8–41.0	30.8–41.0	62, 146	β_{EFISH}	31–34	1.91	115
[Fe(η^5-C_5Me_5)(η^5-C_5Me_4-(E)-CH=CHC_6H_4-NO_2)]	34.7–45.4	34.7–45.5	62, 146	β_{EFISH}	40	1.91	115
[Fc-(Z)-CH=CHC_6H_4-4-CN]	5.49	5.92	62	β_{EFISH}	4.0	1.91	115
[Fc-(E)-CH=CHC_6H_4-4-CN]	9.72	9.76	62	β_{EFISH}	10	1.91	115
[Fc-(E)-CH=CHC_6H_4-4-CHO]	15.1	17.2	62	β_{EFISH}	12	1.91	115
[Fc-(E)-CH=CH-(E)-CH=CHC_6H_4-4-NO_2]	47.4	47.5	62	β_{EFISH}	66	1.91	115
[Fc-(E)-CH=CH-(E)-CH=CH-(E)-CH=CHC_6H_4-4-NO_2]	66.6	66.8	62				
[Fc-(E)-CH=C(CN)C_6H_4-4-NO_2]	25.6–31.9	29.0–37.1	62, 146	β_{EFISH}	21–22	1.91	116
[Fe(η^5-C_5Me_5)(η^5-C_5Me_4=(E)-CH=C(CN)C_6H_4-4-NO_2)]	28.4–35.3	31.9–40.4	62, 146	β_{EFISH}	35	1.91	116
[Fc-(E)-CH=CHC_6H_4-2,4-$(NO_2)_2$]	40.0–51.0	44.8–56.4	62, 146	β_{EFISH}	21–23	1.91	116
[Fc-(E)-CH=CH-(E)-CH=CH-(E)-CH=CHC_6H_4-4-NO_2]	73.7	73.8	146	β_{EFISH}	52	1.91	116
[$Cr(CO)_3$(η^6-C_6H_6)]	−1.30–−1.9	1.30–1.9	62, 146	β_{EFISH}	−0.8	1.91	116
[$Cr(CO)_3$(η^6-C_6H_5OMe)]	−1.36–−1.5	1.50–1.6	62, 146	β_{EFISH}	−0.7–−0.9	1.91	116
[$Cr(CO)_3$(η^6-$C_6H_5NH_2$)]	−1.1–−1.27	1.42–1.9	62, 146	β_{EFISH}	−0.6	1.91	116
[$Cr(CO)_3$(η^6-$C_6H_5NMe_2$)]	−0.51–−0.91	1.12–3.0	62, 146	β_{EFISH}	−0.4	1.91	116
[$Cr(CO)_3$(η^6-$C_6H_5CO_2Me$)]	−1.79–−3.1	4.28–4.8	62, 146	β_{EFISH}	−0.7	1.91	116
[$Cr(CO)_3$(η^6-C_6H_5-(E)-CH=CHPh)]	−3.93–−4.0	4.80	62, 146	β_{EFISH}	−2.2	1.91	116
[$Cr(CO)_3$(η^6-C_6H_5-(E)-CH=CHC_6H_4-4-NO_2)]	8.3	23.5	62				
[$Cr(CO)_3$(η^6-C_6H_5-(E)-CH=CH-(E)-CH=CHC_6H_4-4-NO_2)]	11.5	28.4	62				
[$Cr(CO)_3$(η^6-C_6H_5-(E)-CH=CH-(E)-CH=CH-(E)-CH=CHC_6H_4-4-NO_2)]	15.6	34.5	62				
[$Cr(CO)_3$(η^6-$H_2NC_6H_4$-4-(E)-CH=CHC_6H_4-4-NO_2)]	10.7	23.6	62				
[$Cr(CO)_3$(η^6-$O_2NC_6H_4$-4-(E)-CH=CHC_6H_4-4-NH_2)]	14.8	26.6	62				

(*continues*)

TABLE XIV (*continued*)

Complex	β_{EFISH} (10^{-30} esu)	β_{tot} (10^{-30} esu)	Ref.	Experimental			
				Effect	β (10^{-30} esu)	Freq. (μm)	Ref.
$[Cr(CO)_3(\eta^6\text{-}H_2NC_6H_4\text{-}4\text{-}(E)\text{-}CH{=}CH\text{-}(E)\text{-}CH{=}CHC_6H_4\text{-}4\text{-}NO_2)]$	16.2	31.2	62				
$[Cr(CO)_3(\eta^6\text{-}O_2NC_6H_4\text{-}4\text{-}(E)\text{-}CH{=}CH\text{-}(E)\text{-}CH{=}CHC_6H_4\text{-}4\text{-}NH_2)]$	27.5	41.4	62				
$[Cr(NH_3)_3(\eta^6\text{-}C_6H_5\text{-}(E)\text{-}CH{=}CHC_6H_4\text{-}4\text{-}NO_2)]$	780		62				
$[Mo(CO)_5(NC_5H_4\text{-}4\text{-}NH_2)]$	−3.60		146	β_{EFISH}[a]	−2.1	1.91	116, 126
$[Mo(CO)_5(NC_5H_4\text{-}4\text{-}Bu^n)]$	−5.91		146	β_{EFISH}[a]	−3.4	1.91	116, 126
$[Mo(CO)_5(NC_5H_5)]$	−7.66		146	β_{EFISH}[a]	−4.4	1.91	116, 126
$[Mo(CO)_5(NC_5H_4\text{-}4\text{-}Ph)]$	−5.28		146	β_{EFISH}[a]	−4.5	1.91	116, 126
$[Mo(CO)_5(NC_5H_4\text{-}4\text{-}COMe)]$	−16.80		146	β_{EFISH}[a]	−9.3	1.91	116, 126
$[Cr(CO)_5(NC_5H_4\text{-}4\text{-}NH_2)]$	−1.32		45	β_{EFISH}[a]	−2.1	1.91	116, 126
$[Cr(CO)_5(NC_5H_4\text{-}4\text{-}NO_2)]$	−31.8		45				
$[Cr(CO)_5(NC_5H_5)]$	−5.91		45	β_{EFISH}[a]	−4.4	1.91	116, 126
$[Cr(CO)_5(NC_5H_4\text{-}4\text{-}Ph)]$	−4.49		45	β_{EFISH}[a]	−4.5	1.91	116, 126
$[Cr(CO)_5(NC_5H_4\text{-}4\text{-}COMe)]$	−13.3		45	β_{EFISH}[a]	−9.3	1.91	116, 126
$[Cr(CO)_5(NC_5H_4\text{-}4\text{-}CHO)]$	−15.5		45	β_{EFISH}[a]	−12	1.91	126
$[Cr(CO)_5(NC_5H_4\text{-}4\text{-}N(Me)_2)]$	0.55		45				
$[Cr(CO)_5(NC_5H_4\text{-}4\text{-}Bu^n)]$	−3.98		45	β_{EFISH}[a]	−3.4	1.91	116
$[RuCl(PPh_3)_2(\eta^5\text{-}C_5H_5)]$	1		49	β_{HRS}	<7	1.06	49
				$\beta_{o,HRS}$	<4		
$[RuCl(PMe_3)_2(\eta^5\text{-}C_5H_5)]$	0		49				
$[Ru(C{\equiv}CPh)(PPh_3)_2(\eta^5\text{-}C_5H_5)]$	2		46	β_{HRS}	16	1.06	49
				$\beta_{o,HRS}$	10		
$[Ru(C{\equiv}CPh)(PMe_3)_2(\eta^5\text{-}C_5H_5)]$	5		46				
$[Ru(C{\equiv}CC_6H_4\text{-}4\text{-}NO_2)(PPh_3)_2(\eta^5\text{-}C_5H_5)]$	29		46	β_{HRS}	468	1.06	47
				$\beta_{o,HRS}$	96		

$[Ru(C{\equiv}CC_6H_4\text{-}4\text{-}NO_2)(PMe_3)_2(\eta^5\text{-}C_5H_5)]$	31	46	β_{HRS}	248	1.06	47
			$\beta_{o,HRS}$	39		
$[Ru(C{\equiv}CC_6H_4\text{-}4\text{-}C_6H_4\text{-}4\text{-}NO_2)(PPh_3)_2(\eta^5\text{-}C_5H_5)]$	36	49	β_{HRS}	560	1.06	49
			$\beta_{o,HRS}$	134		
$[Ru(C{\equiv}CC_6H_4\text{-}4\text{-}(E)\text{-}CH{=}CHC_6H_4\text{-}4\text{-}NO_2)(PPh_3)_2(\eta^5\text{-}C_5H_5)]$	45	47	β_{HRS}	1455	1.06	47
			$\beta_{o,HRS}$	232		
$[Ru(C{\equiv}CC_6H_4\text{-}4\text{-}C{\equiv}CC_6H_4\text{-}4\text{-}NO_2)(PPh_3)_2(\eta^5\text{-}C_5H_5)]$	36	49	β_{HRS}	865	1.06	49
			$\beta_{o,HRS}$	212		
$[Ru(C{\equiv}CC_6H_4\text{-}4\text{-}N{=}CHC_6H_4\text{-}4\text{-}NO_2)(PPh_3)_2(\eta^5\text{-}C_5H_5)]$	55	47	β_{HRS}	840	1.06	47
			$\beta_{o,HRS}$	86		
$[Ru(C{\equiv}CC_6H_4\text{-}4\text{-}CH{=}NC_6H_4\text{-}4\text{-}NO_2)(PPh_3)_2(\eta^5\text{-}C_5H_5)]$	52	47				
$[Ru(C{\equiv}CC_6H_4\text{-}4\text{-}N{=}NC_6H_4\text{-}4\text{-}NO_2)(PPh_3)_2(\eta^5\text{-}C_5H_5)]$	89	47				
trans-$[RuCl_2(dppm)_2]$	0	37				
cis-$[RuCl_2(dppm)_2]$	−8	37				
trans-$[RuCl(C{\equiv}CPh)(dppm)_2]$	−13	37				
trans-$[RuCl(C{\equiv}CC_6H_4\text{-}4\text{-}NO_2)(dppm)_2]$	34	37				
trans-$[RuCl(C{\equiv}CC_6H_4\text{-}4\text{-}C_6H_4\text{-}4\text{-}NO_2)(dppm)_2]$	45	37				
trans-$[RuCl(C{\equiv}CC_6H_4\text{-}4\text{-}(E)\text{-}CH{=}CHC_6H_4\text{-}4\text{-}NO_2)(dppm)_2]$	60	37				
trans-$[Ru(C{\equiv}CC_6H_4\text{-}4\text{-}NO_2)_2(dppm)_2]$	0	37				
trans-$[Ru(C{\equiv}CPh)(C{\equiv}CC_6H_4\text{-}4\text{-}NO_2)(dppm)_2]$	32	37				
1,2-bis(phenyl)-1,2-dicarbadodecaborane	0.62	147				
1,2-bis(4-nitrophenyl)-1,2-dicarbadodecaborane	11.1	147				
1,2-bis(3-nitrophenyl)-1,2-dicarbadodecaborane	5.0	147				
1,2-bis(4-aminophenyl)-1,2-dicarbadodecaborane	4.1	147				

[a] Experimental values were measured using the analogous tungsten complex.

ferrocenyl analogues, and complexes with strong acceptor groups have larger nonlinearities than those with weaker acceptor groups [i.e., β(nitro) > β(formyl) > β(cyano)]. The increased nonlinearities calculated upon increasing the length of the delocalized π-electron system in these organometallic complexes is also in accordance with traditional organic NLO material design. The *Z* configuration leads to lower nonlinearities than the *E* geometry for isomers of complexes with stilbene-containing groups, a result observed with organic compounds. The efficiency of the coupling of iron to the substituted cyclopentadienyl ligand has also been probed. MO analysis suggests that the iron does couple efficiently and that optical nonlinearities of these ferrocenyl complexes can be readily understood by traditional organic structure/NLO property relationships, with the ferrocenyl moiety having a similar donor strength to the methoxyphenyl group.

In chromium and molydenum arene complexes of the type [$M(CO)_3(\eta^6$-arene)], the $M(CO)_3$ group acts as an electron acceptor. Good agreement is seen with experimental data where available, and results for some hypothetical molecules have been reported in an effort to facilitate efficient NLO chromophore design. In comparison with the ferrocenyl and ruthenium acetylide complexes presented in this section, the arene complexes that have also been assessed experimentally have reasonably low hyperpolarizabilities despite having intense MLCT bands in the near UV.[62] The reason for the low nonlinearities has been investigated; the orientation of the charge-transfer axis (away from the ground state dipole direction) was discarded as a source of the small nonlinearities, as β_{EFISH} contributes a significant amount to β_{tot} (~75%). The small β values have therefore been attributed to the modest contributions from the MLCT transitions, examined in detail for the complex [$Cr(CO)_3(\eta^6$-$C_6H_6)$]. A comparatively small change in dipole moment for this MLCT transition limits β. The ground state dipole moment is virtually unchanged upon derivatization. When electron-withdrawing substituents are attached to the ring, the arene π^* orbital is lowered in energy, the MLCT band moves to lower energy, and a larger β_{tot} is calculated. Attaching an electron-donating substituent reverses the effect. The sign of β_{EFISH} is negative for each of these complexes.

The nonlinearities of these complexes are too small for any practical applications. In order to find ways of enhancing β_{EFISH} in the chromium tricarbonyl system, a series of hypothetical complexes with increased π conjugation have been designed and their nonlinearities calculated. For these model complexes, the β_{EFISH} values are substantially larger than those of the real complexes, β_{EFISH} values are significantly less than β_{tot}, and β_{EFISH} values are positive. It has been suggested that the much smaller β_{EFISH} with respect to β_{tot} is due to the ground-state dipole moments not being parallel to the charge transfer direction.[62] The uniformly positive β_{EFISH} provides evidence that the charge transfer direction is now in the plane of the chromophore and not along the arene–chromium axis as before. Removing the ligated metal from the hypothetical complexes generates organic compounds for which the calculated nonlinearities are significantly larger than those of the metal-bound analogues. The $(OC)_3Cr(\eta^6\text{-}C_6H_5X)$ unit is the donor group in a classic donor–bridge–acceptor composition. From this viewpoint, electron-withdrawing carbonyl ligands are not favorable. This can be seen clearly by replacing them with electron-donating ammine ligands: The highly asymmetric $[(H_3N)_3Cr(\eta^6\text{-}C_6H_5\text{-}(E)\text{-}CH{=}CHC_6H_4\text{-}4\text{-}NO_2)]$ has (by far) the largest computationally derived quadratic optical nonlinearity thus far.

The $M(CO)_5$ derivatives[45,146] illustrate several points. The computed NLO properties of the pyridine complexes are in good agreement with experiment. The responses of the pyridine-coordinated complexes are larger than those of the free ligand, with a low-energy MLCT transition suggested as the major contributor in these complexes.[45,62] The stilbazole complexes have larger discrepancies between the computed and experimental responses, a result assigned to inaccuracies in input geometry. The pyridine ring in these stilbazole complexes acts as an acceptor; as a consequence, calculated hyperpolarizabilities of the stilbazole complexes do not change appreciably upon substitution at the stilbazole 4-position, as the charge transfer transition is essentially invariant upon substitution. A further consequence of the pyridine ring behaving as the primary charge acceptor is that no dramatic increase in β upon conjugation lengthening is observed.

Ruthenium acetylide complexes are the fourth class of organometallics to be investigated computationally for their second-order NLO properties. For these ruthenium complexes with very large experimental nonlinearities and significant resonance enhancement, ZINDO does not reproduce experimental data in absolute terms. The experimentally observed nonlinearities are substantially larger than those of the ferrocenyl, (arene)tricarbonylmetal, or pentacarbonyl(pyridine)metal complexes, and the agreement with computationally derived values much poorer. With that caveat in mind, it is possible to examine results in more detail. The calculated optical nonlinearity of $[Ru(C{\equiv}CC_6H_4\text{-}4\text{-}(E)\text{-}CH{=}CHC_6H_4\text{-}4\text{-}NO_2)(PPh_3)_2(\eta^5\text{-}C_5H_5)]$ (45×10^{-30} esu) is far greater than the sum of those of its precursors $RuCl(PPh_3)_2(\eta\text{-}C_5H_5)$ (1×10^{-30} esu) and $(E)\text{-}4,4'\text{-}HC{\equiv}CC_6H_4CH{=}CHC_6H_4NO_2$ (11×10^{-30} esu).[47] More importantly, it is about 50% larger than $Ru(C{\equiv}CC_6H_4NO_2\text{-}4)(PPh_3)_2(\eta\text{-}C_5H_5)$; as with organic molecules, chain lengthening leads to an increase in calculated nonlinear response. A small *increase* is observed in calculated response on replacing the ene linkage with an imino group. Despite the small increase on replacing one CH by N, there is a dramatic increase on the second substitution, with the computed response for the azo-linked complex almost double that of $[Ru(C{\equiv}CC_6H_4\text{-}4\text{-}(E)\text{-}CH{=}CHC_6H_4\text{-}4\text{-}NO_2)(PPh_3)_2(\eta^5\text{-}C_5H_5)]$. By contrast, ene-linkage replacement by azo-linkage in the organic system leads to a dramatic *decrease* in molecular first hyperpolarizability.[148] Comparison of the ZINDO-derived β_{EFISH} with the two-level corrected β_{EFISH} reveals that the former underestimates the experimental nonlinearities, an underestimation that would probably be exacerbated if damping were factored into the experimental results.[47] Computationally determined values that approximate gas-phase nonlinearities are expected to be smaller than solution-phase measurements, as dipolar solvents such as $CHCl_3$ help to stabilize the MLCT excited state for the latter. The calculated nonlinearities are consistent with an increase in quadratic optical nonlinearity upon chain lengthening of the organometallic chromophore. Computationally derived molecular quadratic hyperpolarizabilities increase on increasing the donor strength of the co-ligand (progressing from PPh_3 to PMe_3), the opposite trend to that observed experimentally. Computational data also increase with the acceptor strength of the aryl substituent (consistent with experimental data), and decreasing the M—C(acetylide) bond length, although little dependence on orientation of the acetylide aryl group was found (Fig. 14); the last two are difficult to test experimentally.

Results from the bis{bis(diphenylphosphino)methane}ruthenium acetylide complexes are consistent with those of the (cyclopentadienyl)bis (phosphine)ruthenium acetylide data; the presence of a strong acceptor (the nitro group) leads to a marked increase in computed nonlinearity, and chromophore chain-lengthening results in enhanced calculated NLO response. Rotating at the phenyl–phenyl dihedral angle in *trans*-$[RuCl(C{\equiv}CC_6H_4\text{-}4\text{-}C_6H_4\text{-}4\text{-}NO_2)(dppm)_2]$ decreases the nonlinearity with increasing angle. The smaller nonlinearity for the biphenylene-containing complex *trans*-$[RuCl(C{\equiv}CC_6H_4\text{-}4\text{-}C_6H_4\text{-}4\text{-}NO_2)(dppm)_2]$ compared to that for the stilbenylacetylide complex *trans*-$[RuCl(C{\equiv}CC_6H_4\text{-}4\text{-}(E)\text{-}CH{=}CHC_6H_4\text{-}4\text{-}NO_2)(dppm)_2]$, coupled to the difficulty in enforcing coplanarity in the former and the sharp decrease in response when rings are not oriented most favorably, suggests that ene-linkage between phenylene rings is the most effective for constructing efficient metal acetylide chromophores for nonlinear optical responses.

Carborane complexes have been investigated briefly, showing modest calculated responses. An acceptor in the *para*-position gives the highest response, significantly better than *para*-donor, showing that the carborane cage/acceptor combination is a better charge transfer pair than cage/donor.[147]

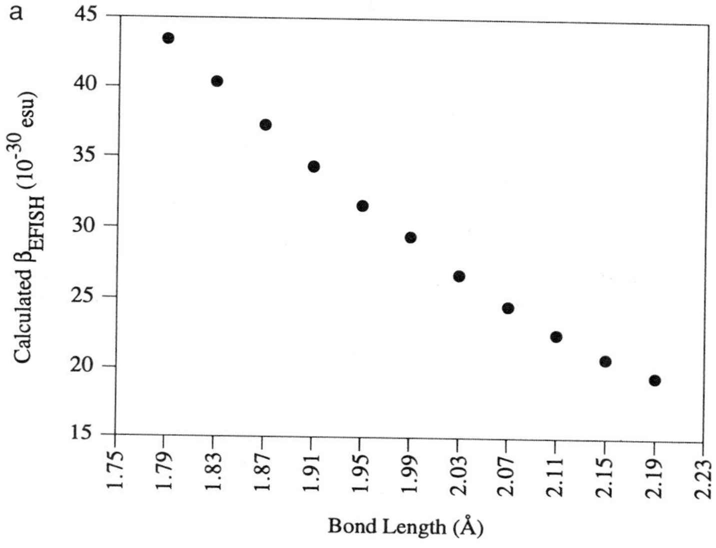

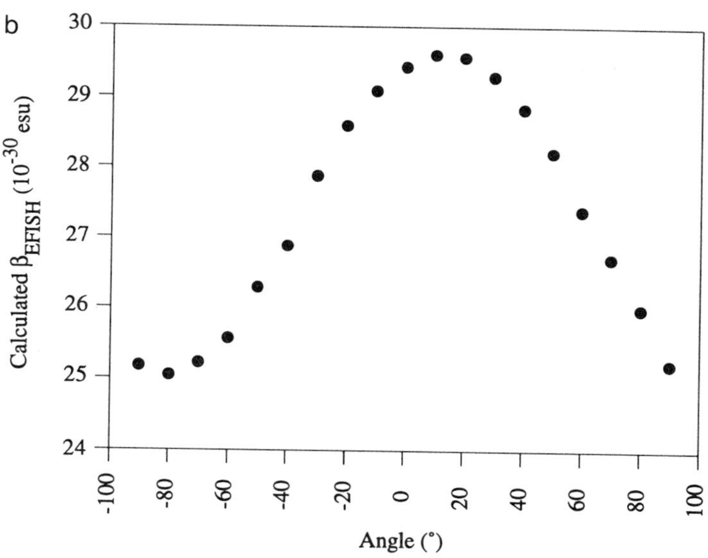

FIG. 14. ZINDO-derived β_{EFISH} for [Ru(C≡CC$_6$H$_4$-4-NO$_2$)(PMe$_3$)$_2$(η^5-C$_5$H$_5$)] showing (a) the effect of Ru–C bond length variation and (b) the effect of acetylide phenyl ring rotation. (Adapted with permission from I. R. Whittall, M. G. Humphrey, D. C. R. Hockless, B. W. Skelton, and A. H. White, *Organometallics* **1995**, *14*, 3978. Copyright © 1995 American Chemical Society.)

IV

CONCLUDING REMARKS

The studies we summarized have begun to establish structure–NLO property relationships for organometallic systems, but a great deal of work investigating variation in metal, charge-transfer ligands, oxidation state, co-ligands, and coordination geometry remains to be done. The molecular nonlinearities obtained for some of the complexes are extremely large, suggesting that the potential for application of organometallics in photonic devices remains. Before this can become a reality, though, a range of materials-related aspects that have been little considered, including optical transparency, scattering loss, processability, one- and two-photon optical stabilities, thermal stability, and (for poled systems) orientational stability, must be addressed.

V

APPENDIX: ABBREVIATIONS

bipy	2,2′-bipyridyl
Bu^n	normal butyl
Bu^t	tertiary butyl
c.c.	complex conjugate
CI	configuration interaction
CPHF	coupled perturbed Hartree-Fock
dc	direct current
DFM	density functional methods
diop	(1,2-dimethyl-1,3-dioxolane-4,5-diyl)bis(methylene) bis(diphenylphosphine)
dmpz	3,5-dimethylpyrazolyl
dppm	bis(diphenylphosphino)methane
EFISH	electric-field-induced second harmonic generation
Et	ethyl
Fc	ferrocenyl
FF	finite field
HOMO	highest occupied molecular orbital
HRS	hyper-Rayleigh scattering
INDO	intermediate neglect of differential overlap
KDP	potassium dihydrogen phosphate
KTP	potassium titanyl phosphate
LMCT	ligand-to-metal charge transfer

LUMO	lowest unoccupied molecular orbital
Me	methyl
MECI	mono-excited configuration interaction
mes	mesityl
MLCT	metal-to-ligand charge transfer
NLO	nonlinear optical
nmdpp	neomenthyldiphenylphosphine
Ph	phenyl
phen	1,10-phenanthroline
RHF	restricted Hartree–Fock
SCF	self-consistent field
SHG	second harmonic generation
SOS	sum over states
STO	Slater-type orbitals
THF	tetrahydrofuran
THG	third harmonic generation
tpph	tetraphenylporphyrinyl
UHF	unrestricted Hartree–Fock
ZINDO	Zerner's intermediate neglect of differential overlap

Acknowledgment

We thank the Australian Research Council for support. I. R. W. was the recipient of an Australian Postgraduate Research Award (Industry), A. M. M. is the recipient of an Australian Postgraduate Award, and M. G. H. holds an ARC Australian Research Fellowship.

References

(1) Nie, W. *Adv. Mater.* **1993**, *5*, 520.
(2) *Nonlinear Optical Properties of Organic and Polymeric Materials;* Williams, D. J., Ed.; American Chemical Society: Washington, D.C., 1983.
(3) *Nonlinear Optical Properties of Organic Molecules and Crystals I;* Chemla, D. S.; Zyss, J., Eds.; Academic Press: Orlando, FL, 1987.
(4) *Nonlinear Optical Properties of Organic Molecules and Crystals II;* Chemla, D. S.; Zyss, J., Eds.; Academic Press: Orlando, FL, 1987.
(5) *Nonlinear Optical Properties of Polymers;* Heeger, A. J.; Orenstein, J.; Ulrich, D. R., Eds.; Materials Research Society: Pittsburgh, PA, 1988.
(6) *Organic Materials for Non-linear Optics;* Hann, R. A.; Bloor, D., Eds.; Royal Society of Chemistry: London, 1989.
(7) *Nonlinear Optical Effects in Organic Polymers;* Messier, J.; Kajzar, F.; Prasad, P.; Ulrich, D., Eds.; Kluwer Academic Publishers: Dordrecht, The Netherlands, 1989.
(8) *Nonlinear Optics of Organics and Semiconductors;* Kobayashi, T., Ed.; Springer-Verlag: Berlin, 1989.
(9) *Materials for Non-linear and Electro-optics 1989;* Lyons, M. H., Ed.; Institute of Physics: Bristol, 1989.

(10) *Materials for Nonlinear Optics, Chemical Perspectives;* Marder, S. R.; Sohn, J. E.; Stucky, G. D., Eds.; American Chemical Society: Washington, D.C., 1991.
(11) *Organic Materials for Non-linear Optics II;* Hann, R. A.; Bloor, D., Eds.; Royal Society of Chemistry: London, 1991.
(12) *Organic Molecules for Nonlinear Optics and Photonics;* Messier, J.; Kajzar, F.; Prasad, P., Eds.; Kluwer: Dordrecht, The Netherlands, 1991.
(13) *Organic Materials for Non-linear Optics III;* Ashwell, G. J.; Bloor, D., Eds.; Royal Society of Chemistry: Cambridge, U.K., 1993.
(14) Williams, D. J. *Angew. Chem., Int. Ed. Engl.* **1984**, *23*, 690.
(15) Pugh, D.; Sherwood, J. N. *Chem. Brit.* **1988**, 544.
(16) Allen, S. *New Scientist* **1989**, *1 July*, 31.
(17) Garito, A.; Fang Shi, R.; Wu, M. *Physics Today* **1994**, *May*, 51.
(18) Marder, S. R.; Perry, J. W.; Bourhill, G.; Gorman, C. B.; Tiemann, B. G.; Mansour, K. *Science* **1993**, *261*, 186.
(19) Nalwa, H. S. *Appl. Organomet. Chem.* **1991**, *5*, 349.
(20) Marder, S. R. In *Inorganic Materials;* Bruce, D. W.; O'Hare, D., Eds.; Wiley: Chichester, U.K., 1992; p. 116.
(21) Long, N. J. *Angew. Chem., Int. Ed. Engl.* **1995**, *34*, 21.
(22) Shen, Y. R. *The Principles of Nonlinear Optics;* Wiley: New York, 1984.
(23) Butcher, P. N.; Cotter, D. *The Elements of Nonlinear Optics;* Cambridge University Press: New York, 1990.
(24) Boyd, R. W. *Nonlinear Optics;* Academic Press: New York, 1992.
(25) Sutherland, R. L. *Handbook of Nonlinear Optics;* Marcel Dekker: New York, 1996.
(26) Zyss, J.; Chemla, D. S. In *Nonlinear Optical Properties of Organic Molecules and Crystals I;* Chemla, D. S.; Zyss, J., Eds.; Academic Press: New York, 1987; p. 23.
(27) Kurtz, S. K.; Perry, T. T. *J. Appl. Phys.* **1968**, *39*, 3798.
(28) Bosshard, C.; Knöpfle, G.; Prêtre, P.; Günter, P. *J. Appl. Phys.* **1992**, *71*, 1594.
(29) Paley, M. S.; Harris, J. M.; Looser, H.; Baumert, J. C.; Bjorklund, G. C.; Jundt, D.; Twieg, R. J. *J. Org. Chem.* **1989**, *54*, 3774.
(30) Lequan, M.; Lequan, R. M.; Ching, K. C. *J. Mater. Chem.* **1991**, *1*, 997.
(31) Maker, P. D.; Terhune, R. W.; Nisenoff, M.; Savage, C. M. *Phys. Rev. Lett.* **1962**, *8*, 21.
(32) Chemla, D. S.; Oudar, J. L.; Jerphagnon, J. *Phys. Rev. B* **1975**, *12*, 4534.
(33) Clays, K.; Persoons, A. *Phys. Rev. Lett.* **1991**, *66*, 2980.
(34) Clays, K.; Persoons, A. *Rev. Sci. Instrum.* **1992**, *63*, 3285.
(35) Hendrickx, E.; Dehu, C.; Clays, K.; Brédas, J. L.; Persoons, A. In *Polymers for Second-Order Nonlinear Optics;* Lindsay, G. A.; Singer, K. D., Eds.; American Chemical Society: Washington, D.C., 1995; p. 82.
(36) Lentzner, H.; Moulik, A. *J. Chem. Inf. Comput. Sci.* **1994**, *34*, 162.
(37) McDonagh, A. M.; Whittall, I. R.; Humphrey, M. G.; Skelton, B. W.; White, A. H. *J. Organomet. Chem.* **1996**, *519*, 229.
(38) Kanis, D. R.; Ratner, M. A.; Marks, T. J. *Chem. Rev.* **1994**, *94*, 195.
(39) Di Bella, S.; Ratner, M. A.; Marks, T. J. *J. Am. Chem. Soc.* **1992**, *114*, 5842.
(40) Di Bella, S.; Marks, T. J.; Ratner, M. A. *J. Am. Chem. Soc.* **1994**, *116*, 4440.
(41) Hehre, W. J.; Radom, L.; Schleyer, P. v. R.; Pople, J. A. *Ab Initio Molecular Orbital Theory;* Wiley: New York, 1986.
(42) Waite, J.; Papadopoulos, M. G. *Z. Naturforsch., A* **1987**, *42*, 749.
(43) Guan, J.; Duffy, P.; Carter, J. T.; Chong, D. P.; Casida, K. C.; Casida, M.; Wrinn, M. *J. Chem. Phys.* **1993**, *98*, 4753.
(44) Levine, I. N. *Quantum Chemistry;* Prentice-Hall: Englewood Cliffs, NJ, 1991.

(45) Kanis, D. R.; Lacroix, P. G.; Ratner, M. A.; Marks, T. J. *J. Am. Chem. Soc.* **1994**, *116*, 10089.
(46) Whittall, I. R.; Humphrey, M. G.; Hockless, D. C. R.; Skelton, B. W.; White, A. H. *Organometallics* **1995**, *14*, 3970.
(47) Whittall, I. R.; Humphrey, M. G.; Persoons, A.; Houbrechts, S. *Organometallics* **1996**, *15*, 1935.
(48) McDonagh, A. M.; Whittall, I. R.; Humphrey, M. G.; Hockless, D. C. R.; Skelton, B. W.; White, A. H. *J. Organomet. Chem.* **1996**, *523*, 33.
(49) Whittall, I. R.; Cifuentes, M. P.; Humphrey, M. G.; Luther-Davies, B.; Samoc, M.; Houbrechts, S.; Persoons, A.; Heath, G. A.; Hockless, D. C. R. *J. Organomet. Chem.*, in press.
(50) Waite, J.; Papadopoulos, M. G. *J. Phys. Chem.* **1991**, *93*, 5426.
(51) Streitweiser, A. *Molecular Orbital Theory for Organic Chemists;* Wiley: New York, 1962.
(52) Hoffmann, R. *J. Chem. Phys.* **1963**, *39*, 1397.
(53) Colwell, S. M.; Murray, C. W.; Handy, N. C.; Amos, R. D. *Chem. Phys. Lett.* **1993**, *210*, 261.
(54) Labanowski, J. K.; Andzelm, J. *Density Functional Methods in Chemistry;* Springer-Verlag: New York, 1991.
(55) Zeigler, T. *Chem. Rev.* **1991**, *91*, 651.
(56) Ward, J. F. *Rev. Mod. Phys.* **1965**, *37*, 1.
(57) Anderson, W. P.; Edwards, W. D.; Zerner, M. C. *Inorg. Chem.* **1986**, *25*, 2728.
(58) Zerner, M. C.; Loew, G. H.; Kirchner, R. F.; Mueller-Westerhoff, U. T. *J. Am. Chem. Soc.* **1980**, *102*, 598.
(59) Bacon, A. D.; Zerner, M. C. *Theoret. Chim. Acta* (*Berl.*) **1979**, *53*, 21.
(60) Oudar, J. L.; Chemla, D. S. *J. Chem. Phys.* **1977**, *66*, 2664.
(61) Oudar, J. L. *J. Chem. Phys.* **1977**, *67*, 446.
(62) Kanis, D. R.; Ratner, M. A.; Marks, T. J. *J. Am. Chem. Soc.* **1992**, *114*, 10338.
(63) Marder, S. R.; Perry, J. W.; Tiemann, B. G.; Schaefer, W. P. *Organometallics* **1991**, *10*, 1896.
(64) Marder, S. R.; Tiemann, B. G.; Perry, J. W.; Cheng, L. T.; Tam, W.; Schaefer, W. P.; Marsh, R. E. In *Materials for Nonlinear Optics: Chemical Perspectives;* Marder, S. R.; Sohn, J. E.; Stucky, G. D., Eds.; American Chemical Society: Washington, D.C., 1991; p. 187.
(65) Marder, S. R.; Perry, J. W.; Schaefer, W. P.; Tiemann, B. G.; Groves, P. C.; Perry, K. J. *Proc. SPIE-Int. Soc. Opt. Eng.* **1989**, *1147*, 108.
(66) Marder, S. R.; Perry, J. W.; Schaefer, W. P.; Ginsburg, E. J.; Gorman, C. B.; Grubbs, R. H. *Mat. Res. Soc. Symp. Proc.* **1990**, *175*, 101.
(67) Green, M. L. H.; Marder, S. R.; Thompson, M. E.; Bandy, J. A.; Bloor, D.; Kolinsky, P. V.; Jones, R. J. *Nature* **1987**, *330*, 360.
(68) Bunting, H. E.; Green, M. L. H.; Marder, S. R.; Thompson, M. E.; Bloor, D.; Kolinsky, P. V.; Jones, R. J. *Polyhedron* **1992**, *11*, 1489.
(69) Bandy, J. A.; Bunting, H. E.; Green, M. L. H.; Marder, S. R.; Thompson, M. E.; Bloor, D.; Kolinsky, P. V.; Jones, R. J. In *Organic Materials for Nonlinear Optics;* Hann, R. A.; Bloor, D., Eds.; Royal Society of Chemistry: London, 1989; p. 219.
(70) Green, M. L. H.; Qin, J.; O'Hare, D.; Bunting, H. E.; Thompson, M. E.; Marder, S. R.; Chatakondu, K. *Pure Appl. Chem.* **1989**, *61*, 817.
(71) Tiemann, B. G.; Marder, S. R.; Perry, J. W.; Cheng, L.-T. *Chem. Mater.* **1990**, *2*, 690.
(72) Bandy, J. A.; Bunting, H. E.; Garcia, M.-H.; Green, M. L. H.; Marder, S. R.; Thompson, M. E.; Bloor, D.; Kolinsky, P. V.; Jones, R. J.; Perry, J. W. *Polyhedron* **1992**, *11*, 1429.

(73) Yamazaki, Y.; Hosono, K.; Matsuda, H.; Minami, N.; Asai, M.; Nakanishi, H. *Biotechnol. Bioeng.* **1991**, *38*, 1218.
(74) Coe, B. J.; Kurek, S. S.; Rowley, N. M.; Foulon, J. D.; Hamor, T. A.; Harman, M. E.; Hursthouse, M. B.; Jones, C. J.; McCleverty, J. A.; Bloor, D. *Chemtronics* **1991**, *5*, 23.
(75) Coe, B. J.; Jones, C. J.; McCleverty, J. A.; Bloor, D.; Kolinsky, P. V.; Jones, R. J. *J. Chem. Soc., Chem. Commun.* **1989**, 1485.
(76) Coe, B. J.; Foulon, J. D.; Hamor, T. A.; Jones, C. J.; McCleverty, J. A.; Bloor, D.; Cross, G. H.; Axon, T. L. *J. Chem. Soc., Dalton Trans.* **1994**, 3427.
(77) Coe, B. J.; Hamor, T. A.; Jones, C. J.; McCleverty, J. A.; Bloor, D.; Cross, G. H.; Axon, T. L. *J. Chem. Soc., Dalton Trans.* **1995**, 673.
(78) Anderson, A. G.; Calabrese, J. C.; Tam, W.; Williams, I. D. *Chem. Phys. Lett.* **1987**, *134*, 392.
(79) Eaton, D. F.; Anderson, A. G.; Tam, W.; Wang, Y. In *Polymers for High Technology: Electronics and Photonics;* Bowden, M. J.; Turner, S. R., Eds.; American Chemical Society: Washington, D.C., 1987; p. 381.
(80) Eaton, D. F.; Anderson, A. G.; Tam, W.; Wang, Y. *J. Am. Chem. Soc.* **1987**, *109*, 1886.
(81) Tam, W.; Eaton, D. F.; Calabrese, J. C.; Williams, I. D.; Wang, Y.; Anderson, A. G. *Chem. Mater.* **1989**, *1*, 128.
(82) Tam, W.; Calabrese, J. C. *Chem. Phys. Lett.* **1988**, *144*, 79.
(83) Tam, W.; Wang, Y.; Calabrese, J. C.; Clement, R. A. *Proc. SPIE-Int. Soc. Opt. Eng.* **1988**, *971*, 107.
(84) Dias, A. R.; Garcia, M. H.; Robalo, M. P.; Green, M. L. H.; Lai, K. K.; Pulham, A. J.; Klueber, S. M.; Balavoine, G. *J. Organomet. Chem.* **1993**, *453*, 241.
(85) Dias, A. R.; Garcia, M. H.; Rodrigues, J. C.; Green, M. L. H.; Kuebler, S. M. *J. Organomet. Chem.* **1994**, *475*, 241.
(86) Marder, S. R.; Perry, J. W.; Schaefer, W. P. *Science* **1989**, *245*, 626.
(87) Meredith, G. R. In *Nonlinear Optical Properties of Organic and Polymeric Materials;* Williams, D. J., Ed.; American Chemical Society: Washington, D.C., 1983; p. 27.
(88) Frazier, C. C.; Harvey, M. A.; Cockerham, M. P.; Hand, H. M.; Chauchard, E. A.; Lee, C. H. *J. Phys. Chem.* **1986**, *90*, 5703.
(89) Kobayashi, T. *IEICE Trans. Electron.* **1992**, *E75*, 36.
(90) Bandy, J. A.; Bunting, H. E.; Garcia, M.-H.; Green, M. L. H.; Marder, S. R.; Thompson, M. E.; Bloor, D.; Kolinsky, P. V.; Jones, R. J. In *Organic Materials for Nonlinear Optics;* Hann, R. A.; Bloor, D., Eds.; Royal Society of Chemistry: London, 1989; p. 225.
(91) Whittall, I. R.; Cifuentes, M. P.; Costigan, M. J.; Humphrey, M. G.; Goh, S. C.; Skelton, B. W.; White, A. H. *J. Organomet. Chem.* **1994**, *471*, 193.
(92) Houlton, A.; Miller, J. R.; Silver, J.; Jassim, N.; Ahmet, M. J.; Axon, T. L.; Bloor, D.; Cross, G. H. *Inorg. Chim. Acta* **1993**, *205*, 67.
(93) Calabrese, J. C.; Tam, W. *Chem. Phys. Lett.* **1987**, *133*, 244.
(94) Eaton, D. F.; Anderson, A. G.; Tam, W.; Wang, Y. *J. Am. Chem. Soc.* **1987**, *109*, 1886.
(95) Benito, A.; Cano, J.; Martinez-Manez, R.; Paya, J.; Soto, J.; Julve, M.; Lloret, F.; Marcos, M. D.; Sinn, E. *J. Chem. Soc., Dalton Trans.* **1993**, 1999.
(96) Coe, B. J.; Jones, C. J.; McCleverty, J. A.; Bloor, D.; Cross, G. *J. Organomet. Chem.* **1994**, *464*, 225.
(97) Kimura, M.; Abdel-Halim, H.; Robinson, D. W.; Cowan, D. O. *J. Organomet. Chem.* **1991**, *403*, 365.
(98) Pollagi, T. P.; Stoner, T. C.; Dallinger, R. F.; Gilbert, T. M.; Hopkins, M. D. *J. Am. Chem. Soc.* **1991**, *113*, 703.
(99) Marder, T. B.; Lesley, G.; Yuan, Z.; Fyfe, H. B.; Chow, P.; Stringer, G.; Jobe, I. R.; Taylor, N. J.; Williams, I. D.; Kurtz, S. K. In *Materials for Nonlinear Optics: Chemical*

Perspectives; Marder, S. R.; Sohn, J. E.; Stucky, G. D., Eds.; American Chemical Society: Washington, D.C., 1991; p. 605.

(100) Fyfe, H. B.; Mlekuz, M.; Stringer, G.; Taylor, N. J.; Marder, T. B. In *Inorganic and Organometallic Polymers with Special Properties;* Laine, R. M., Ed.; Kluwer: Dordrecht, The Netherlands, 1992; p. 331.

(101) Coe, B. J.; Jones, C. J.; McCleverty, J. A. *Polyhedron* **1994**, *13*, 2107.

(102) O'Reilly, S. A.; White, P. S.; Templeton, J. L. *Chem. Mater.* **1996**, *8*, 93.

(103) Copley, R. C. B.; Lambeth, C.; Machell, J.; Mingos, D. M.; Murphy, D. M.; Powell, H. *J. Mater. Chem.* **1991**, *1*, 583.

(104) Wu, X.; Qin, J.; Liu, D.; Wu, B.; Chen, C. *Nonlinear Optics* **1994**, *7*, 153.

(105) Lesley, G.; Yuan, Z.; Stringer, G.; Jobe, I. R.; Taylor, N. J.; Koch, L.; Scott, K.; Marder, T. B. In *Organic Materials for Non-linear Optics II;* Hann, R. A.; Bloor, D., Eds.; Royal Society of Chemistry: London, 1991; p. 197.

(106) Wang, K. Z.; Huang, C. H.; Zhou, D. J.; Xu, G. X.; Xu, Y.; Liu, Y. Q.; Zhu, D. B.; Zhao, X. S.; Xie, X. M. *Solid State Commun.* **1995**, *93*, 189.

(107) Gao, L. H.; Wang, K. Z.; Huang, C. H.; Zhao, X. S.; Xia, X. H.; Xu, J. M.; Li, T. K. *Chem. Lett.* **1995**, *11*, 1049.

(108) Kezhi, W.; Chunhui, H.; Guangxian, X.; Xinsheng, Z.; Xiaoming, X.; Lingge, X.; Tiankai, L. *Thin Solid Films* **1994**, *247*, 1.

(109) Richardson, T.; Roberts, G. G.; Polywka, M. E. C.; Davies, S. G. *Thin Solid Films* **1988**, *160*, 231.

(110) Wright, M. E.; Toplikar, E. G.; Kubin, R. F.; Seltzer, M. D. *Macromolecules* **1992**, *25*, 1838.

(111) Wright, M. E.; Cochran, B. B.; Toplikar, E. G.; Lackritz, H. S.; Kerney, J. T. In *Inorganic and Organometallic Polymers II: Advanced Materials and Intermediates;* Wisian-Neilson, P.; Allcock, H. R.; Wynne, K. J., Eds.; American Chemical Society: Washington, D.C., 1994; p. 456.

(112) Wright, M. E.; Toplikar, E. G.; Lackritz, H. S.; Kerney, J. T. *Macromolecules* **1994**, *27*, 3016.

(113) Lacroix, P. G.; Lin, W.; Wong, G. K. *Chem. Mater.* **1995**, *7*, 1293.

(114) Corriu, R. J. P.; Douglas, W. E.; Yang, Z.; Karakus, Y.; Cross, G. H.; Bloor, D. *J. Organomet. Chem.* **1993**, *455*, 69.

(115) Calabrese, J. C.; Cheng, L.-T.; Green, J. C.; Marder, S. R.; Tam, W. *J. Am. Chem. Soc.* **1991**, *113*, 7227.

(116) Cheng, L. T.; Tam, W.; Meredith, G. R.; Marder, S. R. *Mol. Cryst. Liq. Cryst.* **1990**, *189*, 137.

(117) Marder, S. R.; Beratan, D. N.; Tiemann, B. G.; Cheng, L. T.; Tam, W. In *Organic Materials for Nonlinear Optics II;* Hann, R. A.; Bloor, D., Eds.; Royal Society of Chemistry: London, 1991; p. 165.

(118) Cheng, L.-T. In *Organic Molecules for Nonlinear Optics and Photonics;* Messier, J.; Kajzar, F.; Prasad, P., Eds.; Kluwer: Dordrecht, The Netherlands, 1991; p. 121.

(119) Behrens, U.; Brussaard, H.; Hagenau, U.; Heck, J.; Hendrickx, E.; Kornich, J.; van der Linden, J. G. M.; Persoons, A.; Spek, A. L.; Veldman, N.; Voss, B.; Wong, H. *Chem. Eur. J.* **1996**, *2*, 98.

(120) Yuan, Z.; Taylor, N. J.; Sun, Y.; Marder, T. B. *J. Organomet. Chem.* **1993**, *449*, 27.

(121) Doisneau, G.; Balavoine, G.; Fillebeen-Khan, T.; Clinet, J.; Delaire, J.; Ledoux, I.; Loucif, R.; Puccetti, G. *J. Organomet. Chem.* **1991**, *421*, 299.

(122) Loucif-Saïbi, R.; Delaire, J. A.; Bonazzola, L.; Doisneau, G.; Balavoine, G.; Fillebeen-Khan, T.; Ledoux, I.; Puccetti, G. *Chem. Phys.* **1992**, *167*, 369.

(123) Ledoux, I. *Synth. Met.* **1993**, *54*, 123.

(124) Blanchard-Desce, M.; Runser, C.; Fort, A.; Barzoukas, M.; Lehn, J.-M.; Bloy, V.; Alain, V. *Chem. Phys.* **1995**, *199*, 253.
(125) Tam, W.; Cheng, L. T.; Bierlein, J. D.; Cheng, L. K.; Wang, Y.; Feiring, A. E.; Meredith, G. R.; Eaton, D. F.; Calabrese, J. C.; Rikken, G. L. J. A. In *Materials for Nonlinear Optics: Chemical Perspectives;* Marder, S. R.; Sohn, J. E.; Stucky, G. D., Eds.; American Chemical Society: Washington, D.C., 1991; p. 158.
(126) Cheng, L.-T.; Tam, W.; Eaton, D. F. *Organometallics* **1990**, *9*, 2856.
(127) Maiorana, S.; Papagni, A.; Licandro, E.; Persoons, A.; Clay, K.; Houbrechts, S.; Porzio, W. *Gazz. Chim. Ital.* **1995**, *125*, 377.
(128) Bruce, D. W.; Thorton, A. *Mol. Cryst. Liq. Cryst.* **1993**, *231*, 253.
(129) Houbrechts, S.; Clays, K.; Persoons, A.; Cadierno, V.; Gamasa, M. P.; Gimeno, J.; Whittall, I. R.; Humphrey, M. G. *Proc. SPIE-Int. Soc. Opt. Eng.* **1996**, *2852*, 98.
(130) Houbrechts, S.; Clays, K.; Persoons, A.; Cadierno, V.; Gamasa, M. P.; Gimeno, J. *Organometallics* **1996**, *15*, 5266.
(131) Tamm, M.; Grzegorzeweski, A.; Steiner, T.; Jentzsch, T.; Werncke, W. *Organometallics* **1996**, *15*, 4984.
(132) Hagenau, U.; Heck, J.; Hendrickx, E.; Persoons, A.; Schuld, T.; Wong, H. *Inorg. Chem.* **1996**, *35*, 7863.
(133) Whittall, I. R.; Humphrey, M. G.; Houbrechts, S.; Persoons, A.; Hockless, D. C. R. *Organometallics* **1996**, *15*, 5738.
(134) Whittall, I. R.; Cifuentes, M. P.; Humphrey, M. G.; Luther-Davies, B.; Samoc, M.; Houbrechts, S.; Persoons, A.; Heath, G. A.; Begsány, D. *Organometallics* **1997,** *16*, 2681.
(135) Whitaker, C. M.; Patterson, E. V.; Kott, K. L.; McMahon, R. J. *J. Am. Chem. Soc.* **1996**, *118*, 9966.
(136) Laidlaw, W. M.; Denning, R. G.; Verbiest, T.; Chauchard, E.; Persoons, A. *Nature* **1993**, *363*, 58.
(137) Laidlaw, W. M.; Denning, R. G.; Verbiest, T.; Chauchard, E.; Persoons, A. *Proc. SPIE-Int. Soc. Opt. Eng.* **1994**, *2143*, 14.
(138) Mignani, G.; Kramer, A.; Puccetti, G.; Ledoux, I.; Soula, G.; Zyss, J.; Meyrueix, R. *Organometallics* **1990**, *9*, 2640.
(139) Mignani, G.; Barzoukas, M.; Zyss, J.; Soula, G.; Balegroune, F.; Grandjean, D.; Josse, D. *Organometallics* **1991**, *10*, 3660.
(140) Lequan, M.; Branger, C.; Simon, J.; Thami, T.; Chauchard, E.; Persoons, A. *Adv. Mater.* **1994**, *6*, 851.
(141) Zyss, J.; Ledoux, I. *Chem. Rev.* **1994**, *94*, 77.
(142) Yuan, Z.; Taylor, N. J.; Marder, T. B.; Williams, I. D.; Kurtz, S. K.; Cheng, L. T. In *Organic Materials for Nonlinear Optics II;* Hann, R. A.; Bloor, D., Eds.; Royal Society of Chemistry: London, 1991; p. 190.
(143) Yuan, Z.; Taylor, N. J.; Marder, T. B.; Williams, I. D.; Kurtz, S. K.; Cheng, L. T. *J. Chem. Soc., Chem. Commun.* **1990**, 1489.
(144) Lequan, M.; Lequan, R. M.; Ching, K. C.; Barzoukas, M.; Fort, A.; Lahoucine, H.; Bravic, G.; Chasseau, D.; Gaultier, J. *J. Mater. Chem.* **1992**, *2*, 719.
(145) Murphy, D. M.; Mingos, D. M. P.; Forward, J. M. *J. Mater. Chem.* **1993**, *3*, 67.
(146) Kanis, D. R.; Ratner, M. A.; Marks, T. J. *J. Am. Chem. Soc.* **1990**, *112*, 8203.
(147) Murphy, D. M.; Mingos, D. M. P.; Haggitt, J. L.; Powell, H. R.; Westcott, S. A.; Marder, T. B.; Kanis, D. R. *J. Mater. Chem.* **1993**, *3*, 139.
(148) Tsunekawa, T.; Yamaguchi, K. *J. Phys. Chem.* **1992**, *96*, 10268.

ADVANCES IN ORGANOMETALLIC CHEMISTRY, VOL. 42

Catalytic Dehydrocoupling: A General Strategy for the Formation of Element–Element Bonds

FRANÇOIS GAUVIN

Chemistry Department
Bishops University
Lennoxville, Quebec, Canada J1M 1Z7

JOHN F. HARROD

Chemistry Department
McGill University
Montréal, Quebec, Canada H3A 2K6

HEE GWEON WOO

Chonnam National University
Kwangju 500-757, Korea

I

INTRODUCTION

A. *The Formation of E–E Bonds*

The large-scale synthesis of *p*-group E–E bonds relies almost completely on reactions of covalent halides. The increasing severity of restrictions on the disposal of salt, or acid wastes, particularly when contaminated with heavy elements, places severe constraints on the potential for growth in the use of main group organometallics. The heavy organic chemical industry responded long ago to a similar situation by turning to the highly efficient transformation of hydrocarbon feedstocks with processes that do not produce objectionable side products. Both the efficiency and the specificity of these transformations were largely achieved through the application of catalysis.

0065-3055/98 $25.00

The possibility of developing a broader process chemistry based on hydrides of the heavier main group elements has been one of our major preoccupations since the mid-1980s.[1] The challenge is daunting and the prospect of profit distant. However, there are some industrial landmarks that give cause for optimism. As an example we cite the industrial synthesis of SiH_4 by reduction of $SiCl_4$ with H_2.[2] Since the direct reduction process totally recycles halogen, it furnishes the basis (at least on paper) for a clean, hydride-based process industry for silicon chemistry. Furthermore, the fluctuating demand for semiconductor-grade silicon, which periodically leads to excess SiH_4 capacity, has stimulated interest in finding new value-added uses for SiH_4. A prerequisite to the realization of this goal is a massive increase in our knowledge of the catalytic reactions of silanes. A similar scenario could be drawn for PH_3 (easy to make without halogen) and other *p*-group hydrides.

Materials science applications have driven a vigorous search for new organometallic compounds, both molecular[3] and polymeric,[4] as precursors for new materials. In the case of polymers, this interest is driven by two different end applications. The first is the intrinsic properties of the materials, such as their electronic, photonic, or thermal/mechanical properties. The second is their use as precursors to purely inorganic materials. Organometallic polymers are particularly attractive precursors because of the inherent ease and versatility that they exhibit with regard to processing.

In the case of the heavier *p*-group elements, the synthetic options for the formation of E–E bonds are quite limited and there is a pressing need for new general methods. Synthetic routes to organometallic polymers are also far more limited than those available for the synthesis of organic polymers. This is in large part due to the essential absence of analogues of vinyl compounds involving atoms from the third and subsequent periods. Although such unsaturated compounds are now well established, they can only be made as long-lived species by loading with sterically demanding substituents,[5] which coincidentally inhibits their polymerization. An interesting strategy to bypass this problem is the use of masked unsaturates,[6] but this method has not yet been widely explored.

Dehydrocoupling (also referred to as dehydrocondensation) provides a versatile route to the synthesis of E–E bonds. The generalized dehydrocoupling reaction is shown in Eq. (1).

$$2\,R_nEH \longrightarrow R_nE\text{–}ER_n + H_2 \qquad R = \text{organyl or H} \tag{1}$$

In the case of the heaviest main group hydrides, such as PbH_4, Eq. (1) occurs spontaneously under ambient conditions. However, most such reactions require either more forcing conditions (thermal or photochemical) or catalysis. In the ensuing article, we describe some of the recent advances that have been made with respect to the catalytic homo- and heterodehydrocoupling of main group organohydrides.

B. *Thermodynamic Aspects of Dehydrocoupling*

Although a thorough analysis of the thermodynamics of Eq. (1) is precluded by the relative paucity of data for the bond dissociation energies of $R_nE{-}H$, $R_nE{-}ER_n$, and $R_nE{-}E'R_n$ bonds, some general trends are readily discernible. Some estimated bond disruption enthalpies for a number of main group binary hydrides are shown in Table I. From this data, it is obvious that the general trend is for Eq. (1) to become more exothermic on descending a group. It is also clear from this data that thermodynamically feasible dehydrocoupling reactions are fairly numerous.

Although the data in Table I are useful to illustrate general trends,

TABLE I

SOME THERMODYNAMIC PARAMETERS FOR THE DEHYDROCOUPLING OF SOME MAIN GROUP HYDRIDES (ENERGIES IN $kJmol^{-1}$)[a]

E	$D_{E\text{-}H}$	$D_{E\text{-}E,\ or\ E\text{-}E'}$	D_R[b]
C	415	331	+64
Si	320	197	+8
Ge	289	163	−20
N	389	159	+184
P	318	214	−13
As	247	178	−119
O	465	138	+357
S	364	264	+29
Se	314	193	0
C O	415 465	343	+102
Si O	320 465	368	−18
Si S	320 364	226	+23

[a] Data from Ref. 7.

[b] The quantity D_R is the sum of the bond enthalpy changes for the reaction, and it is expected to approximate the reaction enthalpy. The values of D_R are based on a value of 435 $kJmol^{-1}$ for $D_{H\text{-}H}$.

they should be treated with some caution when attempting to assess the thermodynamic viability of a particular reaction. The E–H enthalpies in the table are average bond enthalpies for the binary hydrides and the E–E enthalpies are for the pure elements. The numbers can be quite different for molecules carrying organic substituents, but there are very few quantitative data for such cases. The case of silicon is an exception since the industrial importance of silanes has encouraged a number of high-quality thermodynamic studies.

Some data relevant to the dehydrocoupling of organosilanes are presented in Table II. From these data two important conclusions may be drawn. In the first place, the dehydrocoupling of silanes is thermodynamically favored by organic substitution at the silicon. Second, substitution at the silicon can cause a large variation in the thermodynamic viability of the reaction. The effect of substitution with silyl groups is particularly marked and is in the opposite sense from substitution with alkyl groups at carbon. Such trends are likely to be the case for other *p*-group elements, but this remains to be demonstrated. Hence, it is clear that absolute conclusions based on data obtained from binary hydrides and the pure elements can be misleading if applied to the case of an organosubstituted hydride.

In conclusion, we draw attention to the fact that the preceding discussion considers only enthalpic factors. It is not expected that the entropy contributions would be large in a dehydrocoupling reaction, but the high fugacity of hydrogen and its low solubility could contribute to driving reactions that are thermodynamically marginal. For example, although the data in the tables indicates that the dehydrocoupling of SiH_4 is not thermodynamically favored, we have in fact observed this reaction in the laboratory.[10]

TABLE II

SOME BOND ENERGIES ($kJmol^{-1}$) FOR E–H AND E–E BONDS (E = C or Si)[a]

Molecule	$D_{C\text{-}H}$	$D_{Si\text{-}H}$	$D_{C\text{-}C}$	$D_{Si\text{-}Si}$	$H_{R(Si)}$(calc.)
EH_4	437	376	—	—	+9
$MeEH_3$	408	373	~350[c]	—	−9
Me_3EH	383	375	291	335	−20
$PhEH_3$	366	367	~350[c]	~320	−21
E_n(diamond)	—	—	354	225	—
$Me_3SiSiMe_3SiH$	—	355[b]	—	~330[c]	−55
$(Me_3Si)_3SiH$	—	325[b]	—	~300[c]	−85

[a] All data from Ref. 8, except those labeled *b*.

[b] Data from Ref. 9.

[c] Estimated by interpolation.

II

HOMODEHYDROCOUPLING

A. *Synthesis of Polymers*

1. *Polysilanes*

a. *Introduction.* Polysilanes with unusual optical and electronic properties due to σ electron conjugation along the silicon backbone have received particular attention as SiC ceramic precursors,[11] luminescent materials,[12] deep-UV photoresists,[13] electroconductors,[14] and photoinitiators.[15,16] Polysilanes are most commonly synthesized by the Wurtz-type coupling reaction of organodichlorosilanes using an alkali metal dispersion in refluxing toluene. The method is intolerant of many interesting functional groups, is difficult to carry out reproducibly, often gives large amounts of low molecular weight products, and is hazardous for large-scale operations.[16–18] Other synthetic methods include anionic polymerization of masked disilenes,[6] ring-opening polymerization of cyclic oligosilanes,[19] electroreductive polymerization of organodichlorosilanes,[20] sonochemical polymerization of organodichlorosilanes,[21] electrochemical polymerization of hydrosilanes,[22] catalytic dehydropolymerization of hydrosilanes by activated transition metals,[23] and redistributive polymerization of hydrodisilanes catalyzed by potassium hydride.[24]

The dehydropolymerization of hydrosilanes catalyzed by Group 4 metallocenes has provided a useful alternative and fruitful approach to polysilanes.[25–27] A major drawback of this method up until the present is the relatively low molecular weights of the polysilanes produced. This precludes the achievement of optimal mechanical and optical properties.[28–30] Intensive efforts have been dedicated to elucidating the dehydropolymerization mechanism, to finding the most efficient catalyst, to increasing the polymer molecular weights, to preparing stereoselective polysilanes, and to dehydropolymerizing various types of hydrosilanes to functional polysilanes.

Several comprehensive reviews on the preparation, structure, properties, photochemistry, and redistribution of polysilanes prepared by the Wurtz coupling method have been published.[15,16,31–33] There are also several excellent reviews on the transition-metal-complex-catalyzed dehydrocoupling of hydrosilanes to polysilanes.[34–36] The present section reviews the more recent work on the dehydropolymerization of hydrosilanes to polysilanes, catalyzed by transition-metal complexes under homogeneous conditions. Extensions of the methodology to the dehydropolymerization of hydrogermanes, hydrostannanes, and hydrophosphanes to the corresponding polymers are also reviewed. The literature is covered up to early 1997.

b. *Dehydropolymerization Catalysts.* Corey has extensively reviewed all the types of early and later transition-metal complexes, including lanthanide and actinide metal complexes, that had been reported as dehydropolymerization catalysts for hydrosilanes up to late 1990.[36] Analysis of the catalytic activity of dehydrocoupling catalysts shows some general trends. The reactivity of hydrosilanes decreases in the order primary > secondary > tertiary. Catalysts that exhibit good dehydrocoupling activity for primary hydrosilanes are often ineffective for the dehydrocoupling of secondary and tertiary hydrosilanes. The large rate difference usually observed for the dehydrocoupling of hydrosilanes with varying degrees of substitution is generally attributed to the steric character of the hydrosilane.

Transition-metal complex catalysts that dehydrocouple silanes can be divided into two broad classes. The first group of catalysts has the general formula Cp'_xMR_y, where M is an early transition metal (Ti, Zr, V), a lanthanide, or an actinide (U, Th). R is usually methyl, but can also be a hydride (y = 0–2), and Cp′ is Cp (η^5-C_5H_5), MeCp, or Cp* (η^5-C_5Me_5) (x = 1–2). Harrod *et al.* reported that although mixtures of moderate molecular weight polymers and lower molecular weight oligomers are formed from primary silanes at ambient temperature with Cp_2MMe_2 (M = Ti, Zr) catalysts, Cp_2VMe_2 and Cp_2V produce only dimer and trimer from $PhSiH_3$ at 100–120°C.[27] $Cp'MMe_3$ (M = Ti, Zr) showed very little catalytic activity with the formation of a trace amount of oligomers.[27] Actinide complexes such as $Cp^*_2UMe_2$ and $Cp^*_2ThMe_2$ only dehydrocouple $PhSiH_3$ to low molecular weight oligomers and dimer, respectively, at ambient temperature.[27] Whereas $Cp^*_2MH_2$ (M = Zr, Hf) and $Cp^*_2Zr(SiPh_3)Me$ catalyzed the dehydrocoupling of $PhSiH_3$ to give dimer, trimer, tetramer, and higher oligomers sequentially, $Cp^*_2HfMe_2$ did not show any dehydrocoupling activity at all.[37] Other early transition-metal complexes such as Cp_2MH_3 (M = Nb, Ta), Cp_2Cr, and Cp_2MH_2 (M = Mo, W) are inactive for $PhSiH_3$ dehydrocoupling, even after prolonged heating.[27]

Interestingly, Cp_2HfMe_2 did not show any dehydrocoupling catalytic activity,[27] whereas a hafnocene catalyst generated *in situ* from Cp_2HfCl_2 and 2 equiv MeLi, in the presence of $PhSiH_3$, produced a predominantly linear polymer from $PhSiH_3$.[38] Given the fact that reaction of Cp_2HfCl_2 with MeLi is the normal reaction for synthesis of Cp_2HfMe_2, this result appears to be anomalous. However, it is important to note that there is still MeLi present when the $PhSiH_3$ is added and it is believed that the MeLi reacts with the latter to generate a dispersed and highly reactive LiH. This in turn produces Cp_2HfH_2, the true catalyst, *in situ*.

Group 4 metallocenes having benzyl, aryl, alkoxy, or silyl ligands, rather than H or Me, can also be employed as dehydrocoupling catalysts. For example, Cp_2TiPh_2 is an active catalyst for $PhSiH_3$ polymerization at moderately elevated temperature.[39] Bourg *et al.*[40] found that titanocene and zirconocene phenoxides $Cp_2M(OAr)_2$ (Ar = C_6H_5, p-$MeOC_6H_4$, p-MeC_6H_4, p-ClC_6H_4, p-$CNOC_6H_4$) are convenient, stable precursors to dehydrocoupling catalysts for primary silanes regardless of the type of substituent on the phenoxy ring. $CpCp^*Zr[Si(SiMe_3)_3]Me$ is known to rapidly dehydropolymerize $PhSiH_3$ with an activity similar to that of $(CpCp^*ZrH_2)_2$.[29,30,35,37] The Cp′ ring may have substituents other than Me, such as $SiMe_3$ (Cp^S) and $(CH_2)_xNR_2$ (Cp^N; $x = 2, 3$). For instance, $Cp^NCp^STiMe_2$ can dehydropolymerize $PhSiH_3$ to polymer and then can be conveniently recovered from the polymerization mixture.[41] The metallocenes $Cp^N_2MR_2$ (M = Ti, Zr; R = Me, CH_2Ph, OPh) and $Cp^NCp'MCl_2$/n-BuLi are also active dehydrocoupling catalysts for $RSiH_3$.[42,43] Lanthanide complexes Cp^*_2Ln-R [Ln = La, Nd, Sm, Y, Lu; R = H, $CH(SiMe_3)_2$] and Cp^*_2Sm are reported to be highly active dehydrocoupling catalysts for $PhSiH_3$.[44–46] Woo and Song[47] have reported the catalytic dehydropolymerization of 3-aryl-1-silabutanes to polysilanes by $Cr(CO)_6$ in refluxing dioxane, shown in Eq. (2). $Mo(CO)_6$

3-aryl-1-silabutane (SiH_3; aryl substituent X) $\xrightarrow[\text{Cr(CO)}_6\text{, dioxane, reflux}]{-H_2}$ poly(3-aryl-1-silabutane) $-(SiH)_n-$ (2)

X = $(CH_3)_2$; Cl; CH_3,Cl; OPh

and $W(CO)_6$ complexes afford higher molecular weight polymers than $Cr(CO)_6$ in the catalytic dehydropolymerization of 3-aryl-1-silabutanes.[48] Group 6 hexacarbonyl complexes catalyze both dehydropolymerization and redistribution reactions of $PhSiH_3$ in refluxing dioxane. In contrast, (mesitylene)$Cr(CO)_3$ and $(CH_3CN)_3Cr(CO)_3$ slowly produced the dehydrocoupling products in 80% and redistributed products in 20% yield based on $PhSiH_3$ in neat and ambient temperature condition, as shown in Eq. (3).[48]

$$PhSiH_3 + (\text{mesitylene})Cr(CO)_3 \xrightarrow{-H_2} \underset{12\%}{Ph_2SiH_2} + \underset{8\%}{Ph_3SiH} + \underset{80\%}{(\text{-Si(Ph)H-})_x} \qquad (3)$$

These reactions are all appreciably accelerated by illuminating with fluorescent room lighting.[48]

Most of the Group 4 dehydrocoupling catalysts just listed are photochemically and/or thermally unstable with short shelf lives at room temperature. They are also generally not commercially available and have to be synthesized from a metallocene dichloride. There are, therefore, obvious advantages to catalysts that can be generated *in situ.* Corey and co-workers[49,50] pioneered the use of *in situ*-generated catalysts from the reagent combinations Cp_2MCl_2/n-BuLi (M = Ti, Zr, Hf). These catalysts do not exhibit the induction periods characteristic of Cp_2MMe_2 precatalysts and in certain cases they exhibit enhanced reactivity. Corey et al.[51] also successfully dehydrocoupled $PhSiH_3$ with an *in situ*-generated $CpMCl_3/n$-BuLi (M = Ti, Zr) combination catalyst. Although this catalyst only yields oligomers from the dimer to the tetramer, Cp^*MMe_3[27] and $Cp^*Cl_2ZrSi(SiMe_3)_3$[29] are completely ineffective for the dehydrocoupling of $PhSiH_3$. An interpretation of the difference between the dimethylmetallocene precatalysts and the BuLi-generated combination catalysts is presented in Section II,A,1,c.

Woo and co-workers[52–55] studied another type of *in situ*-generated combination catalyst, Cp_2MCl_2/Red-Al (M = Ti, Zr, Hf). Red-Al [or Vitride; sodium bis(2-methoxyethoxy)aluminum hydride], which is known to catalyze the polymerization of lactams and olefins[56,57] and the trimerization of organic isocyanates[58] and to reduce many organic and inorganic species, is commercially available. It is also soluble and reasonably stable in toluene and can be easily mixed as a source of hydride ion with other reagents. The exact nature of the reactive species generated from the Cp_2MCl_2/Red-Al combination is unclear. Like Cp_2MCl_2/n-BuLi combination systems, the Cp_2MCl_2/Red-Al combination functions as a simple dehydrocoupling catalyst for hydrosilane dehydrocoupling.[52] For example, Cp_2MCl_2/Red-Al catalyzes dehydrohomopolymerization and dehydroco-polymerization of 3-aryl-1-silabutanes, as shown in Eqs. (4) and (5).

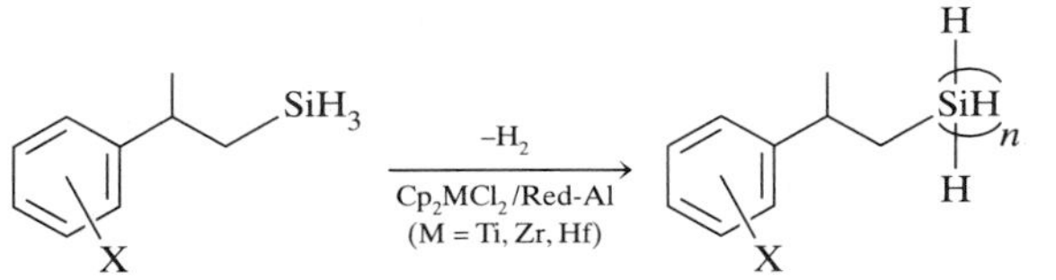

(4)

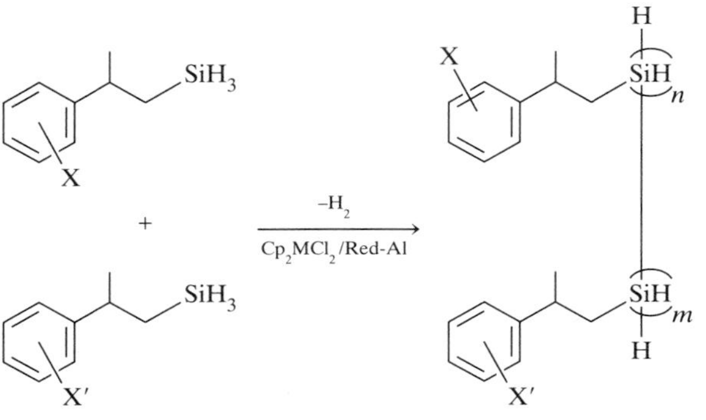

(5)

X, X′ = H; CH_3; CH_3,CH_3; Cl; CH_3,Cl; OPh; *c*-$(CH)_4$; $CH(CH_3)CH_2SiH_3$

Although, to a first approximation, the hydridic Red-Al is expected to generate the hydride Cp_2MH_2 species, it is also expected that the strongly Lewis acidic co-product, (2-methoxyethoxy)aluminum chloride, will be strongly associated with the species involved in the catalytic cycle. The influence of electrophilic species is manifest in that the Cp_2MCl_2/Red-Al combinations catalyze both redistribution and dehydrocoupling of 2-phenyl-1,3-disilapropane, **1**. Neither $[Cp_2MH_2]_2$ nor Cp_2MCl_2/*n*-BuLi combinations catalyze redistribution during dehydrocoupling of **1**, but the Cp_2MCl_2/Red-Al catalysts do, as shown in Eqs. (6) and (7).[54,55] A mechanism proposed by the authors for the catalytic redistribution/dehydrocoupling of **1** by the Cp_2MCl_2/Red-Al is shown in Scheme 1.

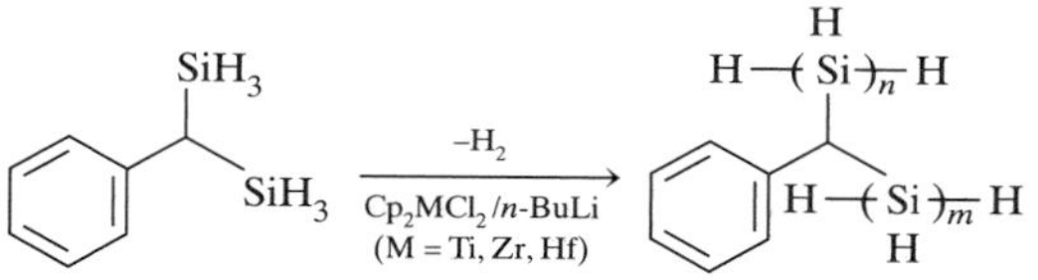
SiH3
SiH3
-H2
Cp2MCl2/n-BuLi
(M = Ti, Zr, Hf)
H
H—(Si)n—H
H—(Si)m—H
H

(6)

1

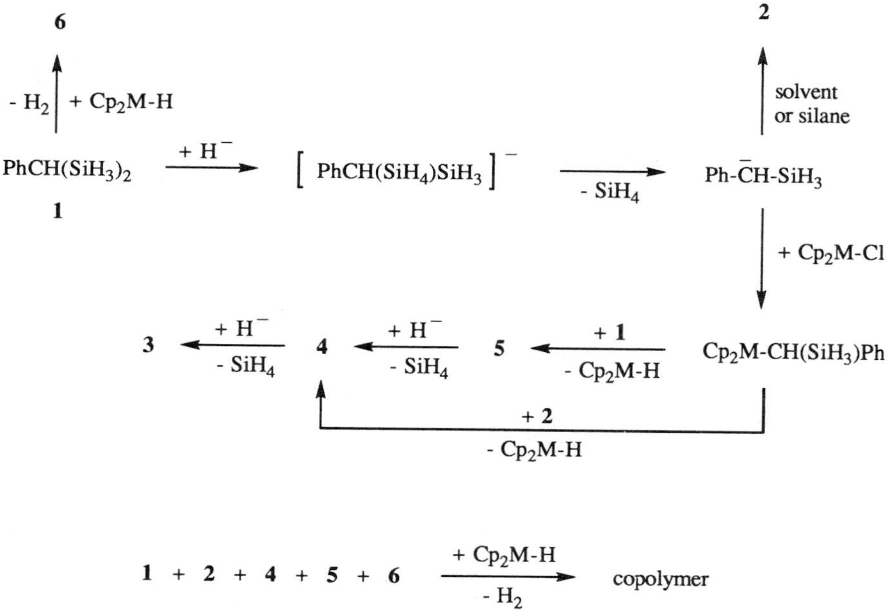
6
- H2
+ Cp2M-H
PhCH(SiH3)2
1
+ H−
[PhCH(SiH4)SiH3]−
- SiH4
Ph-CH-SiH3
2
solvent or silane
+ Cp2M-Cl
Cp2M-CH(SiH3)Ph
+ 1
- Cp2M-H
5
+ H−
- SiH4
4
3
+ 2
- Cp2M-H
1 + 2 + 4 + 5 + 6
+ Cp2M-H
- H2
copolymer

SCHEME 1. Proposed mechanism for the dehydrocoupling/redistribution of $PhCH(SiH_3)$-SiH_3 by Cp_2MCl_2/Red-Al catalysts.

$$\mathrm{PhCH(SiH_3)_2}\ (\mathbf{1}) \xrightarrow[\mathrm{Cp_2MCl_2/Red\text{-}Al}\ (\mathrm{M = Ti, Zr, Hf})]{-\mathrm{H_2}/-\mathrm{SiH_4}} \text{copolymer} + \underset{\mathbf{2}}{\mathrm{PhCH_2SiH_3}} + \underset{\mathbf{3}}{\mathrm{(PhCH_2)_2SiH_2}} + \underset{\mathbf{4}}{\mathrm{H_3SiCH(Ph)SiH_2CH_2Ph}} + \underset{\mathbf{5}}{\mathrm{H_3SiCH(Ph)SiH_2CH(Ph)SiH_3}} + \underset{\mathbf{6}}{\mathrm{H_3SiCH(Ph)SiH_2SiH_2CH(Ph)SiH_3}} \quad (7)$$

This mechanism involves the preferential attack of the hydride on the less hindered silicon with formation of a pentacoordinated anionic species that collapses to give an α-silyl carbanion intermediate and SiH_4 gas. The α-silyl carbanion may then pick up a hydrogen from a hydrogen source (e.g., silane or solvent) to yield $BzSiH_3$, **2**, or it may react with a metallocene chloride to produce its metallocene derivative, which will then undergo a σ-bond metathesis reaction with **1** and **2** to give a Si–C coupling product, **5** and **4**, respectively. **5** may sequentially lose SiH_4 gas by the action of a hydride ion producing **3** and **4**. The resulting silanes might then undergo dehydrocoupling to yield a copolymer.

Unlike the Cp_2MCl_2/n-BuLi catalysts, Cp_2MCl_2/Red-Al catalysts polymerize $PhSiH_3$ selectively to a linear polysilane with higher than usual molecular weight [Eq. (6)]. The molecular weight of the product increases in the order Hf > Zr > Ti.[59]

Another combination catalyst that includes a Lewis acid component is $Cp'_2MCl_2/2n\text{-BuLi}/B(C_6F_5)_3$.[60,61] Interest in such systems came about as a consequence of the successful use of well-characterized homogeneous cationic and cation-like Group 4 and actinide metallocene catalysts for olefin polymerization.[62,63]

$$\mathrm{PhSiH_3} \xrightarrow[-\mathrm{H_2}]{\mathrm{Cp_2HfCl_2/Red\text{-}Al}} \mathrm{H}\!\left[\begin{array}{c}\mathrm{Ph}\\ |\\ \mathrm{Si}\\ |\\ \mathrm{H}\end{array}\right]_n\!\mathrm{H} \quad (8)$$

~95% linear selectivity

The $Cp'_2MCl_2/2n\text{-BuLi}/B(C_6F_5)_3$ catalysts give linear, higher molecular weight poly(phenylsilanes) for M = Zr. For M = Ti and Hf, linear polysilane is produced with ~60% selectivity, the remainder of the product being small rings. This selectivity is similar to that of the $Cp_2ZrCl_2/2n$-BuLi catalyst. The best results were obtained with $CpCp^*ZrCl_2/2n\text{-BuLi}/B(C_6F_5)_3$ (95% linear selectivity; $M_w = 14{,}000$, $M_w/M_n = 1.9$).[61]

The order of reactivity for the various zirconocenes is $Cp^*_2ZrCl_2$ << Cp_2ZrCl_2 ≈ $(MeCp)_2ZrCl_2$ < $CpCp^*ZrCl_2$. None of the combinations $Cp^*_2ZrCl_2/2n$-BuLi/$B(C_6H_5)_3$, $Cp'_2MMe_2/B(C_6F_5)_3$,[61] or $Cp'_2MR_2/[NHR_3]^+[BR_4]^-$ [64] exhibits any significant catalytic activity. The nature of the active species in these systems is discussed further in Section II,A,1,c).

The preceding observations suggest that chain elongation, relative to ring formation and/or chain scissioning, can be controlled by tuning the steric and electronic character of the catalyst. Careful design of the catalyst may provide even higher molecular weight polysilanes.[64] Other experimental parameters such as addition rate of silane, amount of solvent used, reaction temperature, and steric effect of silane also have a significant influence on the rate of the reaction and the molecular weights of the products.[35] To date, the Group 4 metallocene catalysts have proven to be the most effective catalysts for silane polymerization. However, we anticipate that new generations of both early and late transition-metal organometallics will give even better results.

A second major class of dehydrocoupling catalysts has the general formula L_nM, where M is a later transition metal and L is a phosphine, carbonyl, olefin, or halogen ligand, or some combination thereof. Corey has exhaustively reviewed the catalytic dehydrocoupling of hydrosilanes by this type of catalyst.[36] Most such catalysts produce a mixture of oligomers (mainly dimer and trimer) along with significant amount of redistribution products. For example, Wilkinson's catalyst, $RhCl(PPh_3)_3$, catalyzes dehydrocoupling and redistribution of $PhMeSiH_2$ to produce disilane and trisilane in 44% yield as well as Ph_2MeSiH in 30% yield.[65] In the presence of oxygen, later transition-metal catalysts readily give polysiloxanes. For example, quite different results are obtained for the catalytic dehydrocoupling of $RSiH_3$ by platinum complex catalysts in the presence and absence of oxygen.[66,67]

The rate of Group 4 metallocene catalyzed dehydrocoupling is generally increased by the inclusion of a hydrogen acceptor, such as an olefin. Cyclic olefins undergo clean hydrogenation in the titanocene system, but terminal olefins and norbornene give a mixture of hydrogenation and hydrosilation.[26] A mixture of hydrosilation and dehydrocoupling products is also produced with Cp_2ZrCl_2/n-BuLi, with the relative amounts depending on the olefin and the silane.[68] Waymouth[69,70] reported that the Cp_2ZrCl_2/n-BuLi catalyst is efficient for alkene hydrosilation using primary and secondary silane reagents at 90°C.

c. *Mechanistic Considerations.* Mechanisms for dehydrocoupling of hydrosilanes have been extensively scrutinized in the hope that a better understanding of the mechanism will provide clues for the design of catalysts to produce polysilanes with higher molecular weights and better stereotacticity. The clear elucidation of the mechanisms of catalytic dehydrocoupling is difficult since the process is always multistep and involves highly reactive or unstable intermediates, often of unusual chemical structure. A further complication is that the added metal complex is rarely the actual catalytic species and may undergo a complicated sequence of reactions to produce the latter.

To date, six mechanistic schemes have been proposed in the literature, two for catalysis by later transition-metal complexes and four for catalysis by Group 4 metallocenes. The earliest mechanism for catalysis by later transition-metal complexes was proposed by Ojima *et al.*[65] In their scheme they proposed a sequence of oxidative addition/reductive elimination and α-hydride abstractions from Si-H as shown in Scheme 2. No direct evidence for silylene intermediates was obtained. However, since the original proposal by Ojima *et al.*, a number of later transition-metal silylene complexes have been reported in the literature. Although "naked" silylene complexes remain extremely rare, several base-coordinated silylene complexes have been reported, such as $(CO)_5Cr{=}Si(O^tBu)_2 \cdot$(base),[71] $(CO)_4Fe{=}SiX_2 \cdot$(base) (X = Me, O^tBu),[72,73] $(CO)_4Fe{=}Si{=}Fe(CO)_4$(base),[74] Cp(CO)$[SiMe(OMe)_2]Fe{=}SiMe_2$,[75] and $[Cp^*L_2Ru{=}SiPh_2(CH_3CN)]^+[BPh_4]^-$.[76] There have also been a number of reports of other reactions of hydrosilyl complexes in which silylene intermediates are strongly implicated.[77–79] A weakness of mechanisms involving generation of silylene intermediates by α-hydrogen abstraction is that they cannot explain the polymerization of secondary silanes, since in that case the disilanyl ligand has no α-hydrogen to be abstracted.

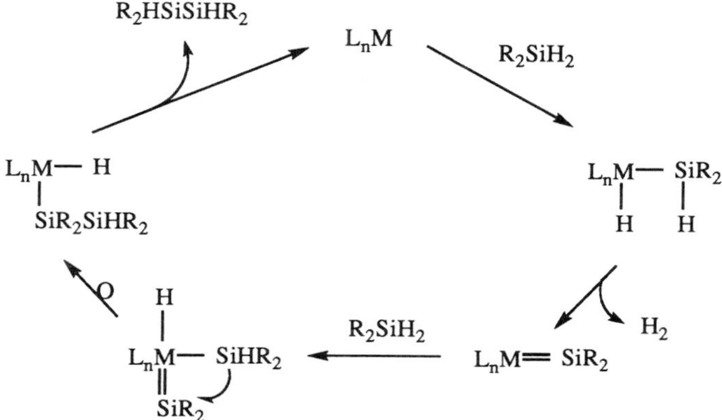

SCHEME 2. The Ojima mechanism for dehydrocoupling/redistribution of R_2SiH_2.

An alternative mechanism for catalysis of Si–Si bond formation by later transition metal complexes was originally proposed by Curtis and Epstein.[33] This mechanism, illustrated in Scheme 3, also invokes a sequence of oxidative addition/reductive elimination steps but does not resort to a silylene intermediate. A serious weakness in this scheme is that, as far as we know, no example of the spontaneous reductive elimination of a disilane from a bis(silyl) complex has yet been observed.

In their early reports on titanocene-catalyzed dehydropolymerization, Harrod and Yun[26] suggested a silylene-based mechanism (Scheme 4). The main justification for this mechanism was the failure to observe low oligomers, particularly the di- and trisilanes, in the early stages of the reaction. This was later shown to be an artifact arising from a rapid equilibrium between these intermediates and the paramagnetic catalyst species.[80] Careful observation of the polymerization indicates that growth of the polymer chain occurs in a stepwise fashion (i.e., condensation type of chain growth).[81]

No terminal silylene complex of a Group 4 metallocene has yet been prepared, which is not surprising since such complexes, if they exist, are expected to be extremely reactive. Although bridging SiR_2 ligands are known for later transition metals, only one example, $[Cp_2Ti(\mu\text{-}SiH_2)]_2$, has been reported for a Group 4 metallocene.[82] There is some doubt about the identity of the latter, which we believe is actually $[Cp_2Ti(\mu\text{-}SiH_3)]_2$. Theoretical calculations suggest that Group 4 metal–silylene complexes can exist.[83] It should also be noted that phosphinidene complexes of Group 4 metallocenes have been reported and that these are isoelectronic with the silylene species discussed here.[84]

Tilley *et al.*[28,29,34,35,85,86] carried out a series of investigations with model

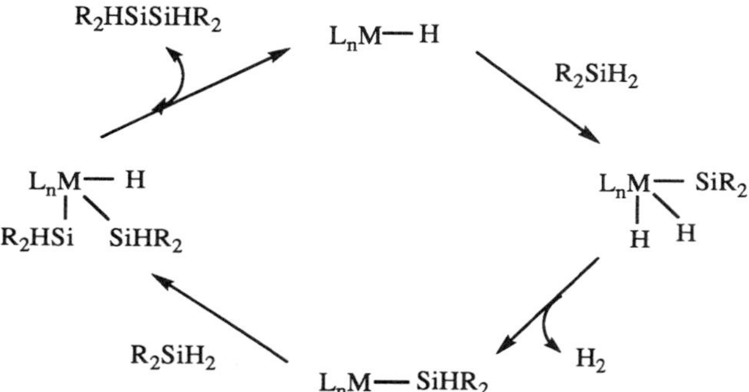

SCHEME 3. The Curtis–Epstein mechanism for catalysis of Si–Si bond formation.

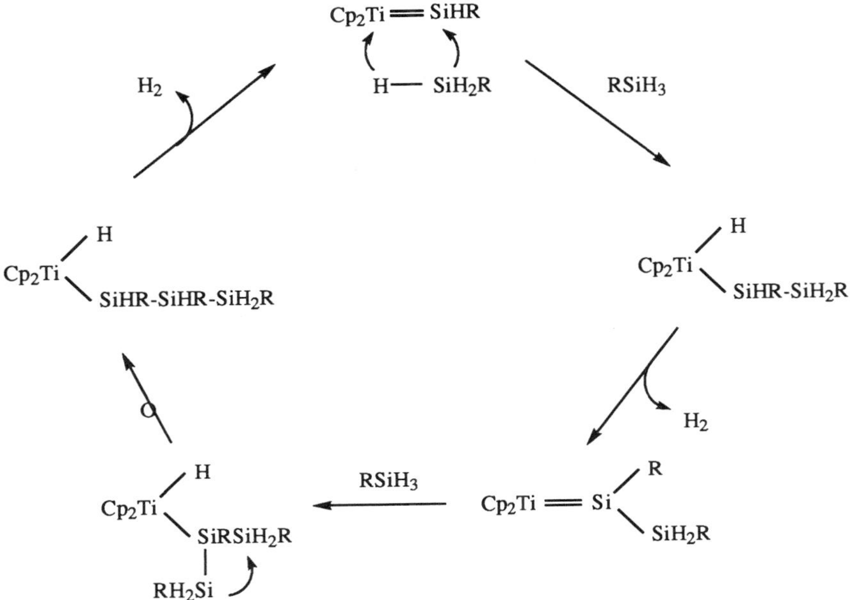

SCHEME 4. A mechanistic scheme for $RSiH_3$ polymerization involving a Group 4 metallocene silylene.

silyl- and hydrido-zirconocene and hafnocene complexes. On the basis of their results they proposed a mechanism involving a sequence of σ-bond metatheses among Si–H, M–H, and M–Si bonds (Scheme 5). Examples of all of the proposed reaction intermediates were isolated and characterized. This mechanism again implies that polymer chain growth occurs in a stepwise fashion rather than by an addition-type growth. Although the Tilley mechanism was developed from the study of zirconocene and hafnocene catalysts, the mechanism is also applicable to titanocene,[38] organolanthanide,[45] and organoactinide[46] catalyzed dehydropolymerization of hydrosilanes.

Marks and co-workers employed organolanthanide complexes of the type Cp'_2Ln-R, where metal-centered redox processes cannot be involved. They studied $PhSiH_3$ polymerization using kinetics and thermochemical measurements and proposed a four-center, heterolytic bond-scission/bond-forming sequence analogous to the Tilley mechanism.[62b]

Corey proposed a scheme that incorporated both σ-bond metathesis and oxidative addition/reductive elimination without resort to a metal–silylene intermediate (Scheme 6).[49–51,68] The authors did not provide detailed kinetic

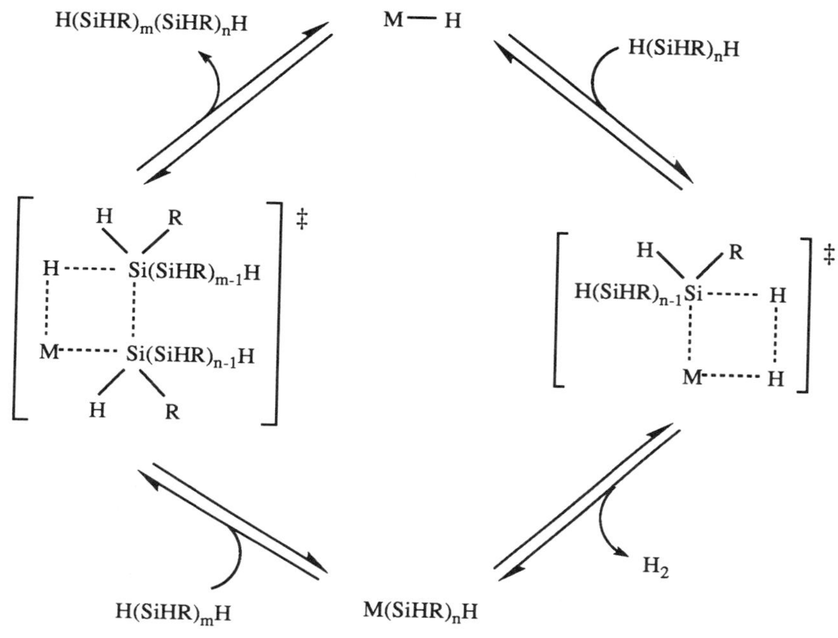

SCHEME 5. The Tilley σ-bond metathesis mechanism.

data or detect any of the proposed intermediates to support this scheme. The assumption that Cp_2Zr is produced in any of these systems is probably incorrect (see later discussion).

Dioumaev and Harrod proposed yet another mechanism for silane dehydropolymerization by the "cation-like" zirconocene catalysts described in Section II,A,1,b.[61] This mechanism (Scheme 7) includes one-electron oxidation/reduction steps as well as σ-bond metathesis steps. This cycle is favored under conditions where Cp_2Zr^{III} is generated in high concentration. Such conditions may be provided by photolysis, or by the use of Cp_2ZrBu_2 as a catalyst precursor.[87] According to this mechanism, the growth of the polymer chain occurs by the coupling of silyl radicals that are generated either by homolysis of a Zr^{IV}–Si bond or by hydrogen atom transfer from Si–H to Zr^{III}.[88] The resulting Zr^{IV}–H then undergoes σ-bond metathesis with Si–H to regenerate the Zr^{IV}–Si and a molecule of H_2. This mechanism is supported by a large body of evidence in the case of the cation-like catalysts, but the extent to which it may be important with neutral catalysts has not been established. It should be noted that both the Zr^{IV}–Si and the Zr^{IV}–H species in this cycle could also be participating in a Tilley-type

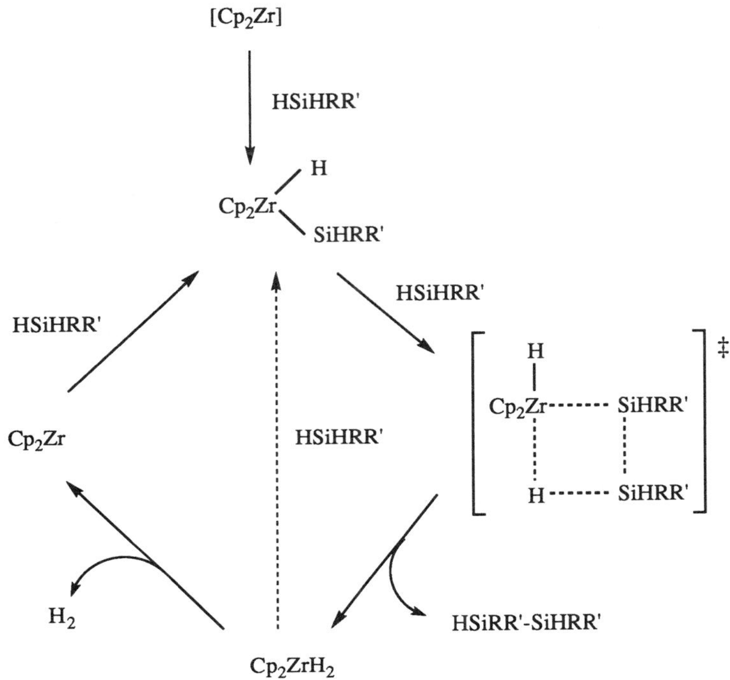

SCHEME 6. The Corey mechanism for $RR'SiH_2$ polymerization.

σ-bond metathesis loop, and that this would be very difficult to establish in the presence of the redox chemistry.

Given the diverse range of catalysts and experimental conditions under which catalytic dehydrocoupling occurs, it is likely that any one, or several, of the mechanistic types described earlier may be important under a particular set of conditions. It is therefore of paramount importance that the particular conditions be carefully considered when making comparisons between different sets of experimental results.

The special behavior of BuLi in the cation-like catalysts encouraged Dioumaev and Harrod to probe the nature of the species present in the Cp_2ZrCl_2/*n*-BuLi combination catalytic system at room temperature. Despite numerous attempts to elucidate it, the true nature of this product has been unclear until recently.[68,69,89–92] Using multidimensional, multinuclear NMR and EPR spectroscopy, Dioumaev and Harrod succeeded in identifying a number of the key intermediates and products in the complicated decomposition of Cp_2ZrBu_2, as shown in Scheme 8.[87] After extended peri-

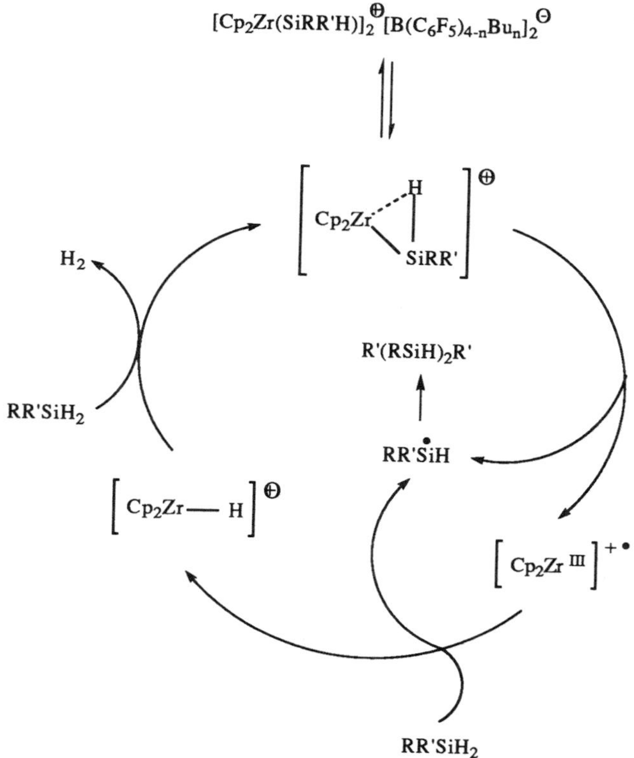

SCHEME 7. A one-electron redox mechanism for $RSiH_3$ polymerization.

ods at room temperature, the main species present are $[Cp_2Zr^{III}-R]_n$, where R = butyl (major product) or crotyl (minor product). The species where R = crotyl was definitively identified by Soleil and Choukroun[92] using EPR spectroscopy.

Given the complexity of the reaction products from the Cp_2ZrCl_2/BuLi reaction and their dependence on temperature of preparation and reaction time, it is likely that several catalytically active species are present under the conditions normally used for dehydrocoupling. However, it is clear that most of the zirconium in these systems is present as Zr^{III}.

d. *Stereoselective Dehydropolymerization of Silanes.* When the two substituents on the backbone silicon atoms of a polysilane are different, each silicon atom is a stereocenter. This is the case with polysilanes produced by

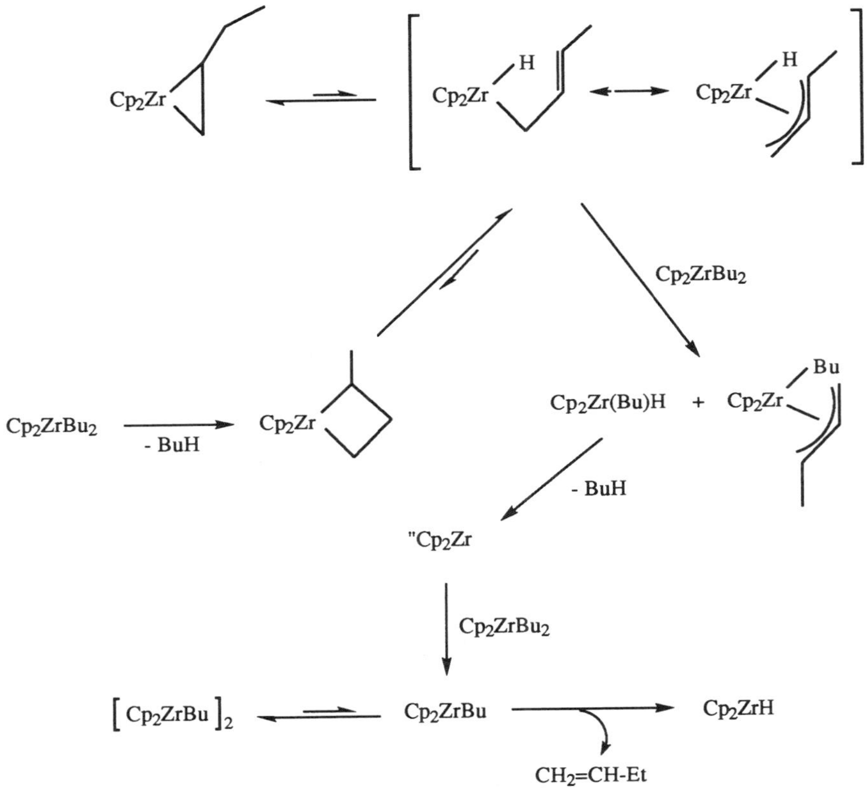

Cp_2Zr Cp_2Zr H Cp_2Zr H

Cp_2ZrBu_2

Cp_2ZrBu_2 - BuH Cp_2Zr $Cp_2Zr(Bu)H$ + Cp_2Zr Bu

- BuH

"Cp_2Zr

Cp_2ZrBu_2

$[Cp_2ZrBu]_2$ Cp_2ZrBu Cp_2ZrH

CH_2=CH-Et

SCHEME 8. The products of Cp_2ZrBu_2 decomposition at room temperature.

dehydrocoupling of primary silanes. A number of preliminary efforts to assign and influence the steroselectivity of such polysilanes have appeared in the literature. Gauvin and Harrod[93] and Gauvin[94] found on the basis of ^1H and ^{29}Si NMR evidence that the sterochemical composition of a polyphenylsilane produced with a dimethyl bis(η^5-indenyl)zirconium catalyst was quite different from that of polymer from a dimethylzirconocene-catalyzed reaction. In light of a later assignment of the ^1H NMR spectra of polyphenylsilanes, it is now evident that the products of this reaction are largely cyclics and the sterochemical significance of the simplified NMR spectrum is not clear. Both Banovetz *et al.*[95] and Corey and Zhu[96] found evidence for diastereoselection in the dehydrocoupling of $PhSiH_3$ with the Cp_2ZrCl_2/BuLi catalyst. Banovetz *et al.* fractionated their polyphenylsilane and found that a high molecular weight fraction gives a ^{29}Si NMR spectrum of unusual simplicity. They assigned the NMR spectrum to a highly syndiotactic microstructure, mainly on the basis of a coincidence of the chemical shift of the principal resonance of the polymer fraction with that of all-*trans* hexaphenylcyclohexasilane. The basis for this assignment is weak since cyclic silicon compounds normally have ^{29}Si resonances at different chemical shifts from their linear analogues. We have also found that all ^{29}Si resonances for Si's more distant than three atoms from a chain end resonate in a narrow chemical shift range, which includes the value for the Banovetz polymer.[100] Indeed, in our experience, the spectrum of Banovetz *et al.* is typical of all high molecular weight, linear polyphenylsilanes that are free of low molecular weight linear and cyclic oligomers, whether as a result of fractionation or directly from the polymerization reaction. Whether it is diagnostic of some kind of stereoregularity remains an open question.

Corey and Zhu[96] examined the sterochemistry of the oligomers (up to Si_4) obtained by dehydrocoupling of $PhMeSiH_2$ with the Cp_2ZrCl_2/BuLi catalyst. They used the distribution and intensities of the ^{1}H NMR resonances due to the Si–Me groups in the trimer and tetramer to estimate the isomeric composition. In the case of the trimer, the 11 resonances anticipated for a statistical distribution of the 2 meso and 1 racemic isomers were all observed, including one with twice the intensity of the other 10. The case was less clear cut for the tetramer, but the cluster of Me resonances did not have the intensity distribution expected for a statistical distribution of stereoisomers.[96]

Group 4 *ansa*-metallocenes have been widely used for the stereoselective polymerization of olefins. They show much reduced activity in dehydrocoupling of silanes, compared to the simple metallocene derivatives.[93,94,97–99] We have used an *S,S*-[bis-1,2-(tetrahydroindenyl)ethane]dichloridotitanium/*n*-BuLi catalyst to dehydrocouple $PhSiH_3$.[100] The reaction is extremely slow and only produces oligomers up to Si_5 and Si_6 after extended reaction times. For much of the time during the reaction, the dominant oligomer is $Ph_4Si_4H_6$. ^{29}Si NMR data show the two diastereomers of the tetrasilane to be present in roughly equal amounts. Since the racemic isomer is expected to be twice as abundant as the meso isomer on purely statistical grounds, this result implies a slight bias toward the meso product.

In summary, although there are some indications that the stereochemistry of silane dehydrocoupling may not be completely random, the effect is not dramatic and can be rationalized by any of the mechanisms discussed previously. In the case of radical coupling, steric interaction between the substituents on the end silicons would tend to favor syndiotactic enchainment. Coupling in the coordination sphere of the catalyst would involve interactions between catalyst substituents and substituents at the end of the silicon chain.

e. *Dehydropolymerization of Hydrosilanes of More Complex Functionality.* A number of investigators have sought ways in which silane dehydrocoupling can be extended to include monomers other than simple alkyl and aryl silanes. Banovetz *et al.*[101] reported the dehydropolymerization of some *p*- and *m*-CF_3- and CH_3-substituted phenylsilanes, using $Cp_2Zr[Si(SiMe_3)_3]Me$ as a catalyst, to give polymers with molecular weight properties similar to those of unsubstituted poly(phenylsilane) ($M_w \approx 1000$–3000 Da and $M_w/M_n \approx 1.5$–2.4). In the case of $CF_3C_6H_4SiH_3$, the Wurtz-type coupling method gives branched polymers due to reaction of the CF_3 groups. This does not occur in the case of the dehydrocoupling reactions.

Hengge and co-workers[102–105] studied the coupling of several silyl-substituted permethylcyclohexasilanes and methylated oligosilanes to give polymers. 1,4-$(SiH_3)_2(Si_6Me_{10})$ undergoes polymerization to yield a polymer with M_w of 4200 (with Cp_2ZrMe_2 catalyst) to 10,500 (Cp_2TiMe_2 catalyst). These materials are presumably highly branched, but the structure was not investigated in any detail. The linear oligosilanes such as 1,2,3-trimethyltrisilane and 1,2,3,4-tetramethyltetrasilane undergo facile polymerization in the presence of Cp_2MMe_2 (M = Ti, Zr) catalysts. A mechanism involving sigma-bond metathesis/β-Si–Si bond cleavage and a metal–silylene intermediate was proposed.[104] The dehydropolymerization of disilanes $Me_3SiSiHMe_2$, $Me_2HSiSiHMe_2$, and $H_2MeSiSiMeH_2$ also proceeds rapidly with Cp_2MMe_2 (M = Ti, Zr) catalysts to yield polymethylsilanes.[104]

Woo and co-workers[106] prepared highly cross-linked polysilanes with an extensive network structure by dehydropolymerization of multisilyl derivatives such as 1,3-$(H_3Si)_2C_6H_4$, 1,4-$(H_3Si)_2C_6H_4$, 1,3,5-$(H_3Si)_3C_6H_3$, 4,4′-$(H_3Si)_2C_6H_4C_6H_4$, and 2,5-$(H_3Si)_2C_4H_2S$ in the presence of a $(CpCp^*ZrH_2)_2$ catalyst. They also prepared lightly cross-linked, soluble polysilanes at room temperature by polymerization of 1,4-$(RH_2Si)_2C_6H_4$ (R = Me, Et, hexyl) with $CpCp^*Zr[Si(SiMe_3)_3]Me$ catalyst.[107] These polymers could then be converted to highly cross-linked insoluble polysilanes with extensive network structure by further heating. Woo *et al.*[52] also prepared a highly cross-linked polysilane with extensive network structure by polymerization of bis(1-trihydrosila-3-butyl)benzene with a Cp_2TiCl_2/Red-Al combination catalyst.

f. *Preparation of Backbone Functionalized Poly(phenylsilanes).* Both the Wurtz-type coupling reaction and the Group 4 metallocene–catalyzed dehydrocoupling reaction suffer the severe limitation that they do not tolerate reactive organic functionality in the monomer. An interesting strategy has been developed by Banovetz *et al.*[108,109] for the free-radical functionalization of polysilanes of the type $[Si(H)R]_n$. Since these polymers can be prepared by dehydrocoupling, this greatly extends the accessibility of polymers with a more varied functionality. Using this approach, chloro derivatives of poly(phenylsilane), $(\text{-Si(Cl)Ph-})_n$, are prepared by reacting polyphenylsilane $(\text{-Si(H)Ph-})_n$ with CCl_4.[109] The highly reactive chloro derivatives are then converted to other derivatives $(\text{Si(X)Ph})_n$ (X = Me, OMe) by classical nucleophilic substitutions. Backbone Si–H functions can also undergo free radical hydrosilation reactions with vinyl and carbonyl compounds. Using this strategy Hsiao and Waymouth[110] prepared other derivatives $(\text{Si(X)Ph})_n$ (X = CH_2CH_2R, OCH_2R, $CH_2CH_2C(O)OH$) by reacting polyphenylsilane with the appropriate vinyl compound or aldehyde in the presence of AIBN radical initiator. The 2-carboxyethyl derivative is the only known water-soluble polysilane.

g. *Applications of Dehydrocoupling in Preceramic Polymers.* The growth of interest in polysilanes after 1980 was driven to a considerable extent by their application as polymer precursors for SiC and Si_3N_4.[111] The ideal polysilane precursor for SiC is poly(methylsilane), which, in principle, can give the maximum theoretical ceramic yield according to Eq. (9).

$$[\text{Si(CH}_3\text{)H}]_n \longrightarrow \text{SiC} + 2\text{H}_2 \tag{9}$$

Early attempts to obtain SiC from a poly(methylsilane) synthesized by Wurtz-type coupling of $Me_2Si(H)Cl$ gave disappointing results.[112] Mu and Harrod[113] subsequently prepared a poly(methylsilane) by Cp_2TiMe_2-catalyzed dehydrocoupling of $MeSiH_3$, which gave SiC in excellent yield and stoichiometry on pyrolysis. Subsequently, Seyferth *et al.*[114–116] prepared and characterized poly(methylsilanes) using Wurtz-type coupling and having the generalized structure $[(\text{MeSiH})_x(\text{MeSi})_y(\text{MeSiH}_2)_z]_n$, where $x + y + z = 1$, $x = 0.8$–0.95. When spiked with various dehydrocoupling catalysts, these polymers give highly cross-linked polysilanes that, when pyrolyzed to 1500°C, produce β-SiC of excellent purity and stoichiometry.[114] The ceramic yield increased from ~20% for uncatalyzed polymer to 70–80% for catalyzed polymer.

A similar improvement in the ceramic yield and stoichiometry in the presence of Cp_2MMe_2 was observed for the pyrolysis of other polysilanes and polycarbosilanes. Seyferth *et al.*[115,116] synthesized poly(vinylsilane), $[\text{CH}_2\text{CH(SiH}_3)]_n$, and pyrolyzed it to 1500°C in the absence and presence of Group 4 metallocene catalysts. The ceramic yield increased from 40% for uncatalyzed polymer to 70–80% for catalyzed polymer. In a similar manner, Corriu polymerized 1,4-disilapentane ($H_3SiCH_2CH_2SiH_2Me$) at room temperature with Cp_2TiMe_2 or Cp_2TiCl_2/n-BuLi.[117] The room-temperature reaction produces a soluble polymer, $[\text{HSi(CH}_2\text{CH}_2\text{SiH}_2\text{Me)}]_n$, with $M_w = 1040$, which converts slowly to a cross-linked partially soluble polymer, $\{\text{HSi}[(\text{CH}_2\text{CH}_2)\text{SiHMe}]_m\}_n$, with $M_w = 32{,}900$. On mild heating, the product becomes completely insoluble with a densely cross-linked network structure. These polymers give β-SiC when pyrolyzed to 1500°C. The ceramic yield is 30% for lightly cross-linked polymer from which the catalyst is removed, but 70% for densely cross-linked polymer that still contains catalyst.

The precise role of the metallocene catalyst in the control of ceramic yield and composition is not understood. There is no doubt that cross-linking by Si–H/Si–H dehydrocoupling is important. In the polymerization of $MeSiH_3$ with Cp_2MMe_2 catalysts, gelation can occur under ambient conditions once the $MeSiH_3$ is used up.[113] In that case it is the backbone tertiary Si–H bonds that react. However, at moderate temperatures (150–300°C) and when only a small proportion of the [Si(Me)H] units have reacted by dehydrocoupling, poly(methylsilane) undergoes a Kumada-type rearrangement as depicted in Eq. (10).[118]

$$\left[\begin{array}{c} CH_3 \\ | \\ Si \\ | \\ H \end{array}\right]_n \xrightarrow[\text{O}]{\Delta} \left[SiH_2CH_2\right]_n \tag{10}$$

This rearrangement produces secondary SiH_2 units that will be much more reactive toward homodehydrocoupling. Another factor that may be of some importance but for which there is no direct evidence is the possibility of heterodehydrocoupling between Si–H and C–H bonds. Such coupling might occur either by a σ-bond metathesis cycle, or by a Berry-type mechanism (see later discussion).[119] Since the activation energy for Si–H/C–H coupling is almost certainly larger than that for Si–H/Si–H, the rate of the former will increase relative to the latter with increasing temperature.

2. *Polygermanes*

Although polygermanes are expected to have similar properties to polysilanes, much less attention has been paid to them. As in the case of polysilanes, polygermanes have been generally prepared by the Wurtz-type coupling of halogermanes.[120,121] Aitken *et al.*[122] first reported the catalytic dehydropolymerization of $PhGeH_3$ with a Cp_2TiMe_2 catalyst. In contrast to the $PhSiH_3$ polymerization, $PhGeH_3$ readily produces an insoluble gel, presumably by further dehydrocoupling of backbone Ge–H bonds. The catalyst chemistry is the same for Si and Ge, and all of the observable Ti in the functioning systems is in the form of Ti^{III}. Ph_2GeH_2 also undergoes dehydrocoupling in the presence of Cp_2TiMe_2.[122] In this case, two parallel reactions are observed. One, which involves Ti^{IV}, selectively gives tetraphenyldigermane. The other, which occurs as the Ti spontaneously reduces to Ti^{III}, gives rapid coupling to a mixture of oligomers.

Our attempts to catalyze the dehydrocoupling of $PhGeH_3$ with Cp_2ZrMe_2 met with failure and we detected no reaction of any kind between these reactants under ambient conditions.[123] However, Choi and Tanaka[124] have now prepared both soluble and gelled poly(phenylgermanes) by the dehydropolymerization of $PhGeH_3$ with $CpCp^*ZrCl_2/n$-BuLi or Cp_2ZrCl_2/n-BuLi combination catalyst.

The mechanism for the dehydropolymerization of $PhGeH_3$ by the Group 4 metallocene catalysts is expected to be similar to the mechanism for the dehydropolymerization of hydrosilanes by the same catalysts. The results with Ph_2GeH_2 and Cp_2TiMe_2 point to a slow σ-bond metathesis mechanism (see Scheme 5) as long as the Ti remains in oxidation state IV. Following the catastrophic reduction to Ti^{III} that occurs on exhaustion of the monomer supply, a much faster dehydrocoupling reaction, possibly occurring by a mechanism of the kind shown in Scheme 7, takes place on replenishing the monomer. The dramatic difference in reactivity between Cp_2ZrMe_2 and the BuLi-derived catalysts points to the importance of the role of Zr^{III} in these reactions. Apparently, the σ-bond metathesis necessary to generate a hydrido/silyl Zr^{IV} species with catalytic activity does not occur with Cp_2ZrMe_2, whereas the products of Cp_2ZrBu_2 decomposition, probably Zr^{III} species, are intrinsically reactive.

Reichl *et al.*[125] reported a remarkable ruthenium-complex-catalyzed "demethanative coupling" to produce polygermanes, as shown in Eq. (11).

$$(n + 1)Me_3GeH \xrightarrow{\text{catalyst}} Me_3Ge[Ge(Me)_2]_nH + n\ CH_4 \tag{11}$$

The authors proposed a mechanism involving an α-Me migration from Ge to Ru, to produce a Ru=Ge species, followed by intramolecular insertion of the germylene into a Ru–$GeMe_3$ bond (see Section III,A,3).

3. *Polystannanes*

Although polystannanes are anticipated to have interesting physical properties, little attention has been paid to them because there has not been, until very recently, a method for making them. Wurtz-type coupling, like most other synthetic approaches, generally gives cyclic oligomers. Sita[126] has prepared homologously pure linear polystannane oligomers by hydrostannolysis between $H(R_2Sn)_nH$ and R_3SnNMe_2 to produce $R_3Sn(R_2Sn)_nSnR_3$ and $HNMe_2$. This was quickly followed by a major breakthrough when Imori and Tilley[127] reported the preparation of linear polystannanes with $M_w \approx 70{,}000$ by dehydropolymerization of R_2SnH_2 (R = ^{n}Bu, ^{n}Hex, ^{n}Oct) with a $CpCp^*Zr[Si(SiMe_3)_3]Me$ catalyst. Subsequently, the Tilley group produced poly(dibutylstannane), $M_w = 2400$, by the Wurtz-type coupling of R_2SnCl_2 by Na/15-crown-5.[128] Babcock and Sita[129] have reported an unusual polymerization of di-*n*-butylstannane with a $RhH(CO)(PPh_3)_3$ catalyst. When the reaction is performed in neat stannane, with all of the catalyst added at once, a vigorous, exothermic dehydrocoupling occurs to produce only the cyclic oligomers $(Bu_2Sn)_5$ and $(Bu_2Sn)_6$. However, slow addition of the catalyst as a dilute solution in toluene to a dilute solution of the monomer gives a 30% yield of a high polymer. Structural analysis of this polymer showed it to be moderately branched and to contain BuSn and Bu_3Sn units in significant amounts, as well as the predominant Bu_2Sn units. In this case, evidently a catalyzed redistribution of Bu groups, as well as dehydrocoupling, occurs. No such redistribution occurs with the Tilley protocol. It should be noted that dilute conditions usually favor the formation of cyclics, the opposite of what is observed in this case.

The preparation of these higher molecular weight polystannanes, and the demonstration that they possess a chemical and mechanical integrity similar to that of the polysilanes (they are, however, much more photosensitive than polysilanes), will no doubt stimulate the further study of their properties and methods for their synthesis.

The mechanism for the dehydropolymerization of R_2SnH_2 by the Group 4 metallocene catalysts is expected to be similar to the mechanism for the dehydropolymerization of hydrosilanes by the same catalysts. The mechanism for the Rh-catalyzed polymerization remains to be determined.

4. *Polyphosphanes*

Since phosphines are isoelectronic with silanes, one might expect that they would also undergo catalytic dehydrocoupling. However, unlike silanes, the phosphines readily reduce and form strong coordinate bonds to the metals that are likely to catalyze their dehydropolymerization. We failed to observe any dehydropolymerization of either $PhPH_2$ or Ph_2PH with Cp_2TiMe_2 or Cp_2ZrMe_2 catalysts and rationalized this failure in terms of the aforementioned particular properties of phosphines.[59] Subsequently, we inadvertently found that $PhPH_2$ yields a mixture of cyclic and linear polyphosphanes [with $(PhP)_6$ being the most abundant of the rings] under the catalytic influence of *rac*-$[(BTHIE)TiH]_2$ [BTHIE = bis-(1,2-tetrahydroindenyl)ethane] at room temperature.[130,131] The catalytic reaction slowly dies because of precipitation of the very insoluble $(BTHIE)Ti(P_2Ph_2)$. The latter contains a diphenylphosphene ligand:

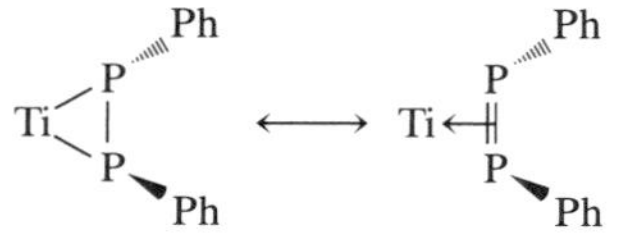

Ti(η²-PPh–PPh) ⟷ Ti←(PPh=PPh)

Spectroscopic evidence shows that the structure shown on the left is closer to the actual structure. The existence of this compound encourages speculation as to the existence of a silicon analogue, or the participation of such ene complexes in catalytic dehydrocoupling cycles. To date, no evidence to this effect has been reported and the best hypothesis is that the phosphine polymerization goes by a σ-bond metathesis cycle, with or without one-electron redox steps.[130]

Breen and Stephan[132,133] reported the dehydrocoupling of $PhPH_2$ in the presence of the anionic hydride complex, $[Cp_2^*ZrH_3][K(THF)_2]$. The dominant product in this case is $(PhP)_5$. This may not be the primary product, since the temperature of the reaction was in the range where oligophosphanes are known to redistribute. Etkin and co-workers[134] prepared an extraordinary polyphosphane by dehydrocoupling of 1,2-$C_6H_4(PH_2)_2$ with the same catalyst. The product in this case is the cyclic octamer $(C_6H_4P_2)_8$, which contains a 16-membered polyphosphorus ring, the largest so far reported.

It is evident that there will be other exciting developments in this area in the near future.

B. *Synthesis of Small Molecules*

Although the bulk of the work carried out on dehydrocoupling has been directed toward the synthesis of polymers, a significant number of useful syntheses of simple molecules with E–E bonds have been discovered along the way. Many of these are mentioned in the previous section. Many catalysts that are too inactive for generation of higher molecular weight polymers can be used to make oligosilanes up to Si_4 in useful amounts.

We found that Cp_2V is an effective catalyst for making 1,2-$Ph_2Si_2H_4$ and 1,2,3-$Ph_3Si_3H_5$ from $PhSiH_3$.[27] The dehydrocoupling is quite slow in this case and the oligosilanes are produced in a stepwise fashion. 1,1,2,2-$Ph_4Si_2H_2$ can be made quantitatively from Ph_2SiH_2 with a Cp_2ZrMe_2 catalyst, since the dimer is completely resistant to coupling.[135] The dimer and trimer of $PhMeSiH_2$ can be made in good yields using the same catalyst.

Corey and co-workers[136,137] prepared the dimers and trimers of heterocyclic dihydrosilanes, such as dihydrosilaanthracene, by using Cp_2MMe_2 (M = Ti, Zr) and $RhCl(PPh_3)_3$ catalysts. Corey's group has also studied the oligomerization of $PhMeSiH_2$ and co-oligomerization of a number of secondary and primary silanes.[138] The Cp_2MCl_2/nBuLi catalyst system yields oligomers up to Si_{5-6} in analytically isolable quantities with $PhMeSiH_2$.

Hengge *èt al.*[102–105] studied the reactions of some (permethyl-*c*-hexasilyl)-silanes with Cp_2MMe_2 (M = Ti, Zr) catalysts. *c*-$(Si_6Me_{11})SiMe_2H$ and

c-(Si_6Me_{11})$SiMeH_2$ react very slowly to produce small amounts of dimers. c-(Si_6Me_{11})SiH_3 dimerizes to give [c-(Si_6Me_{11})-SiH_2]$_2$, whereas 1,4-(SiH_2Me)$_2$(Si_6Me_{10}) undergoes oligomerization to give a mixture of dimer to tetramer.

Dehydrocoupling of primary silanes yields all-*trans* c-$R_6Si_6H_6$ (R = Ph or Bz) with a number of Group 4 metallocene catalysts. The Bz derivative, **7**, is obtained in good yield with Cp_2TiMe_2.[139] The Ph derivative is obtained in ca. 1% yield with $CpCp^*MMe_2$ (M = Ti or Zr)[27] but can be produced almost quantitatively with a $CpCp^*TiMe_2/PhPH_2$ catalyst in dilute solution.[59] Little is known of the mechanism of formation of these cyclics. Their very low solubility is an important factor in the isolation of the all-*trans* isomers, and equilibration by catalytic redistribution of Si–Si bonds is another. Stereoselection by the catalyst and entropic favoring of ring closure in an all-syndiotactic chain may also be important in some cases. Finally, dilution undoubtedly plays an important role. Exploration of these factors in the future will, it is hoped, allow control over the relative rates of linear propagation vs ring closure in these dehydrocoupling reactions.

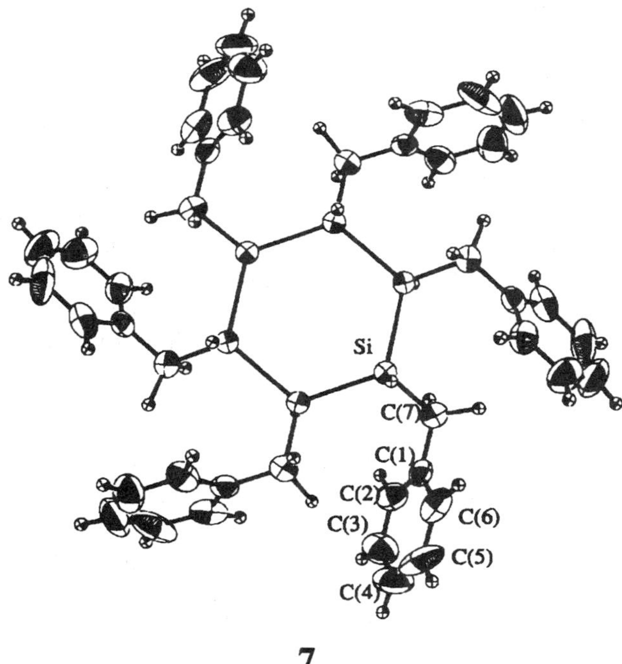

7

Most of the significant examples of synthesis of small molecules with E–E bonds (E $\neq$ Si) have been alluded to in Section II,A. From the relatively small number of examples to date, it is clear that there is potential for developing dehydrocoupling as a synthetic method for other main-group elements. Unfortunately, our understanding of the mechanistic principles is still quite primitive and it is difficult to predict which catalysts will be the most effective for a particular synthetic goal. However, we anticipate that this situation will change rapidly in the near future.

III

Heterodehydrocoupling

A. *B–H/E–H Coupling*

There is a large body of literature on the direct, or acid/base-catalyzed, reactions of boranes with a variety of E–H compounds, particularly water and alcohols. Such reactions are not covered in this article. However, there is to our knowledge only one reported study of metal-catalyzed homo- or heterodehydrocoupling of B–H compounds. Sneddon *et al.*[140] reported both the thermal and catalytic dehydropolymerization of borazine to polyborazylene polymers [Eq. (12)].

$$\mathrm{B_3N_3H_6} \xrightarrow[\text{or catalyst}]{\Delta} \left[\mathrm{B_3N_3H_4}\right]_n + \mathrm{H_2} \qquad (12)$$

Given the general similarities of B–H/Si–H and B–E/Si–E bonds, it is very likely that such chemistry will yield interesting results in the future.

B. *Si–H/H–E Coupling*

The subject of Si–H/H–E coupling was reviewed recently, and the present section covers mainly work reported since 1991.[36]

1. *Si–H/H–O Coupling*

A wide range of catalysts have been reported for dehydrocoupling of OH-containing compounds with silanes. These include acids and bases and both homogeneous and heterogeneous transition-metal catalysts. Such reactions can be useful in organic synthesis (protection of OH groups), in organosilicon synthesis (alkoxysilanes), and in sol–gel chemistry (direct reaction of silanes with H_2O). The subject has been adequately reviewed up to 1990, and we cover only more recent developments.[36,141]

Bedard and Corey[142] carried out a detailed survey of the dehydrocoupling of a range of organosilanes and alcohols under the influence of Cp_2TiCl_2/^{n}BuLi. This catalyst is effective under mild conditions (ambient temperature for $t \leq 1$ h). EtOH and ^{i}PrOH react with primary and secondary organosilanes to give, in most cases, near quantitative yields of the corresponding alkoxy silanes. With secondary and tertiary alcohols more forcing conditions (higher reaction temperatures and/or longer reaction times, t = several hours) are necessary. Some representative data are shown in Table III.

Although partial replacement of the SiH occurs in the more sterically hindered combinations, it is not possible to achieve partial substitution with the less hindered silanes, even using substoichiometric amounts of alcohol and low temperatures. The authors suggested that rapid redistribution of the hydroalkoxysilane intermediates may account for this behavior. This suggestion seems reasonable, since we have observed that the rates of dimethyltitanocene-catalyzed redistributions are strongly dependent on steric encumbrance on the silicon.[143]

A variety of diols also react with secondary silanes to give the 1,3-dioxa-2-silacycles in excellent yields.[141] This method is superior to the alcoholysis

TABLE III

SOME REACTIONS OF ORGANOSILANES WITH ALCOHOLS AND PHENOL CATALYZED BY Cp_2TiCl_2/BuLi[a,b]

Alcohol	Silane	Temperature[c]	Reaction time (h)	Yield % by GPC	Product
EtOH	^{n}HexSiH_3	RT	1	89	^{n}HexSi(OEt)$_3$
	PhMeSiH_2	RT	0.17	95	PhMeSi(OEt)$_2$
	Ph_2SiH_2	RT	0.67	95	Ph_2Si(OEt)$_2$
	Me_2PhSiH	RT	24	85	Me_2PhSi(OEt)
^{i}PrOH	^{n}HexSiH_3	RT	1	86	^{n}HexSi(O^{i}Pr)$_3$
	PhMeSiH_2	RT	1	90	PhMeSi(O^{i}Pr)$_2$
	Ph_2SiH_2	reflux	8	93	Ph_2Si(O^{i}Pr)$_2$
	Me_2PhSiH	reflux	48	No reaction	
^{t}BuOH	^{n}HexSiH_3	reflux	2	76	^{n}HexSiH(O^{t}Bu)$_2$
	PhMeSiH_2	RT	1	90	PhMeSiH(O^{t}Bu)
	Ph_2SiH_2	reflux	1	66	Ph_2SiH(O^{t}Bu)
PhOH	^{n}HexSiH_3	reflux	20	84	^{n}HexSi(OPh)$_3$

[a] From Ref. 142.
[b] Reactions carried out in THF solvent.
[c] RT, room temp.; ref., under reflux.

of the dichlorosilanes, particularly will respect to the absence of polymeric products in the dehydrocoupling reactions. The diol reactions do not work well with primary silanes because of formation of oligomeric products. Attempts to react all three OH groups of *cis,cis*-1,3,5-cyclohexanetriol, or glycerol, intramolecularly with primary silanes were also unsuccessful, but high yields of the silatrane were obtained from triethanolamine.[142]

The reaction of H_2O with hydrosilanes is of interest as an approach to sol–gel materials.[144,145] Alkyl and aryl hydrosilanes do not react at a significant rate with neutral water but are hydrolyzed by aqueous acid or base. Catalytic hydrolysis offers the prospect of control of the outcome of the reaction, particularly with respect to the formation of silanols, or siloxanes. Bedard and Corey[142] briefly studied the reactions of $PhMeSiH_2$ and Ph_2SiH_2 with H_2O using the $Cp_2TiCl_2/^nBuLi$ catalyst. The Ph_2SiH_2 reaction gives the disiloxane, $(Ph_2SiH)_2O$, in quantitative yield, presumably via the silanol. The $PhMeSiH_2$ reaction gives a complicated mixture of oligo(methylphenylsiloxanes), some of which carry PhMe(OH)Si- and/or PhMe(H)Si- end groups.

Schubert and Lorenz[145] explored the activity of $[(Ph_3P)CuH]_6$ as a catalyst for SiH/OH dehydrocoupling. An advantage of this catalyst is that, in contrast to Group 4 metallocene catalysts, it does not catalyze either homodehydrocoupling of SiH or hydrosilation of C=C and C=O bonds. It also gives good to excellent yields with a wide combination of silanes and alcohols, even the combination tertiary alcohol/tertiary silane. The reactivity orders are $Me_2PhSiH > Ph_2MeSiH > Ph_3SiH > Et_3SiH$ and $EtOH > MeOH > {}^iPrOH > {}^tBuOH \approx PhOH$. A notable feature of these reactions is that the yields are usually significantly greater when they are run in air than when they are run under an argon atmosphere, suggesting the possibility of a redox mechanism involving a Cu^I/Cu^{II} cycle.

The $[(Ph_3P)CuH]_6$ catalyst has also been studied for the dehydrocoupling of silanes with water and with acetic acid.[146] Only tertiary silanes were investigated to avoid complications due to polymerization. The water reaction gives simple silanols as the main product with Ph_3SiH, Ph_2MeSiH, and Et_3SiH, whereas $PhMe_2SiH$ gives the disiloxane. Reactions of CH_3COOH followed the same general pattern as those of the alcohols to give excellent yields of the silyl acetates.

$[(Ph_3P)CuH]_6$ did not catalyze reaction of Et_2NH with secondary and tertiary silanes.[145] This is not surprising in view of the results of Liu and Harrod, who showed that CuCl, or CuH catalysts do not activate the second N—H of primary amines, or the third N—H bond of NH_3, toward reaction with silanes.[147,148]

Intramolecular hydrosilation has been extensively used to control regio- and enantioselectivity.[149] This strategy has been applied by Bergens *et al.*[150]

to the synthesis of chiral diols by catalytic intramolecular hydrosilation of alkenyl silyl ethers. These workers originally synthesized the intermediate alkenylsilyl ethers by reaction of the appropriate alcohol with R_2SiClH [Eq. (13)] and in a second stage catalyzed the intramolecular hydrosilation with a rhodium complex catalyst [Eq. (14)]:

$$R_2SiClH + HO(CH_2)_nCMe{=}CH_2 \longrightarrow R_2Si(H)O(CH_2)_nCMe{=}CH_2 + HCl \quad n = 1 \text{ or } 2 \qquad (13)$$

$$R_2Si(H)O(CH_2)_nCMe{=}CH_2 \xrightarrow{\text{catalyst}} \text{cyclic } R_2Si{-}O{-}(CH_2)_n \text{ ring} \qquad (14)$$

$$R_2SiH_2 + HO(CH_2)_nCMe{=}CH_2 \xrightarrow{\text{catalyst}} R_2Si(H)O(CH_2)_nCMe{=}CH_2 + H_2 \qquad (15)$$

Xin and Harrod[151] carried out similar reactions as a one-pot reaction using the titanocene-catalyzed dehydrocoupling of the alcohol with a diorganosilane to generate the intermediate silyl ether [Eq. (15)]. The titanocene-based catalysts are of little synthetic utility since they give both intra- and intermolecular hydrosilation products, as well as products resulting from hydrogenation and redistribution reactions. The one-pot dehydrocoupling/hydrosilation strategy has also been shown to work to better effect with a rhodium-based catalyst.[152] In this case, although an unspecified amount of hydrogenation was reported to take place, it was very minor compared to the hydrogenation and other side reactions observed with the titanocene catalysts. A comparison of the two catalyst systems is presented in Table IV. The prevalence of the bis(alkenyloxy)silane products in the titanocene-catalyzed reactions is due to the high activity of these catalysts for redistributing the mono(alkenyloxy)silanes, rather than to a competitive second dehydrocoupling reaction. In the case of the last entry of Table IV, the hydrosilation reaction occurs at a negligible rate compared to the first dehydrocoupling step and the subsequent redistribution.

2. *Si–H/H–N Coupling*

The heterogeneously catalyzed coupling of hydrosilanes and amines has long been recognized as a useful route to silazanes.[153] More recently, a number of transition-metal-based homogeneous catalysts have been explored and found to give interesting reactivities.[154] Reactions have been effected between tertiary, secondary, and primary organosilanes and either ammonia or amines. An advantage of dehydrocoupling over the classical

TABLE IV

COMPARISON OF CATALYTIC DEHYDROCONDENSATION/HYDROSILATION REACTIONS OF $CH_2{=}C(Me)CH_2OH$ IN THE PRESENCE OF Rh AND Ti-BASED CATALYSTS

Catalyst	Silane	Time	Yield (**a** + **b** + **c**)[a]	**a**:**b**:**c**
$[Rh(diphos)(solv)_2]^{+}$[b]	Ph_2SiH_2	<10 min	>95%	95:5:0
$[Rh(diphos)(solv)_2]^{+}$[b]	Et_2SiH_2	<10 min	>95%	95:5:0
Cp_2TiMe_2[c]	Ph_2SiH_2	49 h	90	29:24:35
Cp_2TiMe_2[c]	$PhMeSiH_2$	19 h	93	68:5:20
rac-[BTHIE]$TiMe_2$[c,d]	Ph_2SiH_2	24 h	82	26:28:28
rac-[BTHIE]$TiMe_2$[c]	$PhMeSiH_2$	6 h	99	0:65:34

[a] **a**, cyclosiloxane; **b**, mono(alkenyloxy)silane; **c**, bis(alkenyloxy)silane.
[b] Catalyst concentration 2–4%.[150]
[c] Catalyst concentration 1 mol% based on substrate.[151]
[d] BTHIE, bis(1,2-tetrahydroindenyl)ethane.

chlorosilane route is the relative ease with which hydrosilazanes can be prepared.

The only homogeneous catalyst that has been systematically studied for the reaction of silanes with ammonia is Cp_2TiMe_2.[147] This catalyst is effective for tertiary, secondary, and primary organosilanes, but the products from secondary and primary organosilanes are quite complex and strongly dependent on the reaction conditions. Reasonable rates were achieved at temperatures of 70 to 100°C, at partial pressures of 1 to 2 atm NH_3, and at a catalyst concentration of ca. 1 mol % relative to silane. Some typical results are shown in Table V.

The initially observed products in the reactions of $PhMe_2SiH$ and $PhMeSiH_2$ with NH_3 are the disilazanes [Eq. (16)]. No monosilylazane is observed under any conditions.[147]

$$PhMeSiH_2 + NH_3 \begin{cases} \not\longrightarrow (PhMeSiH)NH_2 \\ \longrightarrow (PhMeSiH)_2NH \end{cases} \tag{16}$$

Since amino hydrosilanes do not undergo rapid redistribution under the reaction conditions (unlike alkoxyhydrosilanes; see later discussion), it may be concluded either that the residual N–H bonds are greatly activated by the first Si–N bond, or that the monosilylamine is never produced. The first possibility is precluded by the fact that -NH_2 terminated oligomers are formed in certain cases under forcing conditions (e.g., see third entry in Table V).

TABLE V

DEHYDROCOUPLING OF AMMONIA AND AMINES WITH ORGANOSILANES[a]

Catalyst	Silane (mmol)	Amine (mmol)	T(°C)	Time (h)	Conversion (%)	Isolated product (%)
Cp_2TiMe_2	Ph_2MeSiH (5)	NH_3 (9)	100	0.5	60	$(Ph_2MeSi)_2NH$ (100)
Cp_2TiMe_2	$PhMeSiH_2$ (7.3)	NH_3[b]	100	0.1	100	$(PhMeSiH)_2NH$ (80)
Cp_2TiMe_2	$PhMeSiH_2$	NH_3[b]	100	1.0	100	$(PhMeSiNH)_3$ (25) $H_2N(PhMeSiNH)_3H$ (67)
Cp_2TiMe_2	$MeSiH_3$ (10)	NH_3 (10)	70	48.0	100	$[MeSi(NH_2)]_n$
CuCl	$PhMeSiH_2$	$PhCH(Me)NH_2$	100	21.0	84	PhCH(Me)NHSiHPhMe
CuCl	$PhMeSiH_2$ (3.8)	$BzNH_2$ (4.3)	110	8.0	87	HPhMeSiNHBz (61) $(HPhMeSi)_2NBz$ (39)
CuCl	$PhMeSiH_2$ (2.6)	$BzNH_2$ (5.9)	105	6.0	88	HPhMeSiNHBz (81)
CuCl	$PhMeSiH_2$ (2.6)	$BzHN_2$ (5.9)	105	67.0	88	$PhMeSi(NHBz)_2$ (100)
CuCl	$PhSiH_3$ (1.0)	$PhNH_2$ (1.0)	110	1.0	95	$PhH_2SiNHPh$ (100)

[a] The ammonia reactions are catalyzed by 1 mol% Cp_2TiMe_2. The amine reactions are catalyzed by 5 mol% $CuCl_2$.

[b] Ammonia was bubbled through the solution at atmospheric pressure.

At long reaction times, both Si–H bonds of $PhMeSiH_2$ react to give products of greater complexity than the disilazane. However, no replacement of the third N–H bond was observed under any circumstances.

An unexpected feature of the Cp_2TiMe_2-catalyzed cross-coupling of silanes with NH_3 is the sequence of reactivity with respect to the substitution on the reactant silane. Organosilanes normally show a sequence of reactivity: 1° > 2° > 3°. In the present case the order is dramatically reversed. The anomaly was attributed to an antagonistic competition between the homo- and heterodehydrocoupling reactions.[147] The Cp_2TiMe_2-catalyzed homodehydrocoupling of silanes follows the normal reactivity sequence, with the primary silanes reacting very rapidly and the tertiary silanes not at all. In the presence of ammonia, the homodehydrocoupling of phenylsilane occurs more rapidly than heterodehydrocoupling and the N-containing products result mainly from amination of oligo(phenylsilanes). However, the rate of homocoupling is still very much lower than in the absence of NH_3.

Cuprous chloride and cuprous hydride are effective catalysts for the coupling of primary amines with silanes.[148] Under the conditions used, CuCl is converted to the hydride by the silane. Some results obtained with this catalyst are shown in Table V. With the relatively encumbered amine $PhCH(Me)NH_2$, only one Si–N bond is formed on reaction with $PhMeSiH_2$. With benzylamine the major product may be HPhMeSiNHBz, (HPhMe-

Si)$_2$NBz, or (BzNH)$_2$SiPhMe, depending on the ratios of the reactants. The behavior of $PhNH_2$ is similar to that of $BzNH_2$. In contrast to the Cp_2TiMe_2-catalyzed reactions of silanes with NH_3, $PhSiH_3$ reacts more rapidly with amines than does $PhMeSiH_2$ in the Cu^I-catalyzed reactions. This may be related to the fact that the Cu^I catalyst does not promote homodehydrocoupling of the silanes.

N_2H_4 and substituted hydrazines are more reactive toward silanes than are NH_3 and amines.[154–156] Primary silanes react spontaneously with hydrazine to give soluble polymers of ill-defined composition, in which most of the Si–H functions have reacted.[155] Model compound studies on cyclic and acyclic dimethylhydrazino-silanes indicate that a dominant motif in the polymer chain is **8**.

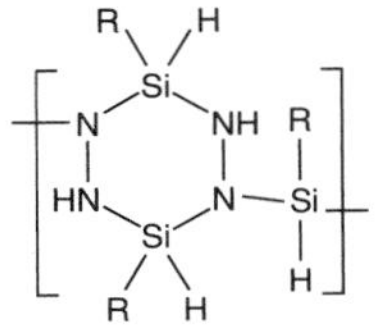

8

In the presence of a 1% Cp_2TiMe_2 catalyst, $PhSiH_3$ reacts violently with N_2H_4 to give an insoluble, cross-linked polymer, presumably of a similar structure to that produced in the absence of catalyst, but with a higher cross-link density resulting from more complete reaction of Si–H functions.[155]

Methyl-substituted hydrazines react at best very slowly with silanes in the absence of catalysts, but they are reactive with primary and secondary silanes in the presence of a Cp_2TiMe_2 catalyst. Even in the presence of catalyst, they react progressively more slowly with increasing methyl substitution. $PhSiH_3$ reacts with $MeNHNH_2$ to give a mixture of oligomers with two types of chain unit, **9** and **10**. As in the case of reactions of primary silanes with NH_3 described earlier, the units **9** arise from direct heterodehydrocoupling of $PhSiH_3$ with $MeNHNH_2$, whereas units **10** arise from competing homodehydrocoupling of $PhSiH_3$, followed by amination of the backbone Si–H units. With unsubstituted NH_2NH_2, the homocoupling of $PhSiH_3$ cannot compete with the faster heterocoupling reaction. With Me_2NNH_2, the rate of heterocoupling is comparable to that of homocoupling of $PhSiH_3$, and a mixture of oligomers with units analogous to **9** and **10** is obtained. With MeNHNHMe, the heterocoupling reaction of the first NH group is faster than the rate of homocoupling of $PhSiH_3$, and a relatively clean reaction occurs to give a mixture of $PhSiH_2$(NMeNHMe) and PhSiH(NMeNHMe)$_2$.

$$\left[\begin{matrix} H & NHMe \\ | & | \\ Si & -N \\ | & \\ Ph & \end{matrix}\right] \quad \text{and} \quad \left[\begin{matrix} NHNHMe \\ | \\ Si \\ | \\ Ph \end{matrix}\right]$$

9 **10**

Ph_2SiH_2 does not react spontaneously with hydrazine, or its methylated derivatives, at temperatures below 100°C. In the presence of 1% Cp_2TiMe_2 these reactions all take place slowly; the reaction rate sequence being $NH_2NH_2 > MeNHNH_2 > Me_2NNH_2$.[155] Some results for these reactions are listed in Table VI. The major products from the NH_2NH_2 and $MeNHNH_2$ reactions were obtained directly as crystalline solids from the cooled reaction mixtures. It should be noted that the rate of heterodehydrocoupling with Me_2NNH_2 is so slow that the major product is the tetraphenyldisilane, the homodehydrocoupling product from the silane, despite the fact that homodehydrocoupling of Ph_2SiH_2 is very slow compared to that of $PhSiH_3$.

An interesting example of dehydrocoupling that permits the synthesis of a simple hydrazinosilane that is not easily synthesized by the classical dehydrochlorination route is the reaction of Me_2SiHCl with NH_2NH_2.[157] A 2:1 molar mixture of these reactants, in the presence of an HCl acceptor, quickly yields $Me_2Si(H)NHNHSi(H)Me_2$. This molecule then undergoes a spontaneous inter-/intramolecular heterodehydrocoupling with a second molecule of NH_2NH_2 to give $Me_2Si(NHNH)_2SiMe_2$ in 50% overall yield. This simple archetypal 1,4-disilatetrazacyclohexane is not formed in isolable

TABLE VI
SOME REACTIONS OF HYDRAZINES CATALYZED BY Cp_2TiMe_2[a,b]

Hydrazine (mole ratio)	T (°C)	t (h)	Conversion (%)	Products (%)	
NH_2NH_2 (1:1)	120	24	100	$Ph_2Si(NHNH)_2SiPh_2$	(40)
				$Ph_2SiH(NHNHSiPh_2H)_2$	(15)
				$Ph_2SiH(NHNH)_2$	(5)
				$Ph_2Si(NHNH_2)_2$	(5)
				$[Ph_2SiNH]_3$	(10)
$MeNHNH_2$ (1:2)	80	20	100	$Ph_2Si(NHNHMe)_2$	(70)
				$Ph_2SiH(NHNHMe)$	(15)
				Higher oligomers	
Me_2NHNH (1:1)	70	20	36	$HPh_2SiSiPh_2H$	(44)
				$Ph_2SiH(NHNMe_2)$	(36)
				$Ph_2Si(NHNMe)_2$	(20)

[a] Catalyst 1 mol% Cp_2TiMe_2 based on silane.
[b] Reactions run in the absence of solvent.

yield by the direct reaction of Me_2SiCl_2 with two moles NH_2NH_2, which gives mainly polymers by intermolecular dehydrochlorination.[158]

3. *Si–H/H–C Coupling*

Relatively little work has been reported on the dehydrocoupling of Si–H/H–C bonds. The thermodynamics of this type of reaction are similar to those of Si–H/Si–H reactions, the stronger Si–C bond in the product (compared to an Si–Si bond) compensating for the stronger C–H bond of the reactant. The kinetic difficulty in realizing this type of reaction is typical of the general difficulty of catalytically activating nonpolar C–H bonds.

It is expected that examples of Si–H/H–C heterodehydrocoupling might be found with substrates that have C–H bonds of unusual weakness, or high polarity. This is in fact the case so far. The earliest example of homogeneous Si–H/H–C heterodehydrocoupling is the reaction of acetylenes with hydrosilanes under the influence of an iodine-modified chloroplatinic acid catalyst. Despite the well-known propensity of chloroplatinic acid to catalyze hydrosilation of acetylenes, the iodine-modified catalyst is highly selective for dehydrocoupling under ambient conditions.[159] In this catalyst system, the I_2 rapidly produces R_3SiI, which is the actual catalyst modifier. PhCCH reacts with trialkylsilanes to give yields of 90–95% and selectivity of 100% in hydrocarbon solvents. No hydrosilation or hydrogenation products are detected. When the reaction is carried out in a coordinating solvent, such as THF or MeCN, hydrosilation is the main reaction. The same is true when $(EtO)_3SiH$ or $EtCl_2SiH$ are used as the hydrosilane.

A CuCl/tertiary amine catalyst was also found to be highly selective for the dehydrocoupling of some arylacetylenes with primary and secondary silanes.[160] With primary silanes the Si–H bonds react in a stepwise sequence with the acetylene, and by appropriate choice of conditions the mono-, di-, or triethynyl silane can be the major product. 1,3-Diethynylbenzene couples with primary and secondary silanes to give polymers of moderate molecular weights of the general structure **11**.

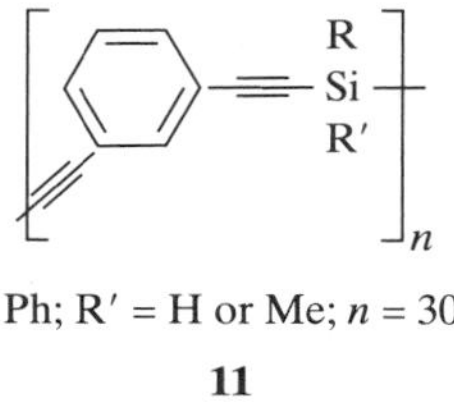

$$\left[\text{C}_6\text{H}_4(\text{C}{\equiv}\text{C})\text{—C}{\equiv}\text{C—Si(R)(R}')\right]_n$$

R = Ph; R′ = H or Me; n = 30–500

11

An apparent Si–H/H–C dehydrocoupling is often observed under hydrosilation conditions. This behavior is illustrated in Eq. (17), which depicts some of the typical products of a hydrosilation reaction. The three organic products shown can be designated as the hydrosilation, the dehydrocoupling, and the hydrogenation product. Depending on conditions, the hydro-

$$R_3SiH + R'CH{=}CH_2 \longrightarrow R_3SiCH_2CH_2R' + R_3SiCH{=}CHR' + R'CH_2CH_3 + H_2 \qquad (17)$$

silation product, or the dehydrocoupling product, may be obtained exclusively, but more often a mixture of the two is obtained. When the alkenylsilane is obtained, it is common to get the hydrogenation product as well.[161] To our knowledge, there is no case of H_2 ever being detected in any of these reactions, even though it should be produced in those cases where the alkenylsilane is the exclusive organic product.

There is some ambiguity as to whether the alkenyl- and alkynylsilane-forming reactions are true dehydrocoupling reactions. According to the definition as embodied in Eq. (1), this would require the direct activation of a C–H bond, such as the $=$C$-$H bond of an olefin. The general consensus is that the alkenyl- and alkynylsilane-forming reactions are more likely to take place via an insertion of the olefin into an M–Si bond, followed by β-hydride elimination [Eqs. (18) and (19)] in the case of electron-rich metal complexes,[161b] or by σ-bond metathesis of M–Si and H–C= in the case of electron-deficient metal complexes [Eq. (20)].[50,138]

$$M{-}Si + CH_2{=}CHR \longrightarrow M{-}CH(R)CH_2Si \qquad (18)$$

$$M{-}CH(R)CH_2Si \longrightarrow M{-}H + CHR{=}CHSi \qquad (19)$$

$$\begin{array}{ccc} M{-}Si & & M\text{-----}Si \\ \vdots \quad \vdots & \longrightarrow & | \qquad | \\ H{-}C{=} & & H\text{-----}C \\ & & \quad \| \end{array} \qquad (20)$$

Although there are many examples of Eq. (17), there is only a single report of the analogous reaction with alkynes.[162] Crabtree and Jun studied the reactions of several tertiary silanes with 1-alkynes using a 7,8-benzoquinolinatoiridium complex. Although hydrosilation generally predominates, some dehydrocoupling is observed in most cases. The amount of dehydrocoupling increases with the steric encumbrance of the silane and the alkyne, with the highest value (60:40, dehydrocoupling:hydrosilation products) being for $Et_3SiH/CH{=}C^tBu$. Again, no H_2 is observed in these reactions, the hydrogen being transferred to either the reactant alkyne or to the alkenylsilane product.

Trimethylsilane undergoes an interesting self-dehydrocondensation in the presence of $Ru(H)_3(SiMe_3)(PMe_3)_3$ as catalyst.[119] The products of this reaction are a mixture of oligo(carbosilanes) [Eq. (21)]:

$$\mathrm{Me_3SiH} \xrightarrow[150^\circ\mathrm{C;C_6H_{12}}]{\text{catalyst}} \mathrm{H_2 + HSiMe_n(CH_2SiMe_3)_{3-n} + H[SiMe_2CH_2]_mSiMe_3}$$
$$\mathrm{+\ Me_3SiCH_2Si(H)MeCH_2SiMe_2CH_2SiMe_3} \quad (21)$$
$$\mathrm{+\ higher\ oligomers}$$
$$n = 1 \text{ or } 2;\ m \text{ mostly} = 2$$

No products with Si–Si bonds were detected and the inertness of Me_4Si to the catalyst excludes a direct C–H bond activation. The authors proposed a mechanism involving β-hydride elimination from a Me_3Si ligand to generate a reactive η^2-silene complex. Subsequent insertion of this η^2-silene into another M–$SiMe_3$ bond forms the new Si–C bond. Recycling the product of reaction (24) through (23) provides a mechanism for chain extension. Branched products, such as $HSiMe_n(CH_2SiMe_3)_{3-n}$, arise from addition of Me_3SiH to higher silenes such as $CH_2{=}SiMe(CH_2SiMe_3)$.

$$L_n\mathrm{Ru(H)_3(SiMe_3)} \longrightarrow L_n\mathrm{RuH(SiMe_3) + H_2} \quad (22)$$

$$\mathrm{H{-}L_nRu{-}SiMe_3} \longrightarrow \mathrm{H_2L_nRu}\leftarrow\|(\mathrm{SiMe_2{=}CH_2}) \xrightarrow[-\mathrm{H_2}]{+\mathrm{Me_3SiH}} \mathrm{H(Me_3Si)L_nRu}\leftarrow\|(\mathrm{SiMe_2{=}CH_2}) \quad (23)$$

$$\mathrm{H_2L_nRu}\leftarrow\|(\mathrm{SiMe_2{=}CH_2}) \xrightarrow{\text{o}} \mathrm{H{-}L_nRu{-}SiMe_2{-}CH_2{-}SiMe_2} \quad (24)$$

4. *Dehydrocoupling with Heavier Elements*

The hydrides of the heavier congeners of the Group 14 to 16 elements have weak E–H bonds and they can be decomposed under mild conditions to yield the pure element or a low-oxidation-state hydride (in many cases of ill-defined chemical composition and structure). This tendency, which also applies to the E–C bonds, underlies the usefulness of hydrides in many gas and vapor phase deposition methods.[3] There is still, however, a need for catalysts, particularly to control the specificity of dehydrocoupling: for example, the ability to make rings of a particular size or isomeric composition, or the ability to avoid cyclic products altogether. In addition, it is desirable to control homo- vs hetero-dehydrocoupling selectivity, something difficult to do by noncatalytic methods.

A nice example of the kind of reaction that can be achieved has been reported by Fisher and co-workers.[163] Their goal was to produce an MOCVD (metallorganic chemical vapor deposition) reagent and they were successful in achieving the reaction shown in Eq. (25).

$$2\ Bu_3SnH + Bu_3P{=}Te \xrightarrow[-Bu_3P,\ 25°C]{Cp^*_2TiH} (Bu_3Sn)_2Te + H_2 \qquad (25)$$

Although at first sight this reaction does not appear to be a formal dehydrocoupling process, the analogy to heterodehydrocoupling can be seen if it is imagined as a reaction between Bu_3SnH and Bu_3SnTeH. Many of the details of the catalytic cycles were uncovered by detecting and isolating several of the key intermediates. The proposed cycles are shown in Scheme 9.

Despite the complexity of Scheme 9, it is still a considerable oversimplification of the full mechanism, and most of the steps depicted must occur by several discrete reactions. This example is a foretaste of the rich chemistry that waits to be uncovered with these heavier element reactions.

IV

CONCLUSION

It is clear that catalytic dehydrocoupling is a versatile technique for the synthesis of E–E and E–E′ bonds. The scope of the methodology, in terms of the types of reactions that can be carried out and the selectivities that are possible, has been broadly established. There is still much to be done in unequivocally establishing the mechanisms and the types of catalytic intermediates involved.

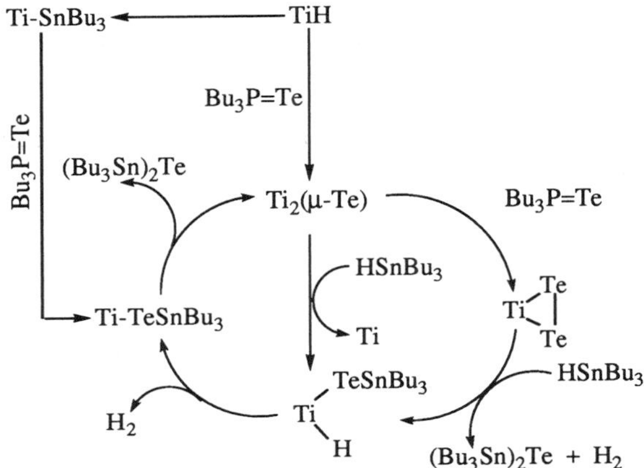

SCHEME 9. The mechanism proposed for Cp^*_2TIH-catalyzed dehydrocoupling of Bu_3SnH and $Ph_3P{=}Te$. (In the scheme, Ti = Cp^*_2Ti.)

There will undoubtedly be many new catalysts described in the near future. In particular, we anticipate that later transition-metal complexes will play a much larger role in this type of chemistry. Although bis-Cp complexes of Groups 3 and 4, the lanthanides, and the actinides have shown exceptional activity and thus far have dominated reported investigations on catalytic homodehydrocoupling, it is most unlikely that they are unique in this property. It is to be expected that other classes of complexes, such as mono-Cp and η^1-alkyl complexes of these elements, will also be active. Other directions for future evolution are the development of catalysts that are air stable, that are tolerant of more functionalities on substrates, and that are more easily manipulated.

Although it is clear from results already obtained that control over ring formation and polymer molecular weight is possible, how this occurs is still very poorly understood.

Acknowledgments

J. F. H. is grateful for the collaboration and hospitality of Dr. Edmond Samuel of the Ecole Nationale Supérieure de Chimie de Paris, in whose laboratory they shared the discovery of the Group 4 metallocene catalyzed dehydrocoupling reactions. H. G. W. is grateful to the Korea Ministry of Education (Basic Science Research Institute program, project no. BSRI-96-3426) and Dow Corning Corporation for support of his research reported herein. J. F. H. and F. G. thank the Natural Sciences and Engineering Research Council of Canada and the Fonds F.C.A.R. du Québec for support of their work.

References

(1) Harrod, J. F. *Stud. Surf. Sci. Catal.* **1992**, *73*, 147.

(2) Breneman, W. C. Can. Pat. 1,151,395, Aug. 9, 1983 (U.S. Pat. Appl. Aug. 23, 1975). C.A. **1983**, 99.

(3) Cowley, A. H.; Jones, R. A. *Angew. Chem., Int. Ed. Eng.* **1989**, *28*, 1208.

(4) *Inorganic and Organometallic Polymers with Special Properties;* R. M. Laine, Ed.; Kluwer Academic Publishers: Dordrecht, 1992.

(5) West, R. *Angew. Chem., Int. Ed. Eng.* **1987**, *26*, 1201; Cowley, A. H.; Lasch, J.; Norman, N. C.; Pakulski, M. *J. Am. Chem. Soc.* **1983**, *105*, 5506; Regitz, M.; Binger, P. *Angew. Chem., Int. Ed. Eng.* **1988**, *27*, 1484; Yoshifuji, M.; Shima, I.; Imamoto, N.; Hirotsu, K.; Higuchi, T. *J. Am. Chem. Soc.* **1981**, *103*, 4587.

(6) Sakamoto, K.; Yoshida, M.; Sakurai, H. *Macromolecules* **1990**, *23*, 4494.

(7) Steudel, R. *Chemistry of the Non-metals;* English edition; Nachod, F. C.; Zuckerman, J. J., Eds.; Walter de Gruyter: Berlin, 1977.

(8) Walsh, R. *Accts. Chem. Res.* **1981**, *14*, 246.

(9) Kanabus-Kaminska, J. M.; Hawari, J. A.; Griller, D.; Chatagilialoglu, C. J. *J. Am. Chem. Soc.* **1987**, *109*, 5267.

(10) Harrod, J. F. In *Design Activation and Transformation of Organometallics into Common and Exotic Materials;* R. M. Laine, Ed.; NATO ASI Series E, No. 141. Martinus Nijhof: Amsterdam, 1988; p. 103.

(11) Yajima, S.; Omori, M.; Hayashi, J.; Okamura, K.; Matsuzawa, T.; Liaw, C. F. *Chem. Lett.* **1976**, 551.
(12) Bianconi, P. A.; Weidman, T. W. *J. Am. Chem. Soc.* **1988**, *110*, 2342.
(13) Griffing, B. F.; West, R. *Polym. Eng. Sci.* **1983**, *23*, 947.
(14) West, R.; David, L. D.; Djurovich, P. I.; Stearley, K. S. V.; Srinivasan, H. Y. *J. Am. Chem. Soc.* **1981**, *103*, 7352.
(15) Miller, R. D.; Michl, J. *Chem. Rev.* **1989**, *89*, 1359.
(16) West, R. *J. Organomet. Chem.* **1986**, *300*, 327.
(17) Trefonas, P.; Damewood, J. R.; West, R.; Miller, R. D. *Organometallics* **1985**, *4*, 1318.
(18) Harrah, L. A.; Zeigler, J. M. *Macromolecules* **1987**, *20*, 601.
(19) Matyjaszewski, K.; Cypryk, M.; Frey, H.; Hrkach, J.; Kim, H. K.; Moeller, M.; Ruehl, K.; White, M. *J. Macromol. Sci.-Chem.* **1991**, *A28*, 1151.
(20) Hengge, E.; Litscher, G. K. *Angew. Chem., Int. Ed. Engl.* **1976**, *15*, 370.
(21) Matyjaszewski, K.; Greszta, D.; Hrkach, J. S.; Kim, H. K. *Macromolecules* **1995**, *28*, 59.
(22) Kimata, Y.; Suzuki, H.; Satoh, S.; Kuriyama, A. *Organometallics* **1995**, *14*, 2506.
(23) Boudjouk, P.; Rajkumar, A. B.; Parker, W. L. *J. Chem. Soc., Chem. Commun.* **1991**, 245.
(24) Becker, B.; Corriu, R.; Guerin, C.; Henner, B. *Polym. Prepr. (Am. Chem. Soc., Div. Polym. Chem.)* **1987**, *28(1)*, 409.
(25) Aitken, C.; Harrod, J. F.; Gill, U. S. *Can. J. Chem.* **1987**, *65*, 1804.
(26) Harrod, J. F.; Yun, S. S. *Organometallics* **1987**, *6*, 1381.
(27) Aitken, C.; Barry, J.-P.; Gauvin, F.; Harrod, J. F.; Malek, A.; Rousseau, D. *Organometallics* **1989**, *8*, 1732.
(28) Woo, H.-G.; Tilley, T. D. *J. Am. Chem. Soc.* **1989**, *111*, 8043.
(29) Woo, H.-G.; Tilley, T. D. *Polym. Prepr. (Am. Chem. Soc., Div. Polym. Chem.)* **1990**, *31(2)*, 228.
(30) Walzer, J. F.; Woo, H.-G.; Tilley, T. D. *Polym. Prepr. (Am. Chem. Soc., Div. Polym. Chem.)* **1991**, *32(1)*, 441.
(31) Kumada, M.; Tamao, K. *Adv. Organomet. Chem.* **1966**, *6*, 19.
(32) Ishikawa, M.; Kumada, M. *Adv. Organomet. Chem.* **1981**, *19*, 51.
(33) Curtis, M. D.; Epstein, P. S. *Adv. Organomet. Chem.* **1981**, *19*, 213.
(34) Tilley, T. D. *Comments Inorg. Chem.* **1990**, *10*, 37.
(35) Tilley, T. D. *Acc. Chem. Res.* **1993**, *26*, 22.
(36) Corey, J. Y. *Adv. Silicon Chem.* **1991**, *1*, 327.
(37) Woo, H.-G. Ph.D. Thesis, University of California at San Diego, 1990.
(38) Li, H.; Gauvin, F.; Harrod, J. F. *Organometallics* **1993**, *12*, 575.
(39) Nakano, T.; Nakamura, H.; Nagai, Y. *Chem. Lett.* **1989**, 83.
(40) Bourg, S.; Corriu, R. J. P.; Enders, M.; Moreau, J. J. E. *Organometallics* **1995**, *14*, 564.
(41) Jutzi, P.; Redeker, T. *Organometallics* **1997**, *16*, 1343.
(42) Jutzi, P.; Redeker, T.; Neumann, B.; Stammler, H.-G. *Organometallics* **1996**, *15*, 4153.
(43) Choi, N.; Onozawa, S.-Y.; Sakakura, T.; Tanaka, M. *Organometallics* **1997**, *16*, 2765.
(44) Sakakura, T.; Lautenschlager, H.-J.; Nakajima, M.; Tanaka, M. *Chem. Lett.* **1991**, 913.
(45) Forsyth, C. M.; Nolan, S. P.; Marks, T. J. *Organometallics* **1991**, *10*, 2543.
(46) Watson, P. L.; Tebbe, F. N. U.S. Patent 4,965,386, 1990; *Chem. Abstr.* **1991**, *114*, 123331p.
(47) Woo, H.-G.; Song, S.-J. *Bull. Korean Chem. Soc.* **1996**, *17*, 494.
(48) Woo, H.-G.; Song, S.-J. Unpublished results.
(49) Corey, J. Y.; Zhu, X.-H.; Bedard, T. C.; Lange, L. D. *Organometallics* **1991**, *10*, 924.
(50) Corey, J. Y.; Zhu, X.-H. *Organometallics* **1992**, *11*, 672.
(51) Corey, J. Y.; Joslin, F. L.; Braddock-Wilking, J.; Thompson, L. S. *Polym. Prepr. (Am. Chem. Soc., Div. Polym. Chem.)* **1992**, *33(1)*, 1239.
(52) Woo, H.-G.; Kim, S.-Y.; Han, M.-K.; Cho, E. J.; Jung, I. N. *Organometallics* **1995**, *14*, 2415.

(53) Woo, H.-G.; Kim, S.-Y.; Kim, W.-G.; Yeon, S. H.; Cho, E. J.; Jung, I. N. *Bull. Korean Chem. Soc.* **1995**, *16*, 1109.
(54) Woo, H.-G.; Song, S.-J.; You, H.; Cho, E. J.; Jung, I. N. *Bull. Korean Chem. Soc.* **1996**, *17*, 475.
(55) Woo, H.-G.; Song, S.-J. *Bull. Korean Chem. Soc.* **1996**, *17*, 1040.
(56) Kralicek, J.; Kubanek, V.; Kondelikova, J. Ger. Pat. 2,301,784, 1973.
(57) Kralicek, J.; Kubanek, V.; Kondelikova, J.; Casensky, B.; Machacek, J. Ger. Pat. 2,445,647, 1976.
(58) Bukac, Z.; Sebenda, J. U.S. Pat. 3,962,239, 1976.
(59) Woo, H.-G.; Harrod, J. F. Unpublished results.
(60) Dioumaev, V. K.; Harrod, J. F. *Organometallics* **1994**, *13*, 1548.
(61) Dioumaev, V. K.; Harrod, J. F. *J. Organomet. Chem.* **1996**, *521*, 133.
(62) (a) Yang, X.; Stern, C. L.; Marks, T. J. *J. Am. Chem. Soc.* **1991**, *113*, 3623; (b) Nolan, S. P.; Porchia, M.; Marks, T. J. *Organometallics* **1991**, *10*, 1450.
(63) Chien, J. C. W.; Tsai, W.-M.; Rausch, M. D. *J. Am. Chem. Soc.* **1991**, *113*, 8570.
(64) Imori, T.; Tilley, T. D. *Polyhedron* **1994**, *13*, 2231.
(65) Ojima, I.; Inaba, S.-I.; Kogure, T.; Nagai, Y. *J. Organomet. Chem.* **1973**, *55*, C7.
(66) Onopchenko, A.; Sabourin, E. T. *J. Org. Chem.* **1987**, *52*, 4118.
(67) Brown-Wensley, K. A. *Organometallics* **1987**, *6*, 1590.
(68) Corey, J. Y.; Huhmann, J. L.; Zhu, X.-H. *Organometallics* **1993**, *12*, 1121.
(69) Kesti, M. R.; Waymouth, R. M. *Organometallics* **1992**, *11*, 1095.
(70) Knight, K. S.; Waymouth, R. M. *Organometallics* **1994**, *13*, 2575.
(71) Zybill, C.; Muller, G. *Angew. Chem., Int. Ed. Engl.* **1987**, *26*, 669.
(72) Zybill, C.; Muller, G. *Organometallics* **1988**, *7*, 1368.
(73) Zybill, C.; Wilkinson, D. L.; Leis, C.; Muller, G. *Angew. Chem., Int. Ed. Engl.* **1989**, *28*, 203.
(74) Zybill, C.; Wilkinson, D. L.; Muller, G. *Angew. Chem., Int. Ed. Engl.* **1988**, *27*, 583.
(75) Ueno, K.; Tobita, H.; Shimoi, M.; Ogino, H. *J. Am. Chem. Soc.* **1988**, *110*, 4092.
(76) Straus, D. A.; Tilley, T. D.; Rheingold, A. L.; Geib, S. J. *J. Am. Chem. Soc.* **1987**, *109*, 5872.
(77) Sharma, H. K.; Pannell, K. H. *Chem. Rev.* **1995**, *95*, 1351.
(78) Procopio, L. J.; Berry, D. H. *J. Am. Chem. Soc.* **1987**, *109*, 3777.
(79) Grumbine, S. K.; Tilley, T. D.; Arnold, F. P.; Rheingold, A. *J. Am. Chem. Soc.* **1994**, *116*, 5495.
(80) (a) Mu, Y.; Aitken, C.; Coté, B.; Harrod, J. F. *Can. J. Chem.* **1991**, *69*, 264; (b) Mu, Y. Ph.D. Thesis, McGill University, 1992; *Diss. Abs. Int. B,* **1993**, *53(12)*, 6307.
(81) Woo, H.-G.; Harrod, J. F.; Hénique, J.; Samuel, E. *Organometallics* **1993**, *12*, 2883.
(82) Hencken, G.; Weiss, E. *Chem. Ber.* **1973**, *106*, 1747.
(83) Harrod, J. F.; Ziegler, T.; Tschinke, V. *Organometallics* **1990**, *9*, 897.
(84) Ho, J.; Dick, D. G.; Stephan, D. W. *Organometallics* **1993**, *12*, 3145.
(85) Woo, H.-G.; Tilley, T. D. *Inorganic and Organometallic Oligomers and Polymers;* Kluwer Academic Publishers: Dordrecht, The Netherlands, 1991; p. 3.
(86) Woo, H.-G.; Walzer, J. F.; Tilley, T. D. *J. Am. Chem. Soc.* **1992**, *114*, 7047.
(87) Dioumaev, V.; Harrod, J. F. *Organometallics,* **1997**, *16*, 1452.
(88) Dioumaev, V.; Harrod, J. F. *Organometallics,* **1997**, *16*, 2798.
(89) Buchwald, S. L.; Watson, B. T.; Huffman, J. C. *J. Am. Chem. Soc.* **1987**, *109*, 2544.
(90) Binger, P.; Muller, P.; Benn, R.; Rufinske, A.; Gabor, B.; Kruger, C.; Blitz, P. *Chem. Ber.* **1989**, *122*, 1035.
(91) Negishi, E.; Holmes, S. J.; Tour, J. M.; Miller, J. A.; Cederbaum, F. E.; Swanson, D. R.; Takahashi, T. *J. Am. Chem. Soc.* **1989**, *111*, 3336.
(92) Soleil, F.; Choukroun, R. *J. Am. Chem. Soc.* **1997**, *119*, 2938.

(93) (a) Gauvin, F.; Harrod, J. F. *Polym. Prepr. (Am. Chem. Soc., Div. Polym. Chem.)* **1991**, *32(1)*, 439; (b) Gauvin, F.; Harrod, J. F. *Can. J. Chem.*, **1990**, *68*, 1638.
(94) Gauvin, F. Ph.D. Thesis, McGill University, 1992; *Diss. Abs. Int. B* **1993**, *54(1)*, 95.
(95) Banovetz, J. P.; Stein, K. M.; Waymouth, R. M. *Organometallics* **1991**, *10*, 3430.
(96) Corey, J. Y.; Zhu, X.-H. *J. Organomet. Chem.* **1992**, *439*, 1.
(97) Shaltout, R. M.; Corey, J. Y.; Rath, N. P. *J. Organomet. Chem.* **1995**, *503*, 205.
(98) Shaltout, R. M.; Corey, J. Y. *Main Group Chem.* **1995**, *1*, 115.
(99) Shaltout, R. M.; Corey, J. Y. *Tetrahedron* **1995**, *51*, 4309.
(100) Rahimian, K.; Dioumaev, V.; Harrod, J. F. Unpublished results.
(101) Banovetz, J. P.; Suzuki, H.; Waymouth, R. M. *Organometallics* **1993**, *12*, 4700.
(102) Hengge, E.; Weinberg, M. *J. Organomet. Chem.* **1993**, *443*, 167.
(103) Hengge, E.; Weinberg, M. *J. Organomet. Chem.* **1992**, *441*, 397.
(104) Hengge, E.; Weinberg, M. *J. Organomet. Chem.* **1992**, *433*, 21.
(105) Hengge, E.; Gspaltl, P.; Pinter, E. *J. Organomet. Chem.* **1996**, *521*, 145.
(106) Woo, H.-G.; Walzer, J. F.; Tilley, T. D. *Macromolecules* **1991**, *24*, 6863.
(107) Imori, T.; Woo, H.-G.; Walzer, J. F.; Tilley, T. D. *Chem. Mater.* **1993**, *5*, 1487.
(108) Banovetz, J. P.; Hsiao, Y.-L.; Waymouth, R. M. *J. Am. Chem. Soc.* **1993**, *115*, 2540.
(109) Banovetz, J. P.; Hsiao, Y.-L.; Waymouth, R. M. *Polym. Prepr. (Am. Chem. Soc., Div. Polym. Chem.)* **1993**, *34(1)*, 228.
(110) Hsiao, Y.-L.; Waymouth, R. M. *J. Am. Chem. Soc.* **1994**, *116*, 9779.
(111) Laine, R. M.; Babonneau, F. *Chem. Mat.* **1993**, *5*, 260.
(112) Seyferth, D. In *Transformation of Organometallics into Common and Exotic Materials: Design and Activation;* R. M. Laine, Ed.; NATO ASI Ser. E: Appl. Sci., No. 141; Martinus Nijhoff: Amsterdam, 1988; p. 133.
(113) Mu, Y.; Harrod, J. F. In *Inorganic and Organometallic Polymers and Oligomers;* Harrod, J. F.; Laine, R. M., Eds.; Kluwer Academic Publishers: Dordrecht, The Netherlands, 1991; p. 23.
(114) Seyferth, D.; Wood, T. G.; Tracy, H. J.; Robison, J. L. *J. Am. Ceram. Soc.* **1992**, *75*, 1300.
(115) Seyferth, D.; Koppetsch, G. E.; Wood, T. G.; Tracy, H. J.; Robison, J. L.; Czubarow, P.; Tasi, M.; Woo, H.-G. *Polym. Prepr. (Am. Chem. Soc., Div. Polym. Chem.)* **1993**, *34(1)*, 223.
(116) Seyferth, D.; Tasi, M.; Woo, H.-G. *Chem. Mater.* **1995**, *7*, 236.
(117) Corriu, R. J. P.; Enders, M.; Huille, S.; Moreau, J. E. *Chem. Mater.* **1994**, *6*, 15.
(118) Scarlete, M.; Brienne, S.; Butler, I. S.; Harrod, J. F. *Chem. Mater.* **1994**, *6*, 977.
(119) Procopio, L. J.; Berry, D. H. *J. Am. Chem. Soc.* **1991**, *113*, 4039.
(120) Trefonas, P.; West, R. *J. Polym. Sci., Polym. Chem. Ed.* **1985**, *23*, 2099.
(121) Szymanski, W. J.; Visscher, G. T.; Bianconi, P. A. *Macromolecules* **1993**, *26*, 869.
(122) Aitken, C.; Harrod, J. F.; Malek, A. J. *J. Organomet. Chem.* **1988**, *249*, 285.
(123) Samuel, E.; Harrod, J. F. Unpublished results.
(124) Choi, N.; Tanaka, M. Submitted for publication. We thank Dr. Tanaka for providing a preprint of this work prior to publication.
(125) Reichl, J. A.; Popoff, C. M.; Gallagher, L. A.; Remsen, E. E.; Berry, D. H. *J. Am. Chem. Soc.* **1996**, *118*, 9430.
(126) Sita, L. R. *Organometallics* **1992**, *11*, 1442.
(127) Imori, T.; Tilley, T. D. *J. Chem. Soc., Chem. Commun.* **1993**, 1607.
(128) Imori, T.; Lu, V.; Cai, H.; Tilley, T. D. *J. Am. Chem. Soc.* **1995**, *117*, 9931.
(129) Babcock, J. R.; Sita, L. R. *J. Am. Chem. Soc.*, **1996**, *118*, 12481.
(130) Xin, S.; Woo, H.-G.; Harrod, J. F.; Samuel, E.; Lebuis, J.-M. *J. Am. Chem. Soc.*, **1997**, *119,* 5307.
(131) Xin, S. Ph.D. Thesis, McGill University, 1995.

(132) Breen, T. L.; Stephan, D. W. *Organometallics* **1996**, *15*, 4509.
(133) Breen, T. L.; Stephan, D. W. *Organometallics* **1996**, *15*, 4223.
(134) Etkin, N.; Fermin, M. C.; Stephan, D. W. *J. Am. Chem. Soc.* **1997**, *119*, 2954.
(135) Mu, Y.; Aitken, C.; Coté, B.; Harrod, J. F. *Can. J. Chem.,* **1991**, *69*, 264.
(136) Corey, J. Y.; Chang, L. S.; Corey, E. R. *Organometallics* **1987**, *6*, 1595.
(137) Chang, L. S.; Corey, J. Y. *Organometallics* **1989**, *8*, 1885.
(138) Corey, J. Y.; Rooney, S. M. *J. Organometal. Chem.* **1996**, *521*, 75.
(139) Li, H.; Butler, I. S.; Harrod, J. F. *Organometallics* **1993**, *12*, 4553.
(140) Sneddon, L. G.; Kai, S.; Fazen, P.; Lynch, A. T.; Remsen, E. E.; Beck, J. S. in *Inorganic and Organometallic Polymers and Oligomers,* Harrod, J. F.; Laine, R. M., Eds.; Kluwer Academic Publishers: Dordrecht, The Netherlands, 1991; p. 199.
(141) Lukevics, E.; Dzintara, M. *J. Organomet. Chem.* **1985**, *295*, 265.
(142) Bedard, T. C.; Corey, J. Y. *J. Organomet. Chem.* **1992**, *428*, 315.
(143) Xin, S.; Harrod, J. F. Unpublished results.
(144) Corriu, R. J. P.; Moreau, J. J.; Wong Chi Man, M. J. *Sol-Gel Sci. Technol.* **1994**, *2*, 87.
(145) Schubert, U.; Lorenz, C. *Inorg. Chem.* **1997**, *36*, 1258.
(146) Lorenz, C.; Schubert, U. *Chem. Ber.* **1995**, *128*, 1267.
(147) Liu, H. Q.; Harrod, J. F. *Organometallics* **1992**, *11*, 822.
(148) Liu, H. Q.; Harrod, J. F. *Can. J. Chem.* **1992**, *70*, 107.
(149) (a) Tamao, K.; Tanaka, T.; Nakajima, T.; Sumiya, R.; Arai, H.; Ito, Y. *Tetrahedron Lett.* **1986**, *27*, 3377; (b) Tamao, K.; Nakajima, T.; Sumiya, R.; Arai, H.; Higuchi, N.; Ito, Y. *J. Am. Chem. Soc.* **1986**, *108*, 6090; (c) Tamao, K.; Yamauchi, T.; Ito, Y. *Chem. Lett.* **1987**, 171; (d) Tamao, K.; Nakagawa, Y.; Ito, Y. *J. Org. Chem.* **1990**, *55*, 3438; (e) Buck, M. S.; Feaster, J. E. *Tetrahedron Lett.* **1992**, *33*, 2099.
(150) Bergens, S. H.; Noheda, P.; Whelan, J.; Bosnich, B. *J. Am. Chem. Soc.* **1992**, *114*, 2121; *ibid.,* 2128.
(151) Xin, S.; Harrod, J. F. *J. Organometal. Chem.* **1995**, *499*, 181.
(152) Wang, X.; Ellis, W. W.; Bosnich, B. *J. Chem. Soc., Chem. Commun.* **1996**, 2561.
(153) Sommer, L. H.; Citron, J. D. *J. Org. Chem.* **1967**, *32*, 3470.
(154) (a) Tillman, N.; Barton, T. J. *Main Group Metal Chemistry* **1987**, *10*, 307; (b) Biran, C.; Blum, Y. D.; Blaser, R.; Tse, D. S.; Youngdahl, K. A.; Laine, R. M. *J. Mol. Catal.* **1988**, *48*, 183; (c) Blum, Y. D.; Laine, R. M. *Organometallics* **1986**, *5*, 2081; (d) Corriu, R. J. P.; Moreau, J. J. E. *J. Organomet. Chem.* **1977**, *127*, 7; (e) Liu, H. Q.; Harrod, J. F. *Organometallics* **1992**, *11*, 822.
(155) He, J.; Liu, H. Q.; Harrod, J. F.; Hynes, R. *Organometallics* **1994**, *13*, 336.
(156) He, J.; Harrod, J. F. *J. Am. Ceram. Soc.* **1995**, *78*, 3009.
(157) He, J.; Harrod, J. F. *Phosphorus, Sulfur and Silicon* **1994**, *93–94*, 349.
(158) Wannegat, U. *Adv. Inorg. Radiochem.* **1964**, *6*, 225.
(159) (a) Voronkov, M. G.; Ushakova, N. I.; Tsikhanskaya, I. I.; Pukhnaravich, V. B. *J. Organometal. Chem.* **1984**, *264*, 39; (b) Voronkov, M. G.; Pukhnaravich, V. B.; Ushakova, N. I.; Tsikhanskaya, I. I.; Albanov, A. L.; Vikovskii, V. Yu. *J. Gen. Chem. USSR* **1985**, *55*, 80.
(160) (a) Ojima, I.; Fuchikamo, T.; Yatabe, M. J. *Organometal. Chem.* **1984**, *260*, 335; (b) Onopchemko, A.; Sabourin, E. T.; Beach, D. L. *J. Org. Chem.* **1983**, *48*, 5101.
(161) (a) Nesmeyanov, A. N.; Friedlina, R. K.; Chukovskaya, E. C.; Petrova, R. G.; Belyavsky, A. B. *Tetrahedron* **1961**, *17*, 61; (b) Harrod, J. F.; Chalk, A. J. In *Catalysis by Metal Carbonyls;* Wender, I.; Pino, P., Eds.; J. Wiley & Sons: New York, 1977; p. 673.
(162) Crabree, R.; Jun, C.-H. *J. Organometal. Chem.* **1993**, *447*, 177.
(163) Fisher, J. M.; Piers, W. E.; Pearce Batchilder, S. D.; Zaworotko, M. *J. Am. Chem. Soc.* **1996**, *118*, 283.

Index

G

H

I

K

M

N

O

P

R

S

T

V

X

Z

Cumulative List of Contributors for Volumes 1–36

Cumulative Index for Volumes 37–42

ISBN 0-12-031142-9